Principles of Engineering Mechanics

Volume 1

Kinematics—The Geometry of Motion

MATHEMATICAL CONCEPTS AND METHODS IN SCIENCE AND ENGINEERING

Series Editor: Angelo Miele
Mechanical Engineering and Mathematical Sciences
Rice University

Recent volumes in this series:

22 **APPLICATIONS OF FUNCTIONAL ANALYSIS IN ENGINEERING** • *J. L. Nowinski*

23 **APPLIED PROBABILITY** • *Frank A. Haight*

24 **THE CALCULUS OF VARIATIONS AND OPTIMAL CONTROL: An Introduction** • *George Leitmann*

25 **CONTROL, IDENTIFICATION, AND INPUT OPTIMIZATION** • *Robert Kalaba and Karl Spingarn*

26 **PROBLEMS AND METHODS OF OPTIMAL STRUCTURAL DESIGN** • *N. V. Banichuk*

27 **REAL AND FUNCTIONAL ANALYSIS, Second Edition**
Part A: Real Analysis • *A. Mukherjea and K. Pothoven*

28 **REAL AND FUNCTIONAL ANALYSIS, Second Edition**
Part B: Functional Analysis • *A. Mukherjea and K. Pothoven*

29 **AN INTRODUCTION TO PROBABILITY THEORY WITH STATISTICAL APPLICATIONS** • *Michael A. Golberg*

30 **MULTIPLE-CRITERIA DECISION MAKING: Concepts, Techniques, and Extensions** • *Po-Lung Yu*

31 **NUMERICAL DERIVATIVES AND NONLINEAR ANALYSIS** • *Harriet Kagiwada, Robert Kalaba, Nima Rasakhoo, and Karl Spingarn*

32 **PRINCIPLES OF ENGINEERING MECHANICS**
Volume 1: Kinematics—The Geometry of Motion • *Millard F. Beatty, Jr.*

33 **PRINCIPLES OF ENGINEERING MECHANICS**
Volume 2: Dynamics—The Analysis of Motion • *Millard F. Beatty, Jr.*

34 **STRUCTURAL OPTIMIZATION**
Volume 1: Optimality Criteria • *Edited by M. Save and W. Prager*

Principles of Engineering Mechanics

Volume 1

Kinematics—The Geometry of Motion

Millard F. Beatty, Jr.
University of Kentucky
Lexington, Kentucky

PLENUM PRESS • NEW YORK AND LONDON

Library of Congress Cataloging in Publication Data

Beatty, Millard F.
Principles of engineering mechanics.

(Mathematical concepts and methods in science and engineering; 32-)
Includes bibliographies and index.
Contents: v. 1. Kinematics—the geometry of motion. 1. Mechanics, Applied. 2. Kinematics. I. Title. II. Series.
TA350.B348 1985 620.1 85-24429
ISBN 0-306-42131-3

A Division of Plenum Publishing Corporation
233 Spring Street, New York, N.Y. 10013

Printed in the United States of America

To my wife and best friend,
NADINE CHUMLEY BEATTY,
and to our children,
LAURA, ANN, and SCOTT

Preface

This book is a vector treatment of the principles of mechanics written primarily for advanced undergraduate and first-year graduate students of engineering. However, a substantial part of the material on kinematics—exclusive of special advanced topics clearly identified within the text—much of the content dealing with particle dynamics, and some selected topics from later chapters, have been used in a first course of a lower level. This introductory course usually is taken by junior students prepared in general physics, statics, and two years of university mathematics through differential equations, which may be studied concurrently. The reader is assumed to be familiar with elementary vector methods, but the essentials of vector calculus are reviewed in the applications and separately in a brief Appendix A, in case this familiarity is occasionally inadequate.

The arrangement of the book into two parts—Volume 1: Kinematics and Volume 2: Dynamics—has always seemed to me the best approach. I have found that students who first master the kinematics have little additional difficulty when finally they reach the free body formulation of the dynamics problem. In fact, this book was conceived initially from a less intensive two-term sequence of introductory courses in kinematics and dynamics that I first taught to beginning undergraduate mechanical engineering students at the University of Delaware in 1963. From this beginning, the current structure has grown from both elementary and intermediate level mechanics courses I have taught for several years at the University of Kentucky.

When used at the beginning graduate level, I envision that both parts may be covered in a single semester course; however, the instructor who prefers to move at a slower pace may use these volumes in consecutive semesters or quarters. In this case, however, I recommend that this material be supplemented by use of selected papers and books that treat the more traditional topics in kinematics of mechanisms, more advanced topics in Lagrangian mechanics, and possibly some elements of continuum mechanics. This advanced pair of courses also should include a variety of meaningful, computer-oriented problems. The limitations of space and my desire to present a fresh development at an intermediate level force the exclusion of these other subjects.

Naturally, the presentation is influenced by my personal interests and background in mechanical engineering, engineering science, and mechanics. Consequently, the approach I have chosen is somewhat more sophisticated and mathematical than is often found in traditional textbooks on engineering mechanics. In keeping with this approach, the aforementioned prerequisite mathematics, largely that of the eighteenth century and earlier, is used without apology. Nevertheless, aware that many readers may not have mastered these prerequisite materials, I have exercised care to reinforce the essential tools indirectly in the illustrations and problems selected for study.

Unusual mathematical topics, such as singularity functions and some elements of tensor analysis, are introduced within the text; and the elements of matrices, nowadays studied by most engineering students, are reviewed in Appendix B. Parenthetical reference to use of these tools is provided along the way, with careful indication where such materials may be omitted from a first course, without loss of continuity. Some elements of set notation are used in Volume 2, but the student usually is familiar with these simple applications. In any event, where familiarity may be lacking, comprehension of the ideas may be readily inferred from the context. Otherwise, the teacher is expected to elaborate upon remedial mathematical topics peculiar to his or her special needs by expanding upon these few areas and by building upon the many examples and problems provided.

The mathematical development and the numerous companion examples are structured to place emphasis on the predictive value of the methods and principles of mechanics, rather than on the often empty and less interesting computational aspects, but not to the exclusion of numerical examples that illustrate the various operations and definitions. In addition, some meaningful introductory computer applications are provided in the problems. Examples have been selected for their instructive value and to help the student achieve understanding of the various concepts, principles, and analytical methods presented. In some instances, experimental data, factual situations, and applications or designs that confirm analytical predictions are described. Numerous assignment problems, ranging from easy and straightforward extensions or reinforcements of the subject matter to more difficult problems that test the creative skills of better students, are given at the end of each chapter. To assist the student in his studies, some answers to the odd-numbered problems are provided at the end.

It is axiomatic that physical intuition or insight cannot be taught. On the other hand, competence in mathematics and physical reasoning may be developed so that these special human qualities may be intelligently cultivated through study of physical applications that mirror the world around us and through practice of the rational process of reasoning from first principles. With these attributes in mind, one objective of this book is to help the engineering student develop confidence in transforming problems into appropriate mathematical language that may be manipulated to derive substantive and useful physical conclusions or specific numerical results. I intend that this treatment should provide a more penetrating look at the elements of classical mechanics and their applications to engineering problems; therefore, the book is designed to deepen and broaden the student's understanding and to develop his or her mastery of the fundamentals. However, to reap a harvest from the seeds sown here, it is important that the student work through many of the problems provided for study. The mere understanding that one may apply theoretical concepts and formulas to solve a particular problem is not equivalent to possession of the knowledge and skills required to produce its solution. These talents grow only from experience in dealing repeatedly with these matters. My view of the importance of solving a lot of problems is expressed further at the beginning of the problem set for Chapter 1, and the attitude emphasized there is echoed throughout the text.

It is unfortunate that the subject of mechanical design analysis has suffered such considerable neglect in engineering curricula in this country. I shall not speculate on the reasons for this decay. On the contrary, it is pleasant to see in some schools rejuvenation of the role of mechanics and mathematics, and innovation of the use of the computer in mechanical design curricula. It is only in recent years that these ingredients have begun to restore life to this important and exciting area of engineering.

But I feel that more needs to be accomplished. Various aspects of mechanical, electrical, and structural design, for example, should be introduced in certain pilot courses, and the content of these premier courses taken earlier must be integrated into the various design sequences. I perceive no reason why problems in mechanical design analysis, for instance, ought not to be introduced as examples in courses prerequisite for a major course in this subject. If this plan were followed, advanced problems and computer-aided applications could be studied in a more carefully planned design curriculum that draws materials from virtually every previous fundamental course in the student's program, namely, statics and dynamics, solid and fluid mechanics, thermodynamics and heat transfer, vibrations and controls, circuits and fields, and so on. Consequently, I have chosen for illustration several examples and problems that illustrate simple introductory applications of kinematics and dynamics in analysis of some problems in mechanical design. It is my hope that this book may provide engineering students with solid mathematical and mechanical foundations for future advanced study of topics in mechanical design analysis, advanced kinematics of mechanisms and analytical dynamics, mechanical vibrations and controls, electomagnetics and acoustics, and continuum mechanics of solids and fluids.

The Contents of This Volume

In any treatment of a classical subject like mechanics, it is difficult to know with certainty what may be new or what is simply an unfamiliar, rediscovered result. On the other hand, a fresh approach usually is not hard to spot; and I believe the reader may find many fresh developments within these pages. The division into kinematics and dynamics, though not unique to this text, is uncommon. Yet this division surely is as logical and pedagogically natural as the separation of statics from dynamics found in virtually every elementary book I have seen.

Volume 1: Kinematics, concerns the geometry of motion from its basic definition for a particle through the general theory of motion of a rigid body and of a particle referred to a moving reference frame. Rectilinear motion, commonly covered in general physics and elementary mechanics, is reviewed only indirectly by illustrations in the text and in assignment problems, so the work herein begins with the spatial description of motion in three dimensions. The reader will find here a consistent, logical, and gradual building of well-known kinematical concepts, theorems, and formulas, beginning from the definitions of motion, velocity, and acceleration of a particle in Chapter 1: Kinematics of a Particle, and extending to the beautiful general relations for the velocity and acceleration of a material point referred to a moving reference frame presented in Chapter 4. And there is much in between that is novel.

The use of singularity functions appears often these days in a good first course in the mechanics of deformable solids, and certainly the subject is useful in courses in mechanical vibrations and electrical circuits, for example. However, I know of no source that provides a thorough and elementary introduction to singularity functions with applications to problems in kinematics. These topics are presented at the close of Chapter 1. Illustrated by several elementary examples, this treatment provides the student with powerful tools to treat discontinuous motions common to many mechanical systems. Therefore, this study shows another useful and important application of singularity functions at an elementary level.

In Chapter 2: Kinematics of Rigid Body Motion, the construction begins with the derivation of the finite rotation of a rigid body about a line and leads ultimately to the fundamental equations for the velocity and acceleration of a rigid body point in terms of the translation and rotation of the body. This unusual approach, in my opinion, provides the clearest and most natural way to arrive at the proper description of the angular velocity and angular acceleration vectors for a rigid body. The chapter includes many worked examples and applications; and it closes with a discussion of the theory of instantaneous screws, including a description of the graphical method of instantaneous centers for a rigid body.

The heuristic and intuitive introduction in Chapter 2 to the elegant primary theorems of Euler and Chasles on finite displacements of a rigid body and the rule of composition of several rotations is expanded analytically in Chapter 3: Finite Rigid Body Displacements. The proofs of these theorems use elementary properties of matrices and tensors. Tensor algebra is introduced to simplify the description of a finite rigid body rotation in terms of the rotation tensor, which is the key to understanding Euler's theorem, the composition rule for several rotations, and Chasles' elegant screw theorem. I know of no other book where one may find a similar treatment of these theorems and their applications. The coordinate invariant construction of the axis and angle of the equivalent Euler rotation given by the simple equations (3.75) and (3.76) provides the essentially unique solution of the eigenvalue problem expressed by (3.88) for the Euler rotation matrix. The simple component equation (3.80) for the unique Euler axis of rotation is not usually found in mechanics texts. Some of these results are buried in the well-known work of Gibbs; but nowhere do I recall having seen the easy, coordinate invariant derivation of (3.146) for the location of the screw axis of a general finite rigid body displacement.

In addition, the transformation laws for vector and tensor components are derived and illustrated in some introductory examples in Chapter 3. We return to these rules again in Volume 2, where some further important properties of symmetric tensors are studied and applied to characterize the rotational dynamics of a rigid body. The use of the tensor representation theorem to study the composition of several finite rotations and to represent the composition of an arbitrary rotation in terms of the classical Euler angles also is carefully described. Many examples, useful applications, and problems for further study also are provided.

The time rate of change of a vector referred to a moving frame, the fundamental kinematic chain and composition rules for several angular velocity and acceleration vectors, and the elegant basic equations for the absolute velocity and absolute acceleration of a particle referred to a moving reference frame are carefully developed in Chapter 4: Motion Referred to a Moving Reference Frame and Relative Motion. The composition rules for angular velocity and angular acceleration appropriate to multiple reference frames are thoroughly and carefully developed. These important results have varied and useful applications in the study of motion of several connected, independently rotating rigid bodies such as occur in the study of robotics and mechanisms. The notation introduced herein to describe rotational chains, in my estimation, is simpler and less awkward than any used elsewhere in the current literature. Motions referred to cylindrical and spherical reference frames in terms of corresponding curvilinear coordinates are derived as special applications of the aforementioned basic relations for the absolute velocity and acceleration. Many examples and applications demonstrate use of all of the concepts and formulas. It is emphasized that the results derived in Chapter 4 include as particular cases virtually all of the fundamental equations encountered throughout the text.

The final section in Chapter 4 concerns special advanced topics that provide preparation for advanced studies in kinematics of rigid body motion and in continuum mechanics of solids and fluids. The time rate of change of a tensor referred to a moving reference frame is derived; and an elementary introduction to frame-indifferent transformations, especially useful in continuum mechanics, is presented. The volume ends where advanced texts on theoretical kinematics of a rigid body and continuum mechanics often begin. A survey of other topics not mentioned above is provided by the table of contents below.

In writing this book, I have appealed over many years to numerous sources, especially for guidance in selection of appropriate problems. While I do not necessarily subscribe to their approach to mechanics, some of the references that I found particularly helpful are listed at the end of each chapter, usually with annotations to describe the substance of the work and to identify specific chapters that may be consulted for collateral study. It is impossible to be precise in citing my specific use of any source, and I apologize if I may have overlooked or forgotten a particular reference. There are, however, two books that I have always found more useful than many others for their excellent problems, and several of the examples and exercises in this book, though redesigned and cast in different language, have originated in some measure with these texts. These are the books by Professors James L. Meriam and Irving H. Shames. However, so far as I am aware, the treatment of problems provided herein usually is quite different and more thorough than found in other comparable sources known to me. Nevertheless, by consulting the listed references or their own favorite books, both the teacher and the student should be able to supplement the many examples and problems to meet their special needs.

Acknowledgment

The final draft of Chapter 4 was completed and the final section dealing with advanced topics on tensor rates was created while I was a Senior Fellow in the Institute for Mathematics and Its Applications at the University of Minnesota during 1984–1985. I am grateful to Professors Jerald L. Ericksen and David Kinderlehrer for inviting my participation in the many workshop activities on continuum physics and partial differential equations held throughout the year. I thank the Institute for providing me the unique opportunity to interact with so many world scholars and to advance my own research interests in continuum mechanics and nonlinear elasticity, and for providing me the use of a personal computer and almost three months of virtually uninterrupted time that enabled me to complete Volume 1 of this work. My fellowship at Minnesota was a most delightful and memorable experience.

I wish also to extend my thanks to others who have helped me. I am grateful to my colleague Professor Oscar W. Dillon for his useful comments and helpful discussions on an early draft. It is a pleasure to thank Dr. Lincoln Bragg for his helpful critical comments and thoughtful remarks on a rough draft of this volume; these served as a valuable guide throughout my writing of this text. I am grateful also to Professor Donald E. Carlson; he read the final manuscript and provided many helpful suggestions, critical comments, and some corrections. Mrs. Laura Robbins deserves special thanks for her pleasant attitude and patience in typing and retyping the manuscript. Finally, I owe much to Mr. Joseph Haas for his good humor, patience, and painstaking care during the tedious task of adding typeset lettering to my many drawings, and for applying his artistic talent to draw from my rough sketches some of the illustrations that appear in this book.

Millard F. Beatty, Jr.
Lexington, Kentucky

Contents

Volume 1

Kinematics

The Geometry of Motion

I cannot too strongly urge that a kinematical result is a result valid forever, no matter how time and fashion may change the "laws" of physics.

Clifford A. Truesdell,
The Kinematics of Vorticity

1

Kinematics of a Particle

> Mathematics deals exclusively with the relations of concepts to each other without consideration of their relation to experience. Physics too deals with mathematical concepts; however, these concepts attain physical content only by the clear determination of their relation to the objects of experience. This in particular is the case for the concepts of motion, space, time.
>
> I believe that the first step in the setting of a "real external world" is the formation of the concept of bodily objects and of bodily objects of various kinds.
>
> Albert Einstein,
> *Essays in Physics*

1.1. Introduction

Dynamics is the branch of analytical mechanics devoted to the study of the motion of bodies and the forces and torques that may produce it. However, since geometric considerations play a principal role in dynamics, we shall consider initially only the purely geometrical aspects of motion. The theory of motion without regard to the agents that produce it is called *kinematics*. Kinematics is the *geometry of motion*. Thus, in this chapter we set the stage for future developments by making precise the idea of motion and its relation to velocity and acceleration. These definitions subsequently are applied to describe the general motion of a material point in terms characterized by the geometry of the curve along which the point moves. After the basic kinematical ideas are explored thoroughly for a rigid body in Chapters 2 and 3 and for moving reference systems in Chapter 4, we shall proceed to investigate the relation of forces and torques to the motion of bodies. Let us begin with a discussion of some primitive terms needed in our work.

1.2. Primitive Terms

It clearly is impossible to define everything. One term is defined by other terms, which are defined by others, and so on; but eventually we must stop at some still undefined, *primitive term* whose interpretation is left to our intuition or to our experience, psychological or otherwise, in relation to the real world and to the context in which the term is used. In classical geometry, for example, "point" is an undefined term, whereas the word "line" is defined as a contiguous set of points. A dictionary, on the other hand, tells us that a "point" is "a dimensionless geometrical object having no property but location." But what meanings shall we assign to the terms "dimensionless" and "location"? Searching the pages, we shall find that "dimensionless" means "without length, width, or height"; and we may add intuitively that "no property but location" implies that a "point" also has no smell, taste, or color. And if any of these terms may be unclear, we continue our march through the lexicon until, finally, our intuitive or psychological needs are satisfied by seemingly more concrete terms, and our understanding of the still primitive, abstract term "point" somehow is rendered more familiar. Thus, we come eventually to perceive the original empirical meaning of certain undefined terms. In any event, the meanings of the primitive terms of any subject ultimately are determined by the use we make of them. In fact, from birth, we learn the meanings of most words in our language in just this way. It is not unusual, therefore, that we must begin our study of mechanics by the enumeration of some primitive terms.

A *material object*, such as an electron, a ball, a river, an aircraft, a planet, etc. is called a *body*. More abstractly, a body $\mathscr{B}$ is defined as a set of material points P called *particles*; we write $\mathscr{B} = \{P\}$. Of course, any part of a body, like a bucket of river water, or any collection of discrete material objects, like the fragments of an exploded shell, is also a body. Thus, a body, like the line of elementary geometry, is a set of points, but it differs in that its points need not be connected. A set of separated particles usually is referred to as a *system of particles*, and a contiguous set of material points often is called a *continuum*. Notice, on the other hand, that a particle, like the point of Euclid's geometry, is a primitive, undefined term that is to be interpreted in the context of its application. Although we shall picture a particle as the familiar geometrical point object, it must be understood that they are not the same thing. A speck of pepper, for example, may be identified as a particle. We all know, of course, that it has properties other than location: it has a color, it makes you sneeze, and it tastes bad when consummed alone. But these additional characteristics are unimportant to the study of the motion of a particle of pepper; rather, only its material content is considered relevant. A material point has mass; a geometrical point does not.

Mass is one of the primitive concepts of dynamics that distinguishes classical mechanics from classical geometry. Its introduction stems from the fact that matter exists in varying degrees of concentration or distribution in the universe. Thus, mass is identified as a positive scalar measure of the material content of a body. The weight of an object, which will be defined precisely later on, is not a suitable measure of its material content, for the weight of a man standing on the surface of the Moon is 1/6 his weight on Earth, yet the material of which he is made is certainly the same and exists in the same bulk. Total volume of an object is not suitable either. We could pack different amounts of matter, like foam rubber, into the same space; but the mass (per unit volume) would differ. The mass of a body is a positive number that measures, relative to some assigned standard reference mass, its *invariant* material content.

Length and *time* are primitive concepts. The concept of length as a positive measure of distance has its foundation in classical geometry, so we shall not go into it again here. But what is time? Time is Now; yet Now is no more. There is a before Now and an after Now. This is the way we characterize time. We say there is a present time, a before the present time, and an after the present time; there exists an earlier than Now and a later than Now. This is how we perceive time; but it does not tell us in mathematical terms, for example, what time is. We employ a clock, any repeatable sequence of events, to keep track of events that take place relative to a certain reference event, like the birth of Christ. But a clock does not tell us what time is; rather, in the same way that a ruler is used to measure length, the distance between two points, a clock is an instrument that measures and records how many events have been repeated since the occurrence of the assigned reference event. Time itself is a primitive concept; it is identified as a numerical measure of the duration of events—past, present, and future.

Force is a primitive concept of mechanics. We pull on the ends of a string and make it taut; stretch a rubber band to several times its unstretched length; bend a straight rod of steel to circular shape; twist our hands; drag our feet; and when we fall down, it hurts. Winds topple buildings; a ball thrown upward returns to strike the ground; a magnet moves an iron bar toward itself or pushes it away without touching it. These actions are the influence of forces. But they do not tell us what force is. Force is an undefined term—a primitive concept. However, these experiences teach us that force is a *vector quantity*. A force is exerted by one body on another with a certain intensity in a certain direction—it is identified in mechanics as a vector measure of the push–pull action between pairs of bodies in the universe.

The primitive terms introduced here will be joined together in what follows; and other basic concepts will be introduced as the need arises. Mass, force, and time are primitive terms that distinguish the analytical structure of classical particle mechanics from classical Euclidean geometry. The concepts of force and mass are essential only to dynamics, so we shall not encounter these specifically until much later on; on the other hand, the concepts of particle, length, and time are required presently to study the kinematics of a particle. We shall see that these three basal concepts are intimately related in the idea of motion.

1.3. Motion and Particle Path

To locate an object in space, we need a reference system. The only reference we have is other objects. Therefore, the physical nature of what we shall call a *reference frame* is an assigned set of objects whose mutual distances do not change with time—at least not very much. For example, the walls of a room, the rotating structure of a carrousel, the cabin framework of a spacecraft, or the remote stars may serve as a frame of reference. It is only necessary that the distance between the objects that define the frame, namely, the walls, the structure, the framework, or the stars and so on, do not change to within the accuracy that distances are being measured, during the time that the frame is in use.

For analytical purposes, our definition of a frame must be more precise, easy to use, and correspond to the physical idea of an assigned set of objects. Therefore, we define a three-dimensional Euclidean reference frame φ as a set consisting of a point O of space, called the *origin* of the reference frame, and a vector basis $\{\mathbf{e}_i\} \equiv \{\mathbf{e}_1, \mathbf{e}_2, \mathbf{e}_3\}$. That is, $\varphi = \{O; \mathbf{e}_i\}$. We shall require for convenience that the basis is an orthonormal basis, i.e., a triple of mutually perpendicular unit vectors. A typical reference frame is shown in Fig. 1.1.

The spatial location of a particle P in the frame φ at time t is given by the *position vector* $\mathbf{x}_\varphi$ of P from O. Of course, as time goes on the place occupied by P generally will vary. The time sequence of positions of P in φ is called a *motion* of P *relative to* φ, and it is defined by the equation

$$\mathbf{x} = \mathbf{x}_\varphi(P, t), \tag{1.1}$$

in which $\mathbf{x}$ is the *place* in frame φ occupied at time t by the particle P in its motion relative to φ, as shown in Fig. 1.1. The locus $\mathscr{L}$ of places occupied by

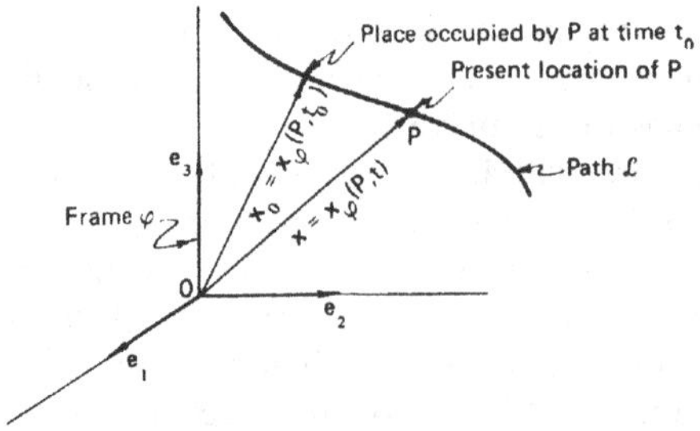

Figure 1.1. A motion of a particle P relative to the frame $\varphi = \{O; \mathbf{e}_k\}$.

P in the motion (1.1) is called the *path* or *trajectory* of P. The place $\mathbf{x}_0$ that P occupies along its path at some instant $t = t_0$, say, which we may sometimes consider as the *initial instant* $t_0 = 0$, is determined by the motion (1.1):

$$\mathbf{x}_0 = \mathbf{x}_\varphi(P, t_0). \tag{1.2}$$

When the particular choice of φ is clear, as it is when only one frame is being considered, for the sake of simplicity we shall descard the subscript φ and write (1.1) and (1.2) as

$$\mathbf{x} = \mathbf{x}(P, t), \tag{1.3a}$$

$$\mathbf{x}_0 = \mathbf{x}(P, t_0). \tag{1.3b}$$

The generic basis $\{\mathbf{e}_k\}$ used to define the reference frame φ may be any convenient vector basis. Frequently, though certainly not always, we shall identify $\{\mathbf{e}_k\}$ as the familiar rectangular Cartesian basis $\{\mathbf{i}_k\} = \{\mathbf{i}_1, \mathbf{i}_2, \mathbf{i}_3\}$ with $\mathbf{i}_1 = \mathbf{i}$, $\mathbf{i}_2 = \mathbf{j}$ and $\mathbf{i}_3 = \mathbf{k}$, as usual. It is understood that different symbols φ, ψ, Φ, μ, etc. may be used to name the reference frame to avoid confusion with other quantities that one or more of these same symbols may be chosen to represent. However, because of frequent use of φ as a frame symbol, we shall reserve the symbol ϕ for use exclusively as an angular placement. Also, for brevity, we shall often write $\mathbf{e}_k$ for the set $\{\mathbf{e}_k\}$.

The foregoing ideas are illustrated in the following example.*

Example 1.1. Let the motion of a particle P relative to an assigned Cartesian reference frame $\varphi = \{O; \mathbf{i}_k\}$ be given by the following position vector function of time t:

$$\mathbf{x}(P, t) = R[\cos \omega t\, \mathbf{i} + \sin \omega t\, \mathbf{j}] + At\mathbf{k}, \tag{1.4}$$

where R, ω, and A are constants and $\{\mathbf{i}_k\} \equiv \{\mathbf{i}, \mathbf{j}, \mathbf{k}\}$, as remarked above. Notice that at $t = 0$ we have from (1.4) and (1.3b)

$$\mathbf{x}_0 = \mathbf{x}(P, 0) = R\mathbf{i}. \tag{1.5}$$

This is the place relative to φ occupied by the particle P initially. Its place at time $t = 2$ units, say, is given by

$$\mathbf{x}_2 \equiv \mathbf{x}(P, 2) = R[\cos 2\omega\, \mathbf{i} + \sin 2\omega\, \mathbf{j}] + 2A\mathbf{k}.$$

Specific measure units for t will depend on the units assigned to ω and A, or conversely.

* The elements of vector algebra are reviewed in Appendix A, Section A.1.

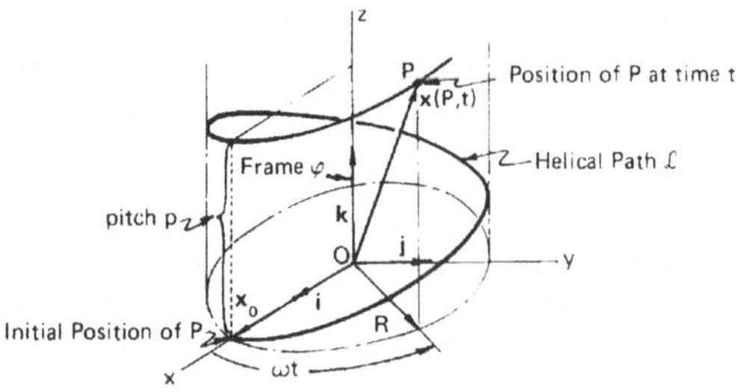

Figure 1.2. The path of a particle having the motion (1.4) is a circular helix.

Of course, in frame φ the place $\mathbf{x}$ occupied by P always may be written in terms of its elementary Cartesian scalar components x, y, z:

$$\mathbf{x} = x(t)\,\mathbf{i} + y(t)\,\mathbf{j} + z(t)\,\mathbf{k}. \tag{1.6}$$

Equating coefficients of base vectors in (1.6) and (1.4), we find that the *time-parametric equations* that define the path of the particle in φ are given by

$$x(t) = R\cos\omega t, \tag{1.7a}$$

$$y(t) = R\sin\omega t, \tag{1.7b}$$

$$z(t) = At. \tag{1.7c}$$

In view of (1.5), it is now clear too that the Cartesian coordinates of P in its initial position are $x_0 = x(0) = R$, $y_0 = y(0) = 0$, $z_0 = z(0) = 0$. We see from (1.7) that $x^2 + y^2 = R^2$, which is the equation of a circle in the xy plane. Thus, the trajectory of P lies on a circular cylinder perpendicular to the xy plane. As shown in Fig. 1.2, the trajectory described by (1.7) is a circular helix, a curve characteristic of the threads of a screw. We see that when the particle has completed one revolution about the cylinder axis, it has also advanced some distance p along that axis, as shown in Fig. 1.2. This unit advance per turn is called the *pitch* of the helix. We recall that a screw that has 10 threads per centimeter (cm) of its length, for example, advances one millimeter (mm) in one revolution, so its pitch is 1 mm or 0.1 cm. Notice also that $\omega t \equiv \theta(t)$ is the angle through which the particle has turned about the $\mathbf{k}$ direction in time t, so the time τ required to complete one turn of 2π radians about the cylinder axis is given by $\omega\tau = 2\pi$. Thus, (1.7c) shows that the pitch is given by $p = z(\tau) = A\tau = 2\pi A/\omega$. It follows, therefore that the constant A depends on the pitch p of the helix and the angular frequency of rotation ω. Angular frequency will be discussed further on.

1.4. Velocity and Acceleration

The *velocity* relative to φ of a particle P at time t is defined by the time rate of change in φ of the position vector (1.1) or (1.3a):

$$\mathbf{v}(P, t) = \dot{\mathbf{x}} \equiv \frac{d\mathbf{x}(P, t)}{dt}. \tag{1.8}$$

The magnitude $v = |\mathbf{v}|$ of $\mathbf{v}$ is called the *speed* of P. We recall that $|\mathbf{v}| = (\mathbf{v} \cdot \mathbf{v})^{1/2}$. The differentiation with respect to time often will be abbreviated by use of a superimposed dot, as indicated in (1.8). Of course, since we are following the particle in its motion relative to φ, neither the identity of the particle P nor any aspect of the reference frame varies with the time. Later on, we shall encounter circumstances where we wish to determine the velocity of a particle relative to a second frame, starting from an equation for the motion relative to the first frame, when the two frames are moving relative to each other; and, in that case, the origin and base vectors of the designated moving frame will vary with time.

The *acceleration* relative to φ of a particle P at time t is defined by the time rate of change in φ of the velocity vector (1.8):

$$\mathbf{a}(P, t) = \dot{\mathbf{v}}(P, t) \equiv \frac{d\mathbf{v}(P, t)}{dt}. \tag{1.9}$$

Substitution of (1.8) into (1.9) gives an equivalent relation in terms of the position vector:

$$\mathbf{a}(P, t) = \ddot{\mathbf{x}}(P, t) = \frac{d^2\mathbf{x}(P, t)}{dt^2}. \tag{1.10}$$

When the position vector is expressed in rectangular Cartesian variables (1.6), the relations (1.8) and (1.10) relative to a Cartesian frame $\varphi = \{O; \mathbf{i}_k\}$ have the specific forms

$$\mathbf{v}(P, t) = \dot{x}(t)\,\mathbf{i} + \dot{y}(t)\,\mathbf{j} + \dot{z}(t)\,\mathbf{k}, \tag{1.11}$$

$$\mathbf{a}(P, t) = \ddot{x}(t)\,\mathbf{i} + \ddot{y}(t)\,\mathbf{j} + \ddot{z}(t)\,\mathbf{k}. \tag{1.12}$$

The Cartesian form of the velocity vector in (1.11) reveals that the speed of P is determined by

$$v = (\dot{x}^2 + \dot{y}^2 + \dot{z}^2)^{1/2} = \frac{ds(t)}{dt}, \tag{1.13}$$

in which we observe that $ds = (dx^2 + dy^2 + dz^2)^{1/2}$ is the elemental arc length along the particle path. Thus, we learn that *the speed is equal to the rate of* *change of the distance $s(t)$ that P moves along its path.* The velocity vector now may be written as the product of its magnitude $v = \dot{s}$ times the unit vector $\mathbf{t} \equiv \mathbf{v}/v$ constructed from $\mathbf{v}$ itself; thus,

$$\mathbf{v}(P, t) = \dot{s}\mathbf{t}. \tag{1.14}$$

It is seen from (1.8) that the velocity vector is in the direction of the infinitesimal particle displacement $d\mathbf{x}(P, t)$ along its path; hence, $\mathbf{t}$ is a unit vector tangent at each point to the particle path. It follows that because the magnitude of a vector is the same in every reference frame, the representation (1.14) holds in every reference system. The relation (1.14) thus shows that *the velocity of a particle always is tangent to its path in every motion.* We shall return to this later on.

A particle P is said to be *fixed* or *at rest* in φ if $x(P, t) = \mathbf{x}_0$ for all times. It is evident from (1.8) that a particle is fixed in φ if and only if $\mathbf{v}(P, t) = \mathbf{0}$, i.e., when and only when it has zero velocity for all times. On the other hand, a particle P is at *instantaneous rest* in φ at a particular time t_0 if and only if $\mathbf{v}(P, t_0) = \mathbf{0}$ at that instant.

The physical dimensions of $\mathbf{v}$ and $\mathbf{a}$, expressed by $[\mathbf{v}]$ and $[\mathbf{a}]$, are derived from the length dimension $[L]$ of the position vector $\mathbf{x}$ and the time dimension $[T]$ of the time variable t on the basis of (1.8)–(1.10). Thus, $[\mathbf{v}] \equiv [V] = [L/T]$, $[\mathbf{a}] = [V/T] = [L/T^2]$. Of course, specific dimensional units will depend upon the choice of units for the fundamental dimensions of length and time employed in a particular problem. If the length is expressed in feet (ft) or meters (m) and the time in seconds (sec), for example, then the velocity will be expressed in ft/sec or m/sec and the acceleration in ft/sec^2 or m/sec^2. Use of specific measure units will arise only in numerical problems.

Example 1.2. The velocity and acceleration of a particle having the motion (1.4) are determined by differentiation in accordance with (1.8) to (1.10). Thus, relative to the frame φ, we find*

$$\mathbf{v} = \dot{\mathbf{x}}(P, t) = \omega R[-\sin \omega t\, \mathbf{i} + \cos \omega t\, \mathbf{j}] + A\mathbf{k}, \tag{1.15}$$

$$\mathbf{a} = \dot{\mathbf{v}}(P, t) = -\omega^2 R[\cos \omega t\, \mathbf{i} + \sin \omega t\, \mathbf{j}], \tag{1.16}$$

in which $A = p\omega/2\pi$. Although these vectors vary with time, we see that both the speed, $v = (R^2\omega^2 + A^2)^{1/2}$, and the magnitude of the acceleration, $|\mathbf{a}(P, t)| = R\omega^2$, are constant in the helical motion (1.4). Also, at $t = 0$, we have $\mathbf{v}_0 \equiv \mathbf{v}(P, 0) = R\omega\mathbf{j} + A\mathbf{k}$ and $\mathbf{a}_0 \equiv \mathbf{a}(P, 0) = -R\omega^2\mathbf{i}$, for example. □

We recall from our earlier Example 1.1 that $\theta(t) \equiv \omega t$ defines the time-varying angle between the two vertical planes that contain the $0z$ axis and the radial lines of R and $\mathbf{x}_0$, as shown in Fig. 1.2. By definition, the time rate of

* The rules for differentiation of a vector function of a scalar variable are reviewed in Appendix A, Section A.2. Integration is outlined in Section A.3.

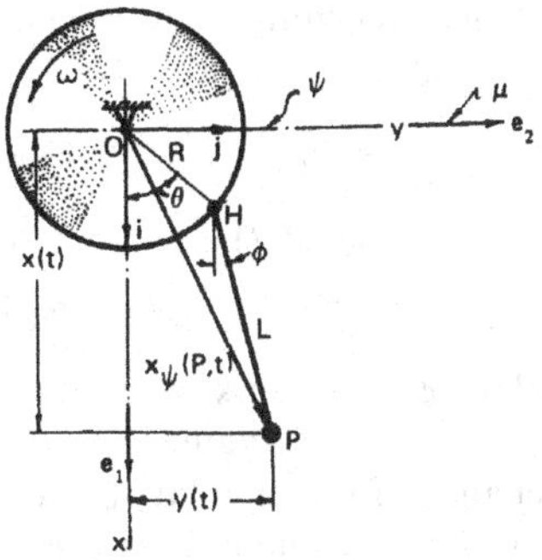

Figure 1.3. A simple mechanical system in which a small body P is modeled as a particle.

change $\dot{\theta}(t)$ of any such *angular placement* $\theta(t)$ is called an *angular speed.** Hence, in the last example $\dot{\theta}(t)=\omega$ is the constant angular speed of rotation of the particle about the positive z axis in φ. Clearly, $[\omega]=[1/T]=[T^{-1}]$; so the dimensional units of ω are expressed in measure units of radians per unit of time. In abbreviated notation, this would be written as rad/sec or rad/min, for example.

In the preceding example, we started from a given motion relation and derived the velocity and acceleration from it. Most of the time, however, we must obtain the motion relation from other data provided in the problem. Let us look at an illustration of a simple mechanical system in which the motion is obtained by construction of the position vector from geometrical considerations.

Example 1.3. The hinged support H of a thin rod of length L to which a small ball P is attached moves with constant angular speed $d\theta(t)/dt \equiv \omega$ on a vertical circle of radius R as shown in Fig. 1.3. We wish to determine the velocity and acceleration of the ball as it moves in the plane relative to frame $\psi=\{O; \mathbf{i}, \mathbf{j}\}$ which is fixed in the plane space at O.

Solution. Since the size of P, though unspecified but finite, is apparently very small compared to the lengths L and R, it is reasonable to model P as a particle attached to the end point of the rod. Then, in terms of the angles θ and ϕ shown in Fig. 1.3, the position vector of P in the fixed Cartesian frame ψ is given by

$$\mathbf{x}_\psi(P, t)=x(t)\,\mathbf{i}+y(t)\,\mathbf{j}=[R\cos\theta(t)+L\cos\phi(t)]\,\mathbf{i} + [R\sin\theta(t)+L\sin\phi(t)]\,\mathbf{j}.$$

* The angular velocity vector $\boldsymbol{\omega}$ will be defined carefully in Chapter 2; its magnitude, the angular speed, has the same physical interpretation illustrated in this simpler intuitive setting.

Hence, recalling that $\dot{\theta}=\omega$, we find with (1.8) and (1.9) that relative to frame ψ

$$\begin{aligned} \mathbf{v}_{\psi}(P, t) &= -(R\omega \sin\theta + L\dot{\phi}\sin\phi)\,\mathbf{i} + (R\omega\cos\theta + L\dot{\phi}\cos\phi)\,\mathbf{j}, \\ \mathbf{a}_{\psi}(P, t) &= -(R\omega^2\cos\theta + L\dot{\phi}^2\cos\phi + L\ddot{\phi}\sin\phi)\,\mathbf{i} \\ &\quad + (-R\omega^2\sin\theta - L\dot{\phi}^2\sin\phi + L\ddot{\phi}\cos\phi)\,\mathbf{j}. \end{aligned} \tag{1.17}$$

Here $\dot{\phi}$ is an unknown angular speed; hence, $\ddot{\phi}\equiv d\dot{\phi}(t)/dt$, the time rate of change of the angular speed, also is unknown. By definition, the time rate of change of the angular speed is called the *simple angular acceleration.** Clearly, its physical dimensions are $[T^{-2}]$; and its usual measure units are rad/sec^2. Notice that the same result would be obtained were we to consider, more precisely, that L was the distance from the support H to the center of the ball at P, or to any other point in the ball. In this case it makes no difference what the dimensions of the ball may be. □

Example 1.4. Determine the velocity and acceleration of the ball P relative to a moving reference frame $\mu=\{O; \mathbf{e}_1, \mathbf{e}_2\}$ fixed in the wheel of the device shown in Fig. 1.3.

Solution. To find the velocity and acceleration of P relative to a moving frame $\mu=\{O; \mathbf{e}_k\}$ fixed in the wheel at O, we first write its position vector relative to frame μ:

$$\mathbf{x}_{\mu}(P, t) = \mathbf{R} + L[\cos\phi(t)\,\mathbf{e}_1 + \sin\phi(t)\,\mathbf{e}_2]$$

where now $\mathbf{R}$ is the constant vector of H from O. Hence, with (1.8) and (1.9), we find relative to frame μ

$$\begin{aligned} \mathbf{v}_{\mu}(P, t) &= L\dot{\phi}[-\sin\phi(t)\,\mathbf{e}_1 + \cos\phi(t)\,\mathbf{e}_2], \\ \mathbf{a}_{\mu}(P, t) &= -L[\ddot{\phi}\sin\phi(t) + \dot{\phi}^2\cos\phi(t)]\,\mathbf{e}_1 \\ &\quad + L[\ddot{\phi}\cos\phi(t) - \dot{\phi}^2\sin\phi(t)]\,\mathbf{e}_2. \end{aligned} \tag{1.18}$$

Notice that in the moving frame μ, both $\mathbf{e}_1$ and $\mathbf{e}_2$ are constant vectors, whereas in the fixed frame ψ their directions vary with the time. Observe also that in this example the angle ϕ is measured relative to the line of $\mathbf{e}_1$ fixed in μ, whereas in the last example ϕ was measured relative to the line of $\mathbf{i}$ fixed in ψ. Hence, these distinct angles coincide only at the instant when the frames coincide. Clearly, the frame μ is turning with the constant angular speed ω relative to the frame ψ, so the directions $\mathbf{e}_1$ and $\mathbf{e}_2$ coincide with $\mathbf{i}$ and $\mathbf{j}$ only at the moment shown in the figure. Comparison of equations (1.17) and (1.18) shows that even at the moment of coincidence the velocity and acceleration

* The angular acceleration vector $\dot{\boldsymbol{\omega}}$ will be defined later on; in general, its magnitude is not equal to the magnitude of the simple angular acceleration.

relative to the two frames differ. We shall learn in Chapter 4 how the velocities and accelerations relative to different frames moving in space are related through the rates of change of direction of their rotating basis vectors. □

Example 1.5. The mechanism shown in Fig. 1.4a consists of a drive crank AB that rotates at a constant angular speed $\dot{\alpha} = \omega$, while the arm OBP oscillates about the hinge pin at O at a varying angular rate $\dot{\beta}$. The slider block at B is attached to the crank AB by a hinge pin and it drives the arm OBP. A pinned pair of slider blocks at P allows the assembly to slide in the arm OBP and in a horizontal guide HP. As AB rotates around the circle BCD, a cutting tool (not shown) attached to P moves in the cutting stroke from its extreme right position to its extreme left position while B moves along the arc DTC. During the return stroke, the tool is idle; but because B moves a shorter distance along the arc CGD, the tool is returned to its starting position in considerably less time. For this reason the device is called a *quick return mechanism*. Find the velocity and acceleration of the point P in $\Phi = \{O; \mathbf{i}_k\}$, and compare the time required for the return stroke to the time expended in the cutting stroke.

Solution. Since P is constrained by the horizontal guide, its position vector in Φ is $\mathbf{x}(P, t) = x\mathbf{i} + 4a\mathbf{j}$. Hence, the velocity (1.11) and the acceleration (1.12) are given by

$$\mathbf{v}(P, t) = \dot{x}\mathbf{i}, \qquad \mathbf{a}(P, t) = \ddot{x}\mathbf{i}. \tag{1.19}$$

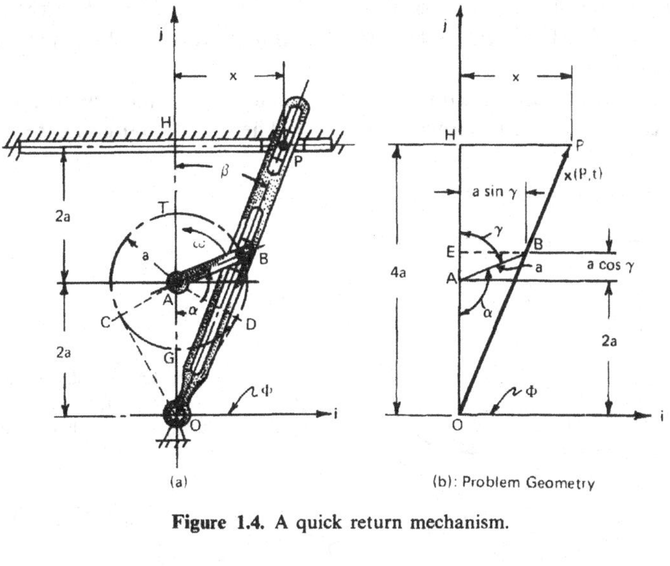

Figure 1.4. A quick return mechanism.

We must now express x in terms of the angle α whose variation is known, using the assigned geometry shown in Fig. 1.4b. By the property of similar triangles OEB and OHP, we see that

$$\frac{x}{4a} = \frac{a \sin \gamma}{2a + a \cos \gamma} = \frac{\sin \alpha}{2 - \cos \alpha},$$

wherein $\gamma = \pi - \alpha$ is used. Hence, $x(t)$ is expressed in terms of the angular position $\alpha(t)$:

$$x(t) = \frac{4a \sin \alpha(t)}{2 - \cos \alpha(t)}. \tag{1.20}$$

A little patience with the differentiation of (1.20) and use of (1.19) yields eventually

$$\mathbf{v}(P, t) = 4a\omega \frac{2 \cos \alpha - 1}{(2 - \cos \alpha)^2} \mathbf{i}, \tag{1.21a}$$

$$\mathbf{a}(P, t) = -8a\omega^2 \frac{\sin \alpha (1 + \cos \alpha)}{(2 - \cos \alpha)^3} \mathbf{i}, \tag{1.21b}$$

in which $\omega = \dot{\alpha}$ denotes the known constant angular speed of the drive crank.

We note from (1.21a) that P is at its extreme positions when $\mathbf{v}(P, t) = \mathbf{0}$. This happens when $\cos \alpha = 1/2$. Thus, in this design C and D are 60° from the vertical line AG. This is also evident from the assigned geometry in Fig. 1.4. Since the 120° arc from C to D through G is one-half as far from D to C through T, and because ω is constant, it is easy to see that the time t_r required for the return stroke is equal to one-half the time t_c expended in the cutting stroke: $t_r = t_c/2$. This characterizes the quick return efficiency of this mechanism.

A graph of the dimensionless cutter velocity $v_P \div (4a\omega)$ as a function of the drive crank angle α is shown in Fig. 1.5. The quick change in the velocity

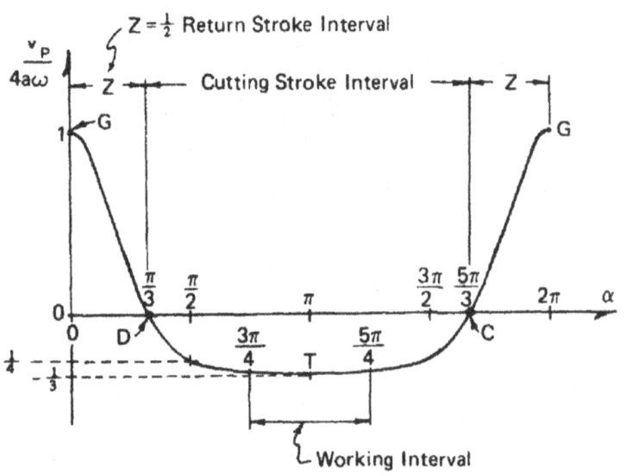

Figure 1.5. Graph of the dimensionless cutter velocity as a function of the drive crank angle α:

$$\frac{v_P}{4a\omega} = \frac{2 \cos \alpha - 1}{(2 - \cos \alpha)^2}.$$

during the return stroke CGD is evident; in fact, the tangent line to the curve at $\alpha=\pi/3$ makes an angle of $-37.6°$ with the α axis. But this rate decreases rather rapidly during the initial phase of the cutting tool stroke, the tangent line to the curve at $\alpha=\pi/2$ making an angle of $-14°$ with the α axis. During the actual tool working interval defined by $\alpha \in [3\pi/4, 5\pi/4]$, say, it is seen that the curve is fairly flat so that the actual cutting operation occurs at a nearly constant rate.

1.5. Some Basic Classifications of Problems

It is impossible to exhaustively classify all varieties of problems involving position, velocity, and acceleration as functions of time; but there are three broad groups of problems that deserve special mention as follows:

Group 1: Given $\mathbf{x}(P, t)$, find $\mathbf{v}(P, t)$, $\mathbf{a}(P, t)$.

This is the basic problem that we have studied above. The solution is explicitly provided by the defining equations (1.8)–(1.10):

$$\mathbf{v}(P, t)=\frac{d\mathbf{x}(P, t)}{dt}, \qquad \mathbf{a}(P, t)=\frac{d\mathbf{v}(P, t)}{dt}=\frac{d^2\mathbf{x}(P, t)}{dt^2}. \tag{1.22}$$

The remaining groups are varieties of this type.

Group 2: Given $\mathbf{v}(P, t)$, find $\mathbf{a}(P, t)$, $\mathbf{x}(P, t)$.

The acceleration is determined from its definition (1.9): $\mathbf{a}(P, t)=\dot{\mathbf{v}}(P, t)$. The position vector is found by integration of the differential equation $d\mathbf{x}(P, t)/dt=\mathbf{v}(P, t)$ by separation of the variables $\mathbf{x}$ and t as indicated below:

$$\int_{t_0}^{t} d\mathbf{x}(P, t)=\mathbf{x}(P, t)-\mathbf{x}_0=\int_{t_0}^{t} \mathbf{v}(P, t)\, dt,$$

wherein $\mathbf{x}_0=\mathbf{x}(P, t_0)$ is the place occupied by P at any given instant t_0. Thus,

$$\mathbf{x}(P, t)=\mathbf{x}_0+\int_{t_0}^{t} \mathbf{v}(P, t)\, dt. \tag{1.23}$$

Group 3: Given $\mathbf{a}(P, t)$, find $\mathbf{v}(P, t)$, $\mathbf{x}(P, t)$.

To determine $\mathbf{v}(P, t)$, we must integrate the differential equation $d\mathbf{v}(P, t)/dt=\mathbf{a}(P, t)$ by separation of the variables $\mathbf{v}$ and t as indicated below:

$$\int_{t_0}^{t} d\mathbf{v}(P, t)=\mathbf{v}(P, t)-\mathbf{v}_0=\int_{t_0}^{t} \mathbf{a}(P, t)\, dt;$$

that is,

$$\mathbf{v}(P, t) = \mathbf{v}_0 + \int_{t_0}^{t} \mathbf{a}(P, t)\, dt, \tag{1.24}$$

in which $\mathbf{v}_0 = \mathbf{v}(P, t_0)$ is the value of $\mathbf{v}(P, t)$ at any given instant t_0. Now that $\mathbf{v}(P, t)$ is known, the rest of the problem falls into group 2, and $\mathbf{x}(P, t)$ is determined by (1.23).

Let us observe that in any problem in which the velocity is known, the distance traveled by the particle along its path is determined by the differential equation (1.13) for the speed. We find by separation of the variables s and t

$$s(t) = \int_{t_0}^{t} v(t)\, dt, \tag{1.25}$$

in which the distance $s(t)$ is measured along the path from the place occupied by the particle at the time t_0.

Finally, we recall that for a time-varying angular placement $\theta(t)$, the angular speed ω and the simple angular acceleration $\dot{\omega}$ are defined by

$$\omega = \dot{\theta}(t), \qquad \dot{\omega} = \ddot{\theta}(t), \tag{1.26}$$

respectively. When $\dot{\omega}(t)$ is given, the angular speed may be found from the second of (1.26) by use of the method of separation of variables; we obtain

$$\omega(t) = \omega_0 + \int_{t_0}^{t} \dot{\omega}(t)\, dt, \tag{1.27}$$

in which $\omega_0 = \omega(t_0)$ denotes the angular speed at the instant t_0. A second integration delivers the angular placement:

$$\theta(t) = \theta_0 + \int_{t_0}^{t} \omega(t)\, dt, \tag{1.28}$$

wherein $\theta_0 = \theta(t_0)$ is the angular placement at time t_0. Thus, in particular, for a constant angular speed $\omega = \omega_0$, we derive from (1.28) the special relation

$$\theta(t) = \theta_0 + \omega_0(t - t_0). \tag{1.29}$$

Let us turn to some sample applications of some of the methods described above.

Example 1.6. The velocity of a particle P which initially is at the place $\mathbf{x}_0 = -2\mathbf{i}$ ft is given by $\mathbf{v}(P, t) = 6t^2\mathbf{i} + 4t\mathbf{k}$ ft/sec. Find the velocity and acceleration of the particle initially, determine the motion of P, and find its path. Compute the distance traveled by the particle in one second.

Solution. It is observed that this problem may be solved by the method for group 2. The acceleration is given for all times by (1.9); therefore,

$$\mathbf{a}(P, t) = \frac{d\mathbf{v}(P, t)}{dt} = 12t\mathbf{i} + 4\mathbf{k} \text{ ft/sec}^2. \tag{1.30}$$

Thus $\mathbf{a}_0 \equiv \mathbf{a}(P, 0) = 4\mathbf{k}$ ft/sec^2 is the initial acceleration of P. Clearly, the initial velocity is $\mathbf{v}_0 = \mathbf{v}(P, 0) = \mathbf{0}$.

The motion of P, hence the path traversed by P, is determined by the differential equation $d\mathbf{x}(P, t)/dt = \mathbf{v}(P, t)$. Application of (1.23) yields

$$\int_0^t d\mathbf{x}(P, t) = \mathbf{x}(P, t)|_0^t = \int_0^t (6t^2\mathbf{i} + 4t\mathbf{k})\, dt = (2t^3\mathbf{i} + 2t^2\mathbf{k})|_0^t.$$

(See Appendix A, Section A.3). Thus, $\mathbf{x}(P, t) = \mathbf{x}_0 + 2t^3\mathbf{i} + 2t^2\mathbf{k}$. With $\mathbf{x}_0 = \mathbf{x}(P, 0) = -2\mathbf{i}$ ft, we have, finally, the motion of P:

$$\mathbf{x}(P, t) = 2(t^3 - 1)\,\mathbf{i} + 2t^2\mathbf{k} \text{ ft}. \tag{1.31}$$

We observe that the motion of P always is in the xz plane.

To determine the path of P, we use (1.6) and (1.31) to obtain the following time-parametric equations:

$$x(t) = 2(t^3 - 1), \qquad y(t) = 0, \qquad z(t) = 2t^2. \tag{1.32}$$

Then elimination of the time parameter t yields the *standard equation* for the path of P in the xz plane:

$$z \equiv \hat{z}(x) = 2\left(\frac{x}{2} + 1\right)^{2/3}. \tag{1.33}$$

Its graph is shown in Fig. 1.6.

It remains to compute the distance traveled in one second. Using the for-

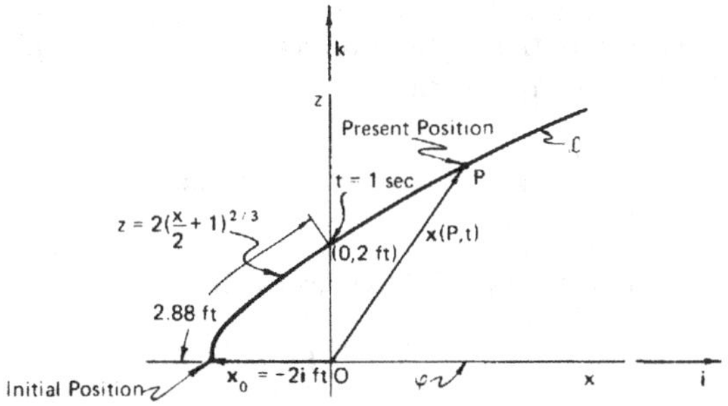

Figure 1.6. Path of a particle having the motion

$\mathbf{x}(P, t) = 2(t^3 - 1)\,\mathbf{i} + 2t^2\mathbf{k}$.

mula provided for $\mathbf{v}(P, t)$, we find that the speed of P is given by $v(t) = 2t(4+9t^2)^{1/2}$ in accordance with (1.13). Then (1.25) becomes

$$s(t) = \int_0^t 2t(4+9t^2)^{1/2}\,dt = \frac{1}{9}\int_0^t (4+9t^2)^{1/2}\,d(4+9t^2)$$
$$= \frac{2}{27}(4+9t^2)^{3/2}\Big|_0^t.$$

Hence, the distance traveled by P in the time t is

$$s(t) = \frac{2}{27}\left[(4+9t^2)^{3/2} - 8\right]. \tag{1.34}$$

In particular, we see from (1.32) and (1.34) that after one second the particle is at the place $x = 0$, $y = 0$, $z = 2$ ft shown in Fig. 1.6, and it has traveled the distance

$$s(1) = \frac{2}{27}\left[13^{3/2} - 8\right] = 2.88 \text{ ft}.$$

We notice in closing that with the help of the path equation (1.32), the distance traveled in (1.34) also may be expressed as a function of z:

$$s = \hat{s}(z) = \frac{2}{27}\left[\left(4 + \frac{9}{2}z\right)^{3/2} - 8\right]. \qquad \square$$

Example 1.7. An electron E is at rest at the position $\mathbf{x}_0 = 2\mathbf{i} + 3\mathbf{j} - \mathbf{k}$ m initially. Subsequently, it is observed that the electron has an acceleration $\mathbf{a}(E, t) = 12t^2\mathbf{i} - 6t\mathbf{j} + 10\mathbf{k}$ m/sec^2. What are the position, velocity, and acceleration of E after 2 sec? Find the speed of E after 1 sec.

Solution. It is clear that this problem belongs to group 3; thus, integration of the differential equation

$$\frac{d\mathbf{v}(E, t)}{dt} = \mathbf{a}(E, t) = 12t^2\mathbf{i} - 6t\mathbf{j} + 10\mathbf{k} \text{ m/sec}^2 \tag{1.35}$$

in the manner demonstrated in (1.24) yields, with $t_0 = 0$,

$$\mathbf{v}(E, t) = \mathbf{v}_0 + 4t^3\mathbf{i} - 3t^2\mathbf{j} + 10t\mathbf{k} \text{ m/sec}.$$

Since $\mathbf{v}_0 = \mathbf{v}(E, 0) = \mathbf{0}$ initially, we have the result

$$\mathbf{v}(E, t) = 4t^3\mathbf{i} - 3t^2\mathbf{j} + 10t\mathbf{k} \text{ m/sec}. \tag{1.36}$$

To find $\mathbf{x}(E, t)$, we integrate the differential equation $d\mathbf{x}(E, t)/dt = \mathbf{v}(E, t)$ as described in (1.23) noting that presently $t_0 = 0$. Thus,

$$\mathbf{x}(E, t) = \mathbf{x}_0 + t^4\mathbf{i} - t^3\mathbf{j} + 5t^2\mathbf{k} \text{ m}.$$

Substitution of the assigned initial value for $\mathbf{x}_0$ delivers the result

$$\mathbf{x}(E, t) = (2 + t^4)\,\mathbf{i} + (3 - t^3)\,\mathbf{j} + (5t^2 - 1)\,\mathbf{k} \text{ m}. \tag{1.37}$$

Finally, we find from (1.35), (1.36), and (1.37) that after 2 sec

$$\mathbf{x}(E, 2) = 18\mathbf{i} - 5\mathbf{j} + 19\mathbf{k} \text{ m}$$
$$\mathbf{v}(E, 2) = 32\mathbf{i} - 12\mathbf{j} + 20\mathbf{k} \text{ m/sec},$$
$$\mathbf{a}(E, 2) = 48\mathbf{i} - 12\mathbf{j} + 10\mathbf{k} \text{ m/sec}^2.$$

The speed of E follows from (1.36): $|\mathbf{v}(E, t)| = (16t^6 + 9t^4 + 100t^2)^{1/2}$ m/sec; hence, after 1 sec, $v(E, 1) = 5^{3/2}$ m/sec. □

More applications of the foregoing methods may be found in the problems given at the end of the chapter. It is evident that the foregoing procedures assume that all quantities are expressed as functions of the time. However, this will not always be so. Later on and in some of the problem assignments, we shall have to develop additional methods to handle other situations. An important special procedure is illustrated below.

Example 1.8. A circular disk of radius a is suspended by a slender rod attached to its center, as shown in Fig. 1.7. The disk is given an angular twist θ_0 from its natural state in frame $\varphi = \{O; \mathbf{i}_k\}$, and released to perform torsional oscillations. The subsequent simple angular acceleration of a particle P on the rim of the disk in its rotation about the vertical axis is determined by the relation $\ddot{\theta} = -K\theta$, in which K is a known constant that depends on certain properties of the disk and the rod. (a) Find the velocity and acceleration of P expressed as functions of θ alone. (b) Determine the maximum magnitude of

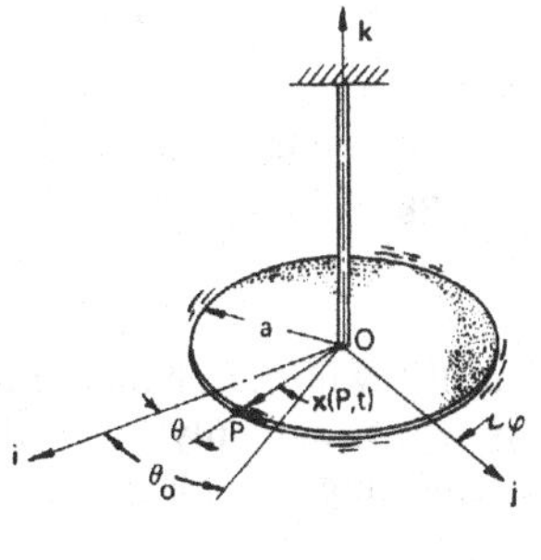

Figure 1.7. Torsional vibrations of a disk.

the angular speed of P. (c) Use these results to show how the angular placement θ may be determined as a function of the time.

Solution. (a) The position vector of P in $\varphi = \{O; \mathbf{i}_k\}$ is

$$\mathbf{x}(P, t) = a(\cos\theta\mathbf{i} + \sin\theta\mathbf{j}); \tag{1.38}$$

and the velocity and acceleration of P are determined by use of (1.8) and (1.9):

$$\mathbf{v}(P, t) = \dot{\mathbf{x}}(P, t) = a\dot{\theta}(-\sin\theta\mathbf{i} + \cos\theta\mathbf{j}), \tag{1.39}$$

$$\mathbf{a}(P, t) = \dot{\mathbf{v}}(P, t) = -a(\ddot{\theta}\sin\theta + \dot{\theta}^2\cos\theta)\,\mathbf{i} + a(\ddot{\theta}\cos\theta - \dot{\theta}^2\sin\theta)\,\mathbf{j}. \tag{1.40}$$

Since $\ddot{\theta} = -K\theta$ is a given function of θ, it remains only to determine $\dot{\theta}$ as a function of θ alone. However, it is evident that because the right-hand side of this differential equation is a function of θ rather than the time, the variables $\dot{\theta}$ and t cannot be separated as we have done in other examples. We must first change from the variable t to the variable θ by use of the chain rule; this yields

$$\ddot{\theta} = \frac{d\dot{\theta}}{d\theta}\frac{d\theta}{dt} = \dot{\theta}\frac{d\dot{\theta}}{d\theta}. \tag{1.41}$$

Substitution of this expression into the given equation $\ddot{\theta} = -K\theta$ yields

$$\dot{\theta}\frac{d\dot{\theta}}{d\theta} = -K\theta. \tag{1.42}$$

The variables $\dot{\theta}$ and θ are now separable.

Bearing in mind the assigned initial conditions that at $t = 0$ the disk is released from rest at $\theta(0) = \theta_0$ so that $\dot{\theta}(0) = 0$, we may now integrate (1.42):

$$\int_0^{\dot{\theta}} \dot{\theta}\, d\dot{\theta} = -K\int_{\theta_0}^{\theta} \theta\, d\theta.$$

Therefore, the angular speed is determined by

$$\dot{\theta} = \pm[K(\theta_0^2 - \theta^2)]^{1/2}. \tag{1.43}$$

The minus sign must be chosen to agree with the initial conditions so that $\theta(t)$ is decreasing in time during the initial part of the motion. Hence,

$$\dot{\theta} = -[K(\theta_0^2 - \theta^2)]^{1/2}. \tag{1.44}$$

The sign of $\dot{\theta}$ will not change until $\dot{\theta} = 0$ again. It is seen from (1.43) that this happens when and only when $\theta = \pm\theta_0$. We recognize $\theta = -\theta_0$ physically as the extreme angular placement of the particle P at which the disk comes momentarily to rest before it reverses its direction to swing back to its initial position of instantaneous rest at $\theta = +\theta_0$. The sign ambiguity that arises from use of the chain rule relation (1.41) usually is easily eliminated by a simple physical or analytical argument of the kind used in this example.

We are now prepared to complete the solution of (a). Substitution of (1.44) and the given relation for $\ddot{\theta}$ into (1.39) and (1.40) yields the velocity and acceleration of P as functions of θ alone:

$$\mathbf{v}(P, t) = -a[K(\theta_0^2 - \theta^2)]^{1/2}(-\sin\theta\mathbf{i} + \cos\theta\mathbf{j}), \tag{1.45}$$

$$\mathbf{a}(P, t) = aK[\theta\sin\theta - (\theta_0^2 - \theta^2)\cos\theta]\,\mathbf{i} - aK[\theta\cos\theta + (\theta_0^2 - \theta^2)\sin\theta]\,\mathbf{j}.$$

(b) It is seen from (1.44) that $\dot{\theta}$ has a maximum magnitude $\omega_{\max}$ at $\theta = 0$:

$$\omega_{\max} = K^{1/2}\theta_0. \tag{1.46}$$

(c) To find θ as a function of the time, we note that a second integration by separation of the variables θ and t in (1.44) yields

$$-K^{1/2}t = \int_{\theta_0}^{\theta} \frac{d\theta}{[\theta_0^2 - \theta^2]^{1/2}}.$$

The right-hand side of this equation is easily integrated following a change of variable defined by $\theta = \theta_0 \cos\psi$ with $\psi = 0$ initially. We find

$$-\int_{\theta_0}^{\theta} (\theta_0^2 - \theta^2)^{-1/2}\, d\theta = \int_0^{\psi} d\psi = \psi.$$

Hence, $\psi = K^{1/2}t$. Therefore, we conclude that the angular placement as a function of time t is given by

$$\theta(t) = \theta_0 \cos(K^{1/2}t). \tag{1.47}$$

Finally, let us observe with the aid of (1.46) and (1.47) that the angular speed also may be expressed as a function of the time:

$$\omega(t) = \dot{\theta}(t) = -\omega_{\max}\sin(K^{1/2}t). \tag{1.48}$$

The results (1.47) and (1.48) may be used to write the velocity and acceleration of P as functions of the time. Notice that the sign in (1.48) agrees with (1.44). □

This completes our study of this problem. Before leaving it behind, let us examine in more general terms the method used to separate the variables in (1.42) above.

Let $f(u)$ be a function of the variable $u(t)$ which depends on the time t, and consider any second-order differential equation of the form

$$\ddot{u}(t) = f(u). \tag{1.49}$$

This equation always has a first integral for $\dot{u}$ expressed as a function of u. Since the right-hand side of this differential equation is a function of u, we must think of $\dot{u}(t)$ as a function $\dot{u}(u)$ depending on u. Then, by use of the chain rule, it follows that

$$\ddot{u}(t) = \frac{d\dot{u}(u)}{du}\frac{du(t)}{dt} = \dot{u}\frac{d\dot{u}}{du}, \tag{1.50}$$

and the equation (1.49) may be rewritten as

$$\dot{u}\frac{d\dot{u}}{du} = f(u). \tag{1.51}$$

Thus, separation of the two variables and integration yields

$$\dot{u}^2 = 2\int f(u)\,du + c_0, \tag{1.52}$$

wherein c_0 denotes the constant of integration.

This development reveals that the same separation of variables may be achieved more directly if in (1.49) one replaces $\ddot{u}$ by $d\dot{u}/dt$ to obtain the first-order equation

$$\frac{d\dot{u}}{dt} = f(u), \tag{1.53}$$

and then multiplies through by $\dot{u}dt = du$ to form

$$\dot{u}d\dot{u} = f(u)\,du, \tag{1.54}$$

in which the variables $\dot{u}$ and u are separated. It is seen that this is the same as (1.51), and integration of (1.54) yields (1.52). Both approaches are based on the chain rule, but the second more direct approach avoids the awkwardness of our having to think of $\dot{u}$ as a function u. This integration procedure will be encountered many times in the solutions of other problems.

1.6. Uniform Motion

The idea of a uniform motion of a particle will be presented below. The important fact that a motion with constant speed is *not necessarily* a uniform motion will be discussed. Afterwards, we shall study an example of uniform motion produced by a cam mechanism.

A motion of a particle for which the velocity vector is a constant vector is called a *uniform motion*. Since $\mathbf{v}(P, t) = \mathbf{v}_0$, a constant vector, it follows that *a motion is uniform if and only if the acceleration is zero* for all times: $\mathbf{a}(P, t) = \mathbf{0}$. Clearly, if the particle moves with constant velocity, then $\mathbf{a}(P, t) = d\mathbf{v}(P, t)/dt = \mathbf{0}$. Conversely, if the acceleration is zero for all times, then $d\mathbf{v}(P, t)/dt = \mathbf{0}$ is a differential equation to be solved for $\mathbf{v}(P, t)$. Its solution by (1.24) is $\mathbf{v}(P, t) = \mathbf{v}_0$, which is a constant vector.

It is natural to ask: What is the path traced out in space by a particle whose velocity is constant? It is easy to see from (1.14) that since the velocity vector at each instant is tangent to the path of the particle, when $\mathbf{v}(P, t)$ has both constant magnitude and constant direction, the path must be a straight line. We can easily demonstrate this geometrical result.

To demonstrate the answer, we need to find $\mathbf{x}(P, t)$ to determine the path of places traced by P as time varies. Given that the velocity $\mathbf{v}(P, t) = \mathbf{v}_0$ is constant, the analysis falls into group 2 above. From (1.23), we get

$$\mathbf{x}(P, t) = \mathbf{x}_0 + \int_{t_0}^{t} \mathbf{v}_0 \, dt = \mathbf{x}_0 + \mathbf{v}_0 \int_{t_0}^{t} dt.$$

That is, the uniform motion is described by

$$\mathbf{x}(P, t) = \mathbf{x}_0 + \mathbf{v}_0(t - t_0). \tag{1.55}$$

This is the time-parametric vector equation of the straight line path of a particle whose place was $\mathbf{x}_0$ at the instant t_0. The result is shown graphically in Fig. 1.8. To see this in other familiar terms, let us recall (1.6); note that $\mathbf{x}_0 = x_0\mathbf{i} + y_0\mathbf{j} + z_0\mathbf{k}$ and write $\mathbf{v}_0 = a\mathbf{i} + b\mathbf{j} + c\mathbf{k}$. Then the vector equation (1.55) may be rewritten as the three scalar equations

$$x = x_0 + a(t - t_0), \qquad y = y_0 + b(t - t_0), \qquad z = z_0 + c(t - t_0).$$

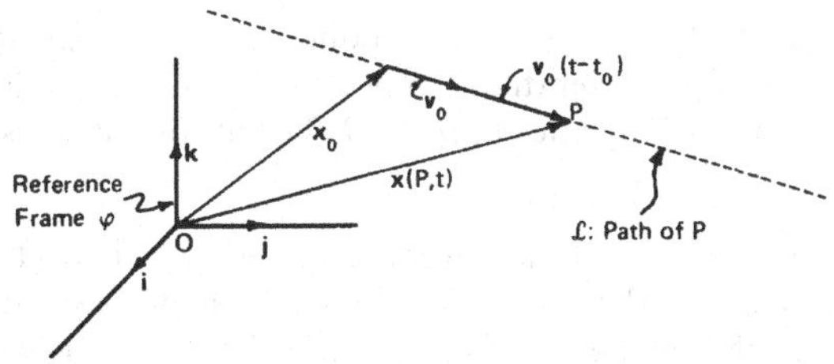

Figure 1.8. The particle path in a uniform motion is a straight line.

The standard equations describing the path are obtained by elimination of the time parameter:

$$t - t_0 = \frac{x - x_0}{a} = \frac{y - y_0}{b} = \frac{z - z_0}{c},$$

The student ought to recognize this set of equations as the intersection of planes defining a straight line whose spatial direction cosines are determined from the components (a, b, c) of the constant velocity vector $\mathbf{v}_0$.

We have shown that a particle can move with a *constant velocity* only if its trajectory is a straight line. The emphasis on *velocity* is important because the velocity vector is constant if and only if both its magnitude, the speed, *and* its direction are constant. In a uniform motion, the speed is always constant. But a constant speed by itself does not constitute a uniform motion because there are infinitely many paths along which a particle may travel with a constant speed. A straight line is only one of them. An example of another one studied earlier is the helical path (1.7) for which the constant speed was found to be $v = (\omega^2 R^2 + A^2)^{1/2}$. For the given initial data $\mathbf{x}_0$ and $\mathbf{v}_0$, there is one and only one uniform motion (1.55); but for the same $\mathbf{x}_0$ and $\mathbf{v}_0$ there are infinitely many motions of P for which only the speed is uniform.

Example 1.9. Suppose that the hinge point H of the system shown in Fig. 1.3 moves with a constant speed $v = 2$ m/sec on a circle of radius $R = 50$ cm. What is the magnitude of the acceleration of H?

Solution. The position vector of H in $\psi = \{O; \mathbf{i}, \mathbf{j}\}$ is given by

$$\mathbf{x}(H, t) = 50(\cos\theta\,\mathbf{i} + \sin\theta\,\mathbf{j})\text{ cm},$$

where $\theta(t)$ denotes the angular placement of H measured from the fixed vertical line shown in Fig. 1.3. Application of (1.8) to the last equation gives $\mathbf{v}(H, t) = 50\dot{\theta}(-\sin\theta\,\mathbf{i} + \cos\theta\,\mathbf{j})$ cm/sec. Therefore, after a change of units, $|\mathbf{v}| = 50\dot{\theta} = v = 200$ cm/sec shows that $\dot{\theta} = 4$ rad/sec, which is the constant angular speed of H about point O in ψ. It is now easy to show with (1.9) that

$$\mathbf{a}(H, t) = -8(\cos\theta\,\mathbf{i} + \sin\theta\,\mathbf{j})\text{ m/sec}^2,$$

whence follows $|\mathbf{a}(H, t)| = 8$ m/sec^2. The point H has a constant speed; but this result shows that its acceleration is not zero because the velocity vector is changing its direction as H rotates around O. The motion of H is not uniform. □

Example 1.10. A *cam* is a mechanical device used to produce a prescribed motion of another body in contact with it. The cam mechanism illustrated in Fig. 1.9 is to be designed to control the oscillatory motion of a push rod so that both its forward motion on $\theta \in [0, \pi]$ and its return motion

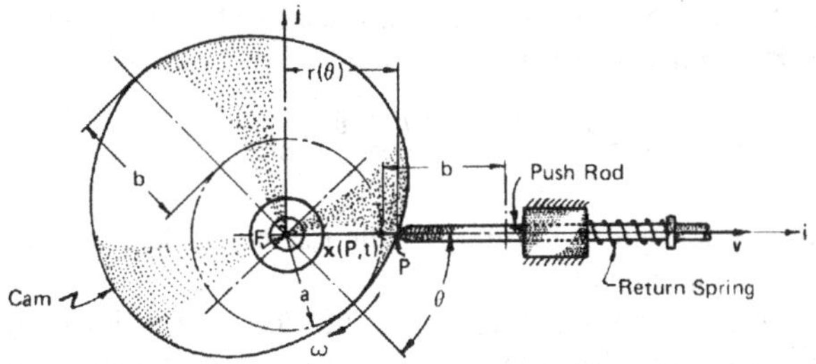

Figure 1.9. Uniform translational motion of a push rod produced by a constant rotary cam motion.

on $\theta \in [\pi, 2\pi]$ are uniform with the same speed v in each direction over the entire stroke b, called the *rise* of the cam. The cam is to rotate with a constant clockwise angular speed $\omega = \dot{\theta}(t)$ about an axle at F. Let $r(\theta)$ denote the variable radial distance from F to the cam surface at the angle θ defined in Fig. 1.9. Find the cam profile $r(\theta)$ that can produce the desired uniform motion.

Solution. We need to relate the shape of the cam defined by $r(\theta)$ to the prescribed uniform motion condition. Since the motion of the push rod is determined by the shape of the cam as it turns through the variable angle $\theta(t)$ about the axle at F, the position vector of the point of contact of the push rod with the cam at P is given by $\mathbf{x}(P, t) = r(\theta)\,\mathbf{i}$. Thus, with the aid of (1.8) and $\omega = \dot{\theta}$, we have

$$\mathbf{v}(P, t) = \frac{d\mathbf{x}(P, t)}{dt} = \frac{dr(\theta)}{d\theta}\,\omega\mathbf{i} \qquad \text{for all} \quad \theta \in [0, 2\pi]. \tag{1.56}$$

In particular, during its uniform forward motion, the push rod has the constant velocity $\mathbf{v}(P, t) = v\mathbf{i}$ for $\theta \in [0, \pi]$. Upon equating this vector to (1.56) we get the relation $v = \omega\, dr(\theta)/d\theta$, in which both v and ω are assigned constants. Then separation of the variables r and θ yields

$$\int_a^{r(\theta)} dr(\theta) = \frac{v}{\omega}\int_0^{\theta} d\theta$$

in which $a = r(0)$ is the value of $r(\theta)$ when $\theta = 0$. Thus, the shape of the cam that will produce the desired uniform forward motion is determined by

$$r(\theta) = a + \frac{v}{\omega}\,\theta \qquad \text{for} \quad \theta \in [0, \pi]. \tag{1.57}$$

The shape function (1.57) also can be expressed in terms of the cam rise b by the condition $r(\pi)=a+b$; we obtain

$$r(\theta)=a+b\frac{\theta}{\pi} \quad \text{with} \quad \frac{b}{\pi}=\frac{v}{\omega} \quad \text{and} \quad \theta\in[0,\pi]. \tag{1.58}$$

The cam profile is a linear function of θ, the rise being determined by assigned design conditions for the ratio v/ω.

To produce the same uniform return motion with velocity $\mathbf{v}(P,t)=-v\mathbf{i}$ on $\theta\in[\pi,2\pi]$, it is clear that the cam must be symmetric about the line $\theta=0$ through F. Nevertheless, let the student show that for the uniform return motion the other half of the profile is determined by

$$r(\theta)=a+b\left(2-\frac{\theta}{\pi}\right) \quad \text{on} \quad [\pi,2\pi]. \tag{1.59}$$

Thus, a cam whose shape is defined by (1.58) and (1.59) converts a constant rotary motion of the cam into a reciprocating, uniform motion of a push rod that maintains contact with it. This design is used often in automatic machine tools where cutter blades are to be moved at constant and relatively slow speeds so that the sudden reversals in the velocity at the beginning and end of the tool stroke are unimportant in the actual cutting operation. Higher operating speeds also may be achieved provided that the return spring is sufficiently stiff to maintain the contact between the push rod and the cam. The sudden jumps that occur in *both* the velocity and the acceleration at the end of each stroke will be discussed later on.

1.7. Velocity and Acceleration Referred to an Intrinsic Frame

Thus far our relations for the motion, velocity, and acceleration of a particle have been applied in situations where a rectangular Cartesian reference frame was adequate. However, this choice frequently proves awkward. In some problems it is more convenient to refer the particle motion to a special moving frame, called the *intrinsic reference frame*, which follows the particle along its tortuous route through space. Thus, instead of decomposing the velocity and acceleration into their usual Cartesian components, we are going to construct their decompositions into *intrinsic components* along directions associated with the path traced by the particle.

1.7.1. Construction of the Intrinsic Velocity and Acceleration

Equations for the velocity and acceleration that are related to certain natural geometrical features of the particle path often are called *intrinsic equations*. The arc length and the unit tangent vector to the particle's path are obvious intrinsic geometrical quantities associated with the velocity vector in (1.14), for example. Therefore, (1.14) is an intrinsic equation for the velocity.

To derive the general formulas for the intrinsic velocity and acceleration, it is natural that we should begin by thinking of the motion as a function of the distance $s(t)$ traveled by the particle along its path. Let us consider a particle P moving along an arbitrary path C in space, as shown in Fig. 1.10. At any instant t its distance along C from any reference point Q on C is $s(t)$. Hence, relative to our Cartesian frame $\Phi = \{O; \mathbf{i}_k\}$ in which the usual coordinates of a point are $\{x_k\} = (x, y, z)$, the position vector of the particle may be written as a function of the time-varying distance traveled along the path:

$$\mathbf{x}(P, t) = \mathbf{x}(s(t)) = x(s)\,\mathbf{i} + y(s)\,\mathbf{j} + z(s)\,\mathbf{k}. \tag{1.60}$$

Then with $d\mathbf{x}(s) = dx(s)\,\mathbf{i} + dy(s)\,\mathbf{j} + dz(s)\,\mathbf{k}$, it is seen that $d\mathbf{x} \cdot d\mathbf{x} = ds^2$, the elemental arc length of the curve. Therefore, the vector $d\mathbf{x}/ds$ is a unit vector. It may be seen in Fig. 1.10 that this unit vector is tangent to the path in the direction of increasing values of s. Thus, the vector $\mathbf{t}(s)$ defined by

$$\mathbf{t}(s) \equiv \frac{d\mathbf{x}}{ds} = \sum_{k=1}^{3} \frac{dx_k}{ds}\,\mathbf{i}_k \tag{1.61a}$$

and

$$\mathbf{t} \cdot \mathbf{t} = 1 \tag{1.61b}$$

is called the *tangent vector* of C. We must insist that C be continuous without corners so that $\mathbf{t}(s)$ is defined uniquely at each point on C. Recalling that the scalar components of a unit vector are the direction cosines of the vector, we see that dx_k/ds are the three direction cosines of the tangent to the path at P. These results are easily visualized for a plane motion, and it may be helpful for the reader to study the equations (1.61) and Fig. 1.11 in relation to the previous remarks.

With (1.60) in mind, we see by (1.8) and use of the chain rule that the

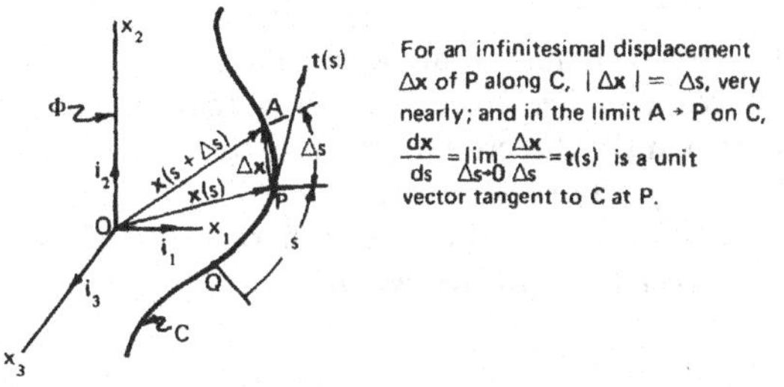

Figure 1.10. Motion of a particle along a space curve.

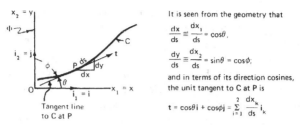

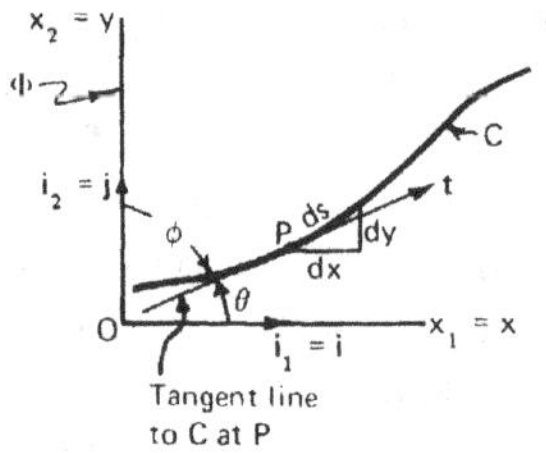

It is seen from the geometry that

$$\frac{dx}{ds} \equiv \frac{dx_1}{ds} = \cos\theta,$$

$$\frac{dy}{ds} \equiv \frac{dx_2}{ds} = \sin\theta = \cos\phi;$$

and in terms of its direction cosines, the unit tangent to C at P is

$$\mathbf{t} = \cos\theta\,\mathbf{i} + \cos\phi\,\mathbf{j} = \sum_{i=1}^{2} \frac{dx_k}{ds}\mathbf{i}_k$$

Figure 1.11. Motion on a plane curve.

velocity of the particle may be expressed by $\mathbf{v}(P, t) = (d\mathbf{x}/ds)(ds/dt)$. Using (1.61a), we now obtain

$$\mathbf{v}(P, t) = \dot{s}\mathbf{t}, \tag{1.62}$$

where $\dot{s} \equiv ds/dt$ denotes the particle speed. This representation of the velocity vector is called the *intrinsic velocity*. We notice that (1.62) is the same result described earlier in (1.14). We see again that *the velocity in every motion is tangent to the particle's path.*

The acceleration is obtained by differentiation of (1.62). Bearing in mind the functional dependence of $\mathbf{t}(s)$ on $s(t)$, we find

$$\mathbf{a}(P, t) = \frac{d\mathbf{v}(P, t)}{dt} = \ddot{s}\mathbf{t} + \dot{s}^2 \frac{d\mathbf{t}(s)}{ds}, \tag{1.63}$$

where $\ddot{s} \equiv d\dot{s}/dt$. Since $\mathbf{t}(s)$ is a unit vector, differentiation of (1.61b) yields [see (A.24) in Appendix A]

$$\mathbf{t} \cdot \frac{d\mathbf{t}}{ds} = 0. \tag{1.64}$$

Hence, C being an arbitrary curve, (1.64) shows that the vector $d\mathbf{t}/ds$ is perpendicular to $\mathbf{t}$. The same result may be seen in more intuitive terms. Because a unit vector has a constant length, the only change it can exhibit is a change in its direction. Since this change can have no component along the invariant unit length of the vector, it must be perpendicular to it, as shown in (1.64). Hence, the unit vector $\mathbf{t}(s)$ must rotate with changes in s, and $d\mathbf{t}/ds$ must be its rate of rotation. We visualize from Fig. 1.12a that $d\mathbf{t}/ds = \lim_{\Delta s \to 0} \Delta\mathbf{t}/\Delta s$ is perpendicular to $\mathbf{t}(s)$ in the direction of the concave side of the path, the direction toward which $\mathbf{t}(s)$ rotates as s varies.

Let $\mathbf{n}$ be a unit vector in the direction of $d\mathbf{t}/ds$ whose magnitude we denote by κ. Then, recalling (1.61a), we may write

$$\frac{d\mathbf{t}}{ds} = \frac{d^2\mathbf{x}}{ds^2} = \kappa\mathbf{n} \tag{1.65a}$$

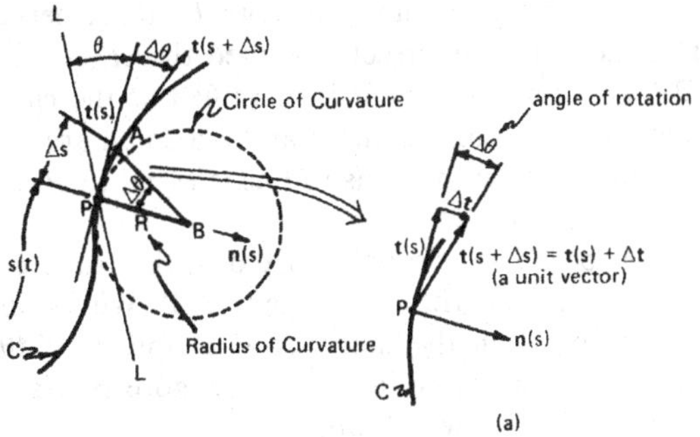

Figure 1.12. Geometry in the principal plane of curvature.

with

$$\kappa \equiv \left|\frac{d\mathbf{t}}{ds}\right| \tag{1.65b}$$

and

$$\mathbf{n}\cdot\mathbf{n} = 1. \tag{1.65c}$$

The unit vector $\mathbf{n}$ is called the *principal normal vector* of C. Since the magnitude κ of $d\mathbf{t}/ds$ is the rate at which the tangent vector rotates its direction with respect to arc length as the particle moves along the curve, κ is called the *curvature* of C. Substitution of (1.65a) in (1.63) yields the basic equation for the *intrinsic acceleration vector*:

$$\mathbf{a}(P, t) = \ddot{s}\mathbf{t} + \kappa\dot{s}^2\mathbf{n}. \tag{1.66}$$

This completes our derivation of the equations for the intrinsic velocity and acceleration. However, some additional useful geometrical details remain to be discussed. Our main results will be summarized afterwards.

1.7.2. Curvature, the Radius of Curvature, and the Intrinsic Frame

The reciprocal of the curvature, denoted by $R \equiv 1/\kappa$, has a simple and useful geometrical interpretation that may be readily visualized from Fig. 1.12. Let us consider the plane defined by the tangent vectors $\mathbf{t}$ and $\mathbf{t} + \Delta\mathbf{t}$ at two points P and A on C separated by an infinitesimal distance Δs. This plane, or, more precisely, its limit as $\Delta s \to 0$, is called the *plane of principal curvature.*

Now let us construct in this plane lines through P and A perpendicular to C. The point B where these lines intersect is named the *center of curvature*. The circle of radius BP with its center at B is identified as the *circle of curvature*. This circle assumes the shape of the curve along a small arc that includes the point P. We are going to show that this radius is equal to R; hence, R will be called the *radius of curvature*.

To see this, let L be a line through P fixed in the plane of principal curvature and making an angle θ with the tangent at P. We see in Fig. 1.12 that the infinitesimal angle between the tangents at P and A is $\Delta\theta$, so $\Delta s = R\Delta\theta$ and $|\Delta t| = |\mathbf{t}|\ \Delta\theta$. These approximations become more precise as Δs is made smaller. In the limit as $\Delta s \to 0$, we obtain

$$\left|\frac{d\theta}{ds}\right| = \frac{1}{R}, \qquad \left|\frac{d\mathbf{t}}{d\theta}\right| = |\mathbf{t}| = 1. \tag{1.67}$$

We now recall (1.65b) and use the chain rule to write $\kappa = |d\mathbf{t}/ds| = |d\mathbf{t}/d\theta|\ |d\theta/ds|$. Then substitution of (1.67) yields the important result

$$\kappa = \left|\frac{d\mathbf{t}}{ds}\right| = \left|\frac{d\theta}{ds}\right| = \frac{1}{R}. \tag{1.68}$$

Thus, as remarked earlier, the curvature κ measures the rate of change in the tangent angle θ with respect to arc length along C; and it is clear that $[R] = [\kappa^{-1}] = [L]$.

Because of the manner in which the plane of principal curvature contacts the path at P, this plane also is called, in more colorful terms, the *osculating plane* of C at P. The principal normal vector $\mathbf{n}$ is perpendicular to the tangent of C and lies in this plane. Therefore, the osculating plane is determined by the vectors $\mathbf{t}$ and $\mathbf{n}$.

Finally, it follows from the properties of $\mathbf{t}$ and $\mathbf{n}$ in (1.61), (1.65), and (1.68) that the vector $\mathbf{b}$ defined by

$$\mathbf{b} \equiv \mathbf{t} \times \mathbf{n} = \frac{d\mathbf{x}}{ds} \times R\frac{d^2\mathbf{x}}{ds^2} \tag{1.69}$$

is a unit vector perpendicular to both $\mathbf{t}$ and $\mathbf{n}$. The vector $\mathbf{b}$ is named the *binormal vector*; it is normal to both C and the osculating plane. The triad $\{\mathbf{t}_k\} = \{\mathbf{t}, \mathbf{n}, \mathbf{b}\}$ of mutually orthogonal unit vectors forms a basis for a moving reference frame $\psi = \{P; \mathbf{t}_k\}$, called the *intrinsic frame*, which follows the particle along its path. In terms of the intrinsic basis, the velocity and acceleration of a particle in a general motion relative to any Cartesian frame

$\Phi = \{O; \mathbf{i}_k\}$ have the simple and useful representations (1.62) and (1.66). These principal results together with the curvature relations (1.68) are summarized with new numbers in the next section.

1.7.3. Velocity and Acceleration Referred to the Intrinsic Frame

We learned in the last section that the *intrinsic velocity* is determined by

$$\mathbf{v} = \dot{s}\mathbf{t}, \tag{1.70}$$

where $\dot{s} \equiv ds/dt$ denotes the particle speed and $\mathbf{t}$ is a unit vector tangent to the trajectory. This shows that the velocity of a particle always is tangent to its path.

We have also found that the *intrinsic acceleration* is given by

$$\mathbf{a} = \ddot{s}\mathbf{t} + \kappa\dot{s}^2\mathbf{n}, \tag{1.71}$$

wherein $\ddot{s} \equiv d\dot{s}/dt$ is called the *tangential acceleration component* and $\kappa\dot{s}^2$ is named the *normal acceleration component*. Since the normal component is directed toward the center of curvature, it also is known as the *centripetal acceleration component*. The result (1.71) shows that *the acceleration vector of P lies always in the osculating plane tangent to its path and containing the center of curvature*. The foregoing description of the intrinsic frame $\psi = \{P; \mathbf{t}_k\}$ and the intrinsic velocity and acceleration vectors in the osculating plane at P are illustrated in Fig. 1.13.

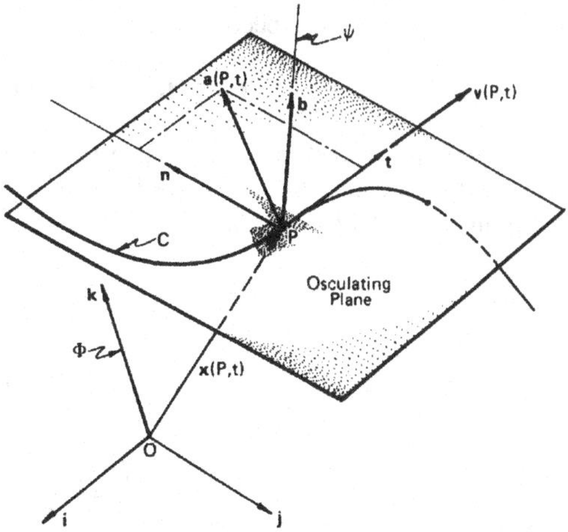

Figure 1.13. Description of the intrinsic reference frame and the intrinsic velocity and acceleration vectors in the osculating plane.

The curvature κ, or the radius of curvature R, is given by

$$\kappa = \frac{1}{R} = \left|\frac{d\mathbf{t}}{ds}\right| = \left|\frac{d\theta}{ds}\right|, \tag{1.72}$$

in which θ is the angle between the tangent vector $\mathbf{t}$ and an arbitrary fixed line in the osculating plane. Two easy applications of (1.72) deserve special mention.

(i) **Rectilinear Motion.** A motion on a straight path is known as a *rectilinear motion.* The tangent vector on a straight path obviously is a constant vector; hence, by (1.72), we have $\kappa = 1/R = 0$. That is, *a straight line path has zero curvature, hence an infinite radius of curvature.* It follows from (1.70) and (1.71) that in every rectilinear motion in the direction $\mathbf{t}$

$$\mathbf{v} = \dot{s}\mathbf{t} \quad \text{and} \quad \mathbf{a} = \ddot{s}\mathbf{t}. \tag{1.73}$$

We recall that a uniform motion is a special rectilinear motion with constant speed; thus, $\mathbf{v} = \dot{s}\mathbf{t} = \mathbf{v}_0$, a constant, and $\mathbf{a} = \mathbf{0}$, as described before.

(ii) **Circular Motion.** It is clear that in a motion on a circle of radius r the tangent vector is tangent to the circle. The principal normal vector at every point around the circle is directed through the center of the circle, so this point is the natural center of curvature of the circle described earlier in Fig. 1.12. Thus, *the radius of curvature of a circle is the radius of the circle*: $R = 1/\kappa = r$.

Notice that the same result follows from the last of (1.72) and the elementary formula $s = r\theta$ for a circle on which θ is the angular placement of an arbitrary radial line from a fixed line through the center. Therefore, we have $\dot{s} = r\dot{\theta}$ and $\ddot{s} = r\ddot{\theta}$. Substituting these relations into (1.70) and (1.71), we obtain the special elementary equations for the motion of a particle on a circle of radius r with angular speed $\omega = \dot{\theta}$ and angular acceleration $\dot{\omega} = \ddot{\theta}$:

$$\mathbf{v} = r\omega\mathbf{t}, \qquad \mathbf{a} = r\dot{\omega}\mathbf{t} + r\omega^2\mathbf{n}. \tag{1.74}$$

Finally, it is easy to verify that the tangential and normal components of the intrinsic acceleration of a particle and the curvature of the path can be computed from the relations

$$a_t \equiv \ddot{s} = \frac{\mathbf{a}\cdot\mathbf{v}}{v}, \tag{1.75a}$$

$$a_n \equiv \kappa\dot{s}^2 = \frac{|\mathbf{a}\times\mathbf{v}|}{v}, \tag{1.75b}$$

$$\kappa = \frac{1}{R} = \frac{|\mathbf{a}\times\mathbf{v}|}{v^3}, \tag{1.75c}$$

in which $v = \dot{s} \neq 0$. The construction of these results is left as an exercise for the student. It should be observed that whereas $\dot{s} = |\mathbf{v}|$, in general $\ddot{s} \neq |\mathbf{a}|$. Rather, by (1.71)

$$|\mathbf{a}| = [\ddot{s}^2 + \kappa^2 \dot{s}^4]^{1/2}. \tag{1.76}$$

We have learned that equations (1.70) and (1.71) are the representations of the velocity and acceleration expressed in terms of a basis that is moving relative to an assigned Cartesian frame $\Phi = \{O; \mathbf{i}_k\}$, as shown in Fig. 1.13. They must not be confused as the velocity and acceleration relative to the intrinsic frame, for it is clear that the particle has no motion relative to an observer situated at the origin of that frame. Rather, the relations (1.70) and (1.71) are representations of the velocity and acceleration of a particle relative to the assigned rectangular Cartesian frame Φ but referred to the moving, intrinsic frame ψ. Said differently, the intrinsic velocity and acceleration components are the instantaneous projections upon the moving, intrinsic frame ψ of the velocity and acceleration as seen by an observer stationed at O in frame Φ. The observer fixed at O first determines $\mathbf{v}$ and $\mathbf{a}$ and afterwards projects them instantaneously upon the axes of the intrinsic frame simply as a matter of convenience. As a consequence, he always finds (1.70) and (1.71). The relation between motion *referred* to an arbitrary moving reference frame and motion *relative* to it will be discussed in greater detail in Chapter 4.

To see this more graphically, let us return to our earlier example of the torsional oscillations of a circular disk. The motion, velocity, and acceleration of a particle P on the rim of the disk are given by (1.38), (1.39), and (1.40) relative to the fixed Cartesian frame $\varphi = \{O; \mathbf{i}_k\}$ shown in Fig. 1.14a. Guided by our previous discussion of motion on a circle, we know that the intrinsic frame $\psi = \{P; \mathbf{t}_k\}$ has the instantaneous orientation shown in Fig. 1.14a. We want to show that the intrinsic velocity and acceleration components are the

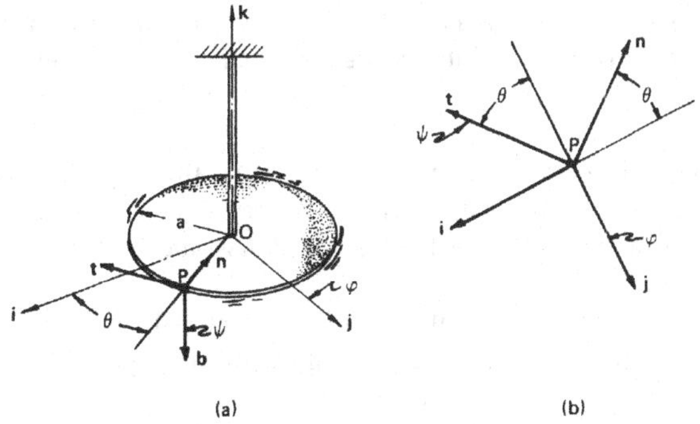

Figure 1.14. Torsional oscillations around a circle referred to the intrinsic frame.

instantaneous projections upon the moving, intrinsic frame ψ of the velocity and acceleration relative to frame φ. The relevant problem geometry is illustrated in Fig. 1.14b. It is seen that the instantaneous projections of the unit vectors $\mathbf{i}$ and $\mathbf{j}$ upon the intrinsic directions $\mathbf{t}$ and $\mathbf{n}$ are given by

$$\mathbf{i} = \sin\theta\mathbf{t} - \cos\theta\,\mathbf{n}, \qquad \mathbf{j} = -\cos\theta\,\mathbf{t} - \sin\theta\,\mathbf{n}. \tag{1.77}$$

Substitution of (1.77) into (1.38), (1.39), and (1.40), and use of a familiar trigonometric identity yields the results desired:

$$\mathbf{x}(P, t) = -a\mathbf{n}, \tag{1.78a}$$

$$\mathbf{v}(P, t) = -a\dot{\theta}\mathbf{t}, \tag{1.78b}$$

$$\mathbf{a}(P, t) = -a\ddot{\theta}\mathbf{t} + a\dot{\theta}^2\mathbf{n}. \tag{1.78c}$$

These equations still describe the motion, velocity, and acceleration of P *relative to* the fixed Cartesian frame $\varphi = \{O; \mathbf{i}_k\}$, but the vectors are now *referred to* the moving, intrinsic frame $\psi = \{P; \mathbf{t}_k\}$. Their new simplicity is evident.

Of course, the equations (1.78b) and (1.78c) may be derived directly from (1.74) for the circular motion of a particle. We must remember, however, that θ in Fig. 1.14a initially is decreasing in time. Therefore, with $\omega = -\dot{\theta}$, $\dot{\omega} = -\ddot{\theta}$, and $r = a$, it is seen that (1.78b) and (1.78c) follow easily from (1.74).

1.7.4. Some Applications of the Intrinsic Velocity and Acceleration

Some examples that illustrate various methods used in the analysis of problems involving intrinsic quantities will be studied next. The formula for the curvature of a plane curve will be reviewed in the solution of the first example, and the formula will be used in two others that follow. Finally, the motion of a particle on a helix will be revisited to find the intrinsic velocity and acceleration; and some interesting properties of this useful curve will be discussed.

Example 1.11. A particle moves along a parabolic path $y = kx^2$ in such a way that the component $\dot{x}$ is a constant C in $\Phi = \{O; \mathbf{i}_k\}$. What are the intrinsic velocity and acceleration of the particle? Assume that $k > 0$.

Solution. To obtain explicit formulas for (1.70) and (1.71), we need to find $\dot{s}$, $\ddot{s}$, and κ. Since $\dot{x} = C$, the equation of the path yields $\dot{y} = 2kx\dot{x} = 2Ckx$. Then with $\dot{s} = (\dot{x}^2 + \dot{y}^2)^{1/2}$ we obtain

$$\dot{s} = C(1 + 4k^2x^2)^{1/2}. \tag{1.79}$$

A second differentiation yields

$$\ddot{s} = 4k^2C^2x(1+4k^2x^2)^{-1/2}. \tag{1.80}$$

It remains to determine the curvature.

To derive the equation for the curvature of any smooth plane curve $y = y(x)$, we recall (1.72) and introduce the angle θ that the tangent to the curve makes with the x axis, as shown in Fig. 1.11. Then $\tan\theta = dy/dx$ and differentiation of each term with respect to s yields

$$\frac{d}{ds}\tan\theta = \frac{d\theta}{ds}\sec^2\theta = \frac{d\theta}{ds}\left(\frac{ds}{dx}\right)^2 \quad \text{and} \quad \frac{d}{ds}\left(\frac{dy}{dx}\right) = \frac{d^2y}{dx^2}\frac{dx}{ds}.$$

Equating the last terms in these expressions and using the relation $ds^2 = dx^2 + dy^2$, we find $d\theta/ds$. Then by (1.72) we get, finally,

$$\kappa = \left|\frac{d\theta}{ds}\right| = \left|\frac{d^2y/dx^2}{[1+(dy/dx)^2]^{3/2}}\right|.$$

This formula gives the curvature for any given smooth, plane curve $y = y(x)$. The derivation of the curvature formula in the case when the plane curve is described by $x = x(y)$ is left to the reader. We find that the curvature of a *plane curve* is determined by

$$\kappa = \left|\frac{d^2y/dx^2}{[1+(dy/dx)^2]^{3/2}}\right| \tag{1.81a}$$

$$= \left|\frac{d^2x/dy^2}{[1+(dx/dy)^2]^{3/2}}\right|. \tag{1.81b}$$

In particular, for the parabola $y = kx^2$ we find $dy/dx = 2kx$, $d^2y/dx^2 = 2k$. Using these values in (1.81a) and recalling that $k > 0$, we obtain

$$\kappa = 2k(1+4k^2x^2)^{-3/2}. \tag{1.82}$$

Thus, collecting the results (1.79), (1.80), and (1.82) in (1.70) and (1.71), we find the intrinsic velocity and acceleration of the particle:

$$\mathbf{v} = C(1+4k^2x^2)^{1/2}\,\mathbf{t}, \qquad \mathbf{a} = 2kC^2(1+4k^2x^2)^{-1/2}(2kx\mathbf{t} + \mathbf{n}). \tag{1.83}$$

Let the student consider what happens to these results if $k < 0$.

Since the path is known, it is an easy geometrical problem to express $\mathbf{t}$

and **n** in terms of **i** and **j** whenever this may be necessary. Unless it is specified that the results must be expressed in $\Phi = \{0; \mathbf{i}_k\}$, it is understood that the answers may be left in terms of $\psi = \{P; \mathbf{t}_k\}$ without our having to write the vectors $\mathbf{t}_k$ as functions of $\mathbf{i}_k$. A sample case where this is required will be illustrated in Example 1.13. □

Example 1.12. Suppose that a particle P moves on the parabola $y = kx^2$ with a constant speed $\dot{s} = 5$ cm/sec and passes the point $(1, 1)$ in $\Phi = \{0; \mathbf{i}, \mathbf{j}\}$. What is the greatest acceleration that the particle P experiences on its path? What is the radius of curvature at the point $(1, 1)$?

Solution. In order that P may pass the point $(1, 1)$ on $y = kx^2$, it is necessary that $k = 1$; hence, $y = x^2$ is the path of interest. Since $\dot{s} = 5$ cm/sec is constant, $\ddot{s} = 0$ and by (1.71) we have

$$\mathbf{a} = 25\kappa\mathbf{n}. \tag{1.84}$$

Thus, **a** is greatest where κ is greatest, i.e., where R is least. Of course, we expect intuitively that this occurs at $(0, 0)$. The result may be established by use of (1.82). For the case $k = 1$, we obtain

$$R = \tfrac{1}{2}(1 + 4x^2)^{3/2}. \tag{1.85}$$

This formula shows that R is least at $x = 0$. Hence, $R_{\min} = 1/2$ cm and $\kappa_{\max} = 2\ \text{cm}^{-1}$. The solution, by (1.84), is $\mathbf{a}_{\max} = 50\mathbf{n} = 50\mathbf{i}$ cm/sec^2 at the origin $(0, 0)$. The radius of curvature at $(1, 1)$ is found from (1.85); we get $R = 5^{3/2} \div 2$ cm. □

Example 1.13. A small guide pin P is attached to a telescopic arm OP of a bell crank mechanism which is hinged at O. The pin must move in a parabolic track as shown in Fig. 1.15a. At point A it has a speed of 10 ft/sec

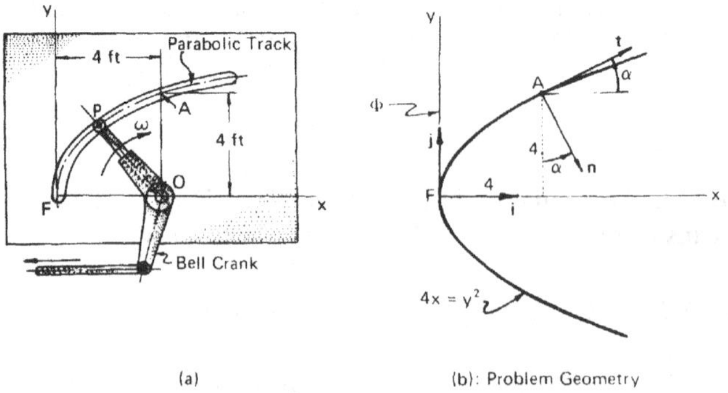

Figure 1.15. Pin motion in a bell crank mechanism.

and a rate of change of speed of 20 ft/sec^2 along the track. What is the acceleration of the pin at point A in the Cartesian frame $\Phi = \{F; \mathbf{i}_k\}$ in Fig. 1.15b?

Solution. The standard equation of the parabolic track shown in Fig. 1.15 is $x = cy^2$. Since the curve contains the point (4, 4), the constant $c = 1/4$ and the actual path equation is

$$4x = y^2. \tag{1.86}$$

The problem data being expressed in terms of intrinsic quantities suggest use of the intrinsic representation for the acceleration. With $\dot{s} = 10$ ft/sec and $\ddot{s} = 20$ ft/sec^2, the acceleration given by (1.71) is $\mathbf{a} = 20\mathbf{t} + 100\kappa\mathbf{n}$ at A. The curvature of the path (1.86) at A may be determined by (1.81b). Using (1.86), we evaluate

$$\frac{dx}{dy} = \frac{y}{2}, \qquad \frac{d^2x}{dy^2} = \frac{1}{2}. \tag{1.87}$$

At the point $A = (4, 4)$ we have $dx/dy = 2$, $d^2x/dy^2 = 1/2$; and by (1.81b) we find $\kappa = \sqrt{5}/50$. Therefore, the acceleration at A is given by

$$\mathbf{a} = 20\mathbf{t} + 2\sqrt{5}\,\mathbf{n} \text{ ft/sec}^2. \tag{1.88}$$

But (1.88) is referred to the intrinsic basis whereas the solution is required in the Cartesian basis. Therefore, a change of basis from $\mathbf{t}_k$ into $\mathbf{i}_k$ is needed.

It is clear from the geometry shown in Fig. 1.15b that

$$\mathbf{t} = \cos\alpha\,\mathbf{i} + \sin\alpha\,\mathbf{j}, \qquad \mathbf{n} = \sin\alpha\,\mathbf{i} - \cos\alpha\,\mathbf{j}, \tag{1.89}$$

wherein α is determined by $\tan\alpha = dy/dx = 2/y$. Evaluating this at $A = (4, 4)$, we get $\tan\alpha = 1/2$; and from this result we determine $\sin\alpha = 1/\sqrt{5}$, $\cos\alpha = 2/\sqrt{5}$. Now (1.89) may be written as

$$\mathbf{t} = \frac{\sqrt{5}}{5}(2\mathbf{i} + \mathbf{j}), \qquad \mathbf{n} = \frac{\sqrt{5}}{5}(\mathbf{i} - 2\mathbf{j}), \tag{1.90}$$

Substitution of (1.90) into the acceleration (1.88) yields

$$\mathbf{a} = 2[1 + 4\sqrt{5}]\,\mathbf{i} + 4[\sqrt{5} - 1]\,\mathbf{j} \text{ ft/sec}^2. \tag{1.91}$$

This is the acceleration of the guide pin at point A in the Cartesian frame Φ. (See Problem 1.53.) □

Example 1.14. The helix is a basic curve found in the design of various kinds of machines used to move solid or granular materials through a screw feed or sorting hopper device. Helices also are used in design of impeller blades of certain air and water pumps, and for drills and screw drive systems. Therefore, it is useful to understand some of the basic geometrical properties of a helix. It is known, for example, that a helix has the unique characteristic that its tangent at each point makes a constant angle with a line parallel to its axis. This property is useful in problems concerning a helical motion similar to the rotation and translation of a nut on a threaded shaft. Establish this result for the helix defined by (1.4). Find the intrinsic velocity and acceleration in this motion.

Solution. We wish to show that the angle γ between the axial unit direction $\mathbf{k}$ and the unit tangent vector $\mathbf{t}$ to the helix described by (1.4) is a constant. Let us recall equation (1.15) for the velocity vector in the helical motion (1.4). Then, according to (1.70), the tangent vector to the helix is given by

$$\mathbf{t} = \mathbf{v}/v = [R\omega(-\sin \omega t\, \mathbf{i} + \cos \omega t\, \mathbf{j}) + A\mathbf{k}]/(R^2\omega^2 + A^2)^{1/2}, \tag{1.92}$$

wherein A, R, and ω are constants. Hence, the speed also is constant:

$$v = \dot{s} = (R^2\omega^2 + A^2)^{1/2}. \tag{1.93}$$

It follows from (1.92) that the angle between $\mathbf{k}$ and $\mathbf{t}$ is given by

$$\mathbf{k} \cdot \mathbf{t} = \cos \gamma = A(R^2\omega^2 + A^2)^{-1/2}, \tag{1.94}$$

which is a constant. Thus, *the tangent at each point on a circular helix makes a constant angle with its axis.* This result is more useful than (1.94) suggests. More generally, the tangent line property shows that when a helix is rolled on a plane, in one revolution its tangent traces a straight line that forms the hypotenuse of a right triangle of altitude p, the pitch of the helix, and base $2\pi R$ as shown in Fig. 1.16. This triangle is called the *pitch triangle.* When a particle moves on a circular helix, it rotates through an angle $\omega t = \theta(t)$ about the helix axis, as described in Fig. 1.2, and it traces in the xy plane a circular arc of length $R\theta(t)$ as it advances a distance $z(t)$ along that axis. We see from the pitch triangle that $\tan \gamma = 2\pi R/p = R\theta(t)/z(t)$. Hence, *the axial advance along a circular helix is proportional to the angle of rotation about its axis,*

$$2\pi z(t) = p\theta(t); \tag{1.95}$$

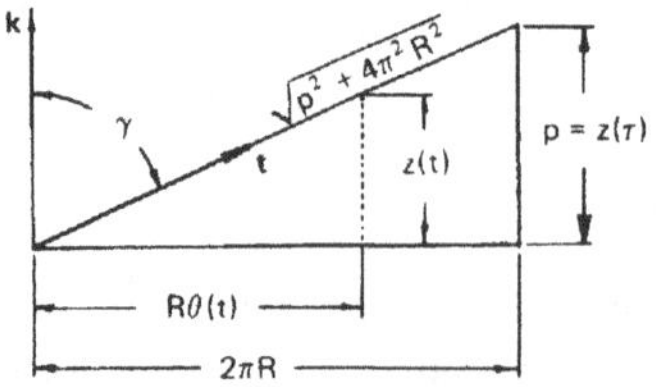

Figure 1.16. Pitch triangle for a circular helix.

and the invariant tangent angle γ *of a circular helix is determined uniquely by the ratio of the circumference of its base circle to its pitch*:

$$\tan \gamma = 2\pi R/p. \tag{1.96}$$

These results will now be applied to find the intrinsic velocity and acceleration for the helical motion (1.4).

We recall from (1.4) or (1.7) that $z(t) = At$ and $\theta = \omega t$. Use of these relations in (1.95) yields $A = \omega p/2\pi$, which is the same result derived differently at the close of Section 1.3. Substitution of this equation into (1.93) and use of (1.96) gives

$$v = R\omega \left[1 + \left(\frac{p}{2\pi R} \right)^2 \right]^{1/2} = R\omega \csc \gamma. \tag{1.97}$$

Since the speed (1.97) is a constant, (1.71) reduces to $\mathbf{a} = \kappa v^2 \mathbf{n}$. But this must be the same as (1.16):

$$\mathbf{a} = \kappa v^2 \mathbf{n} = -R\omega^2 [\cos \omega t\, \mathbf{i} + \sin \omega t\, \mathbf{j}]. \tag{1.98}$$

Therefore, it follows that

$$|\mathbf{a}| = \kappa v^2 = R\omega^2 \tag{1.99a}$$

and

$$\mathbf{n} = -\cos \omega t\, \mathbf{i} - \sin \omega t \mathbf{j}. \tag{1.99b}$$

Substitution of (1.97) and (1.99a) into (1.70) and (1.71) yields the intrinsic velocity and acceleration in the helical motion (1.4):

$$\mathbf{v}(P, t) = R\omega \csc \gamma \mathbf{t}, \qquad \mathbf{a}(P, t) = R\omega^2 \mathbf{n}, \tag{1.100}$$

in which $\mathbf{t}$ and $\mathbf{n}$ are given explicitly by (1.92) and (1.99b). The principal normal vector $\mathbf{n}$ at each point along the helix is directed perpendicular to the helix axis $\mathbf{k}$. Hence, the osculating plane of $\mathbf{t}$ and $\mathbf{n}$ slides along the helix at the constant angle γ defined by (1.96).

Finally, let us observe that the curvature may be found from (1.99a): $\kappa = 1/\rho = R\omega^2/v^2$, where ρ denotes the radius of curvature. This may be simplified further by use of (1.97) and the pitch triangle. We find the interesting result

$$\rho = \frac{1}{\kappa} = \frac{R}{\sin^2 \gamma} = R \left[1 + \left(\frac{p}{2\pi R} \right)^2 \right]. \tag{1.101}$$

Thus, *the radius of curvature of a circular helix is a constant determined by its pitch and the radius of its base circle*. Moreover, since $\sin \gamma < 1$, the radius of curvature is larger than the radius of its base circle. Notice that when $\gamma = \pi/2$, the helix degenerates to its base circle, and then $\rho = R$. When $\gamma = 0$, the helix degenerates to its axis, and in that case $\rho = \infty$. The result (1.101) may be used to rewrite (1.100) in terms of the radius of curvature; we get

$$\mathbf{v}(P, t) = \rho\omega \sin \gamma \mathbf{t}, \qquad \mathbf{a}(P, t) = \rho(\omega \sin \gamma)^2 \, \mathbf{n}. \tag{1.102}$$

Notice that these results are similar to the equations (1.74) for a motion with constant angular speed $\omega \sin \gamma$ on a circle of radius ρ.

The intrinsic velocity and acceleration also may be derived by straightforward application of (1.75) to the Cartesian equations (1.15) and (1.16). This exercise is left for the student.

1.8. Summary of Particle Kinematics

Our study of particle kinematics has evolved naturally from four primary definitions: reference frame, motion, velocity, acceleration.

(i) A reference frame $\varphi = \{O; \mathbf{e}_k\}$ is a set consisting of a suitable origin point O in space and a suitable triple of mutually perpendicular unit vectors $\mathbf{e}_k$.

(ii) A motion of a particle P relative to φ is defined by its time-varying position vector in φ:

$$\mathbf{x} = \mathbf{x}_\varphi(P, t) \qquad [\text{cf. } (1.1)].$$

This specifies the locus of places occupied by the particle as a function of time.

(iii) The velocity of P relative to φ is the time rate of change of its position vector in frame φ:

$$\mathbf{v} = \mathbf{v}_\varphi(P, t) = \dot{\mathbf{x}}_\varphi(P, t) \qquad [\text{cf. } (1.8)].$$

The magnitude of $\mathbf{v}$ is called the speed: $v = |\mathbf{v}| = (\mathbf{v} \cdot \mathbf{v})^{1/2}$.

(iv) The acceleration of P relative to frame φ is the time rate of change of its velocity vector in φ:

$$\mathbf{a} = \mathbf{a}_\varphi(P, t) = \dot{\mathbf{v}}_\varphi(P, t) = \ddot{\mathbf{x}}_\varphi(P, t) \qquad [\text{cf. } (1.9)\text{–}(1.10)].$$

When the identity of φ is clear, the subscript notation may be suppressed.

In a rectangular Cartesian reference frame $\varphi = \{O; \mathbf{i}_k\}$, we have the useful explicit representations

$$\mathbf{x} = x(t)\,\mathbf{i} + y(t)\,\mathbf{j} + z(t)\,\mathbf{k} \qquad [\text{cf. } (1.6)],$$

$$\mathbf{v} = \dot{x}(t)\,\mathbf{i} + \dot{y}(t)\,\mathbf{j} + \dot{z}(t)\,\mathbf{k} \qquad [\text{cf. } (1.11)],$$

$$\mathbf{a} = \ddot{x}(t)\,\mathbf{i} + \ddot{y}(t)\,\mathbf{j} + \ddot{z}(t)\,\mathbf{k} \qquad [\text{cf. } (1.12)].$$

The speed of the particle is the time rate of change of the distance $s(t)$ traveled along its path:

$$v = \dot{s}(t) = (\dot{x}^2 + \dot{y}^2 + \dot{z}^2)^{1/2} \qquad \text{[cf. (1.13)]}.$$

The velocity and acceleration relative to φ have an especially simple and useful representation when referred to the intrinsic reference frame $\psi = \{P; \mathbf{t}_k\}$ that follows the particle:

$$\mathbf{v} = \dot{s}\mathbf{t}, \qquad \mathbf{a} = \ddot{s}\mathbf{t} + \kappa\dot{s}^2\mathbf{n} \qquad \text{[cf. (1.70)–(1.71)]}.$$

The velocity vector is in the direction **t** tangent to the path. The acceleration, if it is not zero, is in the osculating plane and directed toward the concave side of the path, as shown in Fig. 1.13. Among all planes at the point P on the path, the osculating plane lies nearest to the curve at P; it is determined by **t** and the principal normal vector **n** directed from P toward the center of curvature. The curvature κ, or its reciprocal, the radius of curvature R, is a measure of the rate of turning of the tangent line along the path:

$$\kappa = \frac{1}{R} = \left|\frac{d\mathbf{t}}{ds}\right| = \left|\frac{d\theta}{ds}\right| \qquad \text{[cf. (1.72)]},$$

wherein θ is the angle that the tangent vector makes with a fixed line in the osculating plane. All of these properties are independent of the coordinate system used in the spatial frame. See also (1.75) and (1.76).

In a rectangular Cartesian reference frame in which the motion is confined to a plane, the curvature is given by

$$\kappa = \left|\frac{d^2y/dx^2}{[1 + (dy/dx)^2]^{3/2}}\right| = \left|\frac{d^2x/dy^2}{[1 + (dx/dy)^2]^{3/2}}\right| \qquad \text{[cf. (1.81)]}.$$

This short list of important relations must be remembered. The rest, except for a few easy definitions of terms, should be seen as following naturally and logically from these few principal equations rather than by rote memorization of other special formulas. Various methods that are useful in the solution of a wide variety of problems have been described in the examples and expanded further in the selection of problems. Future subject matter builds continuously upon this foundation; therefore, skills developed here will be sharpened further in new applications that lie ahead in the development of other useful kinematical formulas for rigid body motion and for motion relative to a moving reference frame. Some special introductory topics that expand upon our work in this chapter are presented in the next section. However, this matter may be omitted in a first reading with no significant interruption in continuity, if the reader prefers to move forward to Chapter 2.

1.9. Special Topics

Three special topics of an introductory nature will be presented. We begin with the continuation of our study of the intrinsic description of particle motion to account for the rate of rotation of the binormal vector. We shall see that this characterizes the turning motion of the osculating plane as it twists along the path. Our second topic involves the geometrical description of the velocity vector as a "path writer" in the same way that the position vector is the "path writer" in the actual motion of the particle. This velocity vector path is called the hodograph. Finally, we conclude with an introduction to singularity functions and their application in some elementary kinematics problems.

1.9.1. Some Additional Properties of the Intrinsic Basis

We have learned in (1.72) that the rate of rotation of the tangent vector describes the curvature of the path. Similarly, because the unit binormal vector is always perpendicular to the osculating plane, its directional change along the trajectory characterizes the rotational motion of the osculating plane. Since the normal vector must stay in the osculating plane and remain perpendicular to both the tangent and binormal vectors, its rotation is determined by theirs. These additional rotational effects will be studied below.

We begin by recalling that the intrinsic vectors **t**, **n**, **b** are mutually perpendicular unit vectors. Hence,

$$\mathbf{t} \cdot \mathbf{t} = 1, \tag{1.103a}$$

$$\mathbf{n} \cdot \mathbf{n} = 1, \tag{1.103b}$$

$$\mathbf{b} \cdot \mathbf{b} = 1, \tag{1.103c}$$

and

$$\mathbf{t} \cdot \mathbf{n} = 0, \tag{1.104a}$$

$$\mathbf{n} \cdot \mathbf{b} = 0, \tag{1.104b}$$

$$\mathbf{b} \cdot \mathbf{t} = 0. \tag{1.104c}$$

Then differentiation of (1.103c) shows that $\mathbf{b} \cdot d\mathbf{b}/ds = 0$. Therefore, $d\mathbf{b}/ds$ is perpendicular to **b** and lies in the plane of **t** and **n**. Consequently, we may write

$$\frac{d\mathbf{b}}{ds} = \alpha\mathbf{t} + \tau\mathbf{n}, \tag{1.105}$$

where α and τ are certain scalars. These are determined next.

To find α, we form the dot product of (1.105) with $\mathbf{t}$ and use (1.103a) and (1.104a) to get $\alpha = \mathbf{t} \cdot d\mathbf{b}/ds$. But differentiation of (1.104c) and use of (1.65b) and (1.104b) reveals that

$$\frac{d\mathbf{b}}{ds} \cdot \mathbf{t} + \mathbf{b} \cdot \frac{d\mathbf{t}}{ds} = \alpha + \mathbf{b} \cdot \kappa \mathbf{n} = \alpha = 0.$$

Hence, (1.105) shows that $d\mathbf{b}/ds$ is parallel to $\mathbf{n}$:

$$\frac{d\mathbf{b}}{ds} = \tau \mathbf{n} \qquad \text{with} \quad \tau = \pm \left| \frac{d\mathbf{b}}{ds} \right|. \tag{1.106}$$

We see that the scalar τ, which may be positive, negative, or zero, is a measure of the rate of rotation of the binormal vector as the particle moves on its path. Since the change (1.106) in the unit vector $\mathbf{b}$ is parallel to $\mathbf{n}$, and $\mathbf{b}$ is always perpendicular to both $\mathbf{t}$ and $\mathbf{n}$, we see that τ measures the twisting rotation of the osculating plane in either direction about the tangent line. In particular, if $\tau < 0$, $d\mathbf{b}/ds$ has the direction $-\mathbf{n}$; and, in this case, $\mathbf{b}$ revolves around $\mathbf{t}$ in the right-hand sense of a screw advancing along $\mathbf{t}$ as the particle advances on its path. Therefore, τ is called the *torsion* of the curve. Clearly, $[\tau] = [L^{-1}]$ follows from (1.106). When the path is a plane curve, $\mathbf{b}$ is a constant vector and (1.106) shows that *every plane curve has zero torsion.*

Now let us consider the rotation of the normal vector. Since $\mathbf{t}$, $\mathbf{n}$, $\mathbf{b}$ form an orthonormal basis, we have $\mathbf{n} = \mathbf{b} \times \mathbf{t}$. Therefore, with (1.65a) and (1.106), we get $d\mathbf{n}/ds = \mathbf{b} \times \kappa \mathbf{n} + \tau \mathbf{n} \times \mathbf{t}$ [see (A.25) in Appendix A]; that is,

$$d\mathbf{n}/ds = -\kappa \mathbf{t} - \tau \mathbf{b} \qquad \text{with} \quad \left| \frac{d\mathbf{n}}{ds} \right| = (\kappa^2 + \tau^2)^{1/2}. \tag{1.107}$$

This is the scalar rate at which $\mathbf{n}$ rotates with respect to s. Of course, since $\mathbf{n}$ is constrained to follow the rotations of $\mathbf{t}$ and $\mathbf{b}$, its rate of rotation, as shown in (1.107), is determined by κ and τ. For a plane curve, $\tau = 0$ and (1.107) confirms that $\mathbf{n}$ rotates at the same rate κ at which $\mathbf{t}$ turns with respect to s.

Collecting our results (1.65a), (1.106), and (1.107), we have the following set of *intrinsic rotation equations*:

$$d\mathbf{t}/ds = \kappa \mathbf{n}, \tag{1.108a}$$

$$d\mathbf{b}/ds = \tau \mathbf{n}, \tag{1.108b}$$

$$d\mathbf{n}/ds = -\kappa \mathbf{t} - \tau \mathbf{b}. \tag{1.108c}$$

In other books these basic equations often are called the *Serret–Frenet for-*

mulas. It can be shown by aid of (1.108) that the curvature and torsion of any smooth path are given by (cf. Problem 1.81)

$$\kappa = \left(\frac{d^2\mathbf{x}}{ds^2} \cdot \frac{d^2\mathbf{x}}{ds^2} \right)^{1/2}, \qquad \tau = -R^2 \frac{d\mathbf{x}}{ds} \cdot \frac{d^2\mathbf{x}}{ds^2} \times \frac{d^3\mathbf{x}}{ds^3}. \tag{1.109}$$

We shall discover some interesting additional results from the following easy applications of (1.108).

Example 1.15. A particle P moves on a path for which the ratio of its curvature to its torsion at every point is a constant μ. Describe the path and find the torsion.

Solution. Information about the nature of the path may be obtained from the differential equations (1.108) which describe its bending and twisting. We want to characterize the paths for which $\mu = \kappa/\tau$ is a constant. To accomplish this, we notice that elimination of $\mathbf{n}$ between the first two equations in (1.108) yields the general relation

$$\frac{d\mathbf{t}}{ds} = \mu \frac{d\mathbf{b}}{ds} \quad \text{with} \quad \mu = \kappa/\tau. \tag{1.110}$$

Thus, for constant μ, (1.110) yields the simpler equation

$$\frac{d}{ds}(\mathbf{t} - \mu\mathbf{b}) = \mathbf{0},$$

whose general solution is

$$\mathbf{t} - \mu\mathbf{b} = \mathbf{d}, \tag{1.111}$$

wherein $\mathbf{d}$ is a constant vector. It follows from (1.111) that $\mathbf{d}$ has the magnitude

$$d = (1 + \mu^2)^{1/2}. \tag{1.112}$$

In addition, it is seen that the dot product of (1.111) by $\mathbf{t}$ and use of (1.103a) and (1.104c) yield the condition $\mathbf{t} \cdot \mathbf{d} = d \cos \gamma = 1$, where $\gamma \equiv \langle \mathbf{t}, \mathbf{d} \rangle$ denotes the angle between $\mathbf{t}$ and $\mathbf{d}$. That is, with (1.112),

$$\cos \gamma = (1 + \mu^2)^{-1/2}, \tag{1.113}$$

which is a constant. Since $\mathbf{d}$ has a constant direction in the plane of $\mathbf{t}$ and $\mathbf{b}$, this result shows that the tangent to the path at each point makes a constant angle with the fixed direction $\mathbf{d}$. However, as explained in Example 1.14, this property is unique to a helix. We thus learn that *the only twisted paths for which the ratio* $\mu = \kappa/\tau$ *is a constant are helices.*

Although the radius R and pitch p of the helix are unknown, we can find an expression for the torsion of a circular helix in terms of these quantities. The pitch triangle described by (1.113) and (1.96) yields

$$\tan\gamma = \mu = \frac{2\pi R}{p}. \tag{1.114}$$

Since $\kappa = \mu\tau$, (1.101) and the second half of (1.114) show that *the torsion of a circular helix is a constant determined by its pitch and the radius of its base circle*:

$$\tau = \frac{1}{R}\left\{\frac{p}{2\pi R} \div \left[1 + \left(\frac{p}{2\pi R}\right)^2\right]\right\}. \tag{1.115} \quad \square$$

Example 1.16. If the path of a particle is traced on a fixed surface S, the tangent vector to the path also is tangent to S; but the normal vector $\mathbf{n}$ is not necessarily perpendicular to S. This is evident for the circular path 1 on the surface of the sphere shown in Fig. 1.17, for example. If $\mathbf{n}$ is perpendicular to S at every point of the trajectory, the path is called a *geodesic* on S. In particular, the great circle along path 2 shown in Fig. 1.17 is a geodesic on the sphere. Indeed, it is clear that these are the only curves for which $\mathbf{n}$ can be parallel to the normal vector $\mathbf{e}$ along every radial line to the curve from the center of the sphere. Thus, *the only geodesics on the sphere are great circles.* Other examples are less obvious. Find the geodesics on a right circular cylinder.

Solution. The normal vector $\mathbf{e}_r$, say, at each point on the surface of a right circular cylinder is directed along a radial line perpendicular to the cylinder axis $\mathbf{k}$. Therefore, along a geodesic on the cylinder surface the principal normal vector $\mathbf{n} = -\mathbf{e}_r$, so it also is perpendicular to $\mathbf{k}$: $\mathbf{n}\cdot\mathbf{k} = 0$. It follows from (1.108a) that $\mathbf{k}\cdot d\mathbf{t}/ds = d(\mathbf{k}\cdot\mathbf{t})/ds = 0$; hence, $\mathbf{k}\cdot\mathbf{t} = \cos\langle\mathbf{t},\mathbf{k}\rangle = \text{const.}$

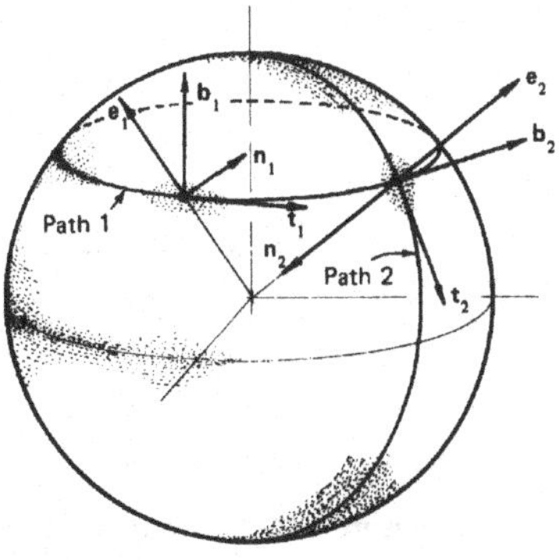

Figure 1.17. Intrinsic motion and the geodesics on the surface of a sphere.

But the only curves whose tangent at each point makes a constant angle with a fixed direction are helices. Thus, *the only geodesic paths on the surface of a cylinder are helices.* Notice that when $\mathbf{k} \cdot \mathbf{t} = 0$, the geodesic on the cylinder is the plane circle normal to $\mathbf{k}$; and when $\mathbf{k} \cdot \mathbf{t} = 1$, the geodesics are axial lines on the cylinder. These are degenerate helices having zero and infinite pitch, respectively.

Finally, it must be mentioned that it is shown in books on differential geometry that among all curves joining two neighboring points on a smooth surface, a geodesic curve is the shortest path connecting them. Thus, the shortest distance between two points on the surface of a sphere is along the arc of a great circle, while on the surface of a cylinder the shortest path is an arc of a helix, its base circle, or a line parallel to its axis.

1.9.2. The Hodograph

The intrinsic equations have revealed interesting geometrical qualities of the velocity and acceleration that are independent of the coordinate system used in the spatial frame. The velocity vector is tangent to the path traced by the position vector, and the acceleration is directed in the osculating plane toward the concave side of the path. In this section, we introduce a simpler kind of geometrical description for the velocity and acceleration that uses the velocity vector as the path writer.

Imagine a fictitious particle P_H whose "position vector" $\mathbf{x}_H$ relative to an origin $0'$ is equal to the velocity vector of the particle P in the actual motion; and let us write $\mathbf{v}_H \equiv \dot{\mathbf{x}}_H$ for the "velocity" of P_H. Then

$$\mathbf{x}_H = \mathbf{v}, \tag{1.116a}$$

$$\mathbf{v}_H = \mathbf{a}. \tag{1.116b}$$

The "motion" $\mathbf{x}_H$ is called the *hodograph motion*; and the path $\mathscr{L}_H$ traced by $\mathbf{x}_H = \mathbf{v}$, as shown in Fig. 1.18b, is called the *hodograph*. In these terms, (1.116)

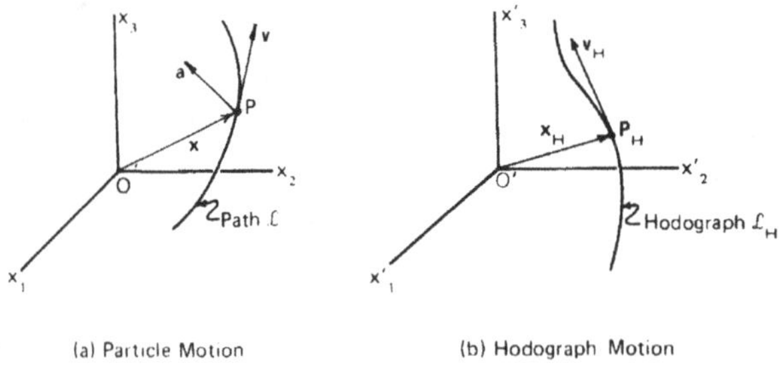

Figure 1.18. Comparison of the hodograph motion and the particle motion.

shows that the "velocity" $\mathbf{v}_H$ in the hodograph motion, called the *hodograph velocity*, is equal to the acceleration in the actual motion. Hence, *the acceleration in the particle motion always is tangent to the hodograph.* This is to be compared with the intrinsic description of the actual motion in Fig. 1.18a. Some examples follow.

Example 1.17. The velocity vector in a uniform motion of a particle is a constant vector $\mathbf{v} = \mathbf{v}_0$. Therefore, the hodograph is a point $\mathbf{x}_H = \mathbf{v}_0$, constant. Notice from (1.116b) that the hodograph velocity is zero: $\mathbf{v}_H = \mathbf{a} = \mathbf{0}$. □

Example 1.18. If a particle has constant acceleration $\mathbf{a} \neq \mathbf{0}$, the hodograph is a path described with constant velocity $\mathbf{v}_H = \mathbf{a}$. The equation of the hodograph is obtained by integrating $\mathbf{a} = \dot{\mathbf{x}}_H$; we get $\mathbf{x}_H = \mathbf{v} = \mathbf{a}t + \mathbf{c}$, where $\mathbf{c}$ is a constant vector. Hence, the hodograph is a straight line. □

Example 1.19. Suppose that the velocity in the hodograph is always perpendicular to the hodograph position vector. Describe the hodograph. What can be said about the actual motion of the particle?

Solution. Since

$$\mathbf{x}_H \cdot \mathbf{v}_H = \frac{d}{dt}\left(\frac{1}{2}\mathbf{x}_H \cdot \mathbf{x}_H\right) = 0,$$

it follows that $\mathbf{x}_H \cdot \mathbf{x}_H =$ constant. Therefore, the hodograph is a curve on the surface of a sphere; and because $\mathbf{x}_H = \mathbf{v}$, the particle has a constant speed. Hence, in the actual motion $\mathbf{a} = \kappa \dot{s}^2 \mathbf{n}$, $\mathbf{v} = \dot{s}\mathbf{t}$ and $\dot{s} =$ const. Nothing more can be said about the actual motion.

1.9.3. Singularity Functions in Particle Kinematics

We have thus far assumed that the motion, velocity, and acceleration are continuous functions of time. But there are numerous applications where this condition is not satisfied by all, or possibly any, of these functions. Our study in Example 1.10 of the reciprocating uniform motion of the push rod showed specifically that the velocity changed abruptly at the end of each stroke but remained constant everywhere else. Behavior of this sort, in which a function is continuous except for a finite number of jump discontinuities at each of which the function has definable limits from both the right and the left, is called *piecewise continuous.*

Problems characterized by piecewise continuous behavior can be solved by ordinary methods applied to the separate continuous parts, and the solutions for adjacent parts eventually can be patched together somehow. In fact, this approach is so common that it was used without special mention in our earlier derivation of (1.58) and (1.59) for the cam shape. This procedure, though straightforward, often proves laborious and cumbersome to do except in simple cases like the symmetric cam operation. In this section, we shall study a more powerful method by which piecewise continuous functions may be treated as if they were continuous functions. To accomplish this trick, two special tools will be needed: the unit step function and the delta function. These functions and their calculus will be described next. Afterwards, the results will be applied in a review of our earlier cam design problem and in the solutions of some other simple kinematics problems.

1.9.3.1. The Unit Step Function

In problems where the position, velocity, or acceleration may change abruptly, it is convenient to introduce special functions to handle these cases. Each of these functions will behave much like a switch that gets turned on or turned off only when the independent variable x, say, takes on certain values. An important example is the *unit step function* $u(x) \equiv \langle x-a \rangle^0$ shown in Fig. 1.19. It is seen that this piecewise continuous function has the value zero for all $x < a$ and the value 1 for all $x > a$; that is, as x just exceeds the value a, the switch $u(x) = \langle x-a \rangle^0 = 1$ is turned on. The limit value of $u(x)$ as x approaches a from the left is zero, while the limit from the right is equal to one. Because the value of $u(x)$ at $x = a$ is undecided, we say that $u(x)$ is undefined at $x = a$. Thus, the unit step function shown in Fig. 1.19 may be defined by

$$u(x) \equiv \langle x-a \rangle^0 = \begin{cases} 0 & \text{if} \quad x < a \\ 1 & \text{if} \quad x > a \\ \text{undefined at } x = a. \end{cases} \tag{1.117}$$

The special angle bracket notation for $\langle x-a \rangle^0$ is used for future convenience. We shall see that the unit step function is a major building block used in construction of other piecewise continuous functions.

This mathematical idea models, among other things, the physical act of turning on a light. In this case x is identified as the time variable t. Prior to

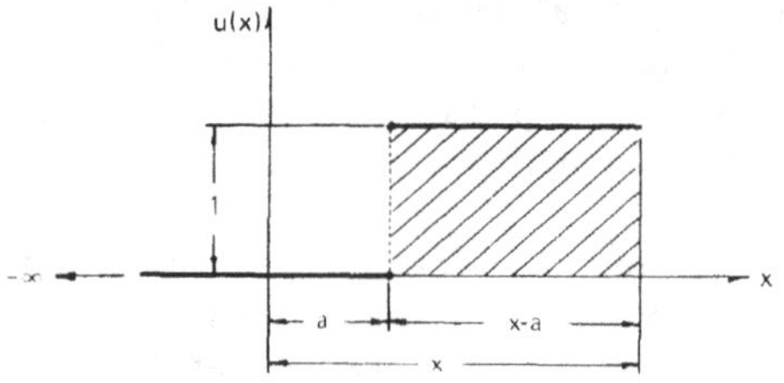

Figure 1.19. Graph of the unit step function $u(x) = \langle x-a \rangle^0$.

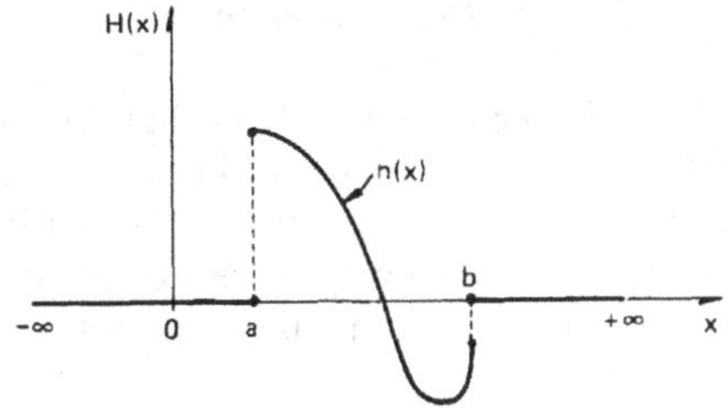

Figure 1.20. Use of the unit step function to turn on a smooth function $h(x)$ at $x=a$ and turn it off at $x=b$.

some instant $t_0 = a$, the light is off; then the switch is thrown and the light is instantaneously turned on—at least it appears so. To see how the unit step function can be used as a similar kind of mathematical switch, let us consider a continuously differentiable function $h(x)$ defined on $a \leqslant x \leqslant b$. To save words, any such function $h(x)$ will be called a *smooth function*. Then the function $H(x)$ defined by

$$H(x) = (\langle x-a\rangle^0 - \langle x-b\rangle^0)\, h(x) \tag{1.118}$$

reveals how the unit step function is used to turn on the function $h(x)$ at $x=a$ and, with use of the minus sign, turn it off again at $x=b$. Indeed, we see from (1.117) that (1.118) implies that

$$H(x) = \begin{cases} 0 & \text{if } x<a \\ h(x) & \text{if } a<x<b \\ 0 & \text{if } x>b \\ \text{undefined at } x=a \text{ and } x=b. \end{cases}$$

This switching effect is pictured in Fig. 1.20. A more specific example is given by the function

$$H(x) = (x-2)^2\langle x-1\rangle^0 \tag{1.119}$$

shown in Fig. 1.21. The parabolic function $h(x) = (x-2)^2$ is turned on at $x=1$, and it remains on indefinitely. If we wish to discontinue its use at $x=3$, say, we simply subtract the term $(x-2)^2\langle x-3\rangle^0$ from the function in (1.119).

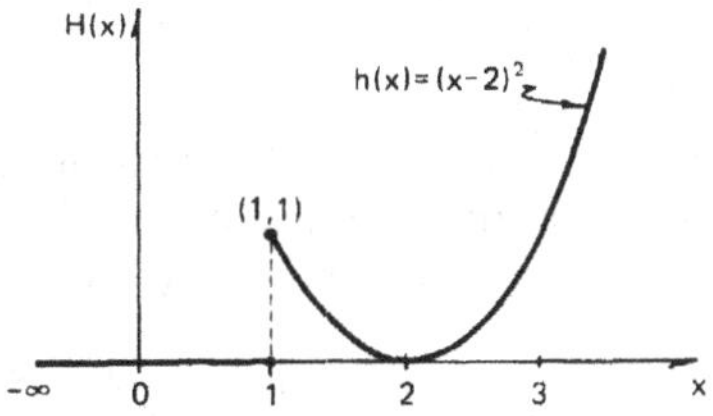

Figure 1.21. Graph of the function $H(x) = (x-2)^2\langle x-1\rangle^0$.

1.9.3.2. The Delta Function

It is seen from (1.119) that to compute the derivative of the function $H(x)$, we shall need to know the derivative of the unit step function $u(x) = \langle x-a \rangle^0$. It is evident from Fig. 1.19 that the slope of the unit step function is zero everywhere except at the discontinuity at $x=a$ where the slope of the step is infinite. This leads us to "define" another special function

$$\delta(x) \equiv \langle x-a \rangle_{-1} = \begin{cases} 0 & \text{if } x \neq a \\ \infty & \text{if } x = a, \end{cases} \tag{1.120}$$

so that $du(x)/dx = \delta(x)$; that is, using the angle bracket notation in (1.117) and (1.120), we have

$$\frac{d}{dx} \langle x-a \rangle^0 = \langle x-a \rangle_{-1}. \tag{1.121}$$

The function described by (1.120) is called the *delta function*. The subscript "-1" is used as a mnemonic device to remind us that the differentiation rule for the unit step function is similar to the usual rule for the derivative of an ordinary power function. Additional motivation for this usage and further interpretation of (1.120) will appear later on. The δ-function is especially useful in the description of concentrated or suddenly applied loads that occur in impact problems, so $\delta(x)$ also is known as the *unit impulse function*. This application will be discussed in Chapter 6.

1.9.3.3. Calculus of Singularity Functions

In view of the jump behavior of the unit step function at $x=a$ and the consequent singular behavior of the delta function at $x=a$, these functions and all others associated with them through products with smooth functions, derivatives and integrals are called *singularity functions*. It is essential that we consider a few rules governing their differentiation and integration. Our first rule (1.121) enables us to compute the derivative of the singularity function (1.119) by use of the usual product rule, for example. More generally, with the help of (1.121), we consider

$$\frac{d}{dx} [h(x)\langle x-a \rangle^0] = \frac{dh(x)}{dx} \langle x-a \rangle^0 + h(x)\langle x-a \rangle_{-1}, \tag{1.122}$$

in which $h(x)$ is any smooth function. In view of the definition (1.120), however, the product of the δ-function and the smooth function $h(x)$, continuous at $x=a$, is defined by

$$h(x)\,\delta(x) = h(x)\langle x-a \rangle_{-1} = h(a)\langle x-a \rangle_{-1}, \tag{1.123}$$

which vanishes whenever $h(a)=0$. Use of this result in (1.122) yields *the general rule for the derivative of the product of a smooth function $h(x)$ and the unit step function $\langle x-a\rangle^0$*:

$$\frac{d}{dx}[h(x)\langle x-a\rangle^0]=\frac{dh(x)}{dx}\langle x-a\rangle^0+h(a)\langle x-a\rangle_{-1}, \tag{1.124}$$

in which the last term vanishes whenever $h(a)=0$.

Let us now recall (1.119) in which $h(x)\equiv(x-2)^2$ is a smooth function whose value at $x=1$ is $h(1)=1$. Application of (1.124) gives the derivative of (1.119):

$$\frac{dH(x)}{dx}=2(x-2)\langle x-1\rangle^0+\langle x-1\rangle_{-1}.$$

More generally, let $h(x)\equiv(x-b)^n$. Then $h(a)=(a-b)^n$ and (1.124) yields

$$\frac{d}{dx}[(x-b)^n\langle x-a\rangle^0]=n(x-b)^{n-1}\langle x-a\rangle^0 + (a-b)^n\langle x-a\rangle_{-1}. \tag{1.125}$$

In particular, when $n\geqslant 1$ and $b=a$, the last term in (1.125) vanishes; and we thereby obtain the *special rule for the derivative of the power function* $\langle x-a\rangle^n$:

$$\frac{d}{dx}\langle x-a\rangle^n=n\langle x-a\rangle^{n-1} \quad \text{for} \quad n\geqslant 1, \tag{1.126}$$

in which, by definition,

$$\langle x-a\rangle^n\equiv(x-a)^n\langle x-a\rangle^0 = \begin{cases}0 & \text{if } x\leqslant a\\ (x-a)^n & \text{if } x>a\end{cases} \quad \text{for} \quad n\geqslant 1. \tag{1.127}$$

The rule (1.121) may be considered as the extension of (1.126) to the case $n=0$. This lends motivation to our earlier use of the subscript notation.

Notice that in (1.127) the value zero is admitted at $x=a$ because the function $\langle x-a\rangle^n$ for $n\geqslant 1$ is continuous and vanishes at $x=a$. In particular, the graph of the function $u_1(x)\equiv\langle x-a\rangle^1$, which is called the *unit slope function*, is shown in Fig. 1.22. It is seen that $u_1(x)$ is equal to zero for $x\leqslant a$ and equal to $(x-a)$ for $x>a$. This function is continuous with value zero at $x=a$, but its derivative $u_1'(x)$ is not. The graph shows that $u_1'(x)=0$ when $x<a$, and $u_1'(x)=1$ for $x>a$; but $u_1'(x)$ is not defined at $x=a$. Therefore, we

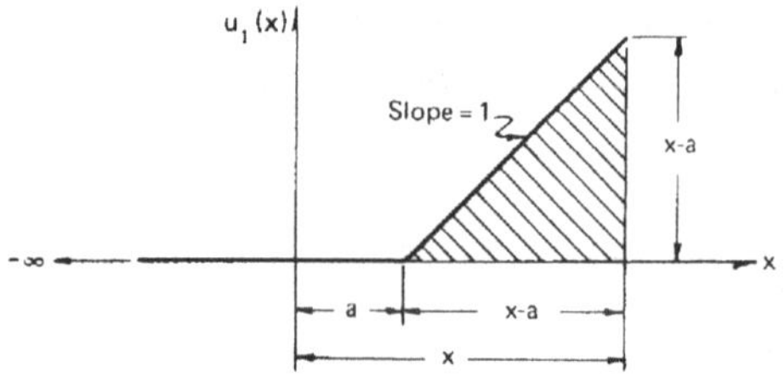

Figure 1.22. Graph of the unit slope functions $u_1(x) = \langle x-a \rangle^1$.

are able to see graphically that the derivative of $u_1(x)$ is equal to the unit step function. Indeed, from (1.126)

$$\frac{d}{dx}\langle x-a\rangle^1 = \langle x-a\rangle^0. \tag{1.128}$$

Higher derivatives of the various singularity functions may be computed similarly. For instance, application of (1.121) in (1.128) yields

$$\frac{d^2}{dx^2}\langle x-a\rangle^1 = \frac{d}{dx}\langle x-a\rangle^0 = \langle x-a\rangle_{-1} = \delta(x). \tag{1.129}$$

The motivation for introduction of the subscript notation in (1.121) is again evident here. Of course, the next derivative of (1.129) will create still another kind of singularity function, namely, $d\delta(x)/dx \equiv \langle x-a\rangle_{-2}$. This mathematical creature, called the *doublet*, will not be studied here.

To invert the differentiation process, we shall need a few easy rules for integration of singularity functions. Let us consider the integral of the product of a smooth function $h(x)$ and the unit step function expressed by

$$F(x) \equiv \int_{-\infty}^{x} h(\xi)\langle \xi - a\rangle^0\, d\xi = \begin{cases} 0 & \text{if } x < a \\ \displaystyle\int_a^x h(\xi)\, d\xi & \text{if } x > a \end{cases} \tag{1.130}$$

for all values of ξ in the interval $-\infty < \xi \leqslant x$. The quantity ξ is the dummy variable of integration, x denotes the problem variable, and a is the fixed position of the step, as usual. When $x < a$, $F(x) = 0$ because the integration runs only to the point $\xi = x < a$ where the switch $\langle \xi - a\rangle^0$ remains off. But if x exceeds a, the unit step switch is turned on, and $\langle \xi - a\rangle^0 = 1$ as before. This leads to the bracketed expression in (1.130). Thus, *the integral of the product of a smooth function and the unit step function is given by*

$$\int_{-\infty}^{x} h(\xi)\langle \xi - a\rangle^0\, d\xi = \langle x-a\rangle^0 \int_a^x h(\xi)\, d\xi. \tag{1.131}$$

In particular, consider $h(\xi) = (\xi - a)^n$ with $n = 0, 1, 2, \ldots,$ and recall

(1.127). It is seen that (1.131) yields the following *special rule for the integral of the power function* $\langle x-a\rangle^n$:

$$\int_{-\infty}^{x} \langle \xi - a\rangle^n \, d\xi = \frac{\langle x-a\rangle^{n+1}}{n+1} \qquad \text{for} \quad n=0, 1, 2,\dots. \tag{1.132}$$

Notice that the graph of the unit step function in Fig. 1.19 shows that the area under the curve is zero if $x<a$ and equal to $(x-a)$ when $x>a$. This agrees with the familiar area interpretation of (1.132) when $n=0$. And, similarly, it is easily seen that for $n=1$ the integral (1.132) is equal to the area of the triangular region shown in Fig. 1.22.

Finally, in view of the rule (1.121), *the integral of the δ-function is defined by*

$$\int_{-\infty}^{x} \delta(\xi)\, d\xi = \int_{-\infty}^{x} \langle \xi - a\rangle_{-1}\, d\xi = \langle x-a\rangle^0. \tag{1.133}$$

With the use of (1.123) and (1.133), we may also derive *the rule for integration of the product of the delta function and a smooth function which is continuous at* $x=a$:

$$\int_{-\infty}^{x} h(\xi)\, \delta(\xi)\, d\xi = \int_{-\infty}^{x} h(\xi)\langle \xi - a\rangle_{-1}\, d\xi = h(a)\langle x-a\rangle^0. \tag{1.134}$$

The behavior of the δ-function certainly is unusual. According to (1.120), $\delta(x)$ vanishes everywhere except at $x=a$ where it becomes indefinite; yet (1.133) shows that the area under its graph for $x>a$ equals one. We appreciate that no ordinary function can have these unusual properties. This behavior can be seen more clearly by review of the derivative concept used in (1.121); namely, the limit definition

$$\delta(x) = \lim_{\Delta x \to 0} \left[\frac{\langle x+\Delta x - a\rangle^0 - \langle x-a\rangle^0}{\Delta x}\right]. \tag{1.135}$$

The expression in square brackets may be interpreted graphically in Fig. 1.23 as a function $1/\Delta x$ which is turned on at $x=a-\Delta x$ and shut off at $x=a$. For a fixed value of Δx, the base of the shaded rectangular region has length Δx and height $1/\Delta x$; hence, the enclosed area is equal to 1. As Δx is made smaller, the height grows larger; but the enclosed area remains the same. This is shown by the dotted rectangles in Fig. 1.23. Thus, as $\Delta x \to 0$, the value of $\delta(x)$ grows indefinitely at $x=a$, as described by (1.120); but the dimensionless area under its graph always is equal to one, a property described by (1.133). Clearly, the physical dimension of $\delta(x)$ must be the reciprocal of that of x: $[\delta(x)] = [x^{-1}]$. We thus see graphically that the properties of the δ-function are not so strange after all.

This completes our introductory study of singularity functions and some of their important properties. In summary, we recall the definitions of the unit step and δ-functions given in (1.117) and (1.120), respectively. These functions

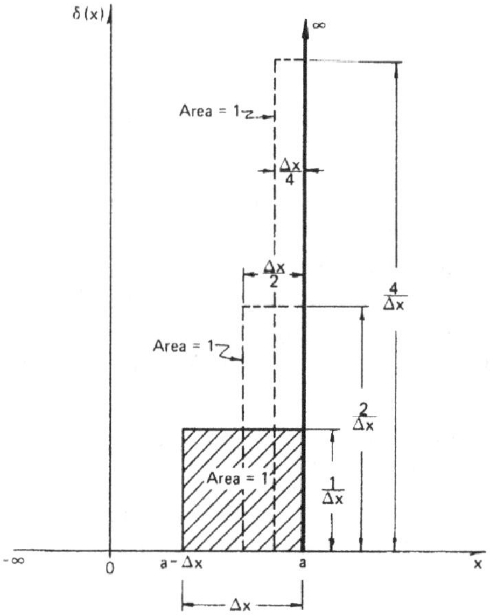

Figure 1.23. Graph of the limit rectangles for the δ-function defined in (1.135).

are related through the first derivative of the unit step function in (1.121). Differentiation and integration of the product of a smooth function and the unit step function are given by (1.124) and (1.131). Some special power rules for differentiation and integration of singularity functions are provided by (1.126) and (1.132), in which the defining relation (1.127) must be remembered. The integral of the δ-function is defined by (1.133), and (1.134) provides its integral with a smooth function. It is emphasized that, in general, products of singularity functions of the same variable are not defined. (However, see Problem 1.99.) In the next part, we shall appeal to these results in their application to some examples.

1.9.3.4. Some Applications of Singularity Functions in Kinematics

It has been emphasized along the way that singularity functions have many useful applications in mechanics and in other areas of engineering that include electrical circuit theory, electromagnetic theory, and heat transfer, for example. Therefore, it is natural that the student may expect to encounter these functions in other work. The elementary analysis of the shear and bending moment functions for beams under various loading situations are noteworthy applications studied in mechanics of deformable solids, for example. In this application, the unit step function turns the load distributions on and off as needed; and the δ-function is used to model a point load on a beam. In other physical applications the δ-function may be used to model the impact of a hammer blow, an instantaneous surge in a shorted circuit, and a lightning bolt striking a tree.

In our present work, the utility of singularity functions in some simple applications to kinematics of a particle will now be illustrated. Our first example of the motion of a mass supported by a spring mounted on a movable support involves the application of the unit step function (1.117) and the rule (1.124) for differentiation of its product with a smooth function that vanishes at the point of discontinuity.

Example 1.20. A particle of mass M is supported vertically by a spring which is fastened to a movable horizontal support. The system initially is at rest when the support experiences a sudden upward displacement of constant amplitude A. The particle is vibrating as a consequence of this disturbance when the spring support experiences at the instant t_0 a sudden downward displacement of the same amplitude. Because of these disturbances, the oscillatory motion of the mass a function of time t is described by

$$\mathbf{x}(M, t) = \mathbf{A}\{(1-\cos\omega t)\langle t-0\rangle^0 - [1-\cos\omega(t-t_0)]\langle t-t_0\rangle^0\}, \tag{1.136}$$

where $\mathbf{A} = A\mathbf{j}$, ω is a constant and the definition (1.117) is to be recalled. Find the velocity and acceleration of M, and show that the motion satisfies the following differential equation:

$$\ddot{\mathbf{x}} + \omega^2\mathbf{x} = \omega^2\mathbf{A}[\langle t-0\rangle^0 - \langle t-t_0\rangle^0]. \tag{1.137}$$

Solution. The velocity is determined by (1.8), as usual; but its application to the motion (1.136) requires use of the rule (1.124) to deal with the singularity functions. First, we identify in (1.136) two smooth vector functions

$$\mathbf{h}_1(t) \equiv \mathbf{A}(1-\cos\omega t) \quad \text{and} \quad \mathbf{h}_2(t) \equiv \mathbf{A}[1-\cos\omega(t-t_0)]$$

for which $\mathbf{h}_1(0) = \mathbf{0}$ at $t = 0$ and $\mathbf{h}_2(t_0) = \mathbf{0}$ at $t = t_0$. As a consequence of these initial conditions, the δ-function terms in the rule (1.124) will vanish in its application to the present case. Thus, with this identification in mind, differentiation of (1.136) in accordance with (1.124) yields the velocity

$$\mathbf{v}(M, t) = \mathbf{A}\omega[\sin\omega t\langle t-0\rangle^0 - \sin\omega(t-t_0)\langle t-t_0\rangle^0]. \tag{1.138}$$

Of course, the acceleration is obtained by (1.9); but in view of the singularity functions in (1.138), we must appeal again to the rule (1.124). By repeating the previous procedure, the reader will find that the acceleration is given by

$$\mathbf{a}(M, t) = \ddot{\mathbf{x}} = -\mathbf{A}\omega^2[\cos\omega t\langle t-0\rangle^0 - \cos\omega(t-t_0)\langle t-t_0\rangle^0]. \tag{1.139}$$

The differential equation of motion (1.137) is easily derived by multiplication of (1.136) by ω^2 and addition of the result to (1.139). Use of (1.117) in (1.137) shows that the motion is governed by two differential equations:

$$\ddot{\mathbf{x}} + \omega^2\mathbf{x} = \omega^2\mathbf{A} \quad \text{for } 0 < t < t_0 \quad \text{and} \quad \ddot{\mathbf{x}} + \omega^2\mathbf{x} = \mathbf{0} \quad \text{for } t_0 < t. \tag{1.140}$$

In the same way, the solutions of these equations may be revealed by expansion of (1.136). As an exercise, the student may derive the equations for the velocity and acceleration from these separate solutions and compare the results with those obtained by expansion of the equations (1.138) and (1.139). Equations of the kind (1.140) will be studied in greater detail in Chapter 6. □

Let us consider another example. This time the δ-function (1.120) and the derivative rule (1.124) for the product of the unit step and a smooth function that *does not* vanish at the point of discontinuity will be applied to study the assembly line motion of a part having a discontinuous velocity. In addition, the impulsive character of the delta function will be observed.

Example 1.21. A small component part P is moved along a straight track in an assembly line operation. The part, previously at rest, begins its journey at $t=0$ by being placed instantaneously upon a conveyor belt that transports it with a constant speed v. At the instant $t=\sigma$, the speed of P is changed suddenly to $\dot{s}=v\sin(2\pi t/\tau)$ when the piece is whisked away by a robot to reach its terminal assembly point at the instant $t=\tau$. There, the robot deposits the part to rest and returns to its starting position for another. What is the jump in the speed of P at $t=\sigma$? Find the acceleration of P as a function of t.

Solution. In general terms, the *jump in the quantity* $Q(x)$ *at* $x=\alpha$ is defined as the difference $\Delta Q \equiv Q(\alpha^+) - Q(\alpha^-)$ in the values of a quantity $Q(x)$ at the value $x=\alpha$ as α is approached from the right, where $x>\alpha$ and $Q(\alpha^+) \equiv \lim_{x\to\alpha} Q(x)$, and from the left, where $x<\alpha$ and $Q(\alpha^-) \equiv \lim_{x\to\alpha} Q(x)$. Of course, the quantity $Q(x)$ is continuous at $x=\alpha$ if and only if $\Delta Q=0$. In the present example, however, the velocity vector of P is a discontinuous function of the travel time t along the straight track. During the first phase of its motion, the part has a constant speed $\dot{s}(t)=v$ that gets turned on at $t=0$, so obviously the jump in the speed at $t=0$ is v. Thereafter, at $t=\sigma$, the speed is changed suddenly to $\dot{s}=v\sin(2\pi t/\tau)$. Thus, the jump $\Delta V \equiv \dot{s}(\sigma^+) - \dot{s}(\sigma^-)$ in the speed at $t=\sigma$ is given by

$$\Delta V = v[\sin(2\pi\sigma/\tau) - 1]. \tag{1.141}$$

The part eventually is brought smoothly to rest at $t=\tau$.

The foregoing description shows that the velocity of P along the entire track may be conveniently expressed in terms of the unit step function (1.117):

$$\mathbf{v}(P, t) = \mathbf{i}v(\langle t-0\rangle^0 - \langle t-\sigma\rangle^0) + \mathbf{i}v \sin\left(\frac{2\pi t}{\tau}\right) \times [\langle t-\sigma\rangle^0 - \langle t-\tau\rangle^0].$$

But the last term may be discarded because t never exceeds τ and the speed vanishes there. We now have

$$\mathbf{v}(P, t) = \mathbf{i}v\langle t-0\rangle^0 + \mathbf{i}v\left[\sin\left(\frac{2\pi t}{\tau}\right) - 1\right]\langle t-\sigma\rangle^0. \tag{1.142}$$

The acceleration is obtained by differentiation of (1.142) by use of the rule (1.124). This time we identify two smooth vector functions

$$\mathbf{h}_1(t) = v\mathbf{i}, \qquad \text{a constant} \quad \text{and} \quad \mathbf{h}_2(t) = \mathbf{i}v\left[\sin\left(\frac{2\pi t}{\tau}\right) - 1\right]$$

with values $\mathbf{h}_1(0) = v\mathbf{i}$ at $t=0$ and, with (1.141), $\mathbf{h}_2(\sigma) = \mathbf{i}\Delta V$ at $t=\sigma$. Therefore, differentiating (1.142) in accordance with (1.124), we derive the acceleration vector

$$\mathbf{a}(P, t) = \mathbf{v}\langle t-0\rangle_{-1} + \frac{2\pi}{\tau}\mathbf{v}\cos\left(\frac{2\pi t}{\tau}\right)\langle t-\sigma\rangle^0 - \Delta\mathbf{V}\langle t-\sigma\rangle_{-1}, \tag{1.143}$$

Wherein $\mathbf{v} \equiv v\mathbf{i}$ and $\Delta\mathbf{V} \equiv \Delta V\mathbf{i}$.

This formula shows that for $\sigma < t \leqslant \tau$ the acceleration is given by

$$\mathbf{a}(P, t) = \frac{2\pi}{\tau}\mathbf{v}\cos\left(\frac{2\pi t}{\tau}\right).$$

Also, in agreement with the uniform motion condition on $0 < t < \sigma$, (1.143) confirms that the acceleration is equal to zero there. The presence of the δ-functions in (1.143) reveals the interesting result that the acceleration has an *impulsive character* whose *strength* or *intensity* may be defined by the coefficients of the δ-functions. Thus, (1.143) shows that these impulsive strengths are proportional to the jumps in the speed at $t=0$ and $t=\sigma$. To fix the proper physical dimensions, we observe in (1.143) that the δ-function must have dimension equal to the reciprocal dimension of t; hence, $[\delta] = [T^{-1}]$. Since τ defines a natural time parameter over the whole interval of interest, it may be used to nondimensionalize the δ-functions in (1.143) by multiplying each coefficient by τ/τ. Now, with the proper physical dimensions, we see, for example, that the acceleration discontinuities at $t=0$ and $t=\sigma$ have shocklike intensities proportional to v/τ and to $\Delta V/\tau$, respectively. □

The same phenomenon occurs in our earlier cam design problem, which is the subject of our third example. The special rule (1.132) for integration of a power function (1.127) and the derivative rule (1.121) for the unit step function will be applied. The strength of the impulsive acceleration at the end of each stroke will be described and the results exhibited graphically.

Example 1.22. Let us reconsider the cam design problem studied in Example 1.10. We recall that two equations (1.58) and (1.59) were needed to describe the entire reciprocating, uniform motion of the push rod as the cam turned with the constant angular speed ω through the angle θ, as shown in Fig. 1.9. This occurs because the velocity is discontinuous at the end of each stroke at $\theta=0$ (or 2π) and $\theta=\pi$. We wish to show that the entire motion can be easily determined with the aid of singularity functions, and we shall also study the acceleration behavior.

Solution. During the forward stroke when $\theta\in[0,\pi]$, the push rod has the uniform forward velocity $\mathbf{v}_f \equiv v\mathbf{i}$. This gets turned on at $\theta=0$ and turned off at $\theta=\pi$. For the return stroke, defined by $\theta\in[\pi, 2\pi]$, the uniform return velocity is $\mathbf{v}_r\equiv -v\mathbf{i}$. This gets turned on at $\theta=\pi$, and it stays on for the rest of the interval. Therefore, recalling (1.117), we see that the velocity on the entire interval $[0, 2\pi]$ may be conveniently expressed by

$$\mathbf{v}(P,t)=\mathbf{v}_f\langle\theta-0\rangle^0-\mathbf{v}_f\langle\theta-\pi\rangle^0+\mathbf{v}_r\langle\theta-\pi\rangle^0-\mathbf{v}_r\langle\theta-2\pi\rangle^0. \qquad (1.144)$$

Since $\mathbf{v}_r = -\mathbf{v}_f$, (1.144) may be rewritten as

$$\mathbf{v}(P,t)=\mathbf{v}_f[\langle\theta-0\rangle^0-2\langle\theta-\pi\rangle^0+\langle\theta-2\pi\rangle^0] \qquad (1.145)$$

Our objective is to derive the cam shape from this expression.

The relation (1.56), in the same way as before, allows us to separate the variables r and θ in (1.145) to obtain

$$\int_a^{r(\theta)} dr\,\mathbf{i}=\frac{\mathbf{v}_f}{\omega}\left[\int_0^\theta\langle\theta-0\rangle^0\,d\theta-2\int_0^\theta\langle\theta-\pi\rangle^0\,d\theta+\int_0^\theta\langle\theta-2\pi\rangle^0\,d\theta\right],$$

wherein $r(0)=a$, as shown in Fig. 1.9. Application of the integration rule (1.132) to the last equation and use of $\mathbf{v}_f=v\mathbf{i}$ yields the cam shape function

$$r(\theta)=a+\frac{v}{\omega}[\langle\theta-0\rangle^1-2\langle\theta-\pi\rangle^1+\langle\theta-2\pi\rangle^1] \qquad (1.146)$$

for all $\theta\in[0, 2\pi]$. But the rule (1.127) shows that the last term vanishes for all $\theta\leqslant 2\pi$, so we may discard it. We recall also that $r(\pi)=a+b$, as shown in Fig. 1.9, whereas (1.146) together with (1.127) gives $r(\pi)=a+(v/\omega)\pi$; hence,

$b=\pi v/\omega$, as before. Therefore, the cam profile (1.146) is determined for all $\theta \in [0, 2\pi]$ by the single equation

$$r(\theta)=a+\frac{b}{\pi}\left[\langle\theta-0\rangle^1-2\langle\theta-\pi\rangle^1\right] \quad \text{with} \quad \frac{b}{\pi}=\frac{v}{\omega}. \tag{1.147}$$

Our earlier equations may now be retrieved by use of (1.127) in (1.147); its expansion yields

$$r(\theta)=\begin{cases} a+\dfrac{b}{\pi}\theta & \text{for} \quad 0\leqslant\theta\leqslant\pi, \\ a+b\left(2-\dfrac{\theta}{\pi}\right) & \text{for} \quad \pi\leqslant\theta\leqslant 2\pi. \end{cases} \tag{1.148}$$

These are the same as (1.58) and (1.59) given before.

The acceleration may be derived from (1.145). With the aid of (1.121), we get

$$\mathbf{a}(P,t)=\dot{\theta}\frac{d\mathbf{v}}{d\theta}=\omega\mathbf{v}_f\left[\langle\theta-0\rangle_{-1}-2\langle\theta-\pi\rangle_{-1}+\langle\theta-2\pi\rangle_{-1}\right]. \tag{1.149}$$

This result, together with (1.120), shows that the acceleration at the end of each stroke changes abruptly with strength $v\omega=b\omega^2/\pi$ at $\theta=0$ and 2π, and $-2v\omega=-2b\omega^2/\pi$ at $\theta=\pi$. Note that $[\delta(\theta)]=[1]$. Thus, the use of singularity functions in this study enables us to obtain a measure of the shocklike, impulsive accelerations that occur instantaneously at the end of each stroke.

Graphs of the motion (1.148) and the components of the velocity (1.145) and the acceleration (1.149) are shown in Fig. 1.24. The unit step character of the speed $v(\theta)$, which is described by (1.145), and the unit slope nature of the push rod motion $x=r(\theta)$, described by (1.147), are evident. The δ-function character of the acceleration $a(\theta)$ found in (1.149) is indicated in Fig. 1.24 by the small open circles at the points of discontinuity, and the corresponding strengths of the impulsive accelerations at these points are indicated by the arrows. We have seen that the latter depend upon the cam rise b and the square of the angular speed ω. □

The next example uses the unit step function to model the oscillograph record of the deceleration of a jet aircraft in its landing on the deck of an aircraft carrier. The velocity and the motion are found by application of the power rule (1.132), and the results are described graphically.

Example 1.23. A jet plane makes a rectilinear landing approach toward an aircraft carrier with a constant velocity $\mathbf{v}_0=v_0\mathbf{i}$. When the arresting hook engages the arresting cable, the plane experiences a sudden deceleration

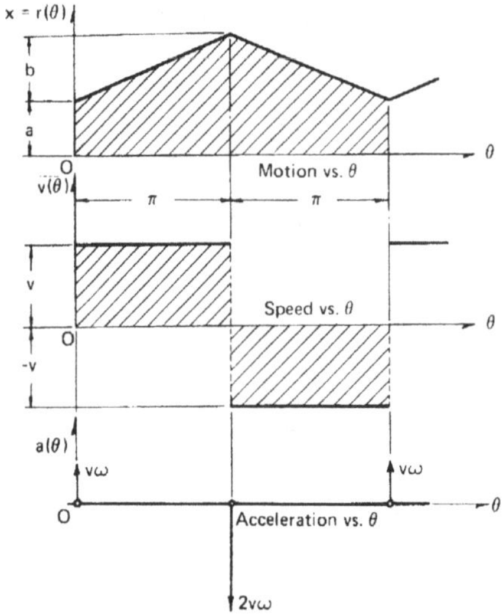

Figure 1.24. Graphs of the motion, speed, and acceleration of the push rod.

$\mathbf{a}(t) = -a(t)\,\mathbf{i}$, which is described by the oscillograph record shown in Fig. 1.25. Find the speed and the distance traveled by the plane during its arrestment expressed as functions of the time, determine their values at the instant $t = 2t_0$, and plot their graphs for $t \leqslant 2t_0$.

Solution. The oscillograph record in Fig. 1.25 indicates properly the zero acceleration of the plane as it approaches the arresting cable with a constant velocity $\mathbf{v}_0$ prior to engagement of its arresting hook at $t = 0$. At the moment of impact, the plane suddenly decelerates at an average constant value a_0 for a time $t = t_0$; then the acceleration returns suddenly to an essentially zero average value up to the time $t = 2t_0$. Therefore, by use of (1.117), the oscillograph record of the deceleration may be sensibly modeled by the function

$$\mathbf{a}(t) = \frac{d\mathbf{v}(t)}{dt} = \mathbf{a}_0\langle t-0\rangle^0 - \mathbf{a}_0\langle t-t_0\rangle^0, \tag{1.150}$$

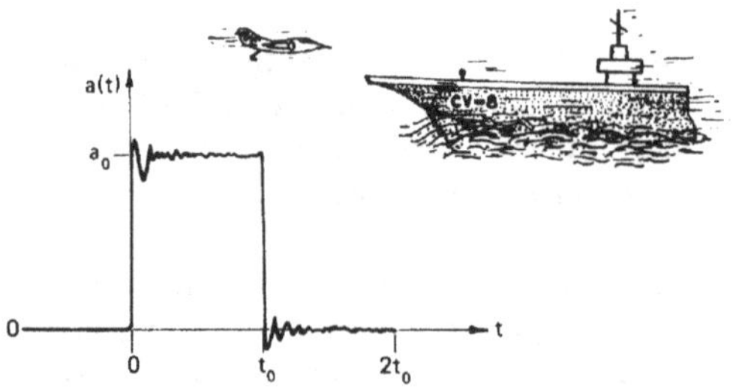

Figure 1.25. Oscillograph record of the deceleration of a carrier jet aircraft during its landing arrestment.

in which $\mathbf{a}_0 = -a_0\mathbf{i}$, a constant. The speed of the plane during its arrestment will now be derived from the first integral of (1.150).

Bearing in mind the condition $\mathbf{v}(t) = \mathbf{v}_0$ for $t \leqslant 0$, we find from (1.150)

$$\int_{\mathbf{v}_0}^{\mathbf{v}(t)} d\mathbf{v}(t) = \mathbf{a}_0\left[\int_0^t \langle t-0\rangle^0\,dt - \int_0^t \langle t-t_0\rangle^0\,dt\right].$$

Application of (1.132) in this equation yields the velocity

$$\mathbf{v}(t) = \dot{s}(t)\,\mathbf{i} = \mathbf{v}_0 + \mathbf{a}_0(\langle t-0\rangle^1 - \langle t-t_0\rangle^1), \tag{1.151}$$

which gives the speed $\dot{s}(t)$ during the plane's arrestment.

The distance $s(t)$ traveled during the landing may be found by integration of (1.151). Recalling the initial condition $\mathbf{x}(0) = s(0)\,\mathbf{i} = \mathbf{0}$ and applying (1.132) to integrate (1.151), we derive

$$\mathbf{x}(t) = s(t)\,\mathbf{i} = \mathbf{v}_0 t + \frac{\mathbf{a}_0}{2}(\langle t-0\rangle^2 - \langle t-t_0\rangle^2). \tag{1.152}$$

To graph the results, the equations (1.150), (1.151), and (1.152) have to be expanded to reveal the separated functions. This yields

$$\mathbf{a}(t) = -a(t)\,\mathbf{i} \tag{1.153a}$$

$$= \begin{cases} \mathbf{0} & \text{for } t \leqslant 0, \\ \mathbf{a}_0 = -a_0\mathbf{i} & \text{for } 0 < t < t_0, \\ \mathbf{0} & \text{for } t_0 < t \leqslant 2t_0, \end{cases} \tag{1.153b}$$

$$\mathbf{v}(t) = \dot{s}(t)\,\mathbf{i} \tag{1.154a}$$

$$= \begin{cases} \mathbf{v}_0 = v_0\mathbf{i} & \text{for } t \leqslant 0, \\ (v_0 - a_0 t)\,\mathbf{i} & \text{for } 0 \leqslant t \leqslant t_0, \\ (v_0 - a_0 t_0)\,\mathbf{i} & \text{for } t_0 \leqslant t \leqslant 2t_0, \end{cases} \tag{1.154b}$$

$$\mathbf{x}(t) = s(t)\,\mathbf{i} \tag{1.155a}$$

$$= \begin{cases} v_0 t\mathbf{i} & \text{for } t \leqslant 0, \\ (v_0 t - \tfrac{1}{2}a_0 t^2)\,\mathbf{i} & \text{for } 0 \leqslant t \leqslant t_0, \\ [(v_0 - a_0 t_0)\,t + \tfrac{1}{2}a_0 t_0^2]\,\mathbf{i} & \text{for } t_0 \leqslant t \leqslant 2t_0. \end{cases} \tag{1.155b}$$

We see that the first equations in (1.153b), (1.154b) and (1.155b) describe the flight conditions before impact; and we notice that the equations (1.154) and (1.155) show that although the acceleration (1.153) is discontinuous at $t = 0$ and $t = t_0$, the motion and velocity are continuous there. It follows from the last of (1.154b) and (1.155b) that the speed of the plane on the flight deck at the time $t = 2t_0$ and the corresponding distance from the impact point are given by

$$\dot{s}(2t_0) = v_0 - a_0 t_0 \quad \text{and} \quad s(2t_0) = t_0(2v_0 - \tfrac{3}{2}a_0 t_0). \tag{1.156}$$

No additional information on further braking is provided beyond $t = 2t_0$. The model results given in (1.153), (1.154), and (1.155) are shown graphically in Fig. 1.26. The unit step character of the acceleration (1.150) and the unit slope nature of the velocity (1.151) are visually evident. The results (1.156) also may be seen there. □

Our final example concerns the motion of a mass supported by a shock absorber. The general rules (1.124) and (1.131) for differentiation and integration of the product of a smooth function with the unit step function are applied.

Example 1.24. A shock absorber is held in a vertical position by a clamp bolted to its lower end. A load M, initially at rest at the top end of the shock absorber in its fully extended position, is moved by a compressive force with a speed $v(t)$ that increases linearly with the time so that $v(\tau) = v^*$ at $t = \tau$. At this instant, the applied force is removed and the load continues its downward compression with the speed $v = v^* \exp(1 - t/\tau)$ until the shock absorber is in

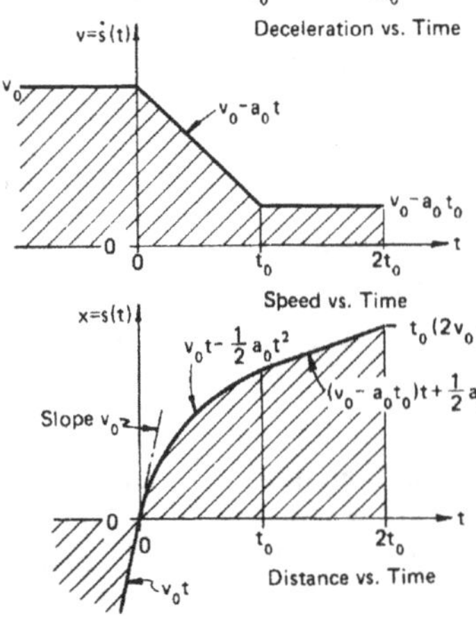

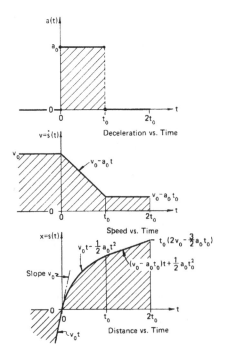

Figure 1.26. Graphs of the results (1.153), (1.154), and (1.155) for the jet aircraft arrestment problem.

its fully compressed position where the load comes to rest. Use singularity functions to derive the displacement and acceleration of the load as functions of the time for $0 \leqslant t < \infty$, and graph the results. What is the fully extended length of the shock absorber?

Solution. During the period $0 \leqslant t \leqslant \tau$, the load moves from its initial rest state with linearly increasing speed $v(t) = kt$. The constant k is determined by the condition $v(\tau) = v^* = k\tau$ at $t = \tau$; therefore, $v(t) = v^* t/\tau$ for $0 \leqslant t \leqslant \tau$. The force that produces this motion is removed at the instant τ, and the load continues its downward motion with the new speed $v(t) = v^* \exp(1 - t/\tau)$ for $\tau \leqslant t$. It is seen that the speed is continuous with value $v(\tau) = v^*$ at $t = \tau$. Hence, with use of (1.117), the speed $v(t)$ for all times $0 \leqslant t < \infty$ may be expressed by

$$v(t) = \frac{v^* t}{\tau} [\langle t - 0 \rangle^0 - \langle t - \tau \rangle^0] + v^* e^{(1 - t/\tau)} \langle t - \tau \rangle^0. \tag{1.157}$$

The acceleration is obtained by differentiation of (1.157) by use of the rule (1.124). We get

$$a(t) = \frac{v^*}{\tau} [\langle t - 0 \rangle^0 - (1 + e^{(1 - t/\tau)}) \langle t - \tau \rangle^0]. \tag{1.158}$$

The rectilinear motion $x(t)$ is obtained by integration of (1.157), as usual. Using the initial condition $x(0) = 0$ and the integration rule (1.131), we obtain

$$\int_0^{x(t)} dx = \frac{v^*}{\tau} \left[\langle t - 0 \rangle^0 \int_0^t t\, dt - \langle t - \tau \rangle^0 \int_\tau^t t\, dt + \tau \langle t - \tau \rangle^0 \int_\tau^t e^{(1 - t/\tau)}\, dt \right].$$

Then, finally, integration of these elementary functions yields the displacement of the load from the initial fully extended state of the shock absorber, namely,

$$x(t) = \frac{v^*}{2\tau} [t^2 \langle t - 0 \rangle^0 - (t^2 - 3\tau^2 + 2\tau^2 e^{(1 - t/\tau)}) \langle t - \tau \rangle^0]. \tag{1.159}$$

Graphs of the results may now be obtained by expansion of (1.159), (1.157), and (1.158). These yield

$$x(t) = \begin{cases} \dfrac{v^*}{2\tau} t^2 & \text{for } 0 \leqslant t \leqslant \tau, & (1.160a) \\ v^* \tau [\tfrac{3}{2} - e^{(1 - t/\tau)}] & \text{for } \tau \leqslant t < \infty, & (1.160b) \end{cases}$$

$$v(t) = \begin{cases} v^* t/\tau & \text{for } 0 \leqslant t \leqslant \tau, & (1.161a) \\ v^* e^{(1 - t/\tau)} & \text{for } \tau \leqslant t < \infty, & (1.161b) \end{cases}$$

$$a(t) = \begin{cases} v^*/\tau & \text{for } 0 \leqslant t \leqslant \tau, & (1.162a) \\ -\dfrac{v^*}{\tau} e^{(1 - t/\tau)} & \text{for } \tau \leqslant t < \infty. & (1.162b) \end{cases}$$

It is seen from (1.160b) that the fully extended length of the shock absorber, denoted by λ, is obtained as the limit value of $x(t)$ as $t \to \infty$. Hence, $\lambda = 3v^*\tau/2$. This is also evident in the graph of the displacement (1.160) shown in Fig. 1.27. The graphs of (1.161) and (1.162) also are sketched there. Notice that the removal of the applied load at $t = \tau$ results in an abrupt drop of magnitude $2v^*/\tau$ in the acceleration, but the displacement and velocity are continuous there. □

This concludes our introductory study of applications of singularity functions to problems in particle kinematics. Although many problems of this kind may be solved by ordinary methods applied to the separate continuous parts of piecewise continuous functions, and sometimes results may be obtained easily by simple geometrical considerations of areas and slopes of linear graphs, these procedures, particularly the latter, prove tedious or impracticable when complicated functions or advanced applications are

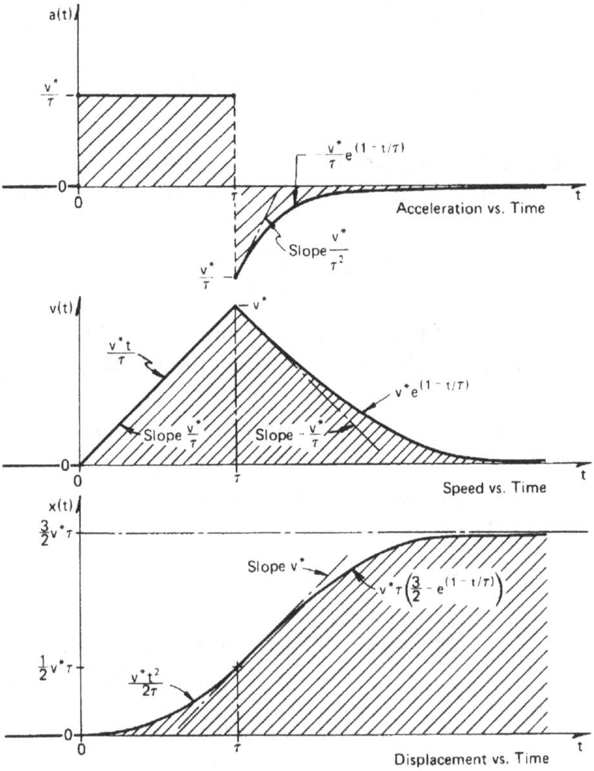

Figure 1.27. Graphical description of (1.59), (1.60), and (1.61) for the motion of a load supported by a viscous shock absorber.

involved. Of course, elementary methods reveal nothing about the strength of an impulsive discontinuity, a concept that proves especially useful in the study of beams bearing concentrated loads and in the study of motion due to impact of two bodies, for example. The latter application will be met in Chapter 6.

References

1. Blanché, R., *Axiomatics*, The Free Press of Glencoe, Macmillan, New York, 1962.
2. Brand, L., *Vector and Tensor Analysis*, Wiley, New York, 1947.
3. Einstein, A., *Essays in Physics*, Philosophical Library, New York, 1950.
4. Kaplan, W., *Operational Methods for Linear Systems*, Addison-Wesley, Reading, Massachusetts, 1962. Chapter 1 provides additional examples and details concerning generalized (singularity) functions.
5. Meriam, J. L., *Dynamics*, 2nd Edition, Wiley, New York, 1975. Chapter 2 contains numerous additional examples and excellent parallel problems for collateral study.
6. Rothbart, H. A., *Cams*, Wiley, New York, 1956. An elementary book on mechanical design of cams.
7. Shames, I., *Engineering Mechanics*, Vol. 2, *Dynamics*, 2nd Edition, Prentice-Hall, Englewood Cliffs, New Jersey, 1966. Chapter 11 provides additional examples and many similar problems useful in collateral study.
8. Synge, J. L., *Science: Sense and Nonsense*, Jonathan Cape Ltd., London, 1951. Reprinted as Euclid and the bright boy, in *The Mathematical Magpie*, a collection of short stories, essays and anecdotes about mathematics, assembled and edited by C. Fadiman, Simon and Schuster, New York, 1962. Synge's fable relates Euclid's tautological attempt to explain to a boy the idea of a point. The story ends in a comical tragedy from which the concept of a point manages to survive.
9. Truesdell, C. A., *The Kinematics of Vorticity*, Indiana University Press, Bloomington, 1954.
10. Weyl, H., *Space-Time-Matter*, Dover, New York, 1922. The concept of time is discussed both philosophically and mathematically in Chapter 1.
11. Yeh, H., and Abrams, J. I., *Principles of Mechanics of Solids and Fluids*, Vol. I, *Particle and Rigid Body Mechanics*, McGraw-Hill, New York, 1960. See Chapter 6 for particle kinematics.

Problems

It is essential that throughout the study of this text the student should work a variety of problems in order to grow familiar with use of the notation, concepts, and definitions; to cultivate, test, and expand his understanding of the subject matter; to learn the general methods of mechanics; and to master various techniques of problem solving. Moreover, in preparation for future work, it also is important that these problems be approached in a spirit and manner similar to that expressed in the sample exercises, namely, by the use of vector methods so far as may be reasonable and, in large measure, without the use of a computing device. Cases where use of a computer is desirable to promote practice with some numerical calculations will be evident, and in a few instances use of a computer will be suggested explicitly. In general, however, numerical values usually will serve only to simplify an analysis and to lay bare the relevant aspects of the example. Therefore, the majority of the problems in this book have been constructed to avoid senseless use of a computer so that the student's skills with direct calculations and with manipulations of algebraic and trigonometric relations may be reinforced and sharpened to develop his ability to handle fundamental aspects of analytical geometry, trigonometry, and calculus, all essential to the modern demands of engineering practice.

1.1. The motion of a grain of sand S for a certain time interval during a wind storm is described by $\mathbf{x}(S, t) = 2t^2\mathbf{u}(S, t)$, where $\mathbf{u}(S, t)$ is a time-dependent vector function that varies with the wind direction. (a) Derive an equation for the velocity of S. (b) Let $\mathbf{u}(S, t) = 5t^2\mathbf{i} + 3\mathbf{j} - 2t\mathbf{k}$ in frame $\varphi = \{O; \mathbf{i}_k\}$. Find in φ the position and the velocity of S initially and at two units later. Appropriate measure units are understood.

1.2. The motion of a particle X in a frame $\varphi = \{O; \mathbf{i}_k\}$ is given by $\mathbf{x}(X, t) = \frac{1}{2}(at^2 + 2bt + c)\,\mathbf{i}_1 + a\cos(pt + q)\,\mathbf{i}_2 + ce^{-pt}\mathbf{i}_3$, where a, b, c, p, and q are constants. Find as functions of the time t the velocity and acceleration of X in φ.

1.3. A particle Q is moving along the x axis with speed $\dot{s} = (4x^2 + 2x + 5)^{1/2}$ cm/sec relative to a frame $\psi = \{O; \mathbf{i}_k\}$. Find the velocity and acceleration of Q at the place $(2, 0, 0)$ cm in ψ.

1.4. Each particle of a fluid may be identified by its unique position vector in an assigned reference state in a frame $\varphi = \{O; \mathbf{i}_k\}$ at an instant t_0, say. This place, whose coordinates are called Lagrangian coordinates, may serve as an identification label for a particle during its subsequent motion in a flow. Let the motion of an arbitrary particle P in a certain flow in frame φ be given by

$$\mathbf{x}(P, t) = Ae^{\mu t}\mathbf{i} + Be^{\lambda t}\mathbf{j} + (C + Dt)\,\mathbf{k},$$

where A, B, C, D, μ, and λ are constants. (a) Determine the place occupied by P initially, identify its Lagrangian coordinates, and find the velocity and acceleration of P in φ. (b) What are the Lagrangian coordinates and the current position, velocity, and acceleration of the particle Q which in the same flow is at the place $\mathbf{X}(Q, 0) = 2\mathbf{i} - 4\mathbf{j}$ initially? (c) What physical interpretation may be assigned to the constant D? Appropriate measure units are understood.

1.5. The motion of a particle P in frame $\psi = \{F; \mathbf{i}_k\}$ is given by

$$\mathbf{x}(P, t) = 2(\sin 2t\,\mathbf{i} + \cos 2t\,\mathbf{j}) + 3t\,\mathbf{k}.$$

(a) Find the speed of the particle and determine the distance it travels as a function of time. (b) Show that the acceleration of P in the given motion is perpendicular to its velocity vector. (c) More generally, suppose that a particle P has an arbitrary motion with a constant speed in frame ψ. Prove that the acceleration of P in ψ is perpendicular to its velocity vector. Does this general result apply to the foregoing special motion of P?

1.6. The guide pin P of a certain machine moves in a groove that is milled in a flat, steel plate. The groove is designed so that the pin motion in frame $\varphi = \{O; \mathbf{i}_k\}$ is given by

$$\mathbf{x}(P, t) = a\cos\omega t\,\mathbf{i}_1 + b\sin\omega t\,\mathbf{i}_2$$

in which a, b, and ω are constants. (a) Derive the standard equation for the geometrical shape of the groove. (b) Show that the acceleration of the pin is directed always toward the origin O, and determine the points along the path of motion at which the acceleration will be greatest.

1.7. The straight path motion of a particle P in the direction $\mathbf{e}$ in a frame ψ is described by $\mathbf{x}(P, t) = (t^3 - 2t^2 + c)\,\mathbf{e}$ ft, in which c is a constant. Find (a) the time at which the particle has attained a speed of 4 ft/sec; (b) the acceleration at this instant; (c) the displacement of the particle during the fifth second; and (d) the value of c if the particle is at the place $\mathbf{x}_0 = 4\mathbf{e}$ ft after 2 sec. Sketch the component graphs of the position $x(t)$, the velocity $v(t)$, and the acceleration $a(t)$ as functions of the time t.

1.8. A point C in an aircraft during a period of its flight moves in a plane rectangular Cartesian frame $\varphi = \{O; \mathbf{e}_k\}$ with a motion

$$\mathbf{x}(C, t) = a \cosh qt\, \mathbf{e}_1 + b \sinh qt\, \mathbf{e}_2,$$

where a, b, q are constants. Find the velocity and acceleration of C, and determine the standard equation of its path. Sketch the path. Recall that $\cosh^2(qt) - \sinh^2(qt) = 1$.

1.9. Two particles P_1 of mass $m_1 = m$ and P_2 of mass $m_2 = 3m$ have motions $\mathbf{x}(P_k, t) = \mathbf{x}_k(t)$ given by

$$\mathbf{x}_1(t) = (4t^3 + 6)\,\mathbf{j} - 4t\mathbf{k}, \qquad \mathbf{x}_2(t) = -3\mathbf{i} + 6t^2\mathbf{k}$$

in $\varphi = \{O; \mathbf{i}, \mathbf{j}, \mathbf{k}\}$. Measure units may be ignored. The center of mass C of the system of two particles in φ is defined by the position vector $\mathbf{x}(C, t) = \mathbf{x}^*(t)$, where

$$(m_1 + m_2)\, \mathbf{x}^*(t) = m_1 \mathbf{x}_1(t) + m_2 \mathbf{x}_2(t).$$

(a) Find the velocity and acceleration of the center of mass of the system. (b) Find the time rate of change of the speed of each particle. How do these compare with the corresponding magnitudes of their accelerations in φ?

1.10. In a frame $\varphi = \{O; \mathbf{i}_k\}$, the moment about point O of the momentum of a particle P of mass m is defined by the vector $\mathbf{h}_O(P, t) = \mathbf{x}(P, t) \times m\mathbf{v}(P, t)$. (a) Find the time rate of change of the moment of momentum of the particles P_1 and P_2 described in the previous problem. (b) How would you define the moment of momentum of this system of two particles? Find it and determine its time rate of change in φ.

1.11. A Scotch crank is a mechanism used to convert a rotary motion into a reciprocating motion or vice versa. If the crank is driven at a constant angular speed $\dot{\psi} = \omega$ as shown in the figure, find the velocity and acceleration of the point P on the piston. Identify your reference frame.

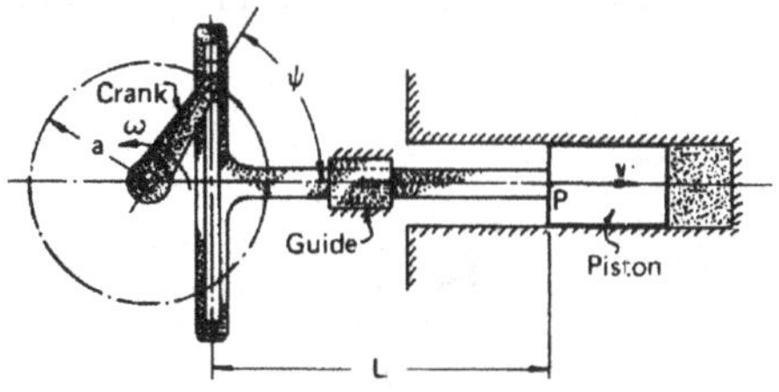

Problem 1.11.

1.12. A small mass M is attached to a rigid rod HM of length $2a$. The rod is hinged at H to a horizontal push rod BH which is driven by a Scotch crank OA of length a. The crank turns with a constant angular speed $\dot{\psi} = \omega$ as shown in the figure, and the system moves in the plane frame $\Phi = \{O; \mathbf{i}, \mathbf{j}\}$ fixed in the engine foundation. Find the velocity and acceleration of M in Φ. What is the speed of M in Φ?

1.13. A device used to damp vibrations in the crankshaft of an engine is modeled in the figure as a small ball B that slides freely in a plane motion on the circumference of a cylindrical cavity within the crankshaft. The cavity has a radius r with center P at

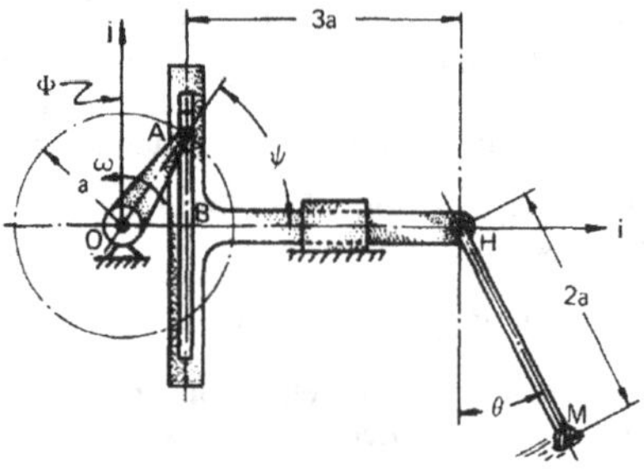

Problem 1.12.

a distance R from the crank axis at O. The shaft rotates relative to the frame $\Phi = \{O; \mathbf{I}_k\}$ with a constant counterclockwise angular speed $\dot{\theta} = \omega$. Find the velocity and acceleration of the ball B in frame Φ expressed as functions of the assigned quantities and the angular measures θ and β.

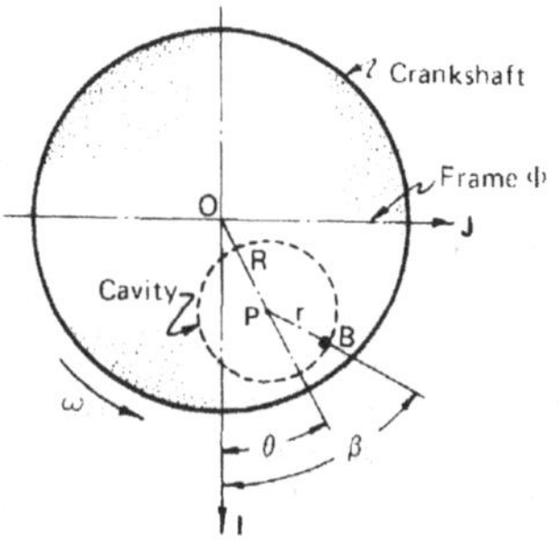

Problem 1.13.

1.14. Two blocks are hinged at the two end A and B of a rod of length L. The blocks slide in slots shown in the figure so that the distance $x(B, t) = a \sin \omega t$, where a and ω denote the constant amplitude and frequency of the oscillations of B, respectively. Find the velocity of the point A as a function of time in frame $\varphi = \{F; \mathbf{e}_k\}$.

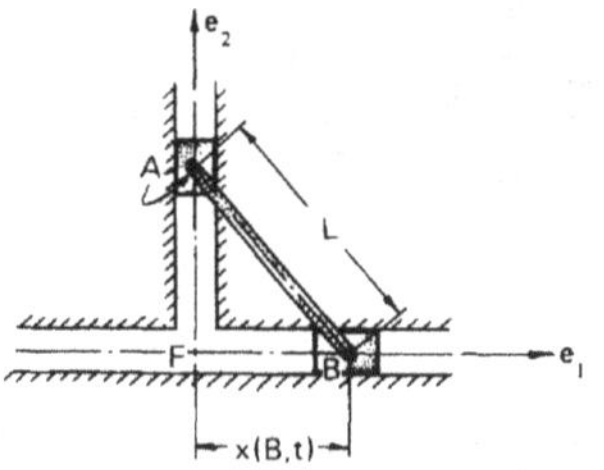

Problem 1.14.

1.15. The Scotch crank shown in the figure has a 60° slanted link driven by a crank of radius a that turns counterclockwise with a constant angular speed $\dot{\theta}(t) = \omega$. Find in frame $\Phi = \{F; \mathbf{i}, \mathbf{j}\}$ the motion, velocity, and acceleration of the point Q on the piston expressed as functions of the time t so that initially $\theta(0) = 0$. What are the crank positions at which the velocity of Q is greatest and least? Find the maximum velocity of Q.

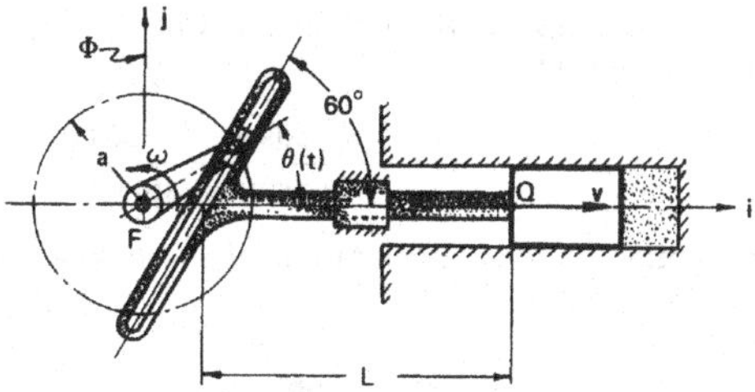

Problem 1.15.

1.16. A pin P is designed to slide in a parabolic slot in a fixed plate shown in the figure. The pin also is designed to move in a vertical slot of a sliding link A which has a constant speed of 2/3 in./sec during a period of its motion toward the right in frame $\psi = \{F; \mathbf{i}, \mathbf{j}\}$. Find the velocity and acceleration of the pin at the position $x = 3$ in.

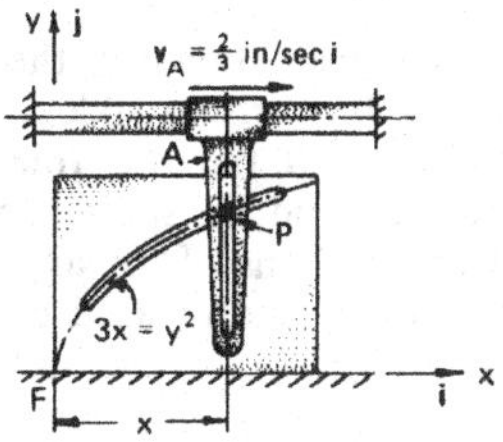

Problem 1.16.

1.17. A cam mechanism shown in the figure is shaped so that the center of the roller R traces a limacon defined by the plane polar equation $r = 20\cos\phi + 30$ cm. The roller center, initially at $\mathbf{x}_0 = 50\mathbf{i}_1$ cm, moves around the fixed cam surface with constant angular speed $\dot{\phi} = \pi/4$ rad/sec. Find the velocity and acceleration of the roller center in $\Phi = \{O; \mathbf{i}_k\}$ after 2 sec. Frame Φ is fixed in the cam.

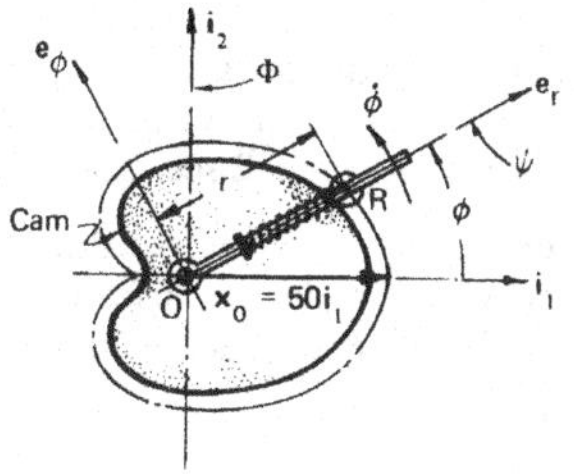

Problem 1.17.

1.18. At each point along the cam contour of the device described in the previous problem, $\mathbf{e}_r$ is a unit vector directed along the radius r from O to R and $\mathbf{e}_\phi$ is a unit vector perpendicular to $\mathbf{e}_r$ in the direction of increasing values of the angle ϕ. Find the velocity and acceleration of the center of the cam roller referred to the moving frame $\psi = \{O; \mathbf{e}_r, \mathbf{e}_\phi\}$. A general method of handling problems of this kind will be described in Chapter 4.

1.19. Two identical Scotch cranks shown in the figure are rotating clockwise about fixed axes with angular speeds ω_1 and ω_2. One crank is hinged at A to a link

AB; the other has a pin P that moves in a slot in AB. (a) Show that the angular speed ω of the link is given by

$$\omega = \frac{3(\omega_2 \cos\phi - \omega_1 \cos\theta)}{9 + (\sin\phi - \sin\theta)^2},$$

(b) Find the simple angular acceleration of the link when the angular speeds of the drive cranks are constant.

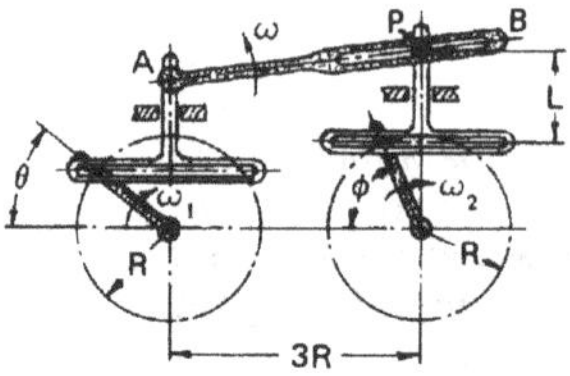

Problem 1.19.

1.20. A trammel mechanism shown in the figure consists of a rod AP of length 6 cm hinged at A and B to blocks that slide in the cross slots. The constant distance between A and B is 4 cm, and the rod rotates counterclockwise with a constant angular speed $\omega = 2$ rad/sec. Find the motion, velocity, and acceleration of the trammel point P expressed as functions of the angle θ in frame $\Phi = \{F; \mathbf{I}, \mathbf{J}\}$, and derive and identify the equation of the path traced by P.

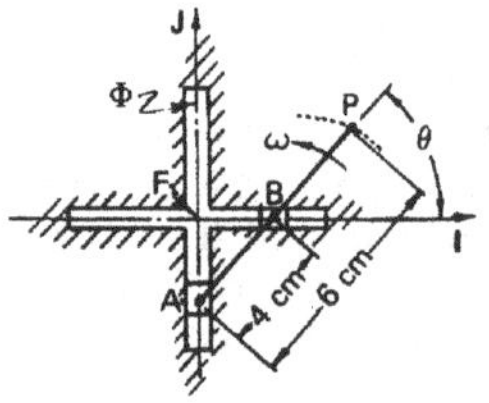

Problem 1.20.

1.21. Suppose that the trammel described in the last problem turns with a constant angular speed $\omega = (\pi/8)$ rad/sec. Find the position and the speed of the trammel point P in frame $\Phi = \{F; \mathbf{I}, \mathbf{J}\}$ after 2 sec, and again 2 sec later.

1.22. Part of an exit ramp for a highway shown in the figure is to be designed as a plane spiral curve OAB whose radius r varies linearly with the angle θ. The design conditions assume that a vehicle P enters the spiral road tangentially at point A with a

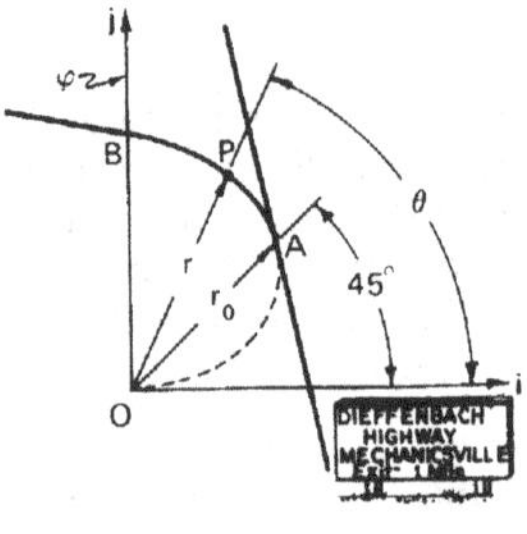

Problem 1.22.

speed v. Afterwards, it maintains a nearly constant but unknown angular speed $\dot{\theta}(t)=\omega$ as it moves toward the exit B where its acceleration is assumed to have magnitude α. Use the reference frame $\varphi=\{O; \mathbf{i}, \mathbf{j}\}$, and determine the radius parameter r_0 at A required for the design expressed in terms of the assigned quantities v and α alone. What is the velocity of P in φ expressed as a function of θ alone?

1.23 Pins A and B are restricted to move in elliptical slots milled in a flat plate as shown in the figure. Their movement is controlled by the motion of the slotted link C, which during a period of its motion moves to the right at a constant speed of 3 in/sec. Find the speed and the acceleration of the pin A when the slot is at $x=3$ in. Determine the motion $\mathbf{x}(A, t)$ of the pin A expressed as a function of time t such that $x=0$ initially.

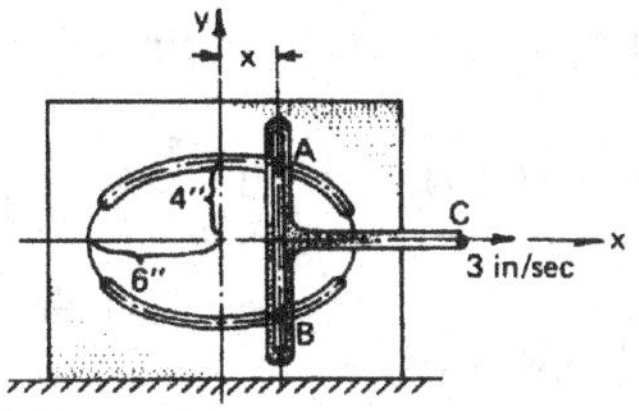

Problem 1.23.

1.24. Find the velocity and acceleration of the center of the piston pin S of the reciprocating engine shown in the figure. Assume that the crank rotates counterclockwise with a constant angular speed $\dot{\theta}=\omega$, and that the length l of the connecting rod is much larger than the length a of the crank so that terms of order larger than a/l may be neglected in the derivation.

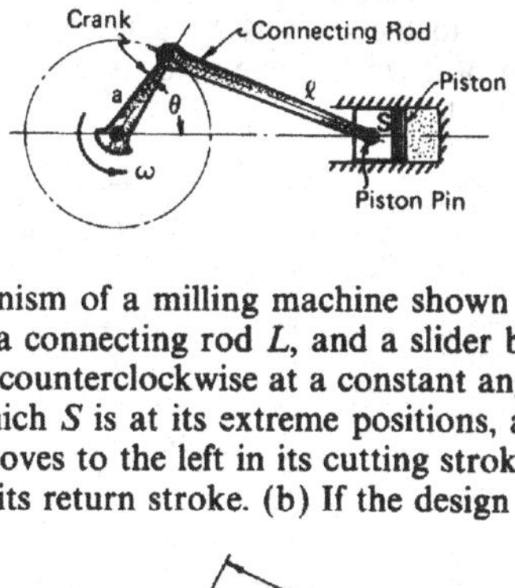

Problem 1.24.

1.25. The quick return mechanism of a milling machine shown in the figure consists of a crank C, a drive arm A, a connecting rod L, and a slider block S that holds the cutting tool. The crank rotates counterclockwise at a constant angular speed $\dot{\theta}=\omega$. (a) Find the crank angles θ for which S is at its extreme positions, and determine the ratio of the time during which S moves to the left in its cutting stroke to the time during which S travels to the right in its return stroke. (b) If the design efficiency E is the

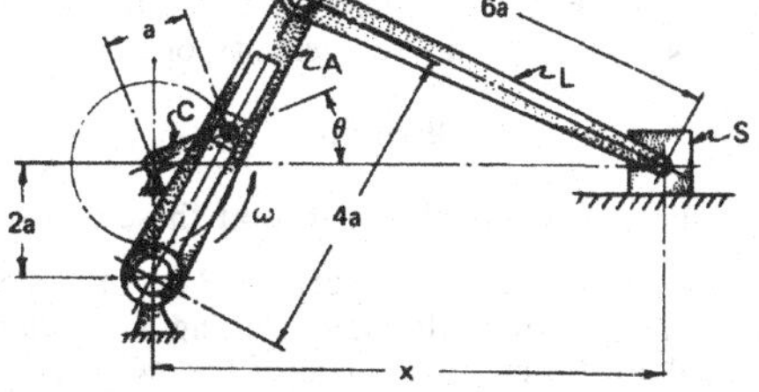

Problem 1.25.

ratio of the cutting stroke time to the full cycle machine stroke time, what is the design efficiency for this machine? (c) Show analytically how the velocity of S as a function of the crank angle θ may be found. Then develop a computer program to evaluate the normalized (nondimensional) motion $X \equiv x/a$ and the normalized velocity $V \equiv \dot{x}/a\omega$ of S in terms of the crank angle θ using a 15° step. Run the program to graph these functions, and construct the so-called phase plane graph of V vs. X. Identify any important features of the graphs. (d) Determine analytically the velocity of S in in./sec when $\theta = 90°$ in a machine for which $a = 6$ in. and $\omega = 240$ rad/min. Does your program check with this result?

1.26. An insect I initially at the place $\mathbf{x}_0 = 10\mathbf{j}$ cm has at time t a velocity $\mathbf{v}(I, t) = 6\mathbf{i} + 5\mathbf{j} + 12t\mathbf{k}$ cm/sec in frame $\psi = \{F; \mathbf{i}_k\}$. Find the motion and acceleration of the insect in ψ. How far has it traveled after 3 sec?

1.27. A particle Q initially at the origin in frame $\varphi = \{Q; \mathbf{i}_k\}$ has an acceleration

$$\mathbf{a}(Q, t) = 6t\mathbf{i} - 4\mathbf{j} + \frac{\pi^2}{9}\cos\left(\frac{\pi t}{3}\right)\mathbf{k}\ \text{m/sec}^2.$$

If the initial velocity is $\mathbf{v}_0 = \mathbf{i} - 4\mathbf{j}$ m/sec, find the acceleration, the velocity and the place occupied by Q after 2 sec.

1.28. A dust particle D at the position $\mathbf{x}_0 = \mathbf{i} + 2\mathbf{j} + 3\mathbf{k}$ at the instant $t_0 = 1$ sec is blown in a storm to a constant acceleration $\mathbf{a} = 3\mathbf{i} + \mathbf{j}$ cm/sec^2 during an interval of its motion in $\Phi = \{F; \mathbf{i}_k\}$. If the velocity of D at t_0 is $\mathbf{v}_0 = 4\mathbf{i} + 5\mathbf{j} - \mathbf{k}$ cm/sec, what are its position and velocity 10 sec later?

1.29. A bullet traveling horizontally pierces three sheets of paper spaced at equal distance d apart, as shown in the figure. If the bullet decelerates at a constant rate with magnitude a so that the measured travel time from the first sheet to the second is t_1 and from the second to the third is t_2, determine the deceleration of the bullet and show that the ratio of its magnitude a to its speed v at the middle sheet is given by $a/v = 2(t_2 - t_1)/(t_1^2 + t_2^2)$.

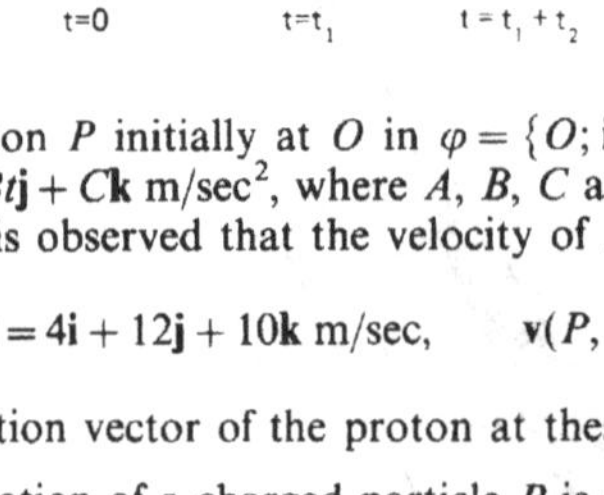

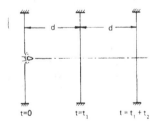

Problem 1.29.

1.30. A proton P initially at O in $\varphi = \{O; \mathbf{i}_k\}$ is in motion with an acceleration $\mathbf{a}(P, t) = At^2\mathbf{i} + Bt\mathbf{j} + C\mathbf{k}$ m/sec^2, where A, B, C are constants. At the instants $t_1 = 1$ sec and $t_2 = 3$ sec it is observed that the velocity of P in φ has the values

$$\mathbf{v}(P, t_1) = 4\mathbf{i} + 12\mathbf{j} + 10\mathbf{k}\ \text{m/sec}, \qquad \mathbf{v}(P, t_2) = 108\mathbf{i} - 48\mathbf{j} + 30\mathbf{k}\ \text{m/sec}.$$

What is the position vector of the proton at these times?

1.31. The motion of a charged particle P is controlled by an electromagnetic field so that it moves, as shown in the figure, along a circular cylindrical helix of radius b and pitch p equal to the circumference of the cylinder base. The particle, initially at

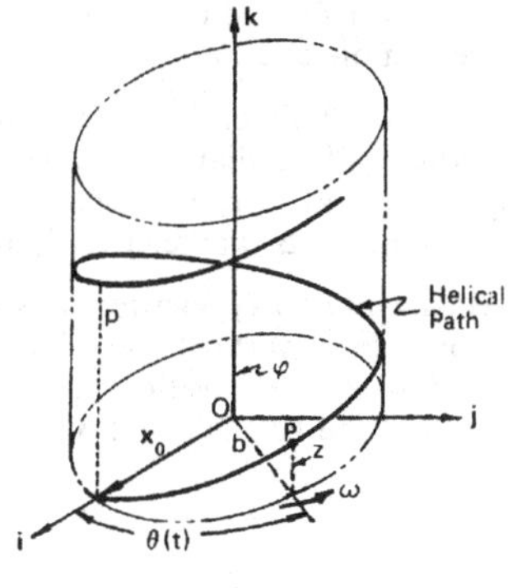

Problem 1.31.

$\mathbf{x}_0 = b\mathbf{i}$, turns about the cylinder axis with a constant angular speed $\dot{\theta}(t) = \omega$ as it advances in the direction of the axis at a constant rate $\dot{z}(t) = A$ in the frame $\varphi = \{O; \mathbf{i}_k\}$. Find the motion, the speed, and the acceleration of P in φ as functions of the time t and in terms of the constants b and ω only.

1.32. A small pin P moves in a straight slot milled in a flat plate of a certain machine in which the pin's motion is programmed to control the motion of a slotted, horizontal link A. In one instance, P is required to move the link from its position at $y = 1$ cm with a velocity $\mathbf{v}_A = 3y\mathbf{j}$, as shown in the figure. Determine the motion, velocity, and acceleration of P in the frame $\varphi = \{O; \mathbf{i}_k\}$, all programmed as functions of the variable y alone. Show how the results may be expressed as functions of the time t alone. Time units are seconds.

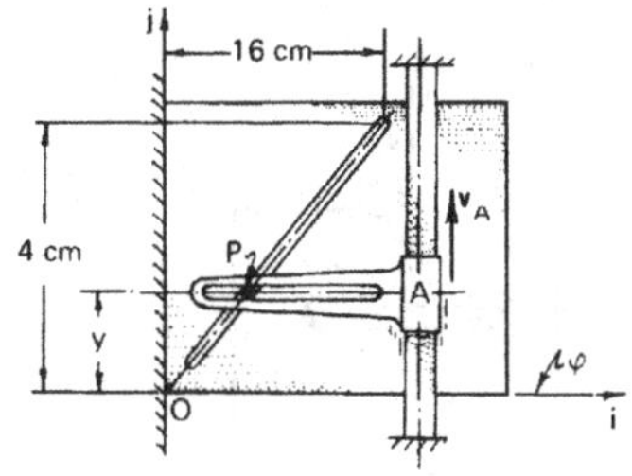

Problem 1.32.

1.33. A material point S initially at rest at the place $\mathbf{x}_0 = -625\mathbf{i} + 35\mathbf{j} + 50\mathbf{k}$ m in $\psi = \{O; \mathbf{i}_k\}$ is given an acceleration $\mathbf{a}(S, t) = 36t^2\mathbf{i} + 42t\mathbf{j} - 4\mathbf{k}$ m/sec^2. Find in ψ the position, velocity, and acceleration of S after 5 sec.

1.34. A particle has the initial velocity $\mathbf{v}_0 = 4\mathbf{i} + \mathbf{j}$ cm/sec at $\mathbf{x}_0 = 1\mathbf{k}$ cm in $\varphi = \{H; \mathbf{i}_k\}$. The acceleration is given by $\mathbf{a}(P, t) = (1 - 6t^2)\mathbf{i} + 8t^3\mathbf{j} + 4\mathbf{k}$ cm/sec^2. Find the velocity and the motion of P in φ.

1.35. A bullet travels through a gun barrel of length l with a constant acceleration from breech to muzzle where it leaves the gun with a speed β. How long does it take the bullet to travel through the barrel? What is the constant acceleration of the bullet relative to the gun?

1.36. An electron E initially at rest at H in $\varphi = \{H; e_k\}$ moves on a straight line with constant acceleration $\mathbf{a}$ so that it attains the velocity $\mathbf{v}_0$ at a distance s_0 from H.

Find the velocity of E as a function of s, the distance traveled, and determine the magnitude of the acceleration of E.

1.37. A tiny spherical bubble of gas ascending through water has a speed that is proportional to the square of its diameter d. The diameter is related to the depth y below the surface by the rule $yd^3 = c$, a constant. Find the time required for the bubble to reach the surface if initially its diameter is d_0 at the depth h where its speed is v_0.

1.38. A shift mechanism of a typewriter consists of a square, steel block B of side $2R$ that slides freely on a fixed vertical rod which passes through the shift lever A. If the block is triggered from its rest position at $y = 0$ with acceleration $\mathbf{a} = c(h - y)^{-2}\mathbf{j}$, where c is a constant, find the velocity with which the block strikes A.

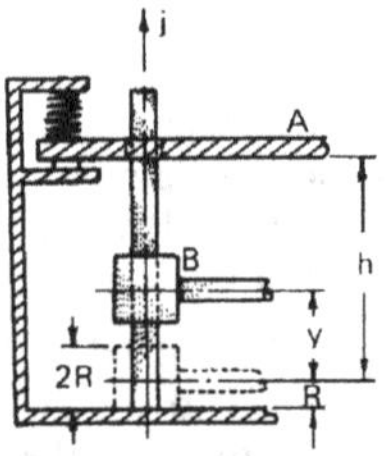

Problem 1.38.

1.39. The slotted link A shown in the figure controls the motion of a small pin P to move in a parabolic groove $3x = y^2 - 9$, which is milled in a flat plate. The link A, initially at rest at $y = 3$ cm, has a controlled acceleration $\mathbf{a}(A, t) = -4y\mathbf{j}$ cm/sec^2. Find the velocity and acceleration of the pin P in the Cartesian frame $\varphi = \{O; \mathbf{i}_k\}$ at the instant when $y = 2$ cm.

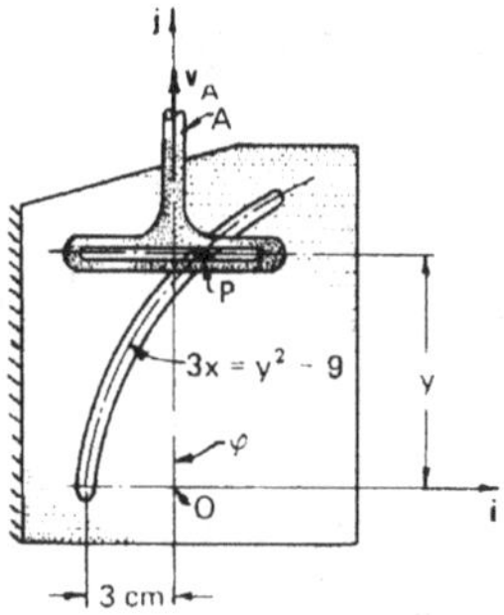

Problem 1.39.

1.40. A jet seaplane has a touchdown speed of 100 mph when making contact with the water. The hull structure of the aircraft fuselage is to be designed so that the landing speed is reduced to 25 mph in a distance of 1/4 mile in time τ. Assume that the deceleration of the seaplane is proportional to the square of its speed through the water: $\ddot{s} = -C\dot{s}^2$, where the design parameter C depends on the shape of the hull, the weight of the aircraft, and average water wave conditions. Determine the value of C, and find the time τ during which the speed is reduced as described.

1.41. A rope ACB is attached to a box B, passes over a pulley of negligible size at C, and slides over a circular surface of radius 15 ft as shown in the figure. The free end

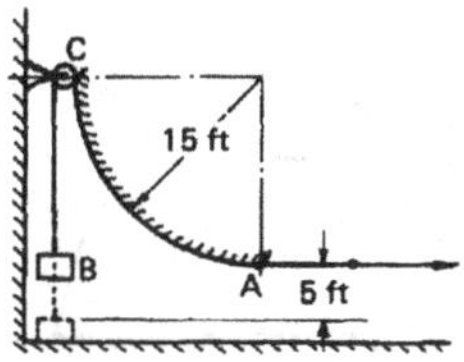

Problem 1.41.

of the rope is at A when the box rests on the ground. If the end A has a constant acceleration of magnitude 10 ft/sec^2, determine the time required for the box to reach C.

1.42. A particle P initially at $\mathbf{x}_0 = b\mathbf{j}$ in $\varphi = \{O; \mathbf{i}_k\}$ has a velocity $\mathbf{v}(P, t) = \omega \sec\theta(a \sec\theta\mathbf{i} + b\tan\theta\mathbf{j})$, where a, b are constants and $\dot{\theta} = \omega$ is a certain constant angular speed. Find the acceleration of P in φ and determine the path along which the particle moves such that $\mathbf{v}(P, 0) = a\omega\mathbf{i}$. Sketch the trajectory of P.

1.43. Two particles P_1 and P_2 are in uniform motion with velocities $\mathbf{v}_1$ and $\mathbf{v}_2$ in space. At a certain instant the vector of P_1 from P_2 is $\mathbf{D}$ and their relative velocity is $\mathbf{v} \equiv \mathbf{v}_2 - \mathbf{v}_1$. If v_1 and v_2 are the respective components of $\mathbf{v}$ parallel and perpendicular to $\mathbf{D}$, show that when P_1 and P_2 are closest together their distance of separation is Dv_2/v and that they arrive in this position after an interval of time Dv_1/v^2, where $v = |\mathbf{v}|$.

1.44. A cam shown in the figure is to be designed with a 3-cm rise to produce two continuous oscillations of a spring-loaded push rod for each revolution of the cam. The cam turns about its axis at F with a constant, clockwise angular speed $\omega = 4$ rad/sec. Construct the equation for the cam profile $r(\theta)$ that will produce a sinusoidal motion of the push rod such that $r(0) = 2$ cm and $\mathbf{v}(0) = \mathbf{0}$ at $\theta = 0$ initially. Determine the maximum velocity and acceleration in frame $\Phi = \{F; \mathbf{i}, \mathbf{j}\}$ of the contact point P on the push rod. How would you design the cam to produce three oscillations of the push rod for each revolution of the cam? And for n oscillations? Sketch the cam profile for $n = 3$.

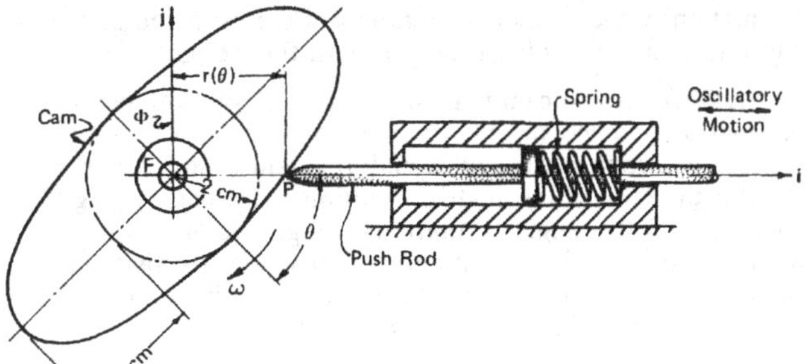

Problem 1.44.

1.45. The tip of a cutting tool has a straight line motion in which the cutter starts from rest and increases its speed from 0 to v^* with a constant acceleration $\mathbf{a}^*$; it retains this constant cutting speed for a time p, as shown in the figure, and ultimately comes to rest at the end of its stroke by a constant deceleration $\mathbf{a}$. The total length of the tool stroke is l. Show that the total time τ required for the full machine stroke is given by

$$\tau = \frac{l}{v^*} + \frac{v^*(a + a^*)}{2aa^*}.$$

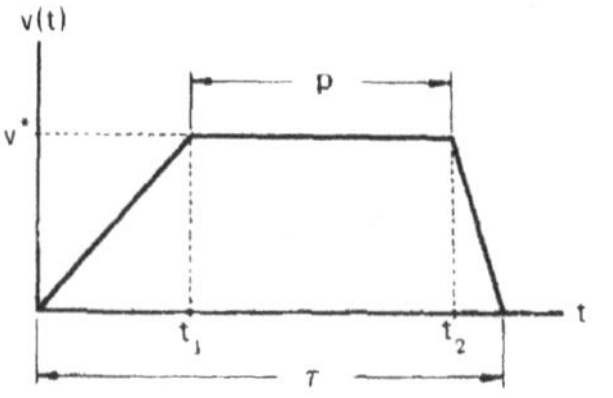

Problem 1.45.

Determine the length λ of the cutting stroke and the time p required to execute it. What is the design efficiency relation (see Problem 1.25) for the case when $a = a^* = 15v^{*2}/l$? Assume that the full cycle machine stroke time is 2τ.

1.46. A cam mechanism similar to that described in Fig. 1.9 is to be designed to control the motion of the spring-loaded push rod to have a constant acceleration $\mathbf{a}$ during its forward motion for $\theta \in [0, \pi]$ and a constant deceleration of the same magnitude during its return motion for $\theta \in [\pi, 2\pi]$. Of course, as a consequence, the acceleration will not be well defined at the stroke extremes. The cam must rotate with constant, clockwise angular speed $\dot{\theta} = \omega$; and the push rod must be at instantaneous rest at $\theta = 0$, where $\mathbf{x}(P, 0) = a\mathbf{i}$, and at $\theta = 2\pi$. (a) Determine the cam rise b at $\theta = \pi$; and find the required continuous cam profile $r(\theta)$ expressed in terms of the geometrical cam design parameters. (b) Sketch the graphs of the motion, velocity, and acceleration of the push rod for $\theta \in [0, 2\pi]$. (c) Sketch the shape of the cam, and discuss briefly any problems you may foresee with this design.

1.47. Suppose that the design specifications for the cam described in Example 1.10 are altered to require that the return motion of the push rod be uniform but only half as rapid as its forward motion, while the radius a and the angular speed ω remain the same as before. (a) What is the ratio of the stroke travel time of the push rod in its forward motion to the travel time in its return motion? How are these times related to the angular rotation of the cam? (b) Find the continuous cam profile $r(\theta)$ that can produce the motion, and determine the rise b. (c) Sketch the cam geometry, and graph the motion, velocity, and acceleration of the push rod as functions of $\theta \in [0, 2\pi]$. Do you foresee any potential problems with this design?

1.48. A smooth cam that rotates with a constant, clockwise angular speed ω is to be used as a quick return device to produce in one revolution a uniform forward motion of a control rod and a uniform return motion which is three times faster. (a) What is the ratio of the angle of rotation of the cam during the forward motion of the control rod to the angle of rotation during its return motion? (b) Find as a function of the rotation angle θ the continuous cam shape $r(\theta)$ that can produce the motion with rise b and $r(0) = a$. (c) Graph the motion, velocity, and acceleration of the control rod for $\theta \in [0, 2\pi]$, and sketch the cam profile.

1.49. A drop cam is used in an automatic hole punch operation for cardboard. The cam turns with a constant, clockwise angular speed ω, and it drives a follower rod which is attached to a spring-loaded punch head having stroke length b. In the clearing interval $0 \leqslant \theta \leqslant \pi$, the spring is compressed further as the punch is raised with a uniform rectilinear motion to the height b at $\theta = \pi$, and the punched work piece is removed from the machine. During the loading interval $\pi \leqslant \theta \leqslant 2\pi$, a new work piece is set for punching as the punch head is lowered to the face of the fresh piece with a uniform motion which is half as rapid as before. At this point, the punch head is at a height h from its initial position at a from the axis of rotation. The hole punch is actuated suddenly by the spring-driven drop action induced at $\theta = 2\pi$ by the continuing cam motion, and the cycle is repeated. (a) Find the height h, the rise b, and the shape $r(\theta)$ of the drop cam, continuous at $\theta=\pi$, which can produce the punch motion. (b) Determine the ratio of the operation time expended in the clearing interval to the time expended in the loading interval. (c) Sketch the cam profile. (d) Graph the motion, velocity, and acceleration of the punch for $0 \leqslant \theta \leqslant 3\pi$. (e) Discuss potential problems that may arise from use of this device.

1.50. In another drop cam design for a hole punch operation similar to that described in the previous problem, the punch is raised into contact with the work piece at a height h during the forward stroke in which the continuing, constant rotational motion ω of the cam in the interval $0 \leqslant \theta \leqslant \pi$ drives the punch through the cardboard. At the end of this punching stroke, which has total length b at $\theta=\pi$, a strong retracting spring is suddenly actuated at $\theta=\pi$ by the drop in the cam shape. Consequently, the punch is cleared instantaneously from the work piece to the height h at $\theta=\pi$, and the spring motion returns the punch control rod to its initial safe starting position at $\theta=2\pi$ where the cam surface is continuous with radius a. The punching motion is to be uniform but twice as rapid as the uniform return motion following its instantaneous retraction from the hole. (a) Find the height h and the stroke length b. (b) What is the ratio of the time expended in the forward stroke to the time expended in the return stroke? (c) Determine the motion, velocity, and acceleration of the punch head, and sketch their graphs and the cam shape.

1.51. The helical motion of a particle P in $\eta=\{O; \mathbf{i}_k\}$ is given by

$$\mathbf{x}(P,t)=2(\sin 2t\,\mathbf{i}+\cos 2t\,\mathbf{j})+3t\mathbf{k}\ \text{cm}.$$

(a) Determine the acceleration of P at $t=\pi/3$ sec. (b) What is the speed of P after 37 days? (c) Find the intrinsic acceleration of P in η, and determine the radius of curvature of the path.

1.52. Consider a particle P in motion along an arbitrary simple curve in the plane Cartesian frame $\Phi=\{O; \mathbf{i}, \mathbf{j}\}$. Show that the velocity $\mathbf{v}=\dot{x}\mathbf{i}+\dot{y}\mathbf{j}$ and the acceleration $\mathbf{a}=\ddot{x}\mathbf{i}+\ddot{y}\mathbf{j}$ of P when projected upon the intrinsic basis directions yield equations (1.70) and (1.71), respectively. Thus, (1.70) and (1.71) are the velocity and acceleration of P in frame Φ, but referred to the plane intrinsic frame $\psi=\{P; \mathbf{t}, \mathbf{n}\}$ following the particle. Hint: Write $\mathbf{i}$ and $\mathbf{j}$ in terms of $\mathbf{t}$ and $\mathbf{n}$.

1.53. (a) Apply (1.70) and (1.71) to derive the formulas in (1.75) for the tangential and normal components of the intrinsic acceleration vector and for the curvature of the path. (b) Suppose that in a plane Cartesian frame $\Phi=\{O; \mathbf{i}, \mathbf{j}\}$ the particle has the following velocity and acceleration at a certain point A on its path:

$$\mathbf{v}=2\sqrt{5}(2\mathbf{i}+\mathbf{j})\ \text{ft/sec}, \qquad \mathbf{a}=2(1+4\sqrt{5})\,\mathbf{i}+4(\sqrt{5}-1)\,\mathbf{j}\ \text{ft/sec}^2.$$

Find the intrinsic velocity and acceleration at A, and determine the curvature of the path at A. The student may check his solution by comparison with Example 1.13.

1.54. The motion of a particle P along a plane curve is given by $\mathbf{x}(P,t)=x(t)\,\mathbf{i}+y(t)\,\mathbf{j}$ in a plane Cartesian frame $\Phi=\{O; \mathbf{i}, \mathbf{j}\}$. Use equation (1.75c) to derive the following relation for the curvature of a plane curve:

$$\kappa=\left|\frac{\dot{y}\ddot{x}-\dot{x}\ddot{y}}{(\dot{x}^2+\dot{y}^2)^{3/2}}\right|.$$

Let the plane curve be expressed as $y=y(x)$ and identify $x=t$. Show that the above formula reduces to the first part of (1.81). Derive the second part of equation (1.81).

1.55. At a certain instant t_0, the velocity and acceleration of a particle P in $\Phi = \{F; \mathbf{i}_k\}$ are given by

$$\mathbf{v}(P, t_0) = 3\mathbf{i} + 4\mathbf{j} + 12\mathbf{k} \text{ ft/sec}, \qquad \mathbf{a}(P, t_0) = 3\mathbf{i} + 5\mathbf{j} + 3\mathbf{k} \text{ ft/sec}^2.$$

Find at the instant t_0 the intrinsic velocity and acceleration of P in Φ, and determine the radius of curvature of the path at the place occupied by P at the instant t_0.

1.56. A particle P moves with a constant speed of 27 m/sec along a parabolic trajectory $y = \sqrt{5}\, x^2$. What is the acceleration of P as it passes the point at $x = 2$ m?

1.57. Find the intrinsic velocity and acceleration of the pin P in the device described in Problem 1.16 when the pin is at $x = 3$ in. What is the curvature at this place?

1.58. The guide pin P of the bell crank device described in Fig. 1.15 has a speed of 10 ft/sec and a rate of change of speed of 20 ft/sec^2 at the point A. What is the angular speed ω of the arm OP when P reaches point A?

1.59. The slotted link A of the device in the figure for Problem 1.39 controls the motion of the pin P to move with a constant speed of 25 cm/sec in the parabolic groove $3x = y^2 - 9$. Find as functions of y the intrinsic velocity and acceleration of P. Determine the velocity and acceleration of the link A in the Cartesian frame $\varphi = \{O; \mathbf{i}_k\}$ at the instant when $y = 2$ cm.

1.60. A small projectile S is fired in the direction $\mathbf{e} = 5/13\mathbf{e}_1 + 12/13\mathbf{e}_2$ with an initial muzzle speed of 130 ft/sec. If the projectile has a constant acceleration $\mathbf{a}(S, t) = -5\mathbf{e}_1 - 30\mathbf{e}_2$ ft/sec^2 in the rectangular Cartesian frame $\varphi = \{F; \mathbf{e}_k\}$, find the radius of curvature of the trajectory after 4 sec. What are the intrinsic velocity and acceleration at this instant?

1.61. A fluid particle which moves along a simple curved path in a certain flow experiment passes the point A with a speed of 18 m/sec; and at 27 m along the curve from A, it decelerates to a speed of 9 m/sec at the point B. The experiment reveals that the deceleration of the particle measured along the path is very nearly proportional to the distance traveled from point A; and measurements show also that the acceleration of the fluid at point B has a magnitude of 90 m/sec^2. Find the radius of curvature of the path at B.

1.62. A crank device shown in the figure moves a small pin P in an elliptical slot milled in a flat plate fixed in the Cartesian frame $\Phi = \{F; \mathbf{I}_k\}$. At the point A, the pin has a speed of 15 cm/sec and a rate of change of speed of 10 cm/sec^2 along the track. Find the acceleration of P referred to Φ at the instant described.

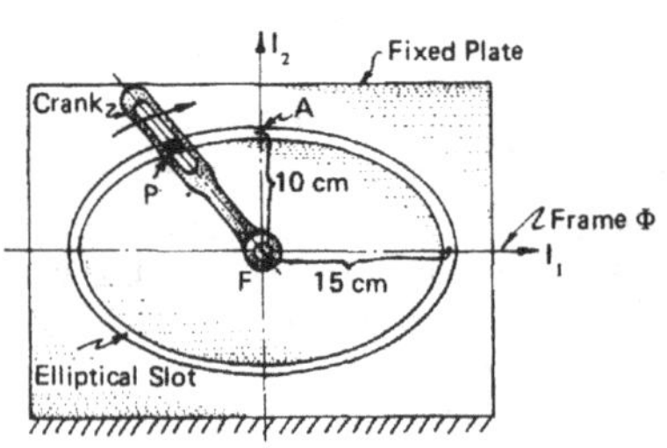

Problem 1.62.

1.63. Suppose that the pin P of the crank device of the previous problem has a constant speed of 15 cm/sec along the track. (a) Where along the slot is the acceleration greatest and least, and what are these extreme values? (b) Find the angular velocity of the crank at these points.

1.64. A particle travels with a constant speed of 4 ft/sec along the path shown in the figure. (a) Find the acceleration at the points 1, 2, 3, and 4 in frame φ. (b) Draw a graph of the magnitude of the acceleration as a function of the distance traveled from the point A to the point B along the path. What can be said about the acceleration at points along a path where the curvature changes abruptly?

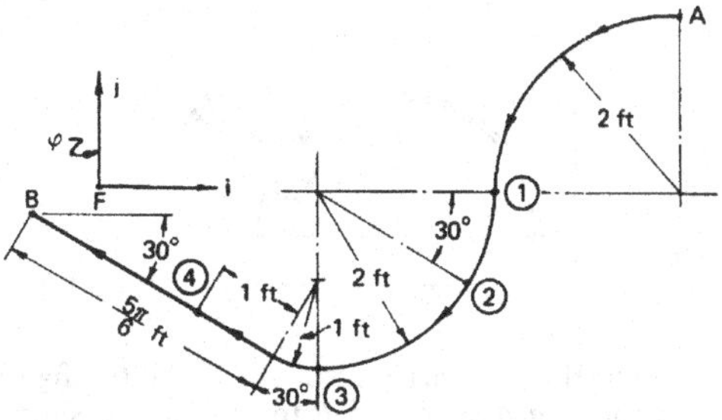

Problem 1.64.

1.65. A particle has a motion given by $12\mathbf{x}(P, t) = t^4\mathbf{i} + 72\mathbf{j} + 2(4t^3 - 9t^2)\mathbf{k}$ in $\varphi = \{B; \mathbf{i}_k\}$. Find the velocity and acceleration of P in φ at the instant $t = 3$. How is the velocity vector related to the acceleration vector at this instant? What does this relationship imply about the curvature of the path? Find $\kappa(t)$, and thereby confirm your conclusion.

1.66. A particle P moves on a twisted cubic trajectory so that

$$\mathbf{x}(P, t) = 6t\mathbf{i} + 3t^2\mathbf{j} + t^3\mathbf{k}$$

in the usual Cartesian frame. Find the velocity and acceleration of P referred to the intrinsic frame.

1.67. An electron emitted from the cathode of a television tube has a helical motion $\mathbf{x}(E, t) = R(\cos\phi\,\mathbf{i} + \sin\phi\,\mathbf{j}) + P\phi\mathbf{k}$, where P is a constant proportional to the pitch, R is the constant radius of the helix, and $\phi(t)$ is the variable angle of rotation about the helix axis. Find the curvature of the helix and the intrinsic velocity and acceleration of E.

1.68. The motion of a fluid particle P is given as $\mathbf{x}(P, t) = 2t^2\mathbf{i} + 3\mathbf{j} - 2t\mathbf{k}$. Find the radius of curvature of the trajectory of P as a function of time t, and determine the intrinsic velocity and acceleration of P.

1.69. A particle has a velocity $\mathbf{v}(P, t) = (2bc)^{1/2}\, t^2\mathbf{i} - bt\mathbf{j} + ct^3\mathbf{k}$. What are the velocity and the acceleration referred to the intrinsic frame?

1.70. An atomic particle P, initially at rest, has an acceleration

$$\mathbf{a}(P, t) = 2t\mathbf{i} - 3t^2\mathbf{j} + 2t^3\mathbf{k}$$

in $\varphi = \{O; \mathbf{i}_k\}$. Find the intrinsic velocity and acceleration of P in φ.

1.71. The center of mass point P of an amusement park vehicle moves on a space curve defined by Cartesian variables $x(t) = 2t$, $y(t) = t^2$, $z(t) = t^3/3$. (a) Find the intrinsic velocity and acceleration of P, and determine the radius of curvature of its path. (b)

What is the distance traveled by the vehicle in time t? (c) Calculate $\mathbf{x}(P, 1)$, $\mathbf{v}(P, 1)$, $\mathbf{a}(P, 1)$, $R(1)$ and $s(1)$, after one second.

1.72. A pin P of a mechanism shown in the figure moves with a constant tangential acceleration α along a circular groove of radius r. (a) Determine as functions of time t the angular speed $\omega(t) = \dot{\theta}(t)$ and the angular position $\theta(t)$ of P such that initially $\omega(0) = \omega_0$ and $\theta(0) = \theta_0$. (b) Show that for $\theta_0 = 0$ and $\beta =$ a constant, $\omega^2 = \omega_0^2 + \beta\theta$ relates the angular speed to the angle θ. Identify β. (c) Find the intrinsic velocity and acceleration of P as functions of θ alone.

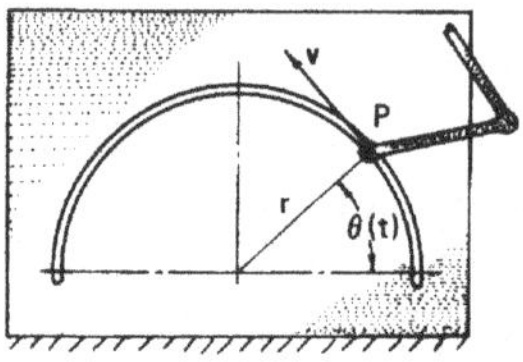

Problem 1.72.

1.73. A trammel valve mechanism shown in the figure consists of a control rod AP of length 8 cm hinged at A and B to blocks that slide in the cross slots. The constant distance between A and B is 2 cm, and the control rod rotates counterclockwise with a constant angular speed $\dot{\theta} = \omega = 2$ rad/sec. Find the velocity and the acceleration of the trammel point P referred to the intrinsic frame $\psi = \{P; \mathbf{t}, \mathbf{n}\}$ when the rod is in the position at $\theta = 45°$. Derive for this position the expressions for $\mathbf{t}$ and $\mathbf{n}$ in terms of $\Phi = \{F; \mathbf{I}, \mathbf{J}\}$.

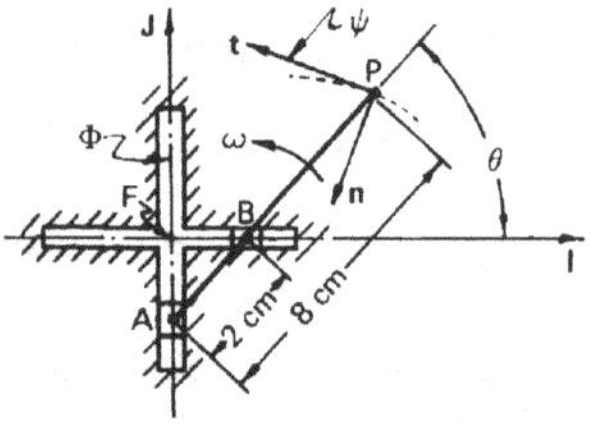

Problem 1.73.

1.74. An electron E emitted from the cathode C of an oscilloscope has a motion $\mathbf{x}(E, t) = a(\cos\omega t\mathbf{i} + \sin\omega t\mathbf{j}) + a\omega t\mathbf{k}$ in frame $\varphi = \{C; \mathbf{i}_k\}$. Here a and ω are constants. Find the velocity and acceleration of E referred to the intrinsic frame $\psi = \{E; \mathbf{t}_k\}$; determine the intrinsic basis $\{\mathbf{t}_k\}$; find the radius of curvature of the trajectory at E at time t; and describe the path.

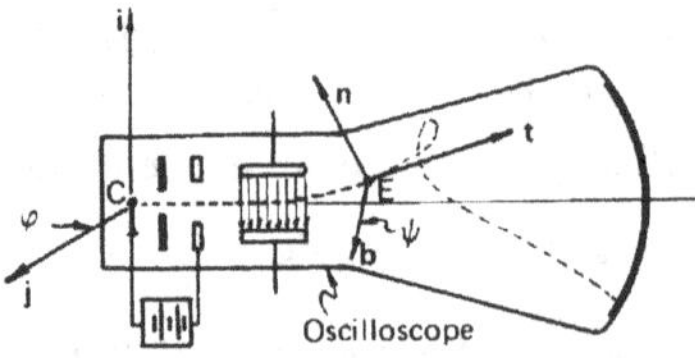

Problem 1.74.

1.75. A particle P, initially at F, has the velocity $\mathbf{v}(P, t) = 3t^2\mathbf{i} + 6t\mathbf{j} + 6\mathbf{k}$ cm/sec in $\Phi = \{F; \mathbf{i}_k\}$. (a) Find the velocity and acceleration of P referred to the intrinsic frame $\psi = \{P; \mathbf{t}_k\}$ at time t. (b) What is the location of P in Φ after 2 sec? What is the radius of curvature of the path at this location? Describe the nature of the path after an infinitely long time.

1.76. A ship S shown in the figure is sailing on a circular arc of radius r with a tangential component of acceleration proportional to its speed, which initially was v_0. Find the intrinsic velocity and acceleration of the ship as functions of the time t. Determine the distance traveled in time t.

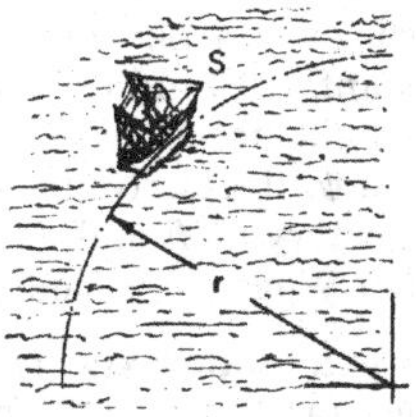

Problem 1.76.

1.77. Find the intrinsic velocity and acceleration of the ship described in the last problem expressed as functions of the distance s traveled along the arc. Determine the time t required to move this distance.

1.78. The center of mass point Q of an object moves parallel to a surface whose profile may be approximated by the curve $y = a \sin \omega x$ shown in the figure. (a) Find the constants a and ω. (b) If Q has a speed of 9 ft/sec and a tangential acceleration of 24 ft/sec^2 at the point A, what is its intrinsic acceleration at A? (c) What is its acceleration at A referred to the fixed frame $\varphi = \{F; \mathbf{i}, \mathbf{j}\}$? (d) Derive expressions for the velocity and acceleration of Q at point B.

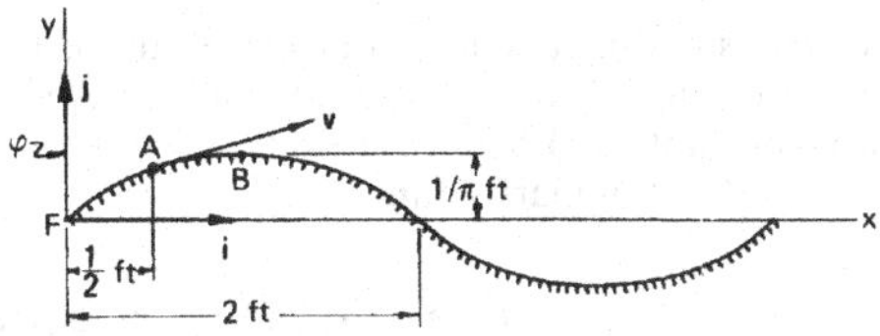

Problem 1.78.

1.79. An electron moves in a plane in such a manner that at each point of its path the intrinsic components of the acceleration vector are constant. Prove that the electron moves on an equiangular spiral $R = \alpha e^{\beta\theta}$ in which α, β are constants, R is the radius of curvature, and θ is the angle that the tangent line to the path makes with the $\mathbf{i}$ direction, namely, $\mathbf{t} \cdot \mathbf{i} = \cos\theta$ in $\varphi = \{O; \mathbf{i}, \mathbf{j}\}$.

1.80. (a) Show that the motion $\mathbf{x}(P, t) = \mathbf{A} \cos pt + \mathbf{B} \sin pt$ with constant vectors $\mathbf{A}$ and $\mathbf{B}$ is the solution of the vector differential equation $\ddot{\mathbf{x}} + p^2\mathbf{x} = \mathbf{0}$. (b) Consider the special case when $\mathbf{A} = a\mathbf{i}$ and $\mathbf{B} = a\mathbf{j}$. Show that the motion $\mathbf{x}(P, t)$ may be written as a similar function $\mathbf{x}(s(t))$ of the arc length parameter $s(t)$; hence, $\mathbf{x}(s)$ is a solution of the differential equation $\mathbf{x}'' + \mathbf{x}/a^2 = \mathbf{0}$, where $' \equiv d/ds$. Determine the curvature and torsion of the path by inspection of the results, and then by differentiation based upon (1.109).

1.81. Use the intrinsic rotation equations (1.108) to show that the torsion of a space curve is given by the formula

$$\tau = \frac{\mathbf{v} \times \mathbf{a} \cdot \dot{\mathbf{a}}}{|\mathbf{v} \times \mathbf{a}|^2}$$

in terms of the physical quantities **v**, **a**, and $\dot{\mathbf{a}}$. Express this relation and (1.75c) in terms of time rates of change of the position vector; then, bearing in mind the intrinsic relation $\mathbf{x}(P, t) = \mathbf{x}(s(t))$, use these formulas to derive (1.109).

1.82. A particle P has a motion $\mathbf{x}(P, t) = t^3/3\mathbf{i} + t^2\mathbf{j} + 2t\mathbf{k}$ in $\Phi = \{F; \mathbf{i}_k\}$. (a) Find the intrinsic velocity and acceleration of P. (b) Use the result of the previous problem to find the torsion of the path, and determine the curvature. (c) Express the intrinsic basis vectors $(\mathbf{t}, \mathbf{n}, \mathbf{b})$ in terms of the basis vectors $(\mathbf{i}, \mathbf{j}, \mathbf{k})$ of Φ.

1.83. The fluid particles in a plane corner flow move in frame $\psi = \{O; \mathbf{i}_k\}$ along the paths $xy = c$, a constant. (a) Find as a function of x the hodograph motion of the fluid paticle P that passes the point (1, 1) with a constant speed of 4 ft/sec and having a positive component $\dot{x}$ in ψ. (b) Derive the standard equation for the hodograph, and sketch graphs of the trajectory of P and its hodograph. (c) Find the intrinsic velocity and acceleration of P. (d) What is the hodograph velocity of P?

1.84. A particle P has a motion $\mathbf{x}(P, t) = \mathbf{l} + \mathbf{m}t^n$, where $\mathbf{l}$, $\mathbf{m}$ are constant vectors and n is a positive integer. (a) Describe the path of P. (b) Find the hodograph motion and velocity, and describe the hodograph.

1.85. An electron E, initially at rest at O in frame $\mu = \{O; \mathbf{i}_k\}$, has an acceleration $\mathbf{a}(E, t) = (k + \omega\dot{y})\,\mathbf{i} - \omega\dot{x}\mathbf{j}$, in which k and ω are constants. (a) Find the velocity and the motion of E as functions of time in μ. Describe the trajectory of the electron in μ. (b) What is the hodograph motion? Describe the hodograph. (c) Write a computer program to graph the particle path and the hodograph for the four combinations of the values $k = 0.4$, 1 and $\omega = 5$, 10.

1.86. A particle moves on the path $y = a \sin \omega x$ with a constant horizontal velocity component. Here a and ω are constants. Find the hodograph motion and its velocity. Describe the hodograph.

1.87. (a) The smooth function $h(x) = x^2$ is to be turned on at $x = 1$. Show that $H(x) \equiv x^2\langle x-1\rangle^0$ may be written as a polynomial $H(x) = \sum_{k=0}^{2} a_k\langle x-1\rangle^k$ of degree 2, where $a_k$are constants. (b) More generally, show that for constants a and b and for positive integers n the singularity function

$$H(x) \equiv (x-b)^n\langle x-a\rangle^0 = \sum_{k=0}^{n} a_k\langle x-a\rangle^k$$

is a polynomial of degree n in the singularity functions $\langle x-a\rangle^k$:

$$\begin{aligned} H(x) = {} & \langle x-a\rangle^n + n(a-b)\langle x-a\rangle^{n-1} + \frac{n(n-1)}{1\cdot 2}(a-b)^2\langle x-a\rangle^{n-2} \\ & + \cdots + \frac{n(n-1)\cdots(n-r+1)}{1\cdot 2\cdot\cdots\cdot r}(a-b)^r\langle x-a\rangle^{n-r} + \cdots \\ & + n(a-b)^{n-1}\langle x-a\rangle^1 + (a-b)^n\langle x-a\rangle^0. \end{aligned}$$

Use this rule to check the special case in (a). (c) Write the polynomial expression for the parabolic function shown in Fig. 1.21.

1.88. Find by two methods the derivatives of the several functions identified as $H(x)$ in the last problem. Show that the results are the same.

1.89. Find by two methods the integrals $F(x) = \int_{-\infty}^{x} H(x)\,dx$ for the several functions identified as $H(x)$ in Problem 1.87. Show that the results are the same.

1.90. The rectilinear speed of a particle P in two situations is expressed as a smooth function $\dot{s} = v\sin(2\pi s/l)$ in which $s(t)$ denotes the distance traveled by P, and v and l are constants. In the first case, the motion begins instantaneously at the origin $s=0$; and in the second instance, it starts suddenly at $s=a$. Express these statements in terms of singularity functions, and find for each case the acceleration of P.

1.91. The acceleration $\mathbf{a}(P, t) = a(t)\,\mathbf{i}$ of a particle P, initially at rest at $\mathbf{x}_0 = 15\mathbf{i} + 6\mathbf{j} + 25\mathbf{k}$ ft, is described by the graph shown in the figure. Use singularity functions to find the motion and velocity of P as functions of time.

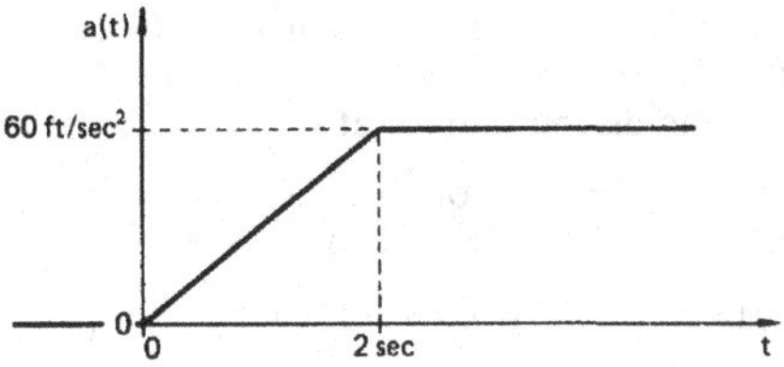

Problem 1.91.

1.92. Starting from rest at the origin, a particle P has a rectilinear motion $x(t)$ with speed given by $v = -2x + g(t)$, wherein the function $g(t)$ is defined by

$$g(t) = \begin{cases} 0 & \text{for } t \leqslant 0 \\ \cos t & \text{for } 0 < t < \pi/2 \\ e^{(\pi/2)-t} & \text{for } t > \pi/2. \end{cases}$$

Find the motion of P and determine its speed and acceleration as functions of time alone. Hint: Multiply the first-order differential equation $dx/dt + 2x = g(t)$ by the integrating factor e^{2t}, and integrate the result. Of course, other methods are possible.

1.93. A spring-loaded push rod of a certain automatic hammer mechanism is driven by a drop cam that turns with a constant, clockwise angular speed $\dot{\theta} = \omega$. The hammer is at rest instantaneously at $\theta = 0$ and at 2π where the hammer motion occurs suddenly due to the spring action at the cam drop. The cam profile function $r(\theta)$ satisfies the conditions $r(0) = a$, $r(\pi) = a + b$, where a and b are constants; and the cam-driven hammer is designed so that it has the acceleration graph $\mathbf{a} = \mathbf{a}(\theta)$ with constant magnitude α as shown in the figure. (a) Find with the aid of singularity functions the motion and the velocity of the hammer as functions of $\theta \in [0, 2\pi]$, draw their graphs, and sketch the shape of the cam needed. (b) Investigate the solution without the use of singularity functions. (c) Discuss the discontinuities in the velocity and acceleration at $\theta = 2\pi$.

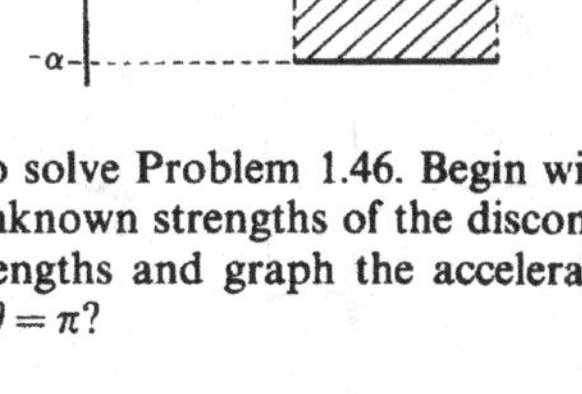

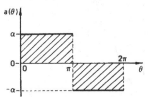

Problem 1.93.

1.94. Use singularity functions to solve Problem 1.46. Begin with the equation for the acceleration accounting for the unknown strengths of the discontinuities at the end points in the motion. Find these strengths and graph the acceleration on $0 \leqslant \theta \leqslant 2\pi$. What is the jump in the velocity at $\theta = \pi$?

1.95. Solve Problem 1.47 with the aid of singularity functions. Show that the results include those obtained before. Find the acceleration in terms of singularity functions and graph the result showing the strengths of the discontinuities.

1.96. Derive the profile of the cam described in Problem 1.48. Use singularity functions.

1.97. Employ singularity functions to investigate Problem 1.49. Discuss the discontinuities that arise at $\theta = 2\pi$, and determine their strengths. What can be said about the acceleration?

1.98. Find with the aid of singularity functions the cam profile required in Problem 1.50. Express the result in terms of the geometrical parameters, and discuss the behavior of the discontinuities at $\theta = \pi$. Describe the acceleration response.

1.99. Consider the function $f(x)\langle x-a\rangle^0 = \{0 \text{ if } x<a;\ f(x) \text{ if } x>a\}$. Now suppose that $f(x) = \langle x-a\rangle^0$ itself. Then for $x>a$, $f(x)=1$. Hence,

$$f(x)\langle x-a\rangle^0 = \langle x-a\rangle^0\langle x-a\rangle^0 = \{0 \text{ if } x<a;\ 1 \text{ if } x>a\}.$$

That is, the square of the unit step function is equal to itself:

$$\langle x-a\rangle^0\langle x-a\rangle^0 = \langle x-a\rangle^0$$

(a) Show that for integers $n, m>0$

$$\langle x-a\rangle^n\langle x-a\rangle^m = \langle x-a\rangle^{n+m}.$$

(b) A particle P is moving with constant speed v along a smooth path having a constant curvature, when a tangential braking force is applied suddenly at time τ to decelerate the motion at a constant rate α along the path. Find in terms of singularity functions the intrinsic velocity and acceleration of P. Determine the time after braking needed to bring P to rest, and find the total distance traveled.

1.100. The oscillograph record of the rectilinear acceleration of a slider block of a certain mechanism during 10 sec of its operation is shown in the figure as a function of time. The block starts from rest at $t=0$. (a) Express the acceleration in terms of singularity power functions, and find the motion and the velocity. (b) Determine the distance traveled by the block after 10 sec. (c) What is its speed after 10 sec? Check this result by relating it to the area under $a(t)$ graph.

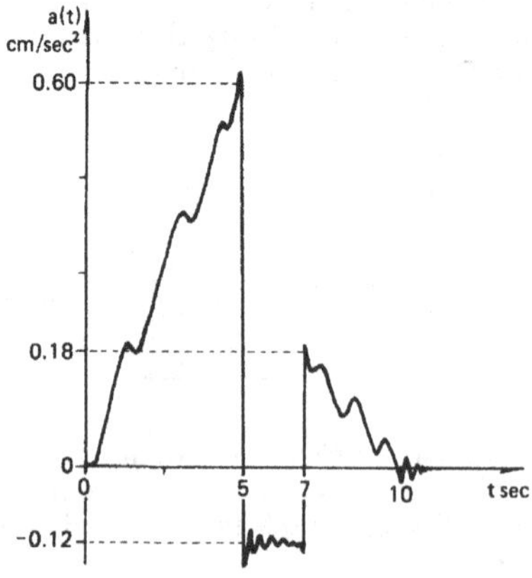

Problem 1.100.

2

Kinematics of Rigid Body Motion

2.1. Introduction

We recall that a body $\mathscr{B}$ is a collection of material points P: $\mathscr{B} = \{P\}$. In general, all of the particles of a body may change their relative positions in $\mathscr{B}$, and in this case the body is said to be *deformable*. At the other extreme, a *rigid body* is a body with the property that the straight line distance between every pair of its particles is constant in time. This idealization of a body that cannot be deformed, however great may be the forces and torques that act upon it, is so intuitively natural that it is often used without mention. The reader surely will recognize that our basic definition of a reference frame embodied the concept of rigidity. In fact, there were several occasions in Chapter 1 where the concept was quietly invoked. In particular, it was tacitly supposed for the mechanical device shown in Fig. 1.3 that the radius R of the wheel and the length L of the hinged rod did not vary with time. These are typical examples of rigid bodies whose motions will be investigated in this chapter. The kinematics of a rigid body in general motion in space will be studied. The main objective will be to learn how the velocity and acceleration of the particles of a rigid body are related to the translational and rotational parts of its motion.

The theory of the motion of a rigid body rests upon a fundamental theorem due to Euler (1775). He proved that when a rigid body is rotated about a fixed point, all particles situated upon some line through that point return to their initial positions upon completion of the displacement. Thus, the infinite variety of displacements by which a rigid body may be brought from one configuration into another consists always of a translation accompanied by a rotation about a line. Chasles (1843) used this result to prove that among these general displacements there exists one of unparalleled simplicity.

He showed that the most general displacement of a rigid body may be represented uniquely by a rotation about a line together with a translation along that line. Study of these important theorems and some other advanced topics concerning finite rigid body motions will be reserved for the next chapter. Our immediate objective in this chapter will be to derive from the exact finite displacement vector equation special representations for the velocity and acceleration of the particles of a rigid body that separate and exhibit clearly the translational and rotational parts of the body's motion. Afterwards, several sample applications of these results will be presented. We shall conclude our study with discussion of some useful theorems related to instantaneous screw motions. Let us begin with a few definitions of terms needed in our study.

2.2. Displacements of a Rigid Body

Let a rigid body $\mathscr{B}$ undergo an arbitrary displacement in space relative to an assigned Cartesian frame $\Phi = \{O; \mathbf{e}_k\}$, as shown in Fig. 2.1; and let $\mathbf{X}$ and $\hat{\mathbf{X}}$ denote the respective position vectors of a particle P in its initial and final positions from O. Then, regardless of the actual motion of P between these positions, the vector $\mathbf{d}(P) \equiv \hat{\mathbf{X}} - \mathbf{X}$ defines the finite displacement of P in Φ. Hence, $\mathbf{d}(P)$ is named the *displacement* of P relative to Φ.

A motion of $\mathscr{B}$ in which every material point sustains the same displacement $\mathbf{d}$ is called a *parallel translation*, or briefly, a *translation*. In this case, $\mathbf{d}(P) = \mathbf{d}$ is the same vector for every particle P in $\mathscr{B}$. Therefore, any arbitrary motion that the body may have suffered in reaching its terminal state is the same as a motion in which its particles traverse parallel, straight line paths, as shown in Fig. 2.1, though the body need not move on that line at each instant. (See Problem 2.1.)

Whenever the initial and final spatial position vectors of at most one point P are the same, then $\mathbf{d}(P) = \mathbf{0}$ only for that one point. The arbitrary motion that the body may have experienced in achieving its end state is indistinguishable from any other having the same end state and for which P is

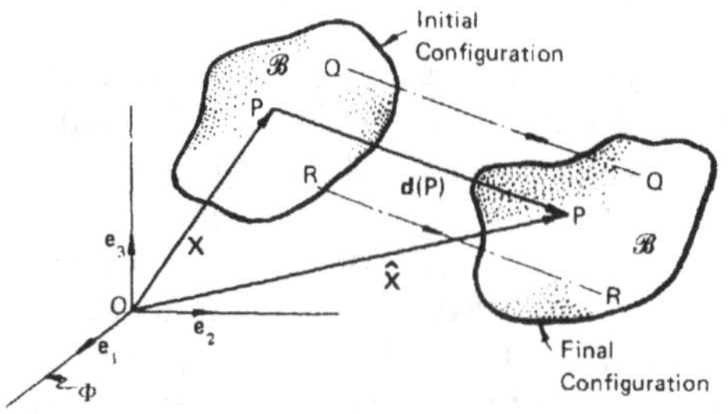

Figure 2.1. Displacement of a particle and parallel translation of a rigid body.

assumed to be fixed in Φ; hence, the displacement of $\mathscr{B}$ is described as a *rotation about a fixed point*. When P is indeed fixed in Φ, the trajectory of any particle of the body is a curved line on the surface of a sphere whose radius is the particle's distance from P.

Similarly, an arbitrary motion in which the spatial position in Φ of every point on some line L is unaltered is the same as a motion in which L is assumed to be fixed in Φ; and the displacement is called a *rotation about a fixed line*. The line L is titled the *axis of rotation*. Our first objective is to describe the displacement $\mathbf{d}(P)$ of a particle P of a rigid body due to a rotation about a fixed line.

2.3. Rotation about a Fixed Line

The vector equation for the displacement of a particle P in a rotation of a rigid body around a fixed line will be derived in terms of three specified quantities: $\mathbf{x}$, the initial position vector of P; $\boldsymbol{\alpha}$, a unit vector that defines the axis of rotation; and θ, the angle of the rotation. Let the body be turned through an angle θ about a fixed line OA in a right-hand sense with respect to a unit axial vector $\boldsymbol{\alpha}$ directed from O to A, as shown in Fig. 2.2; and consider a particle P initially in a plane K which is perpendicular to the axis at A. Since the body is rigid and OA is fixed, the trajectory of P in the plane K is a circle centered on the axis of rotation. The displacement of P is described by

$$\mathbf{d}(P) = \hat{\mathbf{x}} - \mathbf{x} = \hat{\boldsymbol{\rho}} - \boldsymbol{\rho}, \tag{2.1}$$

where $\mathbf{x}\{\hat{\mathbf{x}}\}$ denotes the initial $\{$final$\}$ spatial position vector of P from O, and $\boldsymbol{\rho}\{\hat{\boldsymbol{\rho}}\}$ is the initial $\{$final$\}$ radius vector of P from A in $\Phi = \{O; \mathbf{e}_k\}$. The first equality in (2.1) is just the definition of the displacement of P in any motion whatever; it is the second equality that distinguishes the displacement

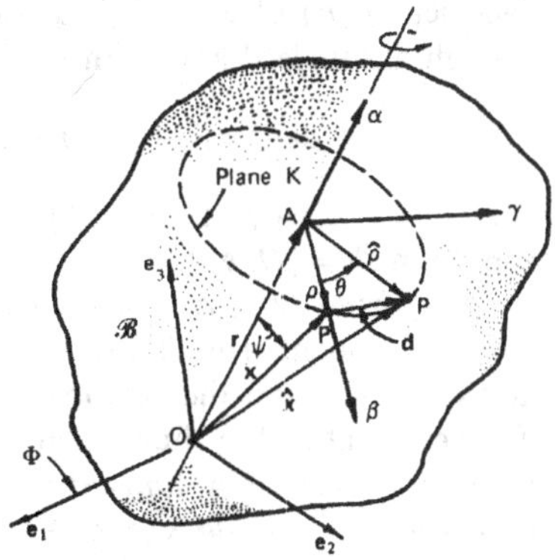

Figure 2.2. Geometry for the description of a finite rotation about a line.

as a rotation about the vector $\boldsymbol{\alpha}$. To exhibit the rotational aspects of the displacement, the last term in (2.1) must be expressed in terms of the specified quantities $\mathbf{x}$, $\boldsymbol{\alpha}$, θ mentioned before.

Since $|\boldsymbol{\rho}| = |\hat{\boldsymbol{\rho}}| = \rho$, the radius of the circular path of P, it is clear from the geometry in Fig. 2.2 that

$$\boldsymbol{\rho} = \rho\boldsymbol{\beta} \tag{2.2a}$$

and

$$\hat{\boldsymbol{\rho}} = \rho(\cos\theta\boldsymbol{\beta} + \sin\theta\boldsymbol{\gamma}), \tag{2.2b}$$

where $\boldsymbol{\beta}$ is the unit vector from A to P and the unit vector $\boldsymbol{\gamma}$ is defined by

$$\boldsymbol{\gamma} = \boldsymbol{\alpha} \times \boldsymbol{\beta}. \tag{2.3}$$

It is plain from their definitions that ρ, $\boldsymbol{\beta}$, and $\boldsymbol{\gamma}$ are independent of θ. Thus, upon putting (2.2) into (2.1), we reach a formula that exhibits exactly the dependence on the angle of rotation:

$$\mathbf{d}(P) = \hat{\mathbf{x}} - \mathbf{x} = \rho\sin\theta\boldsymbol{\gamma} - \rho(1 - \cos\theta)\boldsymbol{\beta}. \tag{2.4}$$

It remains to express $\rho\boldsymbol{\beta}$ and $\rho\boldsymbol{\gamma}$ in terms of the assigned quantities $\mathbf{x}$ and $\boldsymbol{\alpha}$.

Let $\mathbf{r} = |\mathbf{r}|\,\boldsymbol{\alpha}$ denote the vector from O to the center of the circle at A in Fig. 2.2. Since $|\mathbf{r}| = |\mathbf{x}|\cos\psi = \mathbf{x}\cdot\boldsymbol{\alpha}$, where ψ is the angle between $\mathbf{x}$ and $\boldsymbol{\alpha}$, we may write $\mathbf{r} = (\mathbf{x}\cdot\boldsymbol{\alpha})\boldsymbol{\alpha}$. It follows that

$$\boldsymbol{\rho} = \mathbf{x} - \mathbf{r} = \mathbf{x} - (\mathbf{x}\cdot\boldsymbol{\alpha})\boldsymbol{\alpha}. \tag{2.5}$$

Using (2.5) in (2.2a) and (2.3), noting that $\boldsymbol{\alpha}\cdot\boldsymbol{\alpha} = 1$, and recalling the expansion rule for the vector triple product (see (A.14) in Appendix A), we obtain

$$\rho\boldsymbol{\beta} = \boldsymbol{\alpha} \times (\mathbf{x} \times \boldsymbol{\alpha}), \qquad \rho\boldsymbol{\gamma} = \boldsymbol{\alpha} \times \mathbf{x}. \tag{2.6}$$

Finally, substitution of (2.6) into (2.4) yields the desired relation for the displacement $\mathbf{d}(P)$ of a particle P of a rigid body in its right-handed rotation through an angle θ about an axis $\boldsymbol{\alpha}$ fixed in the body:

$$\mathbf{d}(P) = \hat{\mathbf{x}} - \mathbf{x} = \boldsymbol{\alpha} \times \mathbf{x}\sin\theta + (1 - \cos\theta)\boldsymbol{\alpha} \times (\boldsymbol{\alpha} \times \mathbf{x}), \tag{2.7}$$

wherein $\mathbf{x} = \mathbf{x}(P)$ and $\hat{\mathbf{x}} = \hat{\mathbf{x}}(P)$ are the initial and final position vectors of P from O in $\Phi = \{O; \mathbf{e}_k\}$.

The result (2.7) may be visualized graphically. It is seen in Fig. 2.3a, and it is also evident from (2.6), that the vector $\boldsymbol{\alpha} \times \mathbf{x}$ is tangent to the circular path of P at $\mathbf{x}$, and the vector $\boldsymbol{\alpha} \times (\boldsymbol{\alpha} \times \mathbf{x})$ is directed from P at $\mathbf{x}$ to the center of the circle at A. We thus see in Fig. 2.3b that the rotational displacement $\mathbf{d}$ is the vector sum of a *tangential displacement* $\boldsymbol{\alpha} \times \mathbf{x}\sin\theta$ and a *normal displacement* $(1 - \cos\theta)\boldsymbol{\alpha} \times (\boldsymbol{\alpha} \times \mathbf{x})$.

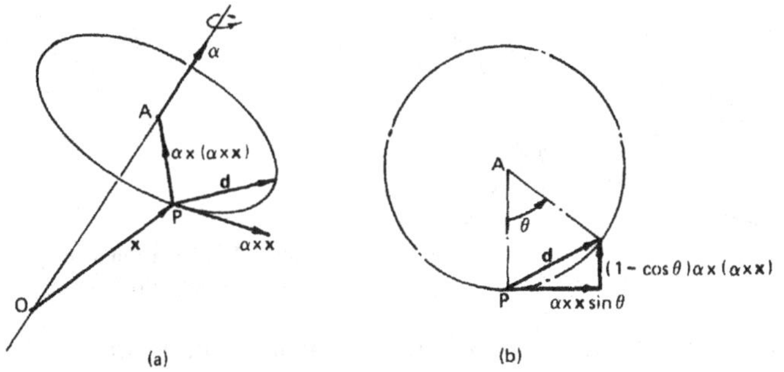

Figure 2.3. Tangential and normal displacement vectors associated with the displacement due to a rotation about an axis.

For future use and simplicity, it is convenient to rewrite (2.7) in the abbreviated form

$$\mathbf{d} = \hat{\mathbf{x}} - \mathbf{x} = \mathbf{T}\mathbf{x}, \tag{2.8}$$

where **T**, which is named the *rotator*, is presently identified as the vector operator

$$\mathbf{T} \equiv \boldsymbol{\alpha} \times [\ \ ] \sin\theta + (1 - \cos\theta)\boldsymbol{\alpha} \times (\boldsymbol{\alpha} \times [\ \ ]). \tag{2.9}$$

Equation (2.8) shows that the rotator transforms the initial position vector of a particle into its displacement vector due to a finite rigid rotation about a fixed line.

If $\mathbf{a} = a\boldsymbol{\alpha}$ is the position vector of a particle on the axis of rotation, then

$$\mathbf{T}\mathbf{a} = \mathbf{0} \tag{2.10}$$

follows from (2.9). This merely confirms the physical constraint that *points on the axis of rotation experience no displacement.*

Let us observe that **T** is a linear operator, namely, it has the basic property that

$$\mathbf{T}(\lambda\mathbf{x} + \mu\mathbf{y}) = \lambda\mathbf{T}\mathbf{x} + \mu\mathbf{T}\mathbf{y}, \tag{2.11}$$

in which λ, μ are scalars and **x**, **y** are vectors. This fact together with (2.10) may be used in an easy demonstration that *the displacement of a* point *P due to a rotation about a fixed line is independent of the choice of reference point O on the axis.* We let $\bar{\mathbf{d}} = \mathbf{T}\bar{\mathbf{x}}$ denote the displacement due to the same rotation of P but with position vector $\bar{\mathbf{x}}$ from any other point $\bar{O}$ on the axis, as shown in Fig. 2.4. Then, with the aid of the rule (2.11) and by use of (2.8) and (2.10), we see that $\mathbf{d} - \bar{\mathbf{d}} = \mathbf{T}\mathbf{a} = \mathbf{0}$ because the vector $\mathbf{a} \equiv \mathbf{x} - \bar{\mathbf{x}}$ is parallel to $\boldsymbol{\alpha}$. Therefore, the previous statement follows. In fact, the same thing is seen more

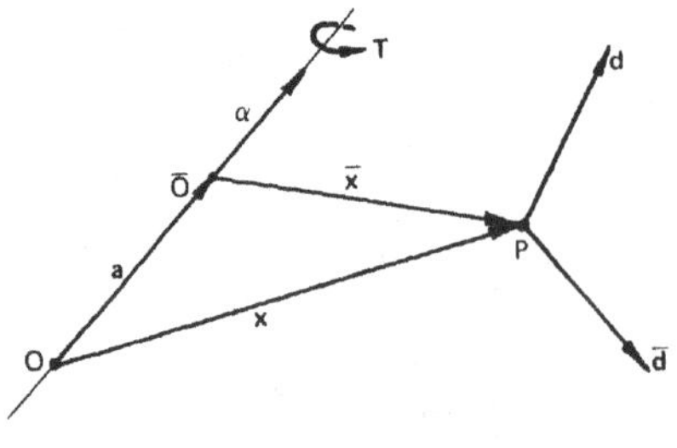

Figure 2.4. Rotational displacements of P for two points on the axis.

directly from (2.1) because $\hat{\boldsymbol{\rho}} - \boldsymbol{\rho}$ is independent of the location of O on the axis.

More important and less obvious applications of the rotator as a linear operator will be encountered in the next chapter. A typical calculation of a displacement vector is illustrated next.

Example 2.1. A satellite tracking antenna fixes on a target by rotating through an angle $\theta = \tan^{-1} 3/4$ about an axis $\boldsymbol{\alpha}$ in the horizontal plane in frame $\Phi = \{F; \mathbf{I}_k\}$. The antenna horn H initially is in the configuration shown in Fig. 2.5. Find the displacement of H and determine its final location in Φ.

Solution. The problem geometry is shown Fig. 2.5a. We see that $\boldsymbol{\alpha} = \cos 45^\circ\, \mathbf{I} + \sin 45^\circ\, \mathbf{J} = (\sqrt{2}/2)(\mathbf{I} + \mathbf{J})$ and $\mathbf{x}(H, t) = 3\boldsymbol{\alpha} + 6\mathbf{K}$ m. Thus, bearing in mind (2.7), we compute

$$\boldsymbol{\alpha} \times \mathbf{x} = \boldsymbol{\alpha} \times 6\mathbf{K} = 3\sqrt{2}(\mathbf{I} - \mathbf{J})\text{ m},$$

$$\boldsymbol{\alpha} \times (\boldsymbol{\alpha} \times \mathbf{x}) = \begin{vmatrix} \mathbf{I} & \mathbf{J} & \mathbf{K} \\ \sqrt{2}/2 & \sqrt{2}/2 & 0 \\ 3\sqrt{2} & -3\sqrt{2} & 0 \end{vmatrix} = -6\mathbf{K}\text{ m}.$$

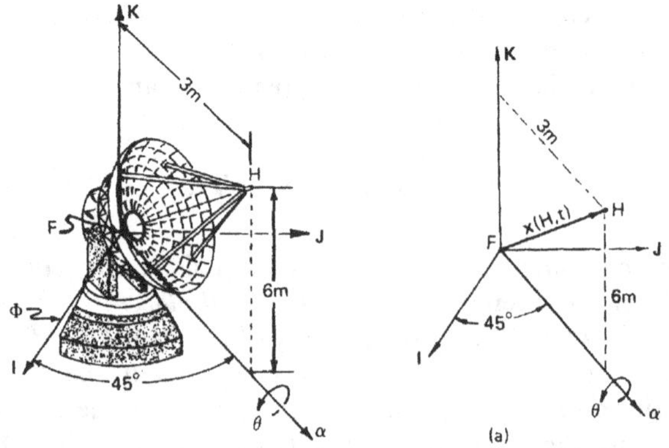

Figure 2.5. Finite rotation of a satellite tracking antenna.

Moreover, $\tan\theta = 3/4$ implies that $\sin\theta = 3/5$, $\cos\theta = 4/5$. Substitution of these data into (2.7) yields the displacement of the antenna horn:

$$\mathbf{d}(H) = \hat{\mathbf{x}} - \mathbf{x} = \frac{9\sqrt{2}}{5}(\mathbf{I} - \mathbf{J}) - \frac{6}{5}\mathbf{K}\ \text{m}.$$

The final position of H in Φ is now given by $\hat{\mathbf{x}}(H) = \mathbf{d}(H) + \mathbf{x}(H)$; we find

$$\hat{\mathbf{x}}(H) = \frac{3\sqrt{2}}{10}(11\mathbf{I} - \mathbf{J}) + \frac{24}{5}\mathbf{K}\ \text{m}.$$

2.4. The Imbedded Reference Frame

Because the distances between the particles of a rigid body are unaltered when the body moves, lines of particles remain lines and angles between all pairs of such lines are preserved. In particular, a triad of perpendicular material lines remains orthogonal; so these imbedded lines may serve as a Cartesian reference frame that moves with the body. More generally, any reference frame whose coordinate lines are fixed relative to a rigid body is known as an *imbedded* or *body reference frame*. The origin of the imbedded frame is named the *base point*.

Although the imbedded frame moves with the body, no part of the imbedded frame actually need be within the body or contain any material part of it. The base point, in particular, belongs to the body only in the sense that it moves with it. A base point fixed at the apex of the frustum of a cone- or pyramid-shaped body, or one chosen at the center of the void of a doughnut-shaped rigid body, are examples of base points that are not material points; but they certainly move with the body.

Obviously, the three coordinates of a particle of the body remain the same when referred to the imbedded frame, whereas the three spatial coordinates of the same particle will vary as the body moves about. It is important to know the number of independent coordinates required to specify the location and orientation of a freely moving rigid body in the spatial frame Φ. This number, which is called the *degrees of freedom* of the body, can be easily determined. We begin by considering an imbedded triangle. Each vertex particle may be located by its three Cartesian coordinates in Φ; but because the sides of the triangle are of fixed length, which we may measure initially, we have also three distance equations of rigid constraint that relate the nine vertex coordinates. Thus, the number of independent coordinates for the three particles is six. Of course, any other particle also is identified by three coordinates. But none of these is free, because there are three additional independent rigid constraint equations that specify its fixed distances from the initial vertex particles. Thus, no more than six independent coordinates are required to specify completely the configuration of a rigid body in Φ. Therefore, *a rigid body has no more than six degrees of freedom.* Naturally, additional constraints on the motion of a body will reduce its degrees of freedom. Suppose, for example, that one point of the body is fixed in Φ; then the degrees of freedom are reduced to three. (How many degrees of freedom are there when two points are fixed?) And, similarly, a rigid disk which is constrained to move in a plane also has at most three degrees of freedom. (How many would it have if its center is also constrained to move on a specified plane curve?)

The six independent coordinates of an unconstrained body may be chosen in a variety of ways. For example, the three coordinates of any particle, or any base point, and any three independent angles that specify the orientation of the body frame relative to the spatial frame Φ may be selected. This natural choice is basic to the following description of the most general displacement of a rigid body.

2.5. The General Displacement of a Rigid Body

The mathematical description of the decomposition of the general displacement of a rigid body into its translational and rotational parts will be outlined here. Let us imagine that the body shown in Fig. 2.6 undergoes an arbitrary motion in space that carries it from a given initial configuration into another configuration relative to an assigned spatial frame $\Phi = \{F; \mathbf{I}_k\}$. Consider an imbedded frame $\varphi' = \{O; \mathbf{i}'_k\}$ which is parallel to Φ when the body is in its initial configuration. The corresponding position vectors of a particle P

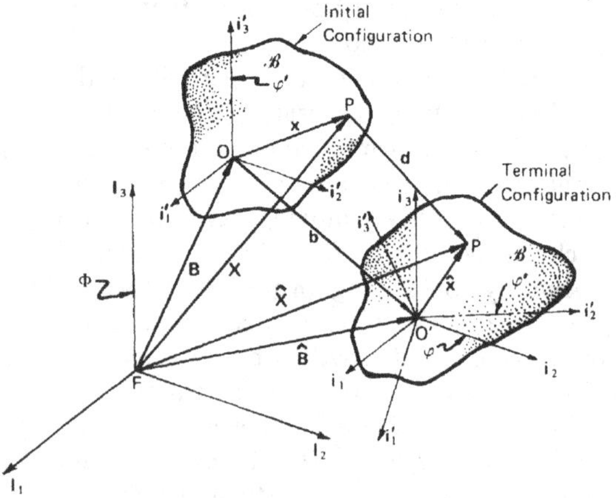

Figure 2.6. Finite displacement of a rigid body.

are denoted by $\mathbf{x}$ in φ' and $\mathbf{X}$ in Φ. After the displacement of the body, P has a new spatial position $\hat{\mathbf{X}}$ in Φ but it retains the same position with respect to the imbedded, body frame. The point O initially at $\mathbf{B}$ in frame Φ is displaced to O' at $\hat{\mathbf{B}}$, as diagrammed in Fig. 2.6. Finally, let us consider in the terminal configuration another auxiliary frame $\varphi = \{O'; \mathbf{i}_k\}$ which is parallel to Φ, hence also to the imbedded frame φ' in the initial configuration. If $\hat{\mathbf{x}}$ denotes the position vector of P from O', relative to φ, then the displacement vector of the particle P is given by

$$\mathbf{d}(P) = \hat{\mathbf{X}} - \mathbf{X} = \mathbf{b} + \hat{\mathbf{x}} - \mathbf{x}, \tag{2.12}$$

where $\mathbf{b} \equiv \mathbf{d}(O) = \hat{\mathbf{B}} - \mathbf{B}$ is the displacement of the base point O.

The foregoing description illustrates our discussion of reference frames in the previous section. Indeed, the angles between the $\mathbf{i}_k$ and $\mathbf{i}'_k$ vectors in the final configuration may be used to characterize the orientation of the body in its final state in Φ. In addition, the auxiliary frames will prove particularly useful in the proof of Euler's theorem and in its applications reserved for advanced study later. However, for our immediate needs, we may ignore their presence and assume that all vectors are referred to the spatial frame Φ alone. We continue with our description of the general displacement of a rigid body.

We note that $\mathbf{b}$ is independent of $\mathbf{x}$; hence, $\mathbf{d}(P) = \mathbf{b}$ for all points P if and only if $\hat{\mathbf{x}} = \mathbf{x}$. In this case, the displacement, by definition, is a parallel translation. On the other hand, if $\mathbf{b} = \mathbf{0}$ for every configuration of the body, the displacement (2.12) is due to a rotation about a fixed point O, and $\mathbf{d}(P) = \hat{\mathbf{x}} - \mathbf{x}$. Therefore, we may draw the following conclusion from (2.12):

(i) *The most general displacement of the points of a rigid body consists of a parallel translation of the base point, hence the body, together with a rotation about the base point.*

It is a remarkable fact that when a rigid body is rotated about a fixed point, all particles on some line through that point return to their original positions upon completion of the displacement. Therefore, the same displacement may be produced by a rotation about a line. We may thus conclude the following:

(ii) *A rotation about a point is equivalent to a rotation about a line through that point.*

This important result is due to Euler (1775). It shows that the equivalent rotation about a line through O is given by (2.8); hence, we now have $\hat{\mathbf{x}} - \mathbf{x} = \mathbf{T}\mathbf{x}$ in (2.12). Collecting our thoughts in (i) and (ii) into (2.12), we have the following:

(iii) *The most general displacement of a rigid body is equivalent to a parallel translation accompanied by a rotation about a line*:

$$\mathbf{d}(P) = \hat{\mathbf{X}} - \mathbf{X} = \mathbf{b} + \mathbf{T}\mathbf{x}. \tag{2.13}$$

Herein, $\mathbf{b}$ *is the displacement of the base point* O, $\mathbf{T}\mathbf{x}$ *is the rotational dis-*

placement for which the rotator **T** *is defined in* (2.9), *and* **x** *is the initial position vector of P from O.*

This is Chasles' corollary (1830). It states, in effect, that the six degrees of freedom of a rigid body consist of three translational degrees of freedom, represented by **b**, together with three rotational degrees of freedom, characterized by **Tx**. The latter may be taken as the two independent direction cosines of a fixed axis of rotation and the angle of rotation about it, for example.

It is easy to visualize the physical content of these major theorems, so the proof of Euler's theorem, which is not trivial, may be omitted in a first reading. However, the interested reader may wish to skip ahead to the following chapter where the proof of Euler's theorem and other results on finite rigid body rotations are presented. Otherwise, with no significant loss of continuity, we may now continue toward our main goal to develop equations for the velocity and acceleration of the points of a rigid body in terms of their translational and rotational parts.

2.6. Infinitesimal Displacement of a Rigid Body

The vector equation describing the infinitesimal displacement of the particles of a rigid body will be derived in this section from the finite displacement relation (2.13). The infinitesimal rotation vector and the instantaneous axis of rotation will be defined. The main result is summarized at the end.

Let a rigid body experience an arbitrary displacement in a time interval Δt so that, relative to the spatial frame Φ, the position vector of any particle P, which was at the place $\mathbf{X}(t)$ at time t, is given by $\hat{\mathbf{X}} = \mathbf{X}(t+\Delta t)$ at the instant $t+\Delta t$. Then $\mathbf{d}(P) = \mathbf{X}(t+\Delta t) - \mathbf{X}(t) \equiv \Delta\mathbf{X}$ is the displacement of P in the interval Δt. Similarly, since $\mathbf{B}(t)$ is the place in Φ at time t of any assigned base point O and $\hat{\mathbf{B}} = \mathbf{B}(t+\Delta t)$ is its place at $t+\Delta t$, then $\mathbf{b} = \mathbf{B}(t+\Delta t) - \mathbf{B}(t) \equiv \Delta\mathbf{B}$ describes the displacement of O during Δt. Of course, the body also experiences during Δt a rotation through an angle $\Delta\theta = \theta(t+\Delta t) - \theta(t)$ about an axis whose direction at time t is $\boldsymbol{\alpha}(t)$, where $\theta(t)$ is the angular placement of P at t. Thus, recalling (2.9), we see that the rotation about the axis $\boldsymbol{\alpha}$ during the time Δt is described by

$$\mathbf{Tx} = \sin \Delta\theta \, \boldsymbol{\alpha} \times \mathbf{x} + (1 - \cos \Delta\theta)\boldsymbol{\alpha} \times (\boldsymbol{\alpha} \times \mathbf{x}), \tag{2.14}$$

wherein $\mathbf{x} = \mathbf{x}(t)$ is the position vector of P from O in frame Φ at the instant t. Using these terms in (2.13), we may write

$$\Delta\mathbf{X} = \Delta\mathbf{B} + \mathbf{Tx}; \tag{2.15}$$

that is, with (2.14),

$$\Delta \mathbf{X} = \Delta \mathbf{B} + \sin \Delta\theta \, \boldsymbol{\alpha} \times \mathbf{x} + (1 - \cos \Delta\theta) \boldsymbol{\alpha} \times (\boldsymbol{\alpha} \times \mathbf{x}). \tag{2.16}$$

The foregoing description of the displacement of the particles of a rigid body actually is valid for a finite displacement that may occur in a possibly finite time interval Δt. However, when the angle of rotation is infinitesimal, the rotational displacement (2.14), hence the total displacement (2.16), may be simplified by use of the series functions

$$\sin u = u - u^3/3! + \cdots, \qquad \cos u = 1 - u^2/2! + \cdots. \tag{2.17}$$

First, we put $u = \Delta\theta$ and retain in (2.17) only terms of first order in the infinitesimal angle $\Delta\theta$ to approximate

$$\sin \Delta\theta = \Delta\theta, \qquad \cos \Delta\theta = 1.$$

Then substitution of these into (2.14) yields the infinitesimal rotational displacement

$$\mathbf{T}\mathbf{x} = \Delta\boldsymbol{\theta} \times \mathbf{x} \qquad \text{with} \quad \Delta\boldsymbol{\theta} \equiv \Delta\theta \, \boldsymbol{\alpha}. \tag{2.18}$$

The vector $\Delta\boldsymbol{\theta}$ is called *infinitesimal rotation vector*. Because the vector $\boldsymbol{\alpha} = \boldsymbol{\alpha}(t)$ is fixed in Φ only momentarily at time t, $\boldsymbol{\alpha}$ is named the *instantaneous axis of rotation*. Thus, when the displacement of the base point is infinitesimal and terms of only the first order in $\Delta\theta$ are retained, the total infinitesimal displacement (2.15), or (2.16), may be written in the form

$$\Delta \mathbf{X} = \Delta \mathbf{B} + \Delta\boldsymbol{\theta} \times \mathbf{x} \qquad \text{with} \quad \Delta\boldsymbol{\theta} = \Delta\theta \, \boldsymbol{\alpha}. \tag{2.19}$$

In words: *Relative to a spatial frame* $\Phi = \{F; \mathbf{I}_k\}$, *the total infinitesimal rigid body displacement* $\Delta\mathbf{X}$ *of a particle P, initially at the place* $\mathbf{x}$ *from an assigned base point O, is equivalent to an infinitesimal displacement* $\Delta\mathbf{B}$ *of the base point together with an infinitesimal rotational displacement* $\Delta\boldsymbol{\theta} \times \mathbf{x}$ *about an instantaneous axis through O.*

2.7. Composition of Infinitesimal Rotations

Successive finite rotations of a rigid body about concurrent axes cannot be compounded by addition, for consecutive rotators generally are neither additive nor commutative. This means that the displacement of a rigid body due to successive finite rotations generally will depend upon the order in which the rotations are performed. On the other hand, consecutive infinitesimal rotations obey the commutative law of vector addition; hence, these are independent of the order of their execution. To visualize this fundamental difference in the properties of finite and infinitesimal rotations, let us begin by considering consecutive finite rotations of a rectangular plate with an edge OP on the x_2 axis and initially oriented in the vertical plane of a spatial frame Φ, as shown in the diagrams of Fig. 2.7. If the plate shown in Fig. 2.7a is rotated first through a right angle about the x_1 axis and then through a right angle about the x_3 axis, while the same plate shown in Fig. 2.7b suffers the same rotations but in reverse order, we see at once that the final position of OP, indeed the orientation of the plate, is not the same. This confirms that the composition of successive finite rotations of a rigid body about concurrent axes is not commutative. The composition of finite rigid body rotations and other related theorems will be studied in the next chapter. Hereafter, we shall focus on the composition of infinitesimal rotations only.

In the derivation of (2.19), we have naturally and correctly represented the infinitesimal rotation by a vector symbol $\Delta\boldsymbol{\theta} = \Delta\theta\, \boldsymbol{\alpha}$, without regard for the noncommutative nature of finite rotations. Nevertheless, these may qualify as vectors only if the commutative law of vector addition is satisfied. Therefore, in light of the character of finite rotations, it is of interest to verify that *successive infinitesimal rotations of a rigid body about concurrent axes, independently of their order of execution, may be added vectorially to form a single equivalent infinitesimal rotation about another concurrent line.*

To establish this result, let $\mathbf{d}_1$ and $\mathbf{d}_2$ denote two displacements due to consecutive infinitesimal rotations through angles $\Delta\theta_1$ and $\Delta\theta_2$ about the respective concurrent axes $\boldsymbol{\alpha}_1$ and $\boldsymbol{\alpha}_2$ through a point fixed at O, as shown in Fig. 2.8a. Then, in view of (2.18), the corresponding infinitesimal displacements given by (2.13) may be written as

$$\mathbf{d}_1 = \Delta\boldsymbol{\theta}_1 \times \mathbf{x}, \tag{2.20a}$$

$$\mathbf{d}_2 = \Delta\boldsymbol{\theta}_2 \times \mathbf{x}_1 = \Delta\boldsymbol{\theta}_2 \times \mathbf{x} + \Delta\boldsymbol{\theta}_2 \times (\Delta\boldsymbol{\theta}_1 \times \mathbf{x}), \tag{2.20b}$$

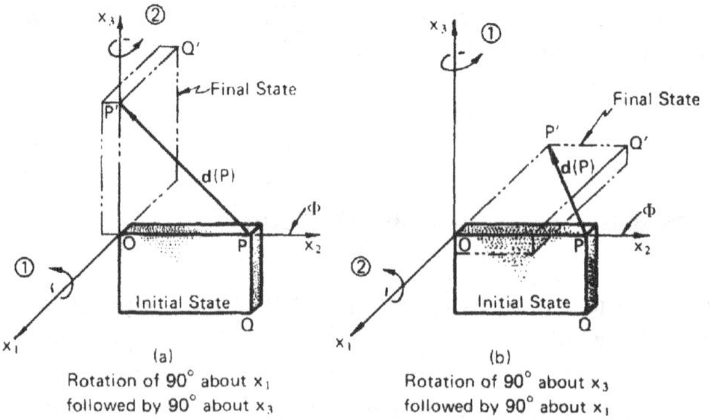

Figure 2.7. Consecutive finite rotations are not commutative, so they cannot be compounded by vector addition.

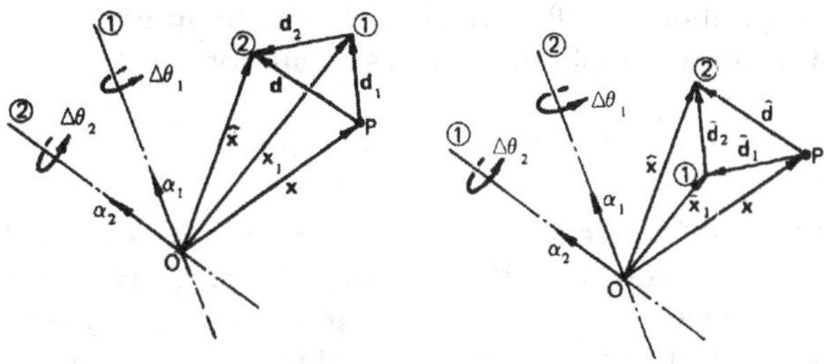

(a) Infinitesimal rotation $\Delta\theta_1$ followed by $\Delta\theta_2$

(b) Infinitesimal rotation $\Delta\theta_2$ followed by $\Delta\theta_1$

Figure 2.8. Composition of successive infinitesimal rotations about concurrent axes.

where $\mathbf{x}_1 = \mathbf{x} + \mathbf{d}_1$ is the position vector of a particle P after the first rotation from its initial place at $\mathbf{x}$. The total displacement is $\mathbf{d} = \mathbf{d}_1 + \mathbf{d}_2$.

Now, starting from the same initial position $\mathbf{x}$ and using identical rotations but performed with $\Delta\theta_2$ followed by $\Delta\theta_1$, as shown in Fig. 2.8b, we see with (2.18) and (2.13) that the corresponding infinitesimal displacement vectors are given by

$$\hat{\mathbf{d}}_1 = \Delta\boldsymbol{\theta}_2 \times \mathbf{x}, \tag{2.21a}$$

$$\hat{\mathbf{d}}_2 = \Delta\boldsymbol{\theta}_1 \times \hat{\mathbf{x}}_1 = \Delta\boldsymbol{\theta}_1 \times \mathbf{x} + \Delta\boldsymbol{\theta}_1 \times (\Delta\boldsymbol{\theta}_2 \times \mathbf{x}), \tag{2.21b}$$

wherein $\hat{\mathbf{x}}_1 = \mathbf{x} + \hat{\mathbf{d}}_1$ is the position vector of P after the first rotation, and now $\hat{\mathbf{d}} = \hat{\mathbf{d}}_1 + \hat{\mathbf{d}}_2$ defines the total displacement.

Upon discarding terms of order larger than the first in $\Delta\theta_1$ and $\Delta\theta_2$ in (2.20b) and (2.21b), which is consistent with our earlier approximation in (2.18), we find

$$\hat{\mathbf{d}}_1 = \mathbf{d}_2 = \Delta\boldsymbol{\theta}_2 \times \mathbf{x} \qquad \text{and} \qquad \hat{\mathbf{d}}_2 = \mathbf{d}_1 = \Delta\boldsymbol{\theta}_1 \times \mathbf{x}.$$

Therefore, regardless of the order of the rotations, the total displacement, $\hat{\mathbf{d}} = \hat{\mathbf{d}}_1 + \hat{\mathbf{d}}_2 = \mathbf{d}_2 + \mathbf{d}_1 = \mathbf{d}$, is the same; and, with the aid of the foregoing relations, it may be written as

$$\mathbf{d} = (\Delta\boldsymbol{\theta}_1 + \Delta\boldsymbol{\theta}_2) \times \mathbf{x} = (\Delta\boldsymbol{\theta}_2 + \Delta\boldsymbol{\theta}_1) \times \mathbf{x}.$$

Consequently, there exists independently of the order of the rotations an equivalent infinitesimal rotation

$$\Delta\boldsymbol{\theta} \equiv \Delta\boldsymbol{\theta}_1 + \Delta\boldsymbol{\theta}_2 = \Delta\boldsymbol{\theta}_2 + \Delta\boldsymbol{\theta}_1 \tag{2.22}$$

about another axis at O so that the total displacement $\mathbf{d} = \Delta\boldsymbol{\theta} \times \mathbf{x}$ is the same.

We recognize that (2.22) shows that consecutive infinitesimal rotations about concurrent lines are indeed compounded by vector addition. The familiar decomposition of $\Delta\boldsymbol{\theta}$ into its vector components in a frame $\Phi = \{F; \mathbf{e}_k\}$ is an important application of this result; for,

$$\Delta\boldsymbol{\theta} = \Delta\theta_1\mathbf{e}_1 + \Delta\theta_2\mathbf{e}_2 + \Delta\theta_3\mathbf{e}_3 \tag{2.23}$$

is the sum of three infinitesimal and conceivably simultaneous rotations about the concurrent coordinate lines. We recall that this property does not hold for finite rigid body rotations whose composition is neither additive nor commutative. But we shall reserve discussion of this advanced topic for later and continue toward our primary objective to learn how the velocity and acceleration of the particles of a rigid body are related to the translational and rotational parts of its motion.

2.8. Velocity and Acceleration of Points of a Rigid Body

We are now prepared to focus on our principal objective to derive from the exact finite displacement vector equation (2.16) special representations for the velocity and acceleration that exhibit the instantaneous translational and rotational parts of the motion in space of the points of a rigid body. The angular velocity and angular acceleration of the body about an instantaneous axis at the base point will enter the results in a natural way; and the relationship of these vectors to certain velocity and acceleration terms having simple geometrical interpretations will be described. The main formulas will show that the usual differentiation operations defined in (1.8) and (1.9) for the velocity and acceleration of any material point whatever, in their application to a rigid body may be replaced almost entirely by simple vector algebraic operations that often are easy to visualize. We begin with the derivation of the velocity equation.

2.8.1. Velocity of a Particle of a Rigid Body

The velocity of a point of a rigid body in motion relative to an assigned frame $\Phi = \{F; \mathbf{e}_k\}$ can be easily derived from either the exact finite displacement vector equation (2.16) or the infinitesimal displacement relation (2.19). We shall use (2.16), leaving the easier application of the latter as an exercise for the reader. We start by rewriting (2.16) in the familiar form of a difference quotient that will allow us to apply the definition of the derivative of a vector function of the scalar variable t:

$$\frac{\Delta\mathbf{X}}{\Delta t} = \frac{\Delta\mathbf{B}}{\Delta t} + \frac{\Delta\theta}{\Delta t}\boldsymbol{\alpha}\times\mathbf{x}\left(\frac{\sin\Delta\theta}{\Delta\theta}\right) + \frac{\Delta\theta}{\Delta t}\boldsymbol{\alpha}\times(\boldsymbol{\alpha}\times\mathbf{x})\left(\frac{1-\cos\Delta\theta}{\Delta\theta}\right). \tag{2.24}$$

Next, we form the limit of (2.24) as $\Delta \mathbf{X}$, $\Delta \mathbf{B}$, and $\Delta\theta$ approach zero with Δt. Then recalling that

$$\lim_{\Delta t \to 0} \frac{\sin \Delta\theta}{\Delta\theta} = 1 \quad \text{and} \quad \lim_{\Delta t \to 0} \frac{1 - \cos \Delta\theta}{\Delta\theta} = 0,$$

writing

$$\mathbf{v}_P \equiv \frac{d\mathbf{X}}{dt} = \lim_{\Delta t \to 0} \frac{\Delta \mathbf{X}}{\Delta t}, \qquad \mathbf{v}_O \equiv \frac{d\mathbf{B}}{dt} = \lim_{\Delta t \to 0} \frac{\Delta \mathbf{B}}{\Delta t}, \tag{2.25}$$

and introducing

$$\boldsymbol{\omega} \equiv \lim_{\Delta t \to 0} \frac{\Delta\theta}{\Delta t} \boldsymbol{\alpha} = \frac{d\theta}{dt} \boldsymbol{\alpha} = \dot{\theta} \boldsymbol{\alpha}, \tag{2.26}$$

we obtain in frame $\Phi = \{F; \mathbf{e}_k\}$ the following important equation for the velocity $\mathbf{v}_P = \mathbf{v}(P, t)$ of a point P of the rigid body:

$$\mathbf{v}_P = \mathbf{v}_O + \boldsymbol{\omega} \times \mathbf{x}, \tag{2.27}$$

wherein $\mathbf{v}_O = \mathbf{v}(0, t)$ is the velocity of the base point O and $\mathbf{x}$ is the position vector of P from O. The velocity $\mathbf{v}_P$ in (2.27) sometimes is called the *total* or *absolute velocity* of P in frame Φ.

The new vector $\boldsymbol{\omega}$ defined by (2.26) is named the *angular velocity* of the body about point O relative to frame Φ. Notice that the angular velocity vector is parallel to the instantaneous axis of rotation defined by $\boldsymbol{\alpha}(t)$ at the base point O; therefore, $\boldsymbol{\omega}$ has the same right-hand sense assigned earlier to $\boldsymbol{\alpha}$. The magnitude $\omega \equiv |\boldsymbol{\omega}|$ of the angular velocity vector is known as the *angular speed.* We see that this coincides with our earlier elementary description of the angular speed as the time rate of change of the increasing angular placement: $\omega = \dot{\theta}$. Moreover, it is seen from (2.23) that the $\lim_{\Delta t \to 0}(\Delta\boldsymbol{\theta}/\Delta t)$ yields

$$\boldsymbol{\omega} = \omega \boldsymbol{\alpha} = \omega_1 \mathbf{e}_1 + \omega_2 \mathbf{e}_2 + \omega_3 \mathbf{e}_3 \tag{2.28}$$

for the angular velocity vector in a frame $\Phi = \{F; \mathbf{e}_k\}$. Thus, the three scalar components $\omega_k \equiv d\theta_k/dt$ of $\boldsymbol{\omega}$ are identified as the instantaneous rates of rotation of the body about the three coordinate directions $\mathbf{e}_k$ in Φ. Clearly, the angular velocity has the physical dimensions $[\boldsymbol{\omega}] = [T^{-1}]$. When the angular measure is expressed in radians and time is in seconds, the units of $\boldsymbol{\omega}$ are rad/sec (radians per second). Unless otherwise stated, use of these units is preferred. On the other hand, in engineering practice the common measure of angular motion often is reported in revolutions and time is in minutes, so in this case the units of $\boldsymbol{\omega}$ are written as rpm (revolutions per minute). The conversion from one set of units to the other is accomplished by recalling that one revolution is equivalent to 2π rad; thus, 1 rpm $= 2\pi/60$ rad/sec.

The vectors introduced in (2.27) are shown in the schematic Fig. 2.9. Additional physical relevance may be assigned to the separate terms in (2.27) by examination of the time derivative in Φ of the position vector $\mathbf{X}=\mathbf{B}+\mathbf{x}$ of the point P. With the aid of (2.25), this yields $\mathbf{v}_P=\mathbf{v}_0+\dot{\mathbf{x}}$; then it follows from (2.27) that

$$\dot{\mathbf{x}}=\mathbf{v}_P-\mathbf{v}_O=\boldsymbol{\omega}\times\mathbf{x}. \tag{2.29}$$

Because $\dot{\mathbf{x}}$ describes the velocity of the point P relative to the base point O, the quantity $\boldsymbol{\omega}\times\mathbf{x}$ is named the *relative rigid body velocity*; it is the velocity that the point P would have if the base point were fixed in Φ. If, in fact, the base point is fixed in Φ, then $\mathbf{v}_O=\mathbf{0}$; and we have $\mathbf{v}_P=\boldsymbol{\omega}\times\mathbf{x}$ for the velocity of P. Therefore, the term $\boldsymbol{\omega}\times\mathbf{x}$ is the contribution to the total velocity of P due to a rotation of P about the base point O.

It is also seen in (2.27) that $\boldsymbol{\omega}\equiv\mathbf{0}$ is a necessary and sufficient condition for which $\mathbf{v}_P=\mathbf{v}_O$ holds for all $\mathbf{x}$. This describes a translation in which $\mathbf{v}_O$ is the instantaneous translational velocity of all points of the body. In this case, the displacement of every particle of the body from an initial configuration to its current configuration is at each instant equivalent to a parallel translation; but the body, as illustrated in Problem 2.1, need not move on a straight line path at each moment.

With these physical descriptions in mind, the result (2.27) may be summarized in more graphic terms: *The velocity of any point P of a rigid body is equal to the velocity of a base point O plus the relative rigid body velocity, the velocity of P due to its rotation about the base point O*:

$$\mathbf{v}_P=\mathbf{v}_O+\boldsymbol{\omega}\times\mathbf{x} \qquad \text{[cf. (2.27)]}.$$

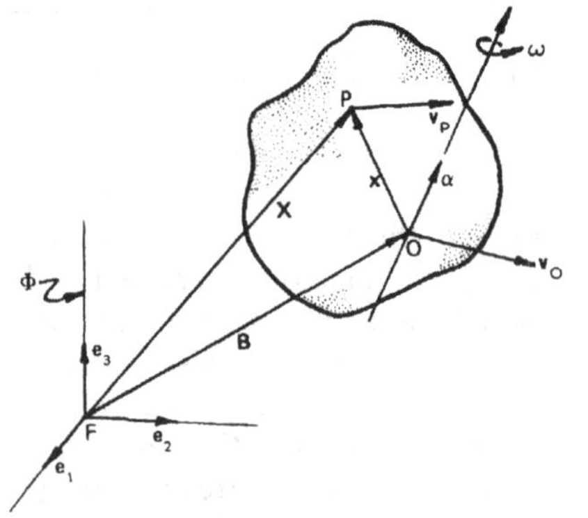

Figure 2.9. Schematic of the vectors associated with the equation $\mathbf{v}_P=\mathbf{v}_0+\boldsymbol{\omega}\times\mathbf{x}$.

2.8.2. Acceleration of a Particle of a Rigid Body

The acceleration $\mathbf{a}_P \equiv \mathbf{a}(P, t) = \dot{\mathbf{v}}_P$ relative to Φ of any point P of a rigid body is obtained by differentiation of (2.27) and use of (2.29). This yields the following important equation for the acceleration of P:

$$\mathbf{a}_P = \mathbf{a}_O + \boldsymbol{\omega} \times (\boldsymbol{\omega} \times \mathbf{x}) + \dot{\boldsymbol{\omega}} \times \mathbf{x}. \tag{2.30}$$

Herein $\mathbf{a}_O \equiv \mathbf{a}(O, t) = \dot{\mathbf{v}}_O$ is the acceleration in Φ of the base point O and $\dot{\boldsymbol{\omega}} \equiv d\boldsymbol{\omega}/dt$, the time rate of change in Φ of the angular velocity of the body, is titled the *angular acceleration* of the body about O relative to frame Φ. The acceleration $\mathbf{a}_P$ in (2.30) sometimes is called the *total* or *absolute acceleration* of P in frame Φ.

It is important to recognize that $\boldsymbol{\omega}$ and $\dot{\boldsymbol{\omega}}$ generally are *not parallel* vectors. Indeed, differentiation of (2.26) shows that the angular acceleration vector is determined by

$$\dot{\boldsymbol{\omega}}(t) = \ddot{\theta}(t)\, \boldsymbol{\alpha}(t) + \dot{\theta}(t)\, \dot{\boldsymbol{\alpha}}(t). \tag{2.31}$$

However, $\boldsymbol{\alpha}(t)$ being a unit vector, it follows that $\boldsymbol{\alpha} \cdot \dot{\boldsymbol{\alpha}} = 0$ must hold for all times in the spatial frame $\Phi = \{F; \mathbf{e}_k\}$; hence, $\dot{\boldsymbol{\alpha}}$ always is perpendicular to $\boldsymbol{\alpha}$. Consequently, $\dot{\boldsymbol{\omega}}(t)$ is situated in the plane of $\boldsymbol{\alpha}$ and $\dot{\boldsymbol{\alpha}}$ and cannot be parallel to $\boldsymbol{\omega}(t) = \dot{\theta}(t)\, \boldsymbol{\alpha}(t)$ unless $\dot{\boldsymbol{\alpha}} = \mathbf{0}$. But this happens if and only if $\boldsymbol{\alpha}$ is a constant vector in Φ; and in this case, we have the simple relations

$$\boldsymbol{\omega}(t) = \dot{\theta}(t)\boldsymbol{\alpha} \quad \text{and} \quad \dot{\boldsymbol{\omega}}(t) = \ddot{\theta}(t)\boldsymbol{\alpha}. \tag{2.32}$$

In addition, it is also clear that the magnitude of the angular acceleration vector is related to the simple angular acceleration introduced in Chapter 1, namely, $|\dot{\boldsymbol{\omega}}| = |\ddot{\theta}|$, but only for this simple case.

The angular acceleration has the physical dimensions $[\dot{\boldsymbol{\omega}}] = [T^{-2}]$. In particular, when the angular speed is expressed in rad/sec, the measure units of the angular acceleration are rad/sec^2 (radians per second, per second). Use of these units is preferred unless explicitly remarked otherwise. In engineering usage, however, the units of angular acceleration often are given as rpm^2 (revolutions per minute, per minute). In any case, the conversion of all units is straightforward.

It is useful to note that the cross-product terms in (2.29) and (2.30) have simple geometrical descriptions. The relative rigid body velocity vector $\boldsymbol{\omega} \times \mathbf{x}$, which at each instant is perpendicular to the plane of $\boldsymbol{\omega}$ and $\mathbf{x}$, instantaneously is tangent to a circle in a plane perpendicular to the axis of rotation. Therefore, the vector $\boldsymbol{\omega} \times (\boldsymbol{\omega} \times \mathbf{x})$ lies in the plane of $\boldsymbol{\omega}$ and $\mathbf{x}$ and is directed toward the instantaneous axis of rotation. This vector is known as the *centripetal acceleration*. Similarly, the vector $\dot{\boldsymbol{\omega}} \times \mathbf{x}$, perpendicular to the plane of $\dot{\boldsymbol{\omega}}$ and $\mathbf{x}$ at each moment, is instantaneously tangent to another circle in a

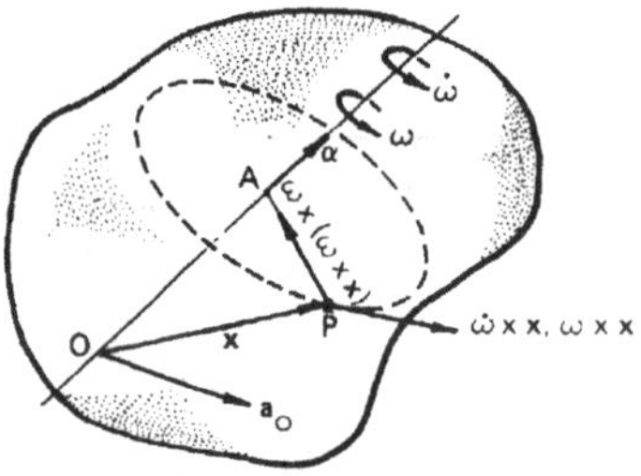

Figure 2.10. Geometrical description of the centripetal and tangential acceleration terms shown for the special case of rotation about a fixed axis.

plane normal to $\dot{\boldsymbol{\omega}}$; hence, this vector is named the *tangential acceleration*. It is easy to picture these terms in the special case when $\boldsymbol{\omega}$ and $\dot{\boldsymbol{\omega}}$ are parallel, as shown in Fig. 2.10.

Finally, differentiation of (2.29) and use of (2.30) shows that the acceleration of the point P of the rigid body relative to the base point O is given by

$$\ddot{\mathbf{x}} = \mathbf{a}_P - \mathbf{a}_O = \boldsymbol{\omega} \times (\boldsymbol{\omega} \times \mathbf{x}) + \dot{\boldsymbol{\omega}} \times \mathbf{x}. \tag{2.33}$$

Hence, this collection of acceleration terms is called the *relative rigid body acceleration*; it is the acceleration that the particle P would have if the base point were fixed in Φ. Therefore, (2.33) is the contribution to the total acceleration of P due to a rotation about the base point O. Moreover, $\boldsymbol{\omega} \equiv \mathbf{0}$ is a necessary and sufficient condition for which $\mathbf{a}_P = \mathbf{a}_O$ holds for all $\mathbf{x}$, so $\mathbf{a}_O$ represents the instantaneous translational acceleration of all points of the body.

Our main result (2.30) may now be summarized in terms of the foregoing physical descriptions: *The acceleration of any point P of a rigid body is equal to the acceleration of a base point O plus the relative rigid body acceleration, the acceleration of P due to its rotation about the base point O; the latter is the sum of the centripetal acceleration and the tangential acceleration*:

$$\mathbf{a}_P = \mathbf{a}_O + \boldsymbol{\omega} \times (\boldsymbol{\omega} \times \mathbf{x}) + \dot{\boldsymbol{\omega}} \times \mathbf{x} \qquad [\text{cf. } (2.30)].$$

2.9. Some Applications of the Basic Equations

The fundamental rules (2.27) and (2.30) are the principal equations we set out to derive for the velocity and acceleration of the points of a rigid body. We have seen that these relations separate and exhibit the translational and rotational parts of the body's motion. In this format, they have wide utility in applications and in further development of other important theoretical results that will be presented later. Presently, however, we shall consider three introductory examples that will illustrate applications of the basic equations.

The first example demonstrates the vector calculations frequently encountered in plane motion applications, and our earlier geometrical interpretation of the various vector product terms is described. The second problem shows how the vector equations may be used in conjunction with previous work to find various unknown, instantaneous quantities in a system that involves the motion of several connected rigid bodies. Our third illustration is an application of (2.27) and (2.30) to an easy mechanical design problem. Some additional examples on the important topic of rolling without slip are reserved for the next section, and analysis of a spatial mechanism will be studied afterwards. We shall conclude with an example concerning the velocity of a rigid body particle referred to an imbedded frame.

Example 2.2. The center O of a circular blade of radius a moves with a constant speed v along the line $Y=2a$ in the machine frame $\Phi=\{F; \mathbf{i}_k\}$. During its return stroke, the blade has an angular velocity $\boldsymbol{\omega}$ and an angular acceleration $\dot{\boldsymbol{\omega}}$, as shown in Fig. 2.11. Find in Φ the absolute velocity and the total acceleration of a point P on the edge of the blade.

Solution. The total velocity and the absolute acceleration of P in Φ are determined by use of (2.27) and (2.30), respectively. We begin construction of the solution by observing that the base point O has the constant velocity

$$\mathbf{v}_O = v\mathbf{i}, \qquad \text{therefore,} \quad \mathbf{a}_O = \mathbf{0}. \tag{2.34}$$

The vector of P from O in Fig. 2.11 is given by $\mathbf{x}=a(\cos\theta\,\mathbf{i}+\sin\theta\,\mathbf{j})$ in Φ; hence, with $\boldsymbol{\omega}=\omega\mathbf{k}$ and $\dot{\boldsymbol{\omega}}=\dot{\omega}\mathbf{k}$, we find

$$\boldsymbol{\omega}\times\mathbf{x}=a\omega(\cos\theta\,\mathbf{j}-\sin\theta\,\mathbf{i}), \qquad \dot{\boldsymbol{\omega}}\times\mathbf{x}=a\dot{\omega}(\cos\theta\,\mathbf{j}-\sin\theta\,\mathbf{i}), \tag{2.35}$$

and

$$\boldsymbol{\omega}\times(\boldsymbol{\omega}\times\mathbf{x}) = -a\omega^2(\cos\theta\,\mathbf{i}+\sin\theta\,\mathbf{j}). \tag{2.36}$$

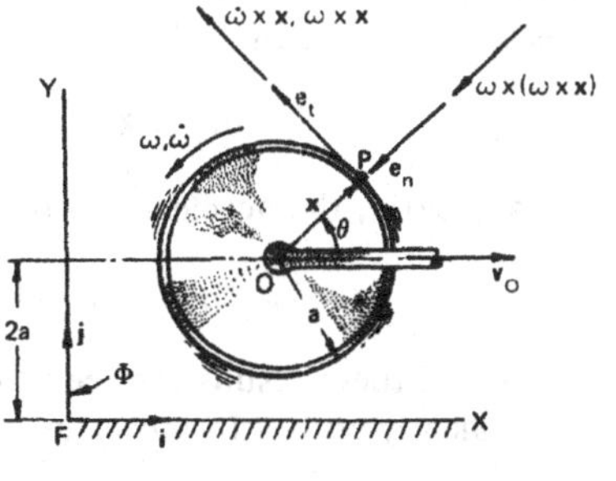

Figure 2.11. Motion of a circular blade during its return stroke.

Collecting the results (2.34)–(2.36) in (2.27) and (2.30), we obtain in frame Φ the absolute velocity

$$\mathbf{v}_P = (v - a\omega \sin\theta)\mathbf{i} + a\omega \cos\theta\, \mathbf{j}, \tag{2.37}$$

and the total acceleration

$$\mathbf{a}_P = (-a\omega^2 \cos\theta - a\dot{\omega} \sin\theta)\mathbf{i} + (a\dot{\omega} \cos\theta - a\omega^2 \sin\theta)\mathbf{j}. \tag{2.38}$$

It is useful to notice in these calculations that because $\dot{\boldsymbol{\omega}}$ is parallel to $\boldsymbol{\omega}$, the product $\dot{\boldsymbol{\omega}} \times \mathbf{x}$ may be obtained immediately from the product $\boldsymbol{\omega} \times \mathbf{x}$ by replacement of ω with $\dot{\omega}$, as illustrated in (2.35).

The geometrical character of the relative rigid body velocity and acceleration terms described earlier may be visualized in this example by our introducing at P an intrinsic frame $\psi = \{P; \mathbf{e}_t, \mathbf{e}_n, \mathbf{e}_b\}$, as shown in Fig. 2.11. It is seen that the vector products $\boldsymbol{\omega} \times \mathbf{x} = a\omega\mathbf{e}_t$ and $\dot{\boldsymbol{\omega}} \times \mathbf{x} = a\dot{\omega}\mathbf{e}_t$ are parallel vectors tangent to the blade at P; and $\boldsymbol{\omega} \times (\boldsymbol{\omega} \times \mathbf{x}) = -a\omega^2\mathbf{e}_n$ is normal to the blade and directed toward its center O. Use of these expressions and (2.34) in (2.27) and (2.30) yields a simpler representation of the solution:

$$\mathbf{v}_P = v\mathbf{i} + a\omega\mathbf{e}_t, \qquad \mathbf{a}_P = a\dot{\omega}\mathbf{e}_t + a\omega^2\mathbf{e}_n, \tag{2.39}$$

wherein the tangent and normal vectors are defined by

$$\mathbf{e}_t = -\sin\theta\, \mathbf{i} + \cos\theta\, \mathbf{j}, \qquad \mathbf{e}_n = -\mathbf{x}/a = -(\cos\theta\, \mathbf{i} + \sin\theta\, \mathbf{j}) \tag{2.40}$$

in frame Φ. Indeed, when these unit vectors are used in (2.39), we recover the results (2.37) and (2.38). The geometrical constructions illustrated above often are valuable aids in the calculation of the cross product terms in (2.27) and (2.30).

Finally, let us recall from Chapter 1 that the velocity of the particle P on the edge of the blade may be obtained in the usual way by differentiation of its position vector in Φ, namely,

$$\mathbf{X}(P, t) = X\mathbf{i} + 2a\mathbf{j} + a(\cos\theta\, \mathbf{i} + \sin\theta\, \mathbf{j}), \tag{2.41}$$

wherein X is the horizontal distance of O from F. Thus, with $\dot{X}\mathbf{i} = \mathbf{v}_O = v\mathbf{i}$ and $\omega = \dot{\theta}$, we obtain the velocity of P in Φ:

$$\mathbf{v}_P = \dot{\mathbf{X}} = v\mathbf{i} + a\omega(-\sin\theta\, \mathbf{i} + \cos\theta\, \mathbf{j}). \tag{2.42}$$

Then, clearly, the acceleration of P in Φ is given by

$$\mathbf{a}_P = \dot{\mathbf{v}}_P = a\dot{\omega}(-\sin\theta\, \mathbf{i} + \cos\theta\, \mathbf{j}) - a\omega^2(\cos\theta\, \mathbf{i} + \sin\theta\, \mathbf{j}). \tag{2.43}$$

Of course, these results are seen to be the same as (2.37) and (2.38) computed algebraically above. In more complex problems in which the geometry is more difficult to perceive, the direct calculations based on the vector algebraic formulas (2.27) and (2.30) will prove especially valuable. □

The relations (2.27) and (2.30) can be applied in circumstances where it may be extremely tedious to produce a solution were one to rely on only the use of the usual differentiation procedure sketched above in (2.41)–(2.43). Situations may arise where only the values of some vector quantities at a particular instant of interest, rather than their functional dependence for all times, are assigned; and, in such cases, it may be possible to obtain additional data desired only for that special moment of concern. This situation is illustrated in the next example involving the motion of several connected rigid bodies.

Example 2.3. A slider block A of an engine mechanism moves in a plane, circular slot of radius 2 ft. At the instant when the links AB and BC are in the perpendicular position shown in Fig. 2.12, the block has a speed of 10 ft/sec in the direction indicated, and a rate of change of speed of 20 ft/sec^2. Find in frame $\varphi = \{F; \mathbf{i}_k\}$ the absolute velocity and acceleration of the hinge pin B, and determine the angular velocities and angular accelerations of the links AB and BC at the instant of interest.

Solution. The point B belongs to both the link AB and the link BC, so the velocity of B as determined by (2.27) may be written as

$$\mathbf{v}_B = \mathbf{v}_A + \boldsymbol{\omega}_1 \times \mathbf{x}_1 = \mathbf{v}_C + \boldsymbol{\omega}_2 \times \mathbf{x}_2, \tag{2.44}$$

wherein $\boldsymbol{\omega}_1 = \omega_1 \mathbf{k}$ and $\boldsymbol{\omega}_2 = -\omega_2 \mathbf{k}$ denote the angular velocities of the links shown in Fig. 2.12. We observe also that $\mathbf{x}_1 = 5\mathbf{j}$ ft and $\mathbf{x}_2 = -2 \sin 45° \, \mathbf{i} = -\sqrt{2}\mathbf{i}$ ft are the position vectors of B from the points A and C whose velocity vectors are

$$\begin{aligned} \mathbf{v}_A &= \dot{s}\mathbf{t} = 10(\cos 45° \, \mathbf{i} + \sin 45° \, \mathbf{j}) \\ &= 5\sqrt{2}(\mathbf{i} + \mathbf{j}) \text{ ft/sec} \quad \text{and} \quad \mathbf{v}_C = \mathbf{0}. \end{aligned} \tag{2.45}$$

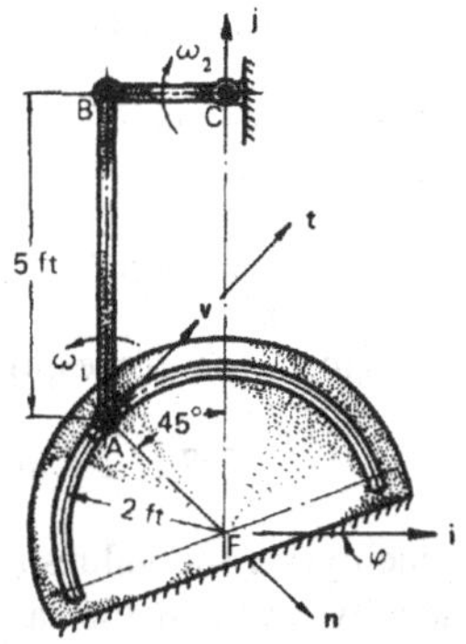

Figure 2.12. Instantaneous motion of several connected rigid parts of an engine mechanism.

We see immediately from the geometry that

$$\boldsymbol{\omega}_1 \times \mathbf{x}_1 = -5\omega_1 \mathbf{i} \text{ ft/sec} \quad \text{and} \quad \boldsymbol{\omega}_2 \times \mathbf{x}_2 = \omega_2\sqrt{2}\mathbf{j} \text{ ft/sec}. \tag{2.46}$$

Substitution of (2.45) and (2.46) into (2.44) yields

$$\mathbf{v}_B = 5(\sqrt{2} - \omega_1)\mathbf{i} + 5\sqrt{2}\mathbf{j} \text{ ft/sec} = \omega_2\sqrt{2}\mathbf{j} \text{ ft/sec}. \tag{2.47}$$

Therefore, the corresponding scalar components must satisfy

$$5(\sqrt{2} - \omega_1) = 0 \quad \text{and} \quad 5\sqrt{2} = \omega_2\sqrt{2}, \tag{2.48}$$

which yield $\omega_1 = \sqrt{2}$ rad/sec and $\omega_2 = 5$ rad/sec. Finally, with the aid of (2.47), we obtain at the instant of interest

$$\mathbf{v}_B = 5\sqrt{2}\mathbf{j} \text{ ft/sec}, \qquad \boldsymbol{\omega}_1 = \sqrt{2}\mathbf{k} \text{ rad/sec}, \qquad \boldsymbol{\omega}_2 = -5\mathbf{k} \text{ rad/sec}, \tag{2.49}$$

for the absolute velocity of B and the angular velocities of the links AB and BC.

The acceleration of B as determined by (2.30) may be similary written in two ways:

$$\begin{aligned} \mathbf{a}_B &= \mathbf{a}_A + \boldsymbol{\omega}_1 \times (\boldsymbol{\omega}_1 \times \mathbf{x}_1) + \dot{\boldsymbol{\omega}}_1 \times \mathbf{x}_1 \\ &= \mathbf{a}_C + \boldsymbol{\omega}_2 \times (\boldsymbol{\omega}_2 \times \mathbf{x}_2) + \dot{\boldsymbol{\omega}}_2 \times \mathbf{x}_2, \end{aligned} \tag{2.50}$$

in which $\dot{\boldsymbol{\omega}}_1 = \dot{\omega}_1 \mathbf{k}$ and $\dot{\boldsymbol{\omega}}_2 = -\dot{\omega}_2 \mathbf{k}$ denote the angular accelerations of the links. The acceleration of the point A is found from $\mathbf{a}_A = \ddot{s}\mathbf{t} + \dot{s}^2/R\mathbf{n}$, where $\ddot{s} = 20$ ft/sec^2, $R = 2$ ft, and $\mathbf{n} = \cos 45°\,\mathbf{i} - \sin 45°\,\mathbf{j} = \sqrt{2}/2(\mathbf{i} - \mathbf{j})$. We now have

$$\mathbf{a}_A = 35\sqrt{2}\mathbf{i} - 15\sqrt{2}\mathbf{j} \text{ ft/sec}^2 \quad \text{and} \quad \mathbf{a}_C = \mathbf{0}. \tag{2.51}$$

Again, with the aid of (2.46), it is easily seen geometrically that

$$\boldsymbol{\omega}_1 \times (\boldsymbol{\omega}_1 \times \mathbf{x}_1) = -5\omega_1^2 \mathbf{j} = -10\mathbf{j} \text{ ft/sec}^2, \qquad \dot{\boldsymbol{\omega}}_1 \times \mathbf{x}_1 = -5\dot{\omega}_1 \mathbf{i}, \tag{2.52}$$

$$\boldsymbol{\omega}_2 \times (\boldsymbol{\omega}_2 \times \mathbf{x}_2) = \omega_2^2\sqrt{2}\mathbf{i} = 25\sqrt{2}\mathbf{i} \text{ ft/sec}^2, \qquad \dot{\boldsymbol{\omega}}_2 \times \mathbf{x}_2 = \dot{\omega}_2\sqrt{2}\mathbf{j}, \tag{2.53}$$

wherein we recall (2.49). Gathering the relations (2.51)–(2.53) in (2.50), we obtain

$$\mathbf{a}_B = 5(7\sqrt{2} - \dot{\omega}_1)\mathbf{i} - 5(3\sqrt{2} + 2)\mathbf{j} = 25\sqrt{2}\mathbf{i} + \dot{\omega}_2\sqrt{2}\mathbf{j}. \tag{2.54}$$

Therefore, the corresponding scalar components provide the relations

$$5(7\sqrt{2} - \dot{\omega}_1) = 25\sqrt{2} \quad \text{and} \quad -5(3\sqrt{2} + 2) = \dot{\omega}_2\sqrt{2}, \tag{2.55}$$

which yield the values $\dot{\omega}_1 = 2\sqrt{2}$ rad/sec^2 and $\dot{\omega}_2 = -5(3 + \sqrt{2})$ rad/sec^2. The negative sign in the last term indicates that the direction of the vector $\dot{\boldsymbol{\omega}}_2$ is opposite to the direction assumed previously. The algebra typically determines both the magnitude and direction of an unknown vector quantity, as shown here for the vector $\dot{\boldsymbol{\omega}}_2$. Use of either of the angular acceleration values in (2.54) completes the solution. At the moment of interest, the angular accelerations of the links AB and BC are

$$\dot{\boldsymbol{\omega}}_1 = 2\sqrt{2}\mathbf{k} \text{ rad/sec}^2, \qquad \dot{\boldsymbol{\omega}}_2 = 5(3+\sqrt{2})\mathbf{k} \text{ rad/sec}^2; \tag{2.56}$$

and

$$\mathbf{a}_B = 25\sqrt{2}\mathbf{i} - 5(3\sqrt{2}+2)\mathbf{j} \text{ ft/sec}^2 \tag{2.57}$$

is the absolute acceleration of the pin B. □

Industrial design problems usually involve integration of several design concepts, technical analyses, and a variety of manufacturing procedures assembled to create some product or system; and it is customary that several parts of a product or a manufacturing system must be analyzed and designed at the same time to accommodate certain specified criteria. The simple mechanical design problem illustrated in the next example integrates an easy cam design analysis with the design of a link mechanism so that both parts satisfy specified design conditions for a particular set of system parameters.

Example 2.4. The shuttle mechanism of a certain sewing machine described in Fig. 2.13 is driven by a cam that rotates with a constant angular velocity $\boldsymbol{\omega} = \omega\mathbf{k}$, while a linkage arrangement moves the shuttle, slider block B, in a fixed vertical slot in the machine. The cam must be designed for the geometry shown in Fig. 2.13 so that the shuttle drive block A has a sinusoidal velocity $\mathbf{v}_A = 6 \sin\theta\, \mathbf{i}$ cm/sec, and the length l of the link AB must be chosen so that the link has no angular acceleration at the instant t_0 when the device is in the configuration where $\phi = 30°$ when $\theta = 60°$. Find the shape of the cam and the angular speed at which it must operate, and determine the link length

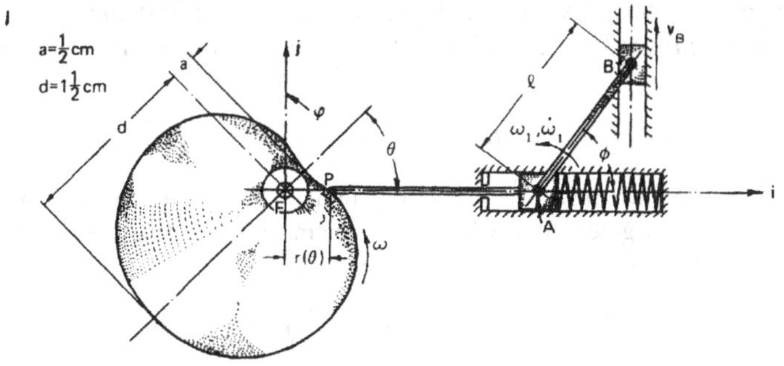

Figure 2.13. A cam-driven linkage mechanism design.

l needed for the design specifications. Determine the velocity and the acceleration of B at the instant t_0. Assume that $\theta = 0$ initially.

Solution. We know from previous experience that the cam shape may be found by integration of the differential equation for the velocity of the contact point P on the push rod. Since the rigid rod PA suffers no rotation, the rule (2.27) reveals the intuitive relation $\mathbf{v}_P = \mathbf{v}_A$ used in Chapter 1. Thus, by aid of the same procedure used in (1.56) and with $\mathbf{v}_A = 6 \sin \theta \, \mathbf{i}$, we obtain

$$\int_a^r dr = \frac{6}{\omega} \int_0^\theta \sin \theta \, d\theta \tag{2.58}$$

in which $r(0) = a = 1/2$ cm is obtained from the given cam geometry, and $\omega = \dot{\theta}$ is the unknown constant angular speed of the cam. Integration of this equation yields

$$r(\theta) = \frac{1}{2} + \frac{6}{\omega}(1 - \cos \theta) \text{ cm}. \tag{2.59}$$

Since $r(\pi) = d = 1.5$ cm, (2.59) determines the constant angular speed and the shape of the cam needed to produce the designated design velocity of A; we obtain

$$\omega = \dot{\theta} = 12 \text{ rad/sec}, \tag{2.60a}$$

$$r(\theta) = 1 - 0.5 \cos \theta \text{ cm}. \tag{2.60b}$$

This completes the cam design analysis.

The motion of the link AB involves four unknown quantities: the velocity $\mathbf{v}_B$ of the slider B; the angular velocity $\boldsymbol{\omega}_1$ and the angular acceleration $\dot{\boldsymbol{\omega}}_1$ of the link AB; and the vector $\mathbf{l}$ of point B from A. These vectors may be related by (2.27) and (2.30) expressed in the form

$$\mathbf{v}_B = \mathbf{v}_A + \boldsymbol{\omega}_1 \times \mathbf{l}, \qquad \mathbf{a}_B = \mathbf{a}_A + \boldsymbol{\omega}_1 \times (\boldsymbol{\omega}_1 \times \mathbf{l}) + \dot{\boldsymbol{\omega}}_1 \times \mathbf{l}. \tag{2.61}$$

The velocity of the base point A is given as $\mathbf{v}_A = 6 \sin \theta \, \mathbf{i}$ cm/sec; hence, using the result (2.60a) derived in the cam analysis, we find the acceleration $\mathbf{a}_A = 72 \cos \theta \, \mathbf{i}$ cm/sec^2. In particular, at the instant of interest t_0 when $\theta = 60°$, we have

$$\mathbf{v}_A = 3\sqrt{3}\mathbf{i} \text{ cm/sec}, \qquad \mathbf{a}_A = 36\mathbf{i} \text{ cm/sec}^2. \tag{2.62}$$

At the same time, we have also $\phi = 30°$; hence, the vector of B from A in frame $\varphi = \{F; \mathbf{i}_k\}$ is given by $\mathbf{l} = l/2(\sqrt{3}\mathbf{i} + \mathbf{j})$ cm. Then, writing $\boldsymbol{\omega}_1 = \omega_1 \mathbf{k}$, we may compute the products

$$\boldsymbol{\omega}_1 \times \mathbf{l} = (\omega_1 l/2)(\sqrt{3}\mathbf{j} - \mathbf{i}), \tag{2.63}$$

$$\boldsymbol{\omega}_1 \times (\boldsymbol{\omega}_1 \times \mathbf{l}) = (-\omega_1^2 l/2)(\sqrt{3}\mathbf{i} + \mathbf{j}). \tag{2.64}$$

Finally, recalling that the design constraint at the instant t_0 demands that $\dot{\boldsymbol{\omega}}_1 = \mathbf{0}$, observing that $\mathbf{v}_B = v_B\mathbf{j}$, $\mathbf{a}_B = a_B\mathbf{j}$, and substituting (2.62), (2.63), and (2.64) into (2.61), we derive the relations

$$\mathbf{v}_B = v_B\mathbf{j} = (3\sqrt{3} - \omega_1 l/2)\mathbf{i} + \omega_1 l\sqrt{3}/2\mathbf{j} \text{ cm/sec},$$

$$\mathbf{a}_B = a_B\mathbf{j} = (36 - \omega_1^2 l\sqrt{3}/2)\mathbf{i} - \omega_1^2 l/2\mathbf{j} \text{ cm/sec}^2.$$

The corresponding scalar components in these identities reveal the results

$$\omega_1 l = 6\sqrt{3} \quad \text{and} \quad \mathbf{v}_B = \omega_1 l\sqrt{3}/2\mathbf{j} = 9\mathbf{j} \text{ cm/sec}; \tag{2.65}$$

$$\omega_1^2 l = 24\sqrt{3} \quad \text{and} \quad \mathbf{a}_B = -\omega_1^2 l/2\mathbf{j} = -12\sqrt{3}\mathbf{j} \text{ cm/sec}^2. \tag{2.66}$$

These equations provide the velocity and acceleration of B required at t_0; and we also may conclude from these equations that

$$\omega_1 = 4 \text{ rad/sec}, \qquad l = 3\sqrt{3}/2 \text{ cm}, \tag{2.67}$$

which is the link length needed for the design specifications. This completes the design analysis for the given set of system parameters. A more general analysis is left as an exercise for the student in Problem 2.33.

2.9.1. Rolling without Slip

An important special application of (2.27) and (2.30) to mechanical design analysis concerns the motion transmitted by bodies through rolling contact. Rotary motion which is transferred between two shafts by disks or gears that roll on one another without slipping is a typical example. Sometimes, however, relative sliding between contacting surfaces in a direction tangent to a contacting line is permissible provided no slipping normal to the line of contact can occur in the direction of the rotational motion. This is a characteristic of helical gear design in which the gear teeth prevent slipping normal to the sliding helicoidal contact surfaces and thus provide the means of transmitting power between the shafts.

The idea of rolling without slip is introduced in the first example below. Afterwards, the criterion for rolling without slip is formulated in general terms, and the equations are illustrated in the study of two further examples of rolling wheels and bevel gears. A final application to helical gear motion provides an example for which the no-slip criterion is invalid.

Example 2.5. A wheel of radius a rolls without slipping on a fixed horizontal surface so that its center moves uniformly with velocity $\mathbf{v}_O$, as shown in Fig. 2.14. (a) Find the angular velocity $\boldsymbol{\omega}$ of the wheel. (b) Let C denote the point on the wheel which instantaneously is in contact with the

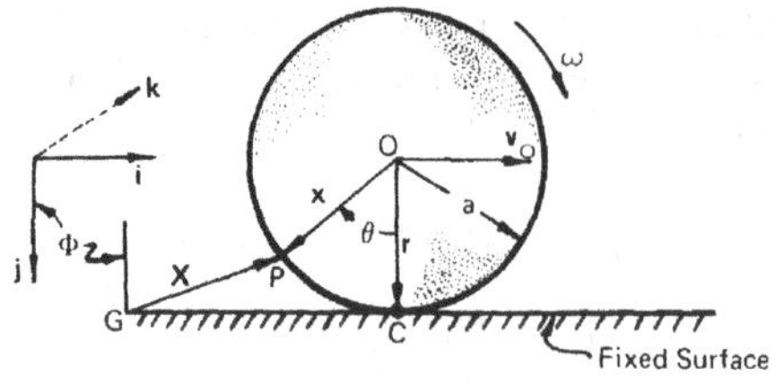

Figure 2.14. Rolling without slip on a fixed, horizontal surface.

fixed surface. Show that its velocity $\mathbf{v}_C$ is zero. (c) Determine the acceleration of a point P on the rim of the wheel. (d) Find the acceleration of the contact point C. (e) Determine the path of P in frame $\Phi = \{G; \mathbf{i}_k\}$.

Solution. (a) The angular velocity of the wheel is related to the condition that the wheel roll without slipping on the fixed surface. This rolling constraint may be characterized in a couple of ways. One definition is that the curves on the surfaces in rolling contact be tangent to each other at each point of contact, and the lengths of the arcs traced on the two curves between successive points of contact be equal. Thus, if a wheel of radius a advances a horizontal distance ds as it turns through an angle $d\theta$ about the direction $\mathbf{k}$ at O, the no-slip rolling condition is expressed by

$$ds = a\, d\theta. \tag{2.68}$$

Then with $v_O = |\mathbf{v}_O| = ds/dt$ and $\omega = |\boldsymbol{\omega}| = d\theta/dt$, we find from (2.68) that $v_O = a\omega$. Therefore, the angular velocity of the wheel about O is given by

$$\boldsymbol{\omega} = \omega\mathbf{k} = (v_O/a)\mathbf{k}. \tag{2.69}$$

(b) The velocity of the point of contact C at $\mathbf{r} = a\mathbf{j}$ from O is determined by $\mathbf{v}_C = \mathbf{v}_O + \boldsymbol{\omega} \times \mathbf{r}$. Thus, with $\mathbf{v}_O = v_O\mathbf{i}$ and use of (2.69), we find

$$\mathbf{v}_C = v_O\mathbf{i} + (v_O/a\mathbf{k}) \times a\mathbf{j} = v_O\mathbf{i} - v_O\mathbf{i} = \mathbf{0}. \tag{2.70}$$

This confirms that the velocity of the rim point C instantaneously in contact with the fixed surface is zero.

(c) The acceleration of an arbitrary point P on the rim of the wheel at $\mathbf{x}$ from O is determined by (2.30). Since $\mathbf{v}_O$ is constant $\mathbf{a}_O = \mathbf{0}$. Also, (2.69) shows that $\boldsymbol{\omega}$ is constant; hence, $\dot{\boldsymbol{\omega}} = \mathbf{0}$. Noting that $\mathbf{x} = a(\cos\theta\,\mathbf{j} - \sin\theta\,\mathbf{i})$ and observing that $\boldsymbol{\omega} \times (\boldsymbol{\omega} \times \mathbf{x})$ is directed along $-\mathbf{x}$ and has the magnitude $a\omega^2$, we now recall (2.30) and compute easily

$$\mathbf{a}_P = -a\omega^2\mathbf{x}/a = -(v_O^2/a)(\cos\theta\,\mathbf{j} - \sin\theta\,\mathbf{i}). \tag{2.71}$$

(d) The point P is in contact with the surface when $\theta = 0$ or 2π; and in this case, (2.71) yields

$$\mathbf{a}_C = -(v_O^2/a)\mathbf{j} \tag{2.72}$$

for the instantaneous acceleration of the contact point C. We thus learn that although the instantaneous velocity of P at C is zero, its acceleration is not. Indeed, (2.72) shows that $\mathbf{a}_C = \mathbf{0}$ if and only if the wheel is at rest in Φ. It should be observed that the acceleration is directed from C toward O.

(e) The equation of the path traced by P in the frame $\Phi = \{G; \mathbf{i}_k\}$ is determined by integration of (2.27) in which

$$\mathbf{v}_P = d\mathbf{X}(P, t)/dt, \qquad \mathbf{v}_O = a\omega\mathbf{i}, \quad \text{and} \quad \boldsymbol{\omega} \times \mathbf{x} = -a\omega(\cos\theta\,\mathbf{i} + \sin\theta\,\mathbf{j})$$

are used to express the right-hand side of (2.27) in terms of θ and ω; namely,

$$d\mathbf{X}(P, t)/dt = a\omega(1 - \cos\theta)\mathbf{i} - a\omega\sin\theta\,\mathbf{j}. \tag{2.73}$$

Assigning $\mathbf{X} = \mathbf{0}$ at $\theta = 0$ and writing $\omega = d\theta/dt$, we determine the motion

$$\mathbf{X}(P, t) = X\mathbf{i} - Y\mathbf{j} = a(\theta - \sin\theta)\mathbf{i} + a(\cos\theta - 1)\mathbf{j} \tag{2.74}$$

whose locus, shown in Fig. 2.15, is a *cycloid* with parametric equations

$$X = a(\theta - \sin\theta), \qquad Y = a(1 - \cos\theta). \tag{2.75}$$

The reader may find it helpful to demonstrate that the motion (2.74) or (2.75) also may be obtained by geometrical construction based on Fig. 2.15. □

2.9.1.1. *Driving Contact and Rolling without Slip*

Two moving rigid bodies that come into contact obviously cannot penetrate one another. In fact, they can maintain their contact with each other only so long as they have the same component of velocity in the direction normal to their surfaces at their instantaneous points of contact; otherwise, the bodies would separate. If their mutual normal velocity component is not

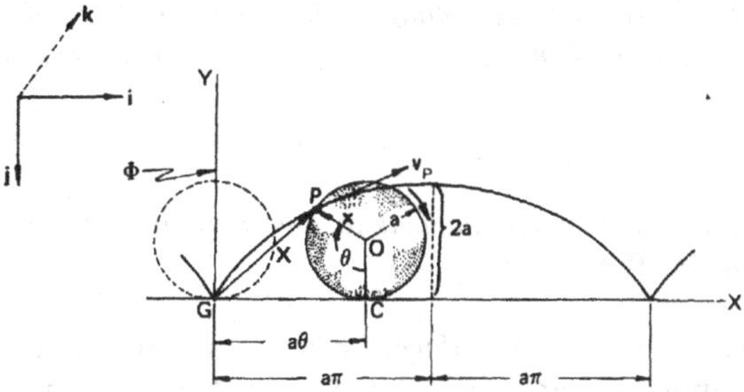

Figure 2.15. The locus generated by the motion of a point P on the circumference of a circle that rolls without slipping along a straight line is a cycloid.

zero, then one body pushes to drive the other in their common normal direction. This driving contact, however, may involve slipping in the tangential plane between the two surfaces, so that the tangential velocity components of their instantaneous contact points will not be equal. In this case, the bodies slide on one another. Thus, in general, two bodies roll on one another without slipping when and only when their instantaneous points of contact have the same velocity vector so that both their corresponding normal and tangential velocity components are equal. In particular, when their mutual normal velocity component vanishes, the rolling contact is called *pure rolling without slip*, or more simply, *rolling without slip*. Otherwise, the contact is identified as *driving contact, with or without slip*. Indeed, the previous example of pure rolling without slip, introduced somewhat differently, showed that the velocity of the rim point instantaneously in contact with the fixed surface is zero. Some additional applications of pure rolling without slip will be studied below. An example of driving contact with partial slip in a specified direction will follow; and other examples may be found in the Problems. (See Problem 2.51, for example.)

Before we turn to the next example, however, let us ask, What acceleration condition must be satisfied at points of instantaneous contact for rolling without slipping? Since the curvatures of the two surfaces of rolling contact generally are different, it is clear that the intrinsic acceleration of their points of instantaneous contact cannot be the same. Indeed, we have seen in the foregoing example that the point on the wheel in contact instantaneously with the fixed, horizontal surface has the acceleration (2.72); but the corresponding point on the *fixed* surface has, of course, no acceleration at all. On the other hand, because the arc lengths traced out by successive points of rolling contact on any two surfaces must be the same in time, in addition to the speed in the direction of their common tangent, the rate of change of speed of the coincident contact points also must be the same. This leads to the following criterion.

Criterion for Rolling without Slip. *The points P and Q of instantaneous contact between any two bodies rolling without slipping on one another must have the same velocity and the same tangential component of acceleration*:

$$\mathbf{v}_P = \mathbf{v}_Q, \qquad \mathbf{a}_P \cdot \mathbf{t} = \mathbf{a}_Q \cdot \mathbf{t}, \tag{2.76}$$

in which $\mathbf{t}$ *is the intrinsic vector tangent to the contacting curves at their coincident contact points P and Q.*

In particular, for a fixed surface, both the speed $\dot{s}$ and the tangential acceleration component $\ddot{s}$ are zero, so the contact point on the fixed surface naturally has no acceleration at all, while the point of contact on the wheel has only a normal component of acceleration, as we have seen before. Since the velocity of the contact point must be zero, we may use the no-slip criterion to determine the angular speed of the wheel in our previous example. Recalling Fig. 2.14, we write

$$\mathbf{v}_C = \mathbf{v}_O + \boldsymbol{\omega} \times \mathbf{r} = v_O \mathbf{i} - a\omega \mathbf{i} = \mathbf{0},$$

from which $\omega = v_O/a$ follows, as before. We shall apply the general conditions (2.76) in the next example.

Example 2.6. A rubber drive wheel of radius r_2 rolls without slipping on the inside surface of a turntable of radius r_1. The turntable has an angular speed ω_1 which is increasing at the rate $\dot{\omega}_1$ about a fixed axis at F, as shown in Fig. 2.16. Find the corresponding angular rates of the drive wheel about its fixed bearing at O.

Solution. The velocities and accelerations of the instantaneous contact points D on the rim of the drive wheel and C on the inside surface of the turntable are determined by

$$\begin{aligned} \mathbf{v}_C &= \mathbf{v}_F + \boldsymbol{\omega}_1 \times \mathbf{r}_1, \qquad & \mathbf{a}_C &= \mathbf{a}_F + \boldsymbol{\omega}_1 \times (\boldsymbol{\omega}_1 \times \mathbf{r}_1) + \dot{\boldsymbol{\omega}}_1 \times \mathbf{r}_1, \\ \mathbf{v}_D &= \mathbf{v}_O + \boldsymbol{\omega}_2 \times \mathbf{r}_2, \qquad & \mathbf{a}_D &= \mathbf{a}_O + \boldsymbol{\omega}_2 \times (\boldsymbol{\omega}_2 \times \mathbf{r}_2) + \dot{\boldsymbol{\omega}}_2 \times \mathbf{r}_2, \end{aligned} \tag{2.77}$$

wherein we have

$$\begin{aligned} \mathbf{r}_1 &= \overline{FC} = -r_1 \mathbf{n}, \qquad & \boldsymbol{\omega}_1 &= \omega_1 \mathbf{b}, \qquad & \dot{\boldsymbol{\omega}}_1 &= \dot{\omega}_1 \mathbf{b}, \\ \mathbf{r}_2 &= \overline{OD} = -r_2 \mathbf{n}, \qquad & \boldsymbol{\omega}_2 &= \omega_2 \mathbf{b}, \qquad & \dot{\boldsymbol{\omega}}_2 &= \dot{\omega}_2 \mathbf{b}, \end{aligned} \tag{2.78}$$

referred to the intrinsic frame $\psi = \{C; \mathbf{t}, \mathbf{n}, \mathbf{b}\}$ shown in Fig. 2.16. Since O and

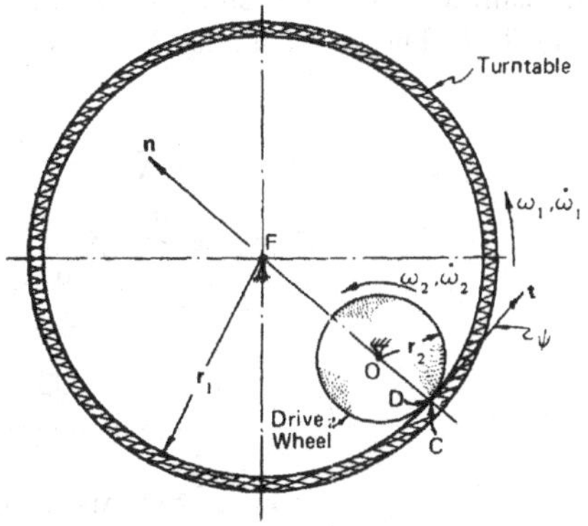

Figure 2.16. A turntable driven without slipping by a rubber wheel.

F are fixed points, $\mathbf{v}_O = \mathbf{v}_F = \mathbf{a}_O = \mathbf{a}_F = \mathbf{0}$; and it is easy to see that the remaining terms are given by

$$\begin{aligned} &\boldsymbol{\omega}_1 \times \mathbf{r}_1 = -\omega_1 \mathbf{b} \times r_1 \mathbf{n} = r_1 \omega_1 \mathbf{t}, \qquad \dot{\boldsymbol{\omega}}_1 \times \mathbf{r}_1 = r_1 \dot{\omega}_1 \mathbf{t}; \\ &\boldsymbol{\omega}_2 \times \mathbf{r}_2 = -\omega_2 \mathbf{b} \times r_2 \mathbf{n} = r_2 \omega_2 \mathbf{t}, \qquad \dot{\boldsymbol{\omega}}_2 \times \mathbf{r}_2 = r_2 \dot{\omega}_2 \mathbf{t}; \\ &\boldsymbol{\omega}_1 \times (\boldsymbol{\omega}_1 \times \mathbf{r}_1) = r_1 \omega_1^2 \mathbf{n}, \qquad \boldsymbol{\omega}_2 \times (\boldsymbol{\omega}_2 \times \mathbf{r}_2) = r_2 \omega_2^2 \mathbf{n}. \end{aligned} \tag{2.79}$$

Substitution of these data into (2.77) yields relations for the velocities and accelerations of the points C and D at the moment of contact:

$$\begin{aligned} \mathbf{v}_C &= r_1 \omega_1 \mathbf{t}, \qquad \mathbf{a}_C = r_1 \dot{\omega}_1 \mathbf{t} + r_1 \omega_1^2 \mathbf{n}, \\ \mathbf{v}_D &= r_2 \omega_2 \mathbf{t}, \qquad \mathbf{a}_D = r_2 \dot{\omega}_2 \mathbf{t} + r_2 \omega_2^2 \mathbf{n}. \end{aligned} \tag{2.80}$$

The criterion (2.76) for rolling without slip may now be applied to these results; we get

$$\begin{aligned} \mathbf{v}_C = \mathbf{v}_D \quad &\text{yields} \quad r_1 \omega_1 = r_2 \omega_2, \\ \mathbf{a}_C \cdot \mathbf{t} = \mathbf{a}_D \cdot \mathbf{t} \quad &\text{yields} \quad r_1 \dot{\omega}_1 = r_2 \dot{\omega}_2. \end{aligned} \tag{2.81}$$

And it follows that the angular rates of rotation of the drive wheel are given by

$$\omega_2 = r_1 \omega_1 / r_2 \quad \text{and} \quad \dot{\omega}_2 = r_1 \dot{\omega}_1 / r_2. \tag{2.82}$$

Notice that the first of (2.81) states that the instantaneous speed $\dot{s}$ of the points of contact are the same, and the second of (2.81) means that their instantaneous rates of change of speed $\ddot{s}$ are the same. Of course, as shown by (2.80), the accelerations of the contact points C and D are not equal. □

Example 2.7. Bevel gears are used to transmit rotary motion between intersecting axles; and their meshed gear teeth assure that this motion will be transmitted without slip. A typical bevel gear arrangement is shown in Fig. 2.17. The bevel angles θ and ϕ are called *pitch angles*; their sum is the

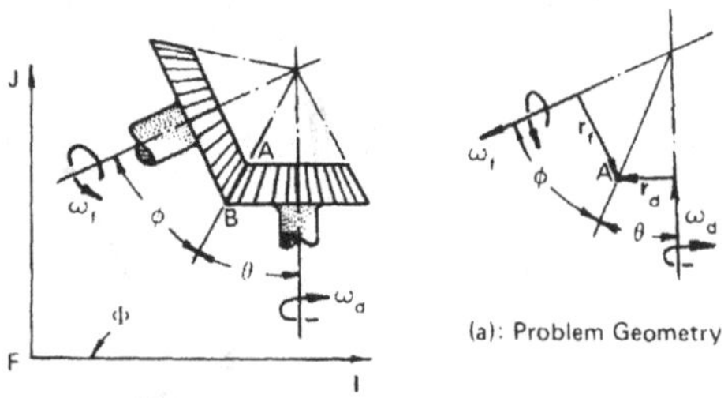

Figure 2.17. Application to bevel gear design.

angle between the gear shafts. If the drive gear with a pitch angle θ has an angular speed ω_d, determine the angular speed ω_f of the driven, follower gear with a pitch angle ϕ. Assume that the shafts are fixed in the machine.

Solution. Because the gears turn about fixed axles and roll on one another without slipping, each pair of points of contact of the gears along their instantaneous contact line AB must have the same velocity. Let r_d and r_f be the radii of the mutual rolling contact point A from points on the fixed axles, as shown in Fig. 2.17. Then, in obvious notation,

$$\mathbf{v}_A = \boldsymbol{\omega}_d \times \mathbf{r}_d = \boldsymbol{\omega}_f \times \mathbf{r}_f \qquad \text{yields} \quad \omega_d r_d = \omega_f r_f. \tag{2.83}$$

Introducing into (2.83) the pitch angle geometry from Fig. 2.17, we obtain the angular speed of the follower gear:

$$\omega_f = \omega_d \sin\theta / \sin\phi. \tag{2.84}$$

We see from this result that *the angular speeds are independent of the gear sizes, so the same rule holds for both large gears and small gears.* Also, (2.84) shows that $\omega_f \geqslant \omega_d$ for $\theta \geqslant \phi$, with $\omega_f = \omega_d$ when and only when $\theta = \phi$. In particular, when the shafts intersect in a right angle, $\theta + \phi = \pi/2$ and (2.84) becomes

$$\omega_f = \omega_d \tan\theta, \qquad 0 < \theta < \pi/2. \tag{2.85}$$

In this case, the angular speed of the follower gear is larger or smaller than the angular speed of the drive gear according as the pitch angle of the drive gear is greater or less than 45°. □

Example 2.8. Helical gears may be used to transmit rotary motion between nonintersecting and nonparallel axles. Their helicoidal shaped teeth allow relative sliding of the gear teeth along their common contact line, but the teeth prevent slipping normal to that line. The sliding motion provides a smooth, quiet operation with less shock than is common with other gears having straight teeth. A typical crossed helical gear arrangement is shown in Fig. 2.18. The angles ψ_d and ψ_f are the constant helix angles of the driver and follower gear teeth, respectively; and their sum is the angle between the gear shafts. Determine the angular speed ω_f of the follower gear when the drive gear has an angular speed ω_d, and the shafts are fixed in the machine.

Solution. Let r_d and r_f denote the radii of a mutual contact point A from points on the fixed axles, as shown in Fig. 2.18. Then we may write

$$\begin{aligned} \mathbf{v}_{Ad} &= \boldsymbol{\omega}_d \times \mathbf{r}_d = \omega_d \mathbf{j} \times r_d \mathbf{k} = \omega_d r_d \mathbf{i}, \\ \mathbf{v}_{Af} &= \boldsymbol{\omega}_f \times \mathbf{r}_f = -\omega_f \mathbf{j}' \times (-r_f \mathbf{k}) = \omega_f r_f \mathbf{i}', \end{aligned} \tag{2.86}$$

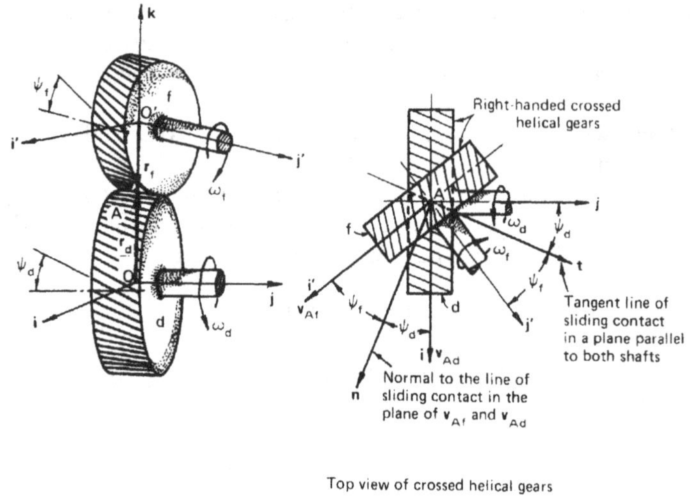

Figure 2.18. Crossed helical gears and their velocity components.

for the respective velocities of the contact point A on the driver and on the follower. The tangent line of the helical teeth at A, denoted by **t**, forms the helix angles ψ_d and ψ_f with the gear axles. Each gear has a sliding component of velocity in the tangential direction **t**, so the criterion for rolling without slip cannot be applied to the velocities (2.86). Rather, only the components of these vectors perpendicular to the teeth are equal:

$$\mathbf{v}_{Ad} \cdot \mathbf{n} = \mathbf{v}_{Af} \cdot \mathbf{n}, \tag{2.87}$$

wherein **n** is the normal vector to **t** as shown in Fig. 2.18. It is seen that the velocity vectors (2.86) make the same angles with **n** that the shafts make with **t**; these are the helix angles. Hence, their components in the direction of the normal **n** satisfy (2.87) when

$$\omega_d r_d \cos\psi_d = \omega_f r_f \cos\psi_f. \tag{2.88}$$

Therefore, the angular speed of the follower gear is

$$\omega_f = \frac{r_d \cos\psi_d}{r_f \cos\psi_f}\,\omega_d. \tag{2.89}$$

Since (2.89) depends on both the helix angles and the gear radii, a variety of angular speed relations is possible. For gears of equal radii, for example, the angular speed of the follower will be greater than that of the driver provided that the helix angle of the follower is the larger. In the special case when the shafts are perpendicular, $\psi_f + \psi_d = \pi/2$ and $\cos\psi_d/\cos\psi_f = \cot\psi_d$;

then, for gears of equal radii, the angular speed of the follower is larger than that of the driver provided that the helix angle of the drive gear is smaller than 45°. Of course, the helix angle of crossed gears may be equal, in which case (2.89) yields the angular speed ratio $\omega_f/\omega_d = r_d/r_f$, which is independent of the angle between the gear shafts. Alternatively, recalling the pitch triangle in Fig. 1.15, we see that $\tan\psi_f = \tan\psi_d$ if and only if $p_f/(2\pi r_f) = p_d/(2\pi r_d)$. Therefore, the angular speed ratio of crossed helical gears having equal helix angles also is equal to the reciprocal ratio of the pitches p_f and p_d of the helices: $\omega_f/\omega_d = p_d/p_f$. Evidently, the pitch triangle may be used to cast (2.89) in terms of ratios of the pitches and the radii of the gears; but we shall omit this relation.

Use of (2.88) in (2.86) provides the velocities of the contact point A on the driver and on the follower:

$$\begin{aligned} \mathbf{v}_{Ad} &= \omega_d r_d \cos\psi_d(\mathbf{n} + \tan\psi_d\,\mathbf{t}), \\ \mathbf{v}_{Af} &= \omega_d r_d \cos\psi_d(\mathbf{n} - \tan\psi_f\,\mathbf{t}). \end{aligned} \tag{2.90}$$

Therefore, the velocities may be equal if and only if $\psi_f = -\psi_d$. This is possible only when the angle between the shafts is zero, that is, when and only when the shafts are parallel. The negative sign means that the helices of parallel helical gears must have opposite hand so that they slope in opposite directions away from the viewer. Crossed helical gears, contrariwise, usually have the same hand so that the helices slope in the same direction away from the viewer, as assumed in Fig. 2.18, which shows right-handed helical gears.

2.9.2. Analysis of a Spatial Mechanism

The problems studied so far have involved only plane motions for which the angular velocity and the angular acceleration are parallel vectors, and for which use of the geometrical interpretations of the vector products often is especially helpful in their calculation. The analysis of spatial motions, because of the increased complexity and the difficulty of visualizing the geometrical aspects of a problem, inherently is more difficult; but the use of vector methods in these applications often simplifies their analysis and eliminates our need to perceive all of the geometrical details. The application of vector methods to the solution of spatial problems demonstrates strikingly the power and utility of this invaluable analytical tool. This is illustrated clearly in the following example of a simple spatial mechanism. Numerous additional examples will be encountered in Chapter 4.

Example 2.9. Two slider blocks A and B are connected by ball joints to a rigid rod, as shown in Fig. 2.19. The motion of A in $\Phi = \{F; \mathbf{i}_k\}$ is controlled so that its translational velocity is constant during an interval of interest. Find

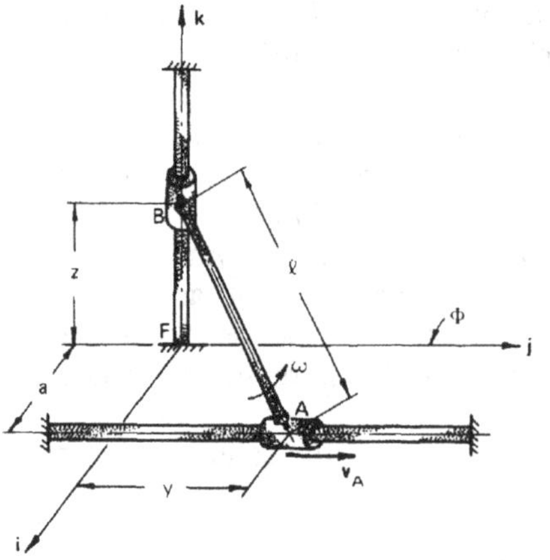

Figure 2.19. A simple spatial mechanism.

in Φ the translational velocity and acceleration of B, and determine the angular velocity and the angular acceleration of the rod in the interval of concern. Assume for simplicity that the ball joints are centered on the axes of the slider guide shafts, and suppose that the joints are ideally smooth so that the rod suffers no angular velocity about its own longitudinal axis. What will be the effect on the rotation of the rod if the ball joint at B is replaced by a hinge pin and yoke assembly?

Solution. The translational velocity of B is given by (2.27). We write

$$\mathbf{v}_B = \dot{z}\mathbf{k} = v_B\mathbf{k} = \mathbf{v}_A + \boldsymbol{\omega} \times \boldsymbol{l}, \tag{2.91}$$

wherein the constant translational velocity of A is

$$\mathbf{v}_A = \dot{y}\mathbf{j} = v_A\mathbf{j} \tag{2.92a}$$

and

$$\boldsymbol{l} = -a\mathbf{i} - y\mathbf{j} + z\mathbf{k}, \tag{2.92b}$$

the vector of B from A, is obtained from the geometry shown in Fig. 2.19 for points on the axes of the guide shafts. Of course,

$$l^2 = a^2 + y^2 + z^2 \tag{2.93}$$

relates z and y. Differentiation of (2.93) yields $2z\dot{z} + 2y\dot{y} = 0$, and use in this expression of the component relation from (2.92a) gives the translational velocity of B in Φ. I.e., in (2.91), we now have

$$\mathbf{v}_B = v_B\mathbf{k} = \dot{z}\mathbf{k} = -(yv_A/z)\mathbf{k}. \tag{2.94}$$

Moreover, because this equation is valid for all times in the interval of interest, the translational acceleration of B in Φ may be obtained most easily by differentiation of (2.94). With the aid of (2.92) and (2.93), we derive

$$\mathbf{a}_B = -v_A^2(l^2 - a^2)/z^3\mathbf{k}. \tag{2.95}$$

Furthermore, since $\dot{y} = v_A$ is a constant, we have by integration $y(t) = y_0 + v_A t$, where y_0 is the initial value of y. Therefore, z is determined by (2.93), and the solution for the velocity and acceleration of B in Φ is now complete.

It remains to determine the angular velocity and the angular acceleration of the rod. The angular velocity may be found from (2.91):

$$\boldsymbol{\omega} \times \boldsymbol{l} = \mathbf{v}_B - \mathbf{v}_A = -v_A(\mathbf{j} + y/z\mathbf{k}), \tag{2.96}$$

in which we recall (2.92) and (2.94). We form the cross product of (2.96) with $\boldsymbol{l}$ and expand the vector triple product to obtain

$$(\boldsymbol{\omega} \times \boldsymbol{l}) \times \boldsymbol{l} = (\boldsymbol{\omega} \cdot \boldsymbol{l})\boldsymbol{l} - (\boldsymbol{l} \cdot \boldsymbol{l})\boldsymbol{\omega}. \tag{2.97}$$

Then we appeal to the ideal ball joint constraint, which requires that the rod shall have no angular velocity about its longitudinal axis, that is,

$$\boldsymbol{\omega} \cdot \boldsymbol{l} = 0. \tag{2.98}$$

Substitution of (2.98) into (2.97) and use of (2.96), as described, yields

$$\boldsymbol{\omega} = v_A(\mathbf{j} + y/z\mathbf{k}) \times \boldsymbol{l}/l^2. \tag{2.99}$$

With the aid of (2.92b), we may compute the remaining cross product to derive the angular velocity of the rod in Φ:

$$\boldsymbol{\omega} = (v_A/zl^2)[(l^2 - a^2)\mathbf{i} - ay\mathbf{j} + az\mathbf{k}]. \tag{2.100}$$

And, finally, differentiation of this result leads to the angular acceleration of the rod in Φ; we get

$$\dot{\boldsymbol{\omega}} = \frac{v_A^2}{z^3 l^2}(l^2 - a^2)(y\mathbf{i} - a\mathbf{j}). \tag{2.101}$$

In the same manner described before, z and y may be expressed as functions of the time alone, and in this way we shall have the complete solution.

The angular velocity and acceleration of the rod depend upon the constraints introduced to fasten the rod to the slider blocks. An ideally smooth ball joint is a simple useful model, but hinged joints often are used too. To see the effect that different types of joint models may have on the angular rates of the rod, let us suppose that the ball joint at B is replaced by a hinge pin and yoke assembly shown in Fig. 2.20. In this case, we see that the rod can rotate about the $\mathbf{k}$ axis along the guide shaft and about the axis $\boldsymbol{\beta}$ parallel to the hinge pin; but it cannot turn about the axis $\boldsymbol{\gamma} \equiv \mathbf{k} \times \boldsymbol{\beta}$, which is perpendicular to these directions. Therefore,

$$\boldsymbol{\omega} \cdot \boldsymbol{\gamma} = 0 \tag{2.102}$$

defines the hinge constraint relation imposed at B on the rotations of the rod.

The vectors $\boldsymbol{\gamma} = \mathbf{k} \times \boldsymbol{\beta}$ and $\boldsymbol{\beta} \equiv \mathbf{k} \times \boldsymbol{l}$, which generally are not unit vectors, may be used to define the directions $\boldsymbol{\beta}$ and $\boldsymbol{\gamma}$ in terms of $\mathbf{k}$ and $\boldsymbol{l}$, which are more easily expressed in Φ. Thus, upon expanding the vector triple product that defines $\boldsymbol{\gamma}$ and recalling (2.92b), we find

$$\boldsymbol{\gamma} = \mathbf{k} \times (\mathbf{k} \times \boldsymbol{l}) = (\mathbf{k} \cdot \boldsymbol{l})\mathbf{k} - \boldsymbol{l} = z\mathbf{k} - \boldsymbol{l} = a\mathbf{i} + y\mathbf{j}. \tag{2.103}$$

Of course, the angular velocity of the rod is still determined by (2.96). Forming its vector product with $\boldsymbol{\gamma}$, expanding the result and introducing the hinge constraint (2.102), we have

$$\boldsymbol{\gamma} \times (\boldsymbol{\omega} \times \boldsymbol{l}) = (\boldsymbol{\gamma} \cdot \boldsymbol{l})\boldsymbol{\omega} - (\boldsymbol{\gamma} \cdot \boldsymbol{\omega})\boldsymbol{l} = (\boldsymbol{\gamma} \cdot \boldsymbol{l})\boldsymbol{\omega} = \boldsymbol{\gamma} \times (\mathbf{v}_B - \mathbf{v}_A). \tag{2.104}$$

Use of (2.103) and (2.92b) yields $\boldsymbol{\gamma} \cdot \boldsymbol{l} = -(a^2 + y^2)$, and substitution of (2.96) into the right-hand side of (2.104) delivers the angular velocity of the rod in Φ:

$$\boldsymbol{\omega} = \frac{v_A}{z(a^2 + y^2)} (y^2\mathbf{i} - ay\mathbf{j} + az\mathbf{k}). \tag{2.105}$$

We thus find that the angular velocity with the hinge constraint is indeed considerably different from the solution for the ideal ball joint case in (2.100); and we notice that

$$\boldsymbol{\omega} \cdot \boldsymbol{l} = v_A az/(a^2 + y^2) \neq 0 \tag{2.106}$$

shows that the hinged joint at B causes the rod to suffer an angular velocity about its own longitudinal axis. Construction of the angular acceleration vector for this case is left for the reader.

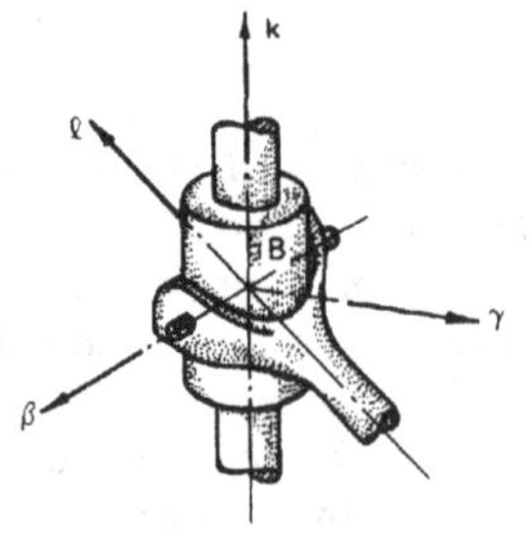

Figure 2.20. Slider block, hinge pin, and yoke assembly.

2.9.3. Velocity Referred to a Moving Reference Frame

We shall learn in future studies that often it is useful to express a vector quantity given in a frame $\Phi = \{F; \mathbf{I}_k\}$ in terms of the basis vectors $\mathbf{i}_k$ of another reference frame $\varphi = \{O; \mathbf{i}_k\}$. When this is done, we shall say briefly that the vector quantity is *referred to frame* φ. In preparation for subsequent development of the use of such auxiliary reference frames, let us consider the following illustration, in which it is helpful to refer the velocity vector of a particle of a rigid body to an imbedded reference frame.

Example 2.10. A thin rigid rod is supported in the vertical plane by the edge of a wall and a horizontal surface along which its end point O is moved with a velocity $\mathbf{v}_O$ in frame $\Phi = \{F; \mathbf{I}_k\}$, as shown in Fig. 2.21. Prove that the velocity in Φ of a particle B of the rod is directed toward O if and only if B is the particle R instantaneously in contact with the edge C over which the rod slides continuously, and find in frame Φ the sliding velocity of R.

Solution. The end-point velocity $\mathbf{v}_O = v_O \mathbf{I} = \dot{X}\mathbf{I}$ is given in frame $\Phi = \{F; \mathbf{I}_k\}$; but it will prove useful to refer $\mathbf{v}_O$ to another frame $\varphi = \{O; \mathbf{i}_k\}$ imbedded in the rod, as shown in Fig. 2.21. To accomplish this, we need to express the basis vectors of Φ in terms of those of frame φ. The geometry in Fig. 2.21 yields

$$\mathbf{I} = \cos\psi\, \mathbf{i} - \sin\psi\, \mathbf{j}, \qquad \mathbf{J} = \sin\psi\, \mathbf{i} + \cos\psi\, \mathbf{j}, \qquad \mathbf{K} = \mathbf{k}. \tag{2.107}$$

In the present case, only the first of (2.107) is needed to write $\mathbf{v}_O$ in terms of the vectors of frame φ; we have

$$\mathbf{v}_O = v_O(\cos\psi\, \mathbf{i} - \sin\psi\, \mathbf{j}), \tag{2.108}$$

and in this form we say that $\mathbf{v}_O$ is *referred to* φ.

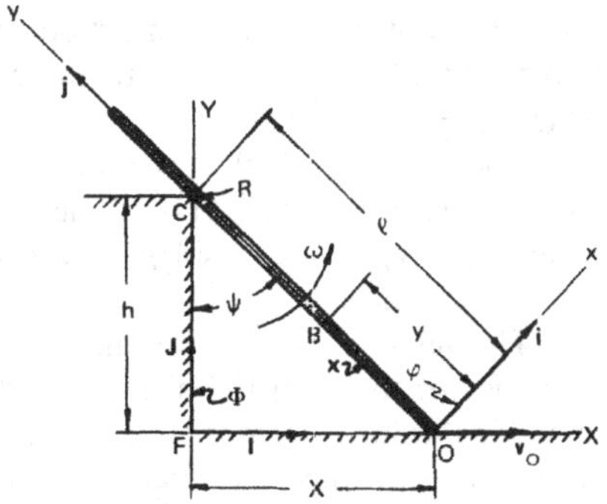

Figure 2.21. Application of an auxiliary, moving reference frame.

The position vector of an arbitrary point B along the thin rod is $\mathbf{x} = y\mathbf{j}$, and the rod has angular velocity $\boldsymbol{\omega} = \dot{\psi}\mathbf{K} = \dot{\psi}\mathbf{k}$ about point O, both vectors being referred to φ. Therefore, with $\boldsymbol{\omega} \times \mathbf{x} = -\dot{\psi}y\mathbf{i}$ and use of (2.108), the velocity of B in frame Φ, but referred to frame φ for our convenience, is given by

$$\mathbf{v}_B = \mathbf{v}_O + \boldsymbol{\omega} \times \mathbf{x} = (v_O \cos\psi - \dot{\psi}y)\mathbf{i} - v_O \sin\psi\,\mathbf{j}. \tag{2.109}$$

The geometry in Fig. 2.21 shows that $X = h\tan\psi$. Therefore, $v_O = \dot{X} = h\dot{\psi}\sec^2\psi$; and with $h = l\cos\psi$, we find

$$v_O \cos\psi = \dot{\psi}l, \tag{2.110}$$

where l is the distance from O the edge point C. Thus, use of (2.110) in (2.109) gives

$$\mathbf{v}_B = \dot{\psi}(l - y)\mathbf{i} - v_O \sin\psi\,\mathbf{j} \tag{2.111}$$

for all positions of the rod. We observe that $l - y$ is the distance from an arbitrary particle B to the particle R which instantaneously is in sliding contact with the edge C. Therefore, (2.111) shows that for $\dot{\psi} \neq 0$ the velocity of B will be directed along the rod toward O when and only when $y = l$. But this means that B must be the unique edge point particle R; hence, (2.111) reveals that the sliding velocity in Φ of the particle R instantaneously in contact with the edge of the wall is given by

$$\mathbf{v}_R = -v_O \sin\psi\,\mathbf{j}. \tag{2.112}$$

It is intuitively clear, of course, that the rod can have no velocity component normal to itself at its point of contact with the edge C; otherwise, the contact constraint would not be maintained. Consequently, it is easy to understand physically why the velocity of the particle R must be directed along the rod; but here we have learned also that R is the only particle having this property. It is instructive to note that the angular speed of the rod and the sliding velocity of R actually may be obtained more directly by application of the sliding contact constraint equation $\mathbf{v}_R \cdot \mathbf{n} = \mathbf{v}_C \cdot \mathbf{n}$, where $\mathbf{n} = \mathbf{i}$ is the mutual normal vector to the contacting surfaces at the contact points. With $\mathbf{v}_C = \mathbf{0}$ and $\mathbf{v}_R = \mathbf{v}_O + \boldsymbol{\omega} \times \mathbf{l} = v_O\mathbf{I} - \dot{\psi}l\mathbf{i}$, we find at once that $\mathbf{v}_R \cdot \mathbf{i} = v_O \cos\psi - \dot{\psi}l = 0$, which states that contact will prevail provided that the rod has the angular speed (2.110). Then the sliding velocity at R is the tangential component $\mathbf{v}_R \cdot \mathbf{j} = -v_O \sin\psi$ in the direction of $\mathbf{j}$ at R, which agrees with (2.112) obtained before.

The velocity vectors of O in (2.108), B in (2.111), and R in (2.112) are the velocities of these particles in frame Φ, but they are referred to the moving, imbedded frame φ for convenience. Clearly, these points have no motion relative to φ; rather, as described in our earlier study of the intrinsic velocity and acceleration, (2.108), (2.111), and (2.112) are the projections onto φ of the velocities of O, B and R relative to frame Φ. Let the reader conclude this example by showing that the velocity (2.112) referred to frame Φ itself is given by

$$\mathbf{v}_B = v_O \sin\psi\,(\sin\psi\,\mathbf{I} - \cos\psi\,\mathbf{J}). \tag{2.113}$$

We shall encounter many situations of this kind in Chapter 4.

2.10. A Basic Invariant Property of the Angular Velocity Vector

It has been emphasized in the derivation and applications of (2.27) and (2.30) that relative to a given reference frame Φ, say, the vector $\boldsymbol{\omega}$ is the angular velocity of a rigid body $\mathscr{B}$ about an instantaneous axis through a specified base point. The body may experience several simultaneous rotations about concurrent axes through the base point, but (2.28) shows that their vector sum is equivalent to a single angular velocity vector about a single instantaneous axis through the same base point. On the other hand, we may ask: What change, if any, occurs in $\boldsymbol{\omega}$ if the base point is shifted arbitrarily to another place in $\mathscr{B}$? To answer this question, we let P be any point of $\mathscr{B}$; and, for the same motion of $\mathscr{B}$, we assume that at the same instant $\boldsymbol{\omega}$ and $\boldsymbol{\omega}^*$ are distinct angular velocities of $\mathscr{B}$ about lines through distinct base points at O and O^*, respectively, as shown in Fig. 2.22. Since the velocity of any particle P, by its definition (1.8), is the same for every base point used in application of (2.27), using the vectors defined in Fig. 2.22, we may write for both O and O^*

$$\mathbf{v}_P = \mathbf{v}_O + \boldsymbol{\omega} \times \mathbf{x} \tag{2.114a}$$

$$= \mathbf{v}_{O^*} + \boldsymbol{\omega}^* \times \mathbf{x}^* \tag{2.114b}$$

$$= (\mathbf{v}_O + \boldsymbol{\omega} \times \mathbf{r}) + \boldsymbol{\omega} \times \mathbf{x}^*, \tag{2.114c}$$

wherein $\mathbf{x} = \mathbf{r} + \mathbf{x}^*$. However, with O as base point, it is clear that the term in

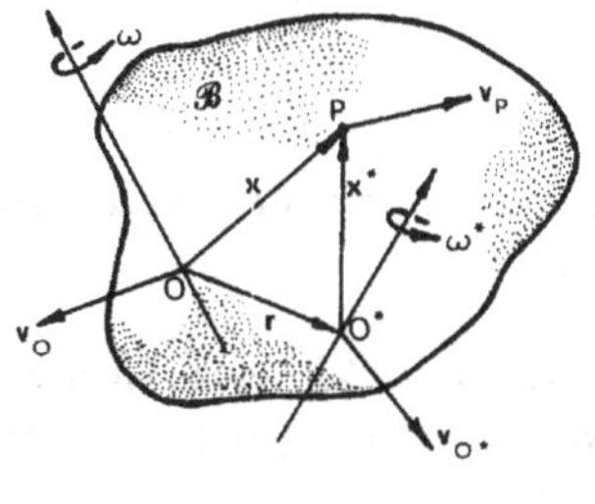

Figure 2.22. A change of base point in a general motion of a rigid body.

parentheses in (2.114c) is equal to $\mathbf{v}_{O^*}$. Consequently, we have $(\boldsymbol{\omega}^* - \boldsymbol{\omega}) \times \mathbf{x}^* = \mathbf{0}$. Since P is an arbitrary particle, this relation must hold for all $\mathbf{x}^*$; therefore, $\boldsymbol{\omega}^* = \boldsymbol{\omega}$ follows. In sum: *The angular velocity vector is the same for every choice of base point associated with a rigid body.*

Let us recall, for example, the description in Fig. 2.14 of a wheel rolling without slipping on a fixed surface. The wheel was naturally assumed to be rotating about its center O. However, since the angular velocity does not change with the choice of base point, we may consider that the same rotation occurs about a parallel line through the point of contact C, for which it is now known that $\mathbf{v}_C = \mathbf{0}$. Then, with C as the base point, we may use (2.27) to write $\mathbf{v}_O = \mathbf{v}_C + \boldsymbol{\omega} \times (-\mathbf{r}) = -\boldsymbol{\omega} \times \mathbf{r}$, which yields the result (2.69) found before. A similar relation may be written for any other base point. The foregoing basic property of the angular velocity vector shows that the instantaneous axes of rotation corresponding to all base points are parallel and the rates of rotation about them are equal.

Thus, the angular velocity vector is a function of time alone; and, without reference to the base point used, we may correctly refer to it as *the* angular velocity of the body. As a consequence of this invariant property of $\boldsymbol{\omega}$, (2.114) becomes

$$\mathbf{v}_P = \mathbf{v}_O + \boldsymbol{\omega} \times \mathbf{x} \tag{2.115a}$$

$$= \mathbf{v}_{O^*} + \boldsymbol{\omega} \times \mathbf{x}^*. \tag{2.115b}$$

The result shows that *a motion of a rigid body due to a translation* $\mathbf{v}_O$ *and a rotation* $\boldsymbol{\omega}$ *about a base point* O *is equivalent to a rotation* $\boldsymbol{\omega}$ *about any other base point* O^* *together with a new translation* $\mathbf{v}_{O^*}$ *given by*

$$\mathbf{v}_{O^*} = \mathbf{v}_O + \boldsymbol{\omega} \times \mathbf{r}, \tag{2.116}$$

where $\mathbf{r} = \mathbf{x} - \mathbf{x}^*$ *is the vector of* O^* *from* O. Thus, a change of base point certainly results in a change of velocity for the new base point, but the angular velocity of the body remains the same. (See Fig. 2.22.)

Two additional easy theorems follow readily from (2.115). Their proof is left for the reader in Problem 2.70.

(i) Invariant Projection Theorem. *The projections upon the instantaneous axis of rotation of the velocities of all points of a rigid body are the same; that is, for all points P*

$$\mathbf{v}_P \cdot \boldsymbol{\alpha} = \mathbf{v}_O \cdot \boldsymbol{\alpha}, \quad \text{or} \quad \mathbf{v}_P \cdot \boldsymbol{\omega} = \mathbf{v}_O \cdot \boldsymbol{\omega}, \tag{2.117}$$

where $\boldsymbol{\alpha} = \boldsymbol{\omega}/|\boldsymbol{\omega}|$ *is the instantaneous axis of rotation and O is any assigned base point.*

(ii) Parallel Axis Theorem. *The velocity of a particle P due to a pure rotation with angular velocity* $\boldsymbol{\omega}$ *about on axis* $\boldsymbol{\alpha}$ *is equivalent to a rotation with* *the same angular velocity about a parallel axis together with a translational velocity perpendicular to that line, and conversely.*

A motion of a rigid body $\mathscr{B}$ in which every particle of $\mathscr{B}$ moves parallel to a fixed plane in frame ψ is called a *plane motion* of $\mathscr{B}$ in ψ. It follows that both $\boldsymbol{\omega}$ and $\dot{\boldsymbol{\omega}}$ must be perpendicular to the plane of the motion. If $\boldsymbol{\omega}=\mathbf{0}$, the plane motion is a pure translation with $\mathbf{v}_P=\mathbf{v}_O$ for all particles P. If there exists a base point O with $\mathbf{v}_O=\mathbf{0}$, perhaps only instantaneously, the plane motion is a pure rotation about an axis at O. It is not difficult to prove that in an *unconstrained* motion of a rigid body in space, there is, in general, no point having zero velocity. On the other hand, it can be shown that *for a plane motion of a rigid body, at any moment, there always exists one and only one base point whose velocity is zero.* (See Problems 2.71, 2.81 and 3.9.)

Notice that (2.117) shows that $\mathbf{v}_P\cdot\boldsymbol{\alpha}=0$ if and only if either O is a fixed base point or it has at each instant a velocity perpendicular to the axis of rotation. In the latter case, each particle P must move at each instant in a plane motion perpendicular to the axis of rotation; therefore, the velocity of P is equivalent to a pure rotation about an axis parallel to $\boldsymbol{\alpha}$ through another base point O^*, say, situated in the fixed plane of the motion. The point O^* whose instantaneous velocity is zero is known as the *instantaneous center of rotation.* (See Problem 2.71.) In general, this unique base point, which often will not be within the body, changes from place to place in the plane as the body moves, unless, of course, the point is a fixed point about which the body is rotating normal to the plane. The locus of instantaneous centers mapped onto the plane is known as the *space centrode.* In particular, when one curved body rolls without slipping on another fixed curved body, the instantaneous center of the moving body is the point of contact, and the space controde is the contact curve of the fixed body. The locus in the plane of corresponding points referred to an imbedded body reference frame is called the *body centrode.* Thus, we may envisage that the body centrode, the curve of instantaneous centers in the body frame, rolls without slipping on the space centrode, the curve of instantaneous centers in the space frame. (See Problems 2.77–2.79.)

We have seen, for example, that the point of contact of a wheel rolling without slipping on a fixed, horizontal surface is the instantaneous center of rotation of the wheel. Therefore, the space centrode is the straight line of the fixed surface, and the body centrode is the moving circumference of the wheel. Every particle of the wheel has at each instant a velocity equivalent to a pure rotation about an axis at the point of contact C, namely, $\mathbf{v}_P=\boldsymbol{\omega}\times\mathbf{x}^*$, where $\mathbf{x}^*$ is the position vector of P from C. When the center O of the wheel is used as the base point, the velocity of the same particle P is determined by the same rotation about the wheel axle together with a translational velocity of O which is perpendicular to it: $\mathbf{v}_P=\mathbf{v}_O+\boldsymbol{\omega}\times\mathbf{x}$, where $\mathbf{x}$ is the vector of P

from O. The parallel axis theorem and existence of an instantaneous center in a plane motion lead to some useful graphical applications included in the exercises at the end of the chapter. (See Problems 2.72–2.80.) Some additional theoretical results are studied in the next section.

2.11. Chasles' Theorem on Screw Motions

The choice of base point to be used in (2.27) and (2.30) is arbitrary, so generally any convenient choice is admissible. In fact, we recall that the base point need not be a material point of the body; so every conceivable point of space is a potential candidate for use as a base point. Because of this arbitrariness in the selection of a base point, the velocity of a rigid body particle may be characterized in an infinite variety of ways. Therefore, we are led to question: Are there any particularly special choices of base point for which the velocity of a rigid body particle may be most simply and uniquely described? The answer is provided by the following remarkable theorem due to Chasles (1830):

Theorem on Screw Motions. *The velocity of any point of a rigid body is equivalent to a screw velocity consisting of a rotation about an axis and a translation along that axis.*

Proof. Let the velocity of a base point O and the angular velocity of the body be assigned so that the velocity of any particle P of a rigid body is determined by (2.115a). We wish to show that there exists a base point O^*, say, whose velocity at each moment is parallel to the instantaneous axis of rotation so that by (2.116)

$$\mathbf{v}_{O^*} = p\boldsymbol{\omega} = \mathbf{v}_O + \boldsymbol{\omega} \times \mathbf{r}, \tag{2.118}$$

where $\mathbf{r}$ is the unknown position vector of the new base point O^* from the assigned base point O and p is an unknown scalar. If this may be done, then with O^* as the base point the theorem follows. Therefore, we shall need to determine p and $\mathbf{r}$ to satisfy (2.118).

The scalar p is determined immediately by the invariant projection theorem (2.117):

$$p \equiv (\mathbf{v}_O \cdot \boldsymbol{\alpha})/\omega = (\mathbf{v}_O \cdot \boldsymbol{\omega})/\omega^2. \tag{2.119}$$

Then $\mathbf{r}$ is to be found from

$$\boldsymbol{\omega} \times \mathbf{r} = p\boldsymbol{\omega} - \mathbf{v}_O = (\mathbf{v}_O \cdot \boldsymbol{\alpha})\boldsymbol{\alpha} - \mathbf{v}_O = \boldsymbol{\alpha} \times (\boldsymbol{\alpha} \times \mathbf{v}_O). \tag{2.120}$$

However, this equation does not determine $\mathbf{r}$ uniquely; for, if $\mathbf{r}$ is a solution of (2.120), so is $\hat{\mathbf{x}} = \mathbf{r} + k\boldsymbol{\omega}$ for all values of the constant k. This is simply the vector equation of the new parallel axis through O^*; hence, O^* may be any point on this line and still satisfy (2.120). Therefore, we may choose O^* to be at the shortest distance from O between the parallel lines so that

$$\mathbf{r} \cdot \boldsymbol{\omega} = 0 \tag{2.121}$$

must hold. Then we form the vector product of (2.120) with $\boldsymbol{\omega}$, expand the result, and use (2.121) to obtain in terms of the originally assigned base point data the unique location $\mathbf{r}$ of the point O^* on the new parallel axis:

$$\mathbf{r} = \frac{\boldsymbol{\omega} \times \mathbf{v}_O}{\omega^2}. \tag{2.122}$$

Thus, shifting the base point to O^* by using (2.118) in (2.115) yields

$$\mathbf{v}_P = p\boldsymbol{\omega} + \boldsymbol{\omega} \times \mathbf{x}^*, \tag{2.123}$$

in which $\mathbf{x}^* = \mathbf{x} - \mathbf{r}$ is the position vector of P from O^* on the new parallel axis. (See Fig. 2.22.)

The equation (2.123) is the content of Chasles' theorem. It shows that the velocity of any particle P of a rigid body may be simply and uniquely characterized by a rotation with angular velocity $\boldsymbol{\omega}$ about a parallel line through a base point O^* at the unique place $\mathbf{r}$ from O given by (2.122) together with a translational velocity (2.118) along that line. This is the motion typical of a nut moved along a threaded screw; it is reminiscent of the helical motion of a particle studied in Chapter 1. In fact, such a motion is known as an *instantaneous screw motion*; and the axis of rotation is named the *screw axis*. The scalar invariant p defined in (2.119) is called the *pitch of the screw*; it is the ratio of the screw translation speed to the angular speed. Hence, the speed of translation along the screw is proportional to the angular speed about the screw axis. (See Problems 2.83 and 2.84.)

Chasles' method of reducing the velocity vector of a particle of a rigid body into a screw motion is unique. To prove this important fact, let us assume that there are two screw motions corresponding to noncollinear base points O' and O^*, respectively. Invariance of the angular velocity vector with respect to the base point shows that $\boldsymbol{\omega}' = \boldsymbol{\omega}^* \equiv \boldsymbol{\omega}$, say; and invariance of the pitch admits $p' = p^* \equiv p$, say. Thus, for the pair of screw motions with respect to distinct base points, we have by (2.123)

$$\mathbf{v}_P = p\boldsymbol{\omega} + \boldsymbol{\omega} \times \mathbf{x}'(P) = p\boldsymbol{\omega} + \boldsymbol{\omega} \times \mathbf{x}^*(P),$$

in which $\mathbf{x}'(P)$ and $\mathbf{x}^*(P)$ are the respective position vectors of P from O' and O^*. We thus find $\boldsymbol{\omega} \times (\mathbf{x}' - \mathbf{x}^*) = \mathbf{0}$. Since in general $\boldsymbol{\omega} \neq \mathbf{0}$, then for distinct base points, this implies that $\mathbf{r} = \mathbf{x}' - \mathbf{x}^*$, the position vector of O^* from

O', contrary to the hypothesis, is parallel to the screw axis. Therefore, $\mathbf{r} = \mathbf{0}$, and we may conclude the following uniqueness theorem: *There is at most one screw motion by which a given motion of a rigid body may be produced.*

Thus, it is now established that among all conceivable motions of a rigid body there exists one of unrivaled simplicity, namely, Chasles' unique screw motion consisting of a rotation about an axis and a translation along that axis. We shall learn in the next chapter that similar results can be proved for finite rigid body displacements. However, study of finite rigid body motions may be omitted from a first course. The reader who may wish to move on to further applications involving relative motions and moving reference frames in Chapter 4 will suffer no serious interruption in the continuity.

References

1. Classical Sources

a. CHASLES, M., Note sur les propriétés générales du système de deux corps semblables entr'eux et placés d'une manière quelconque dans l'espace; et sur le déplacement fini ou infiniment petit d'un corps solide libre, *Bulletin des Sciences Mathématiques par Férussac* **14**, 321–326 (1830). It is interesting that roughly four years later, L. Poinsot published his famous geometrical theory of the rotation of a rigid body about a fixed point, in which he states the screw theorem without reference to Chasles. This preliminary version, published in 1834, was expanded in two subsequent papers that appeared much later: POINSOT, L., Théorie nouvelle de la rotation des corps, *Liouville's Journal de Mathématiques* **16**, 9–129, 289–336 (1851). See also BALL, R. S., *The Theory of Screws*, Cambridge University Press, Cambridge, 1900. This is an advanced treatise on the mathematical theory of screw motion. The thorough annotated bibliography and Ball's parable on the dynamical motions of a rigid body make useful and enjoyable reading for anyone interested in the history of mechanics. Geometrical constructions of the kinematical theorems on rigid body motion may be found in the book by ROUTH, E. J., *Dynamics of a System of Rigid Bodies*, Dover, New York, 1960, Chapter 5, 184–229.

b. EULER, L., *Mémoires de l'Académie de Berlin*, 1750; and *Novi Commentaires de Saint-Pétersbourg* **20**, 189 (1775). See the advanced treatise by WHITTAKER, E. T., *Analytical Dynamics*, Cambridge University Press, Cambridge, 1927, Chapter 1.

2. Sources for Finite Rotations

a. BEATTY, M., Vector representation of rigid body rotation, *American Journal of Physics* **31**, 134–135 (1963).

b. GRUBIN, C., Vector representation of rigid body rotation, *American Journal of Physics* **30**, 416–417 (1962). Equation (2.7) is derived by solving a simple vector differential equation as described in Problem 2.5.

c. SCHWARTZ, H., Derivation of the matrix of rotation about a given direction as a simple exercise in matrix algebra, *American Journal of Physics* **31**, 730–731 (1963). Equation (2.7) is derived by matrix methods.

3. Engineering Applications Sources

a. LONG, R., *Engineering Science Mechanics*, Prentice-Hall, Englewood Cliffs, New Jersey, 1963. Chapter 6 treats the theory of rigid body motion.

b. MERIAM, J. L., *Dynamics*, 2nd Edition, Wiley, New York, 1975. Many excellent collateral problems and additional examples of engineering interest may be found in Chapters 5 and 7.

c. SHAMES, I., *Engineering Mechanics, Vol. 2: Dynamics*, 2nd Edition, Prentice-Hall, Englewood Cliffs, New Jersey, 1966. Chapter 15 is a good source for similar problems and examples useful for parallel study.

Problems

2.1. Prove that if there is no rocking disturbance by its occupants, the displacement of the cabin of a rotating ferris wheel is a parallel translation. Describe the path followed by a typical particle of the cabin.

2.2. (a) Show that the scalar product $\mathbf{Tx}\cdot\boldsymbol{\alpha}=0$ holds for all particles of a rigid body whose finite displacement is a rotation about a fixed line. What is the physical meaning of this result? (b) Prove that for a nonzero vector $\mathbf{c}$, $\mathbf{Tc}=\mathbf{0}$ if and only if $\mathbf{c}$ is parallel to the axis of rotation of the body; otherwise, $\mathbf{T}=\mathbf{0}$.

2.3. Two particles of a rigid body initially are located at

$$\mathbf{X}_1=2\mathbf{i}+3\mathbf{j}-\mathbf{k}\text{ m}\quad\text{and}\quad \mathbf{Y}_1=-3\mathbf{i}+2\mathbf{j}+6\mathbf{k}\text{ m}$$

in $\Phi=\{F;\mathbf{i}_k\}$. After a finite displacement of the body, these particles are, respectively, positioned at

$$\mathbf{X}_2=4\mathbf{i}-3\mathbf{j}+\mathbf{k}\text{ m}\quad\text{and}\quad \mathbf{Y}_2=-\mathbf{i}-4\mathbf{j}+8\mathbf{k}\text{ m}$$

in Φ. Prove that the displacement is not a rotation about a fixed line in Φ. See Problem 2.2.

2.4. Show that equation (2.7) for the displacement of a particle of a rigid body due to a finite rotation about a fixed line may be written as

$$\mathbf{d}(P)=\frac{d}{\rho}\left[\boldsymbol{\alpha}\times\mathbf{x}\cos\frac{\theta}{2}+\boldsymbol{\alpha}\times(\boldsymbol{\alpha}\times\mathbf{x})\sin\frac{\theta}{2}\right],$$

wherein $d=|\mathbf{d}|$, $\rho=|\boldsymbol{\rho}|$ and the other terms are the same as before.

2.5. Let $\mathbf{x}(P,t)$ be the position vector of a rigid body particle P from a point O on a fixed axis with direction $\boldsymbol{\alpha}$; and consider the differential equation $\mathbf{x}'=\boldsymbol{\alpha}\times\mathbf{x}$ subject to the initial condition $\mathbf{x}(P,0)=\mathbf{x}_0$. The prime denotes $d/d\theta$ and the angle $\theta(t)$, with $\theta(0)=0$, is the rotation about $\boldsymbol{\alpha}$. Prove that $\mathbf{x}\cdot\boldsymbol{\alpha}$ is a constant, and derive the relation

$$\mathbf{x}''+\mathbf{x}=(\boldsymbol{\alpha}\cdot\mathbf{x}_0)\boldsymbol{\alpha}.$$

Show that the solution of this vector differential equation is

$$\mathbf{x}(P,t)=\mathbf{A}\cos\theta(t)+\mathbf{B}\sin\theta(t)+(\boldsymbol{\alpha}\cdot\mathbf{x}_0)\boldsymbol{\alpha},$$

in which $\mathbf{A}$ and $\mathbf{B}$ are constant vectors. Find these vectors, and thus show that the solution may be written in the form of (2.7). Hint: Notice that $\mathbf{x}'$ is perpendicular to $\boldsymbol{\alpha}$.

2.6. The figure shows a spatial control device consisting of a drive shaft OA positioned on the line $x_1 = x_2 = x_3$ and connected rigidly to a curved link OB whose end B is constrained by a pin P to slide in a curved door panel of an aircraft. To operate the mechanism, the shaft OA is rotated from its initial position through an angle $\theta = 60°$, as indicated. Find the displacement of P when its initial position from O is $\mathbf{x} = -\mathbf{i} + 2\mathbf{j}$ ft in $\Phi = \{O; \mathbf{i}_k\}$. What is the final location of P in Φ?

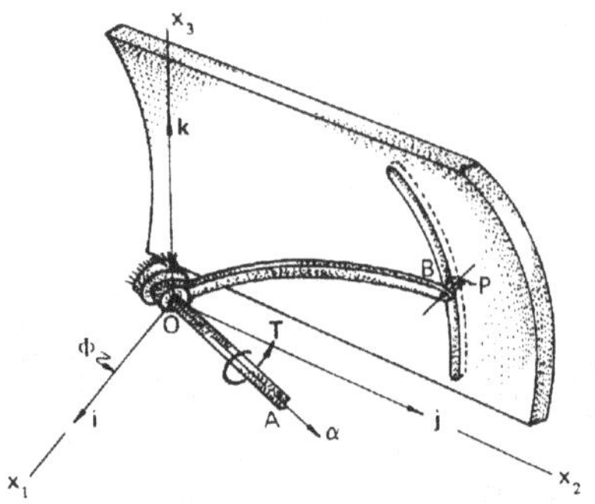

Problem 2.6.

2.7. How many independent coordinates are required to specify the locations of N particles free to move in space? How many degrees of freedom does this system have if one of the particles must remain at the origin of the reference frame?

2.8. Assign independent coordinates adequate to specify the position of a system of four rigid rods hinged together to form a quadrilateral that may move arbitrarily in a plane. (a) How many degrees of freedom does this system have? (b) How many degrees of freedom does the system have when one hinge point is fixed in the plane? (c) If one hinge pin is removed, how many additional degrees of freedom does the system possess?

2.9. Assign independent coordinates that adequately describe plane configurations of the systems shown in the figure, and determine the number of degrees of freedom for each case.

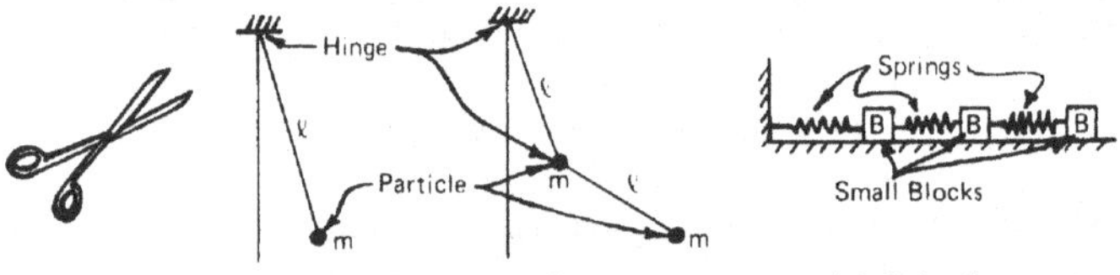

(a): Scissors (b): Simple and Double Pendulums (c): Spring System **Problem 2.9.**

2.10. A rectangular container initially positioned with its center C at the place $(4, 6, 8)$ in a fixed frame Φ is transported to a new configuration so that C is at the place $(6, 8, 10)$ and the container, in its new location, has been rotated 45° about a fixed line through C and the origin of Φ. A point T at the top of the container initially is at the place $(4, 6, 9)$ in Φ. What is the displacement and the final location of T in Φ?

2.11. (a) A unit cube undergoes an infinitesimal rotation $\Delta\theta$ about its diagonal. The displacement of the particle at $A = (1, 1, 1)$ in frame $\Phi = \{F; \mathbf{I}_k\}$ is plainly zero. What is the infinitesimal displacement of the particle at $B = (1, 0, 1)$ in Φ? (b) Suppose that this rotation is viewed as the composition of the corresponding three equal infinitesimal, component rotations of the cube about its three edges through O. What are the resultant displacements of the particles at A and B in Φ? (c) If the rotation in (a) is considered as the composition of three equal infinitesimal rotations about the face diagonals through O and the points (1, 0, 1), (1, 1, 0), and (0, 1, 1), are the displacements of the particles at A and B the same as before? (d) Discuss the results of these three cases.

2.12. If a body is rigid, the velocity of each of its particles P is determined by (2.27). Conversely, suppose that a body moves so that in every motion the velocity of each of its particles is given by (2.27). Prove that the body is rigid.

2.13. Use (1.70), (1.71), and the relation $\dot{s} = \rho\dot{\theta}$ to derive the equations (2.27) and (2.30) for the rigid body velocity and acceleration of a particle P rotating on a circle of radius ρ with angular speed $\dot{\theta}$ about a fixed axis $\boldsymbol{\alpha} = \mathbf{b}$.

2.14. Use the geometrical interpretations of the terms in (2.27) and (2.30) to determine the velocity and acceleration of the rim particle P in Fig. 1.14. Check your solutions against (1.78). See Example 1.8.

2.15. Apply equations (2.27) and (2.30) to determine for the conditions specified in Problem 1.12 the velocity and acceleration of the mass M. Note the geometrical nature of the terms computed.

2.16. Employ (2.27) and (2.30) to find the velocity and acceleration of the ball P in Fig. 1.3. Use the conditions described in Example 1.3, and compare your results with those in (1.17). Observe the geometrical character of the terms computed.

2.17. The slider block of a machine oscillates along a straight line; and at the instant illustrated, it has a speed of 30 ft/sec and is accelerating at 10 ft/sec^2 toward the right. The connecting rod AB has an angular speed of 4 rad/sec and an angular acceleration of 8 rad/sec^2 clockwise about its hinge at A, and the rod is in a vertical position. Determine for this instant the velocity and acceleration of the connecting pin B relative to the slider and relative to the machine foundation.

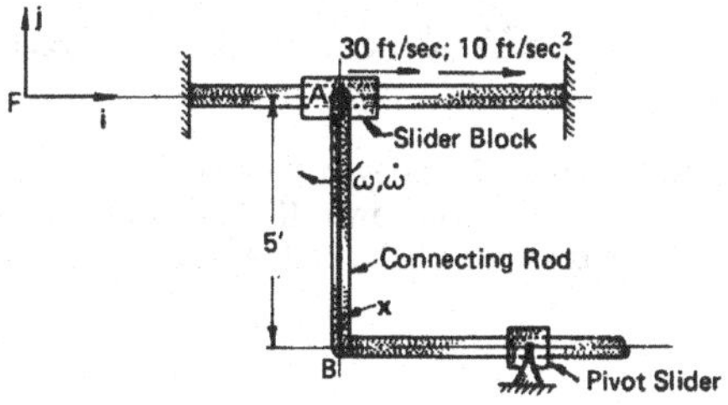

Problem 2.17.

2.18. A hydraulically actuated piston B is connected by a link BC to a gear G that rolls on a fixed ring gear R of radius b, as shown. Show that the ratio of the speed v^* of the slider to the speed v of the center C of the gear is given by

$$\frac{v^*}{v} = \frac{\sin(\theta - \phi)}{\cos\theta}.$$

2.19. The crank device of a metal forming press shown in the figure is used to drive the ram head for the forming dies. The die design requires knowledge of the ram velocity as a function of its distance h above the bottom of the ram stroke. Assume that the crank OA turns with a constant angular velocity $\boldsymbol{\omega}$ and that $r/l \ll 1$, where r is

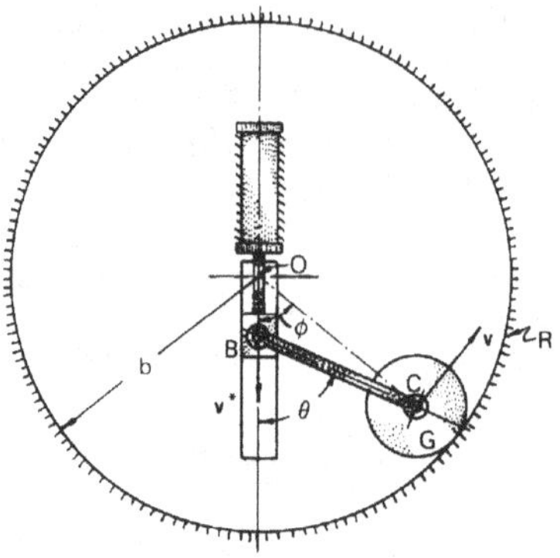

Problem 2.18.

the crank radius and l denotes the length of the connecting rod. Determine the velocity of the ram as a function of h. Find the angular velocity of the connecting rod AB. What is the maximum velocity of the ram head?

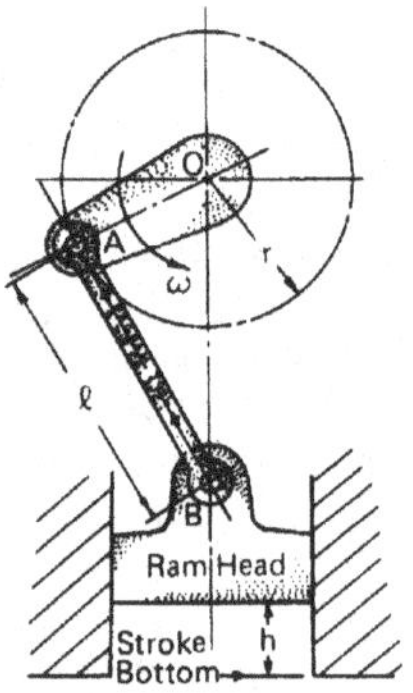

Problem 2.19.

2.20. A rigid of length $2l$ slides in the plane, but its ends maintain contact with the wall at A and ground at B, as shown. If B moves to the right with a constant velocity $\mathbf{v}$, what are the accelerations of the end A and the midpoint C? Determine the trajectory of C.

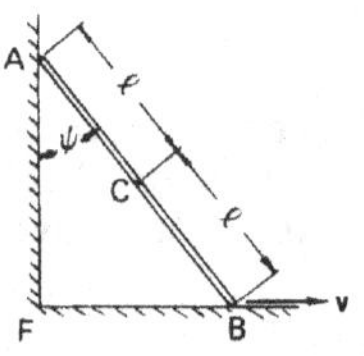

Problem 2.20.

2.21. The ends of a rigid bar AB of length $\sqrt{2}$ ft are constrained to move in guide slots. At the instant t_0 shown in the figure, the end B has a velocity of 2 ft/sec and an acceleration of 4 ft/sec^2 toward the right. Find the velocity and acceleration of the end A, and determine the angular velocity and acceleration of the bar at the moment t_0.

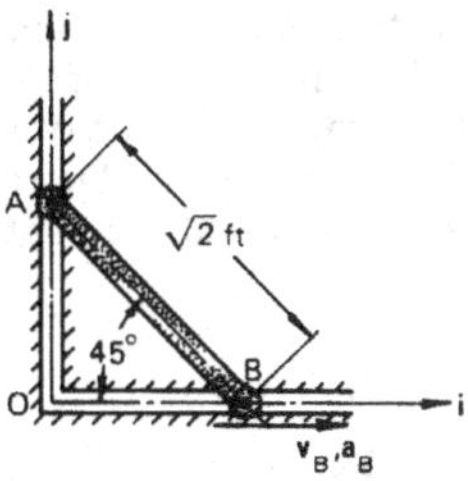

Problem 2.21.

2.22. During an interval of its motion, the slider block S has a constant vertical velocity $\mathbf{v}_S$. Determine the angular acceleration of the link OA when the mechanism is in the position shown.

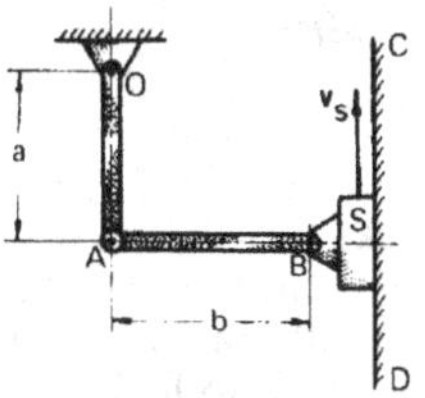

Problem 2.22.

2.23. A slider block B is connected by a hinged link BC to the piston rod HC of a hydraulic actuator A. During an interval of its plane motion, the piston rod has a constant velocity $\mathbf{v}_C = 40\mathbf{j}$ cm/sec in $\Phi = \{C; \mathbf{i}_k\}$. Find the absolute velocity and acceleration of B in Φ when C is at the position shown; and determine at this instant the angular velocity and acceleration of the link.

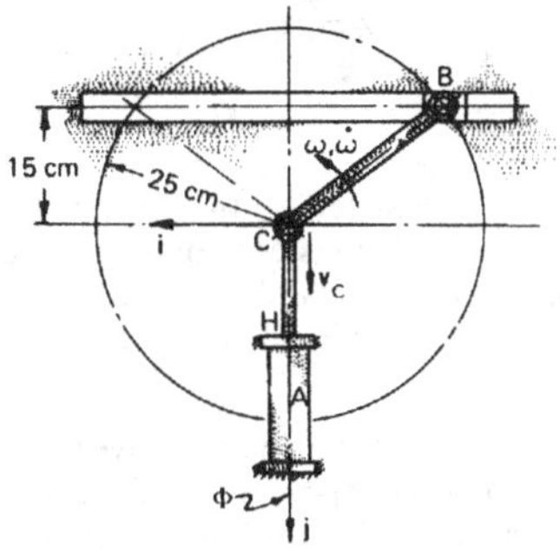

Problem 2.23.

2.24. The control wheel of a crank device is turning clockwise at a rate of $\omega_0 = 4$ rad/sec, which is growing at the rate of 2 rad/sec^2 at the instant t_0 when the device is in the configuration shown. Find the total velocity and the total acceleration of the connecting link pin B; and determine at the moment t_0 the angular velocities and the angular accelerations of the crank links AB about A and BC about C.

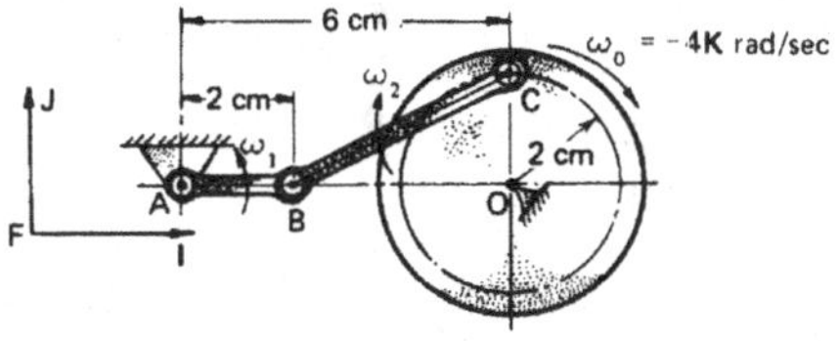

Problem 2.24.

2.25. An equilateral triangular frame that moves freely at A controls the motion of a slider B through a hydraulic actuator D. The piston rod, which is hinged to the frame at C, moves with a constant velocity $v_C = 30\mathbf{i}$ cm/sec during a time interval in which C passes the location shown. Determine at this instant the velocity and acceleration of the slider B, and find the angular velocity and angular acceleration of the triangular frame.

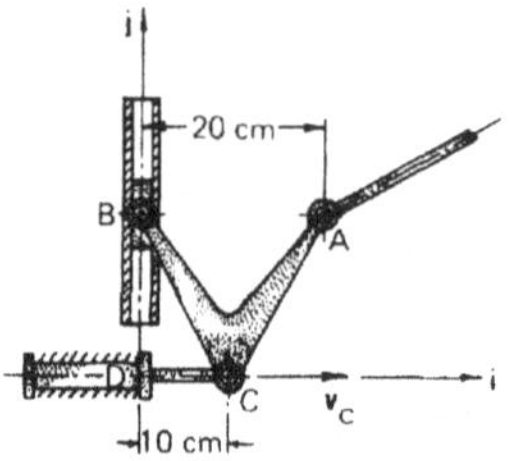

Problem 2.25.

2.26. The trammel shown in the figure consists of a rigid rod AP of length l hinged at A and B to blocks that slide in the cross slots. The distance between A and B is d. (a) Use (2.27) to determine the velocities of particles A, B, and P. Integrate $\mathbf{v}_P$ to derive the equation of the path traced by P. (b) Apply (2.27) to find the velocity of the midpoint C. Integrate the result to find the path of C.

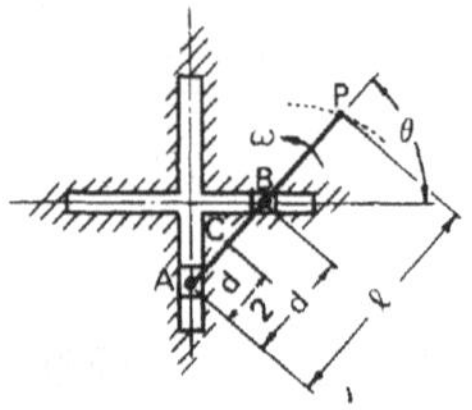

Problem 2.26.

2.27. In the last problem, let $l = 14$ cm and $d = 8$ cm. Suppose that the trammel rod rotates as shown with a constant angular speed of 4 rad/sec. Employ equation (2.30) to compute the accelerations of P and C when $\theta = 30°$. Check your solutions by differentiation of the results obtained in the last problem. Write the equations of the paths traced by P and C.

2.28. The slider A of a mechanism moves in a straight track. In the position shown, the hinge pin B has a speed $v_B = 36$ in./sec and the link BC has an angular speed ω_2 that is decreasing at the rate of 12 rad/sec each second in frame $\Phi = \{F; \mathbf{i}_k\}$. Find for this instant in Φ the acceleration of the slider and the angular acceleration of the link AB.

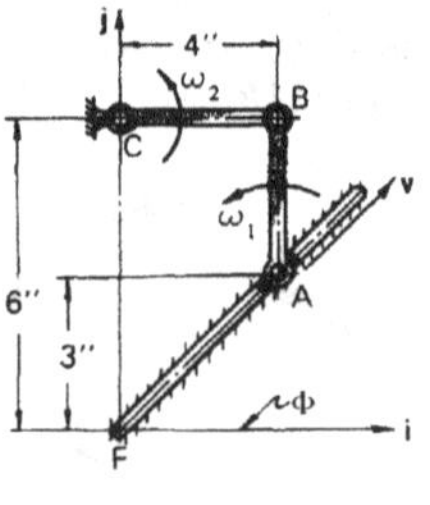

Problem 2.28.

2.29. The link AB of a slider mechanism has an angular velocity $\boldsymbol{\omega}_1 = 0.5\mathbf{k}$ rad/sec and an angular acceleration $\dot{\boldsymbol{\omega}}_1 = 1.5\mathbf{k}$ rad/sec^2 relative to a fixed frame $\varphi = \{A; \mathbf{i}_k\}$ at the moment t_0 shown in the figure. What is the angular acceleration of the rod BC relative to φ at the time t_0?

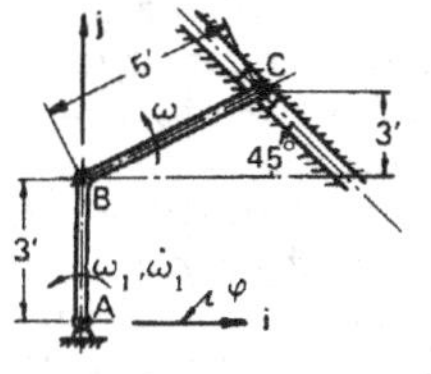

Problem 2.29.

2.30. A mechanism has a slider that moves in a parabolic track with a speed of 10 cm/sec which is changing at the rate of 20 cm/sec each second at the instant when the links AB and BC are in the position illustrated. Find the velocity and acceleration of the hinge pin B in its plane motion at this instant, and determine the angular acceleration of each link in frame $\Phi = \{F; \mathbf{i}_k\}$.

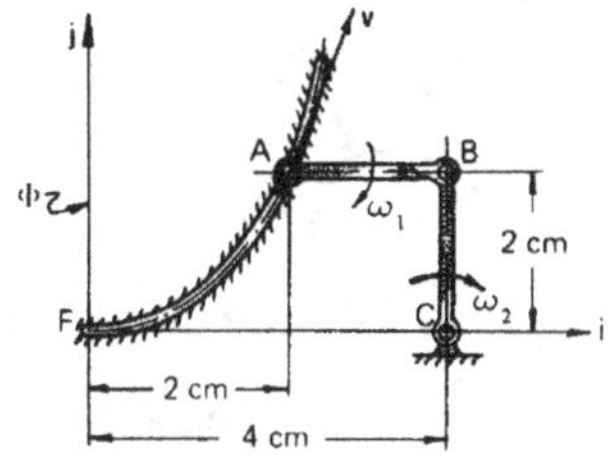

Problem 2.30.

2.31. A four-bar linkage consists of a drive crank AB of length a that rotates with a known angular speed ω, a connecting link BC of length b, a follower link CD of length c, and a fixed link AD that often is part of the foundation of the machine. Find the velocity of joint C as a function of the parameters shown. Evaluate the result for $\omega = 25$ rad/sec, $\theta = 60°$, $\phi = 45°$, $2a = c = 508$ mm, $b = 548$ mm. What is the length of the link AD?

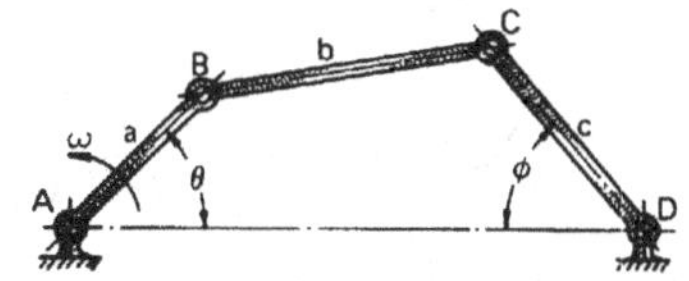

Problem 2.31.

2.32. Two cylinders A and B turn about their respective axes C and D within a rigid housing H. A rod R is hinged at P to the cylinder A, which has a clockwise angular velocity $\boldsymbol{\omega}_A$, and it slides in a block D welded to B so that its center line coin-

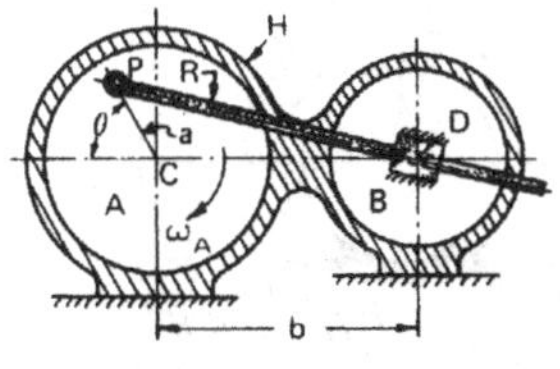

Problem 2.32.

cides with a diameter of B. Find the angular velocity $\boldsymbol{\omega}_B$ of B expressed as a function of the angle θ. Write a program to compute the ratio ω_B/ω_A of the angular speeds as a function of θ with a fixed ratio b/a of the dimensions shown. Graph the result for $b/a=2$, 5, 10 and $0 \leqslant \theta \leqslant 2\pi$. Determine the angles θ for which $\omega_B=0$.

2.33. The boss has decided that a general analysis of the design problem described in Example 2.4 is desirable. The cam is to have rise b and produce a sinusoidal velocity of the shuttle drive block A given by $\mathbf{v}_A = v \cos\theta\,\mathbf{i}$, where v is a constant and $\theta(0)=0$ initially. The cam dimensions a and d are constants. The length l of the link AB must be chosen so that $\dot{\boldsymbol{\omega}}_1=\mathbf{0}$ at the instant t_0 for which $\theta(t_0)=\theta_0$ and $\phi(t_0)=\phi_0$ are assigned. (a) Determine the shape $r(\theta)$ of the cam and the required angular operating speed ω. (b) Find the velocity and acceleration of the link pin B at t_0, and determine the required link length l, all expressed in terms of the assigned data. Check the results for the conditions specified in Example 2.4.

2.34. A wheel of radius 30 cm rolls without slipping down an inclined plane at the controlled rate of 15 m/sec. What are the velocity and acceleration of a point P on the wheel rim at the position shown? What is the angular velocity of the wheel? What is the influence of the inclination of the plane?

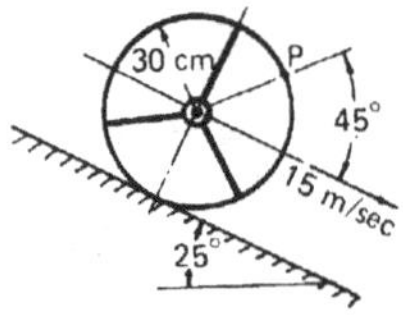

Problem 2.34.

2.35. The drive mechanism for a phonograph turntable of radius R is shown in the figure. The motor shaft turns the gear A with a counterclockwise angular speed ω_1, and A engages an idler gear B which drives the turntable at an angular speed ω_2, as indicated. (a) Find the radius of the drive gear A. What function does the idler gear perform? (b) Compute the drive gear diameter required to move a 9-in.-diam turntable at $33\frac{1}{3}$ rpm with a motor rated at 400 rpm.

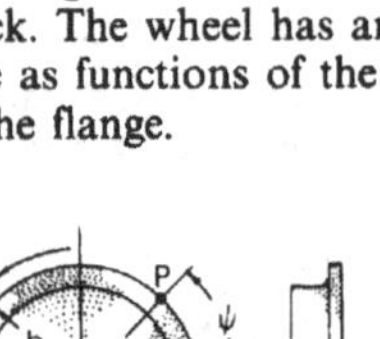

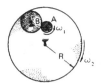

Problem 2.35.

2.36. Guided by a flange of radius b, a train wheel of radius a rolls without slipping on a horizontal track. The wheel has angular velocity $\boldsymbol{\omega}$ and angular acceleration $\dot{\boldsymbol{\omega}}$, as shown. Determine as functions of the angle ψ the velocity and acceleration of a point P on the rim of the flange.

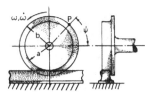

Problem 2.36.

2.37. A flexible, inextensible string is wrapped around a spool core of radius a. The spool, with outer radius b, rolls without slipping along a fixed, horizontal surface as the string is pulled with a constant horizontal velocity $\mathbf{v}$, as indicated. (a) Show that the velocity of the spool center O is in the direction of the pull. (b) Determine the velocity of the points O, P, and Q for the case $b = 2a$. Find the trajectory of the point Q, and describe the curve.

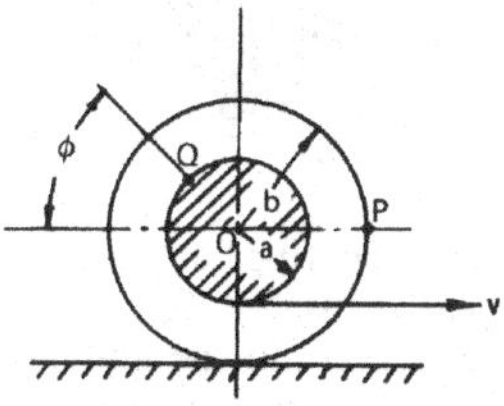

Problem 2.37.

2.38. The segment of a circular cylinder of radius R, initially at rest on a horizontal plane, as illustrated, is slightly disturbed to perform rocking oscillations without slipping so that $\theta = \theta_0 \sin pt$, the constants θ_0 and p denoting the small amplitude and circular frequency of the angular motion, respectively. Find the velocity and acceleration of the point C on the axis of symmetry at the distance h from O in $\varphi = \{O; \mathbf{i}, \mathbf{j}\}$.

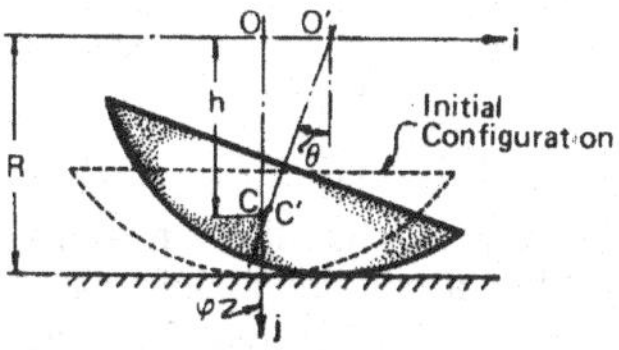

Problem 2.38.

2.39. The face angle ϕ of the bevel gear assembly shown in the figure is chosen so that the gears properly roll on each other at all points of contact along the line AO'. Prove that this arrangement is impossible unless P and O' coincide.

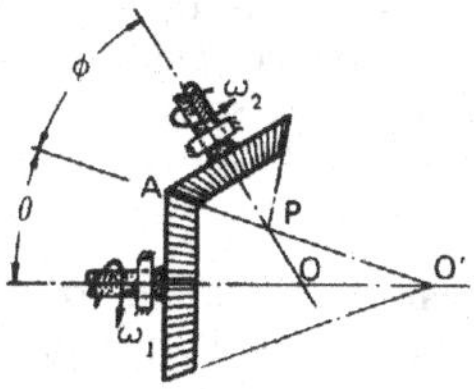

Problem 2.39.

2.40. Two gears A and B are held in rolling contact by a link of length l between their centers, as shown. The gear A of radius R rotates clockwise with angular velocity $\boldsymbol{\omega}_A$ about O in $\Phi = \{O; \mathbf{i}_k\}$, while the gear B of radius r rolls on the periphery of A with counterclockwise angular velocity $\boldsymbol{\omega}_B$ relative to Φ. Find the angular velocity of the link in Φ.

2.41. Planetary gears are used in transmission systems that require gears that can be shifted while remaining interlocked. In the shifting process, power transmission is altered by stopping one or more gears. A typical planetary gear train consists of a sun gear Σ of radius a and three planet gears P of radius b set in a ring gear housing R, as

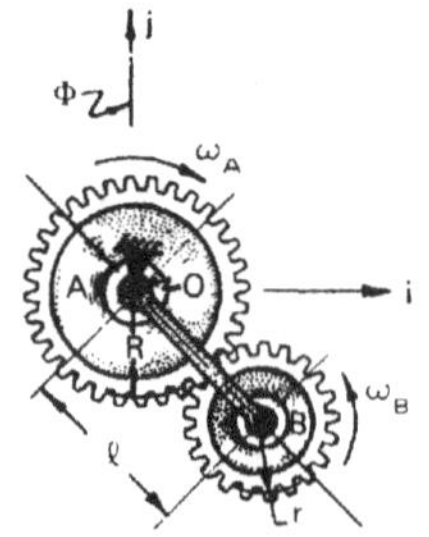

Problem 2.40.

illustrated. The planet gear bearings are mounted in a tripod frame S called the spider. Suppose that the sun gear turns counterclockwise with an angular velocity $\boldsymbol{\omega}$ while the ring gear is fixed in frame Φ. Find the angular velocities in Φ of the planet gears and the spider. How are these related to $\boldsymbol{\omega}$ when $r = 3a$?

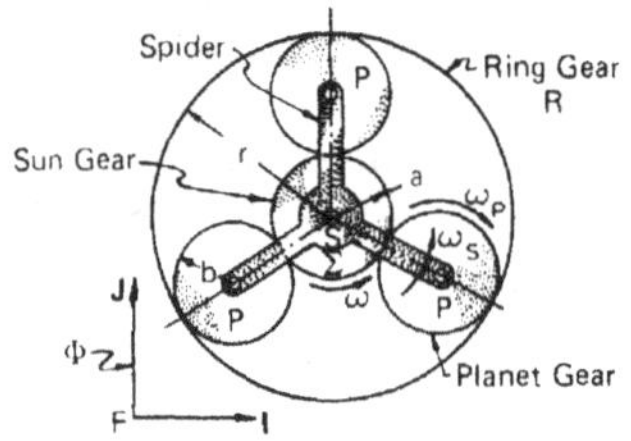

Problem 2.41.

2.42. Determine the angular velocities in Φ of the planet gears and the spider described above for the case when the sun gear is stopped and the system is driven by the ring gear, which has a counterclockwise angular velocity $\boldsymbol{\Omega}$. How are these rates related to $\boldsymbol{\Omega}$ when $r = 3a$; and, additionally, when $\boldsymbol{\Omega} = \boldsymbol{\omega}$, how do these rates compare with those found in the previous problem?

2.43. The sun gear of the planetary gear train described in Problem 2.41 rotates with angular velocity $\boldsymbol{\omega}$ while the ring gear turns with angular velocity $\boldsymbol{\Omega}$, relative to frame Φ. Find the angular velocities of the spider and planet gears in Φ. Check your results for the conditions set in the previous two problems. Discuss conditions necessary and sufficient for (i) the spider to remain at rest while the planets turn, and (ii) the planets to stop while the spider turns. Is it possible for the spider and planet gears to rotate at the same rate? Can they rotate at the same rate about opposite directions?

2.44. The planet gear B rolls on the fixed sun gear A such that the drive link OB turns counterclockwise at 90 rpm, as shown. When the gears are shifted, the ring gear

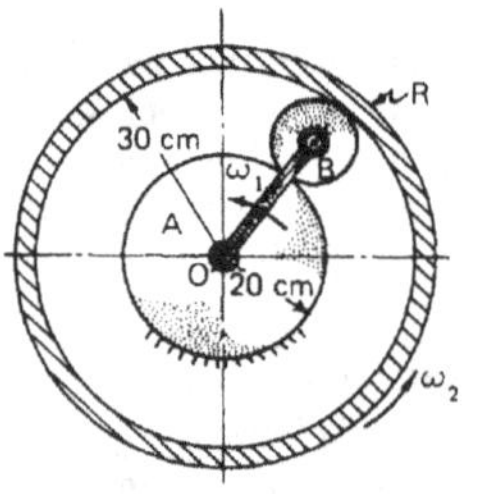

Problem 2.44.

R is held fixed and the sun gear rotates about O, while the rotation of OB is unchanged. Determine the angular velocity of gear B for each case. What is the ratio of the corresponding angular speeds of the sun and ring gears for these two cases?

2.45. Two shafts are mounted in bearings and geared together as shown. A flexible, inextensible cable wound on the drum D, which is attached rigidly to the gear B, supports a crate C. The gear A turns clockwise with angular velocity ω and angular acceleration $\dot{\omega}$. Determine the velocity and acceleration of the crate.

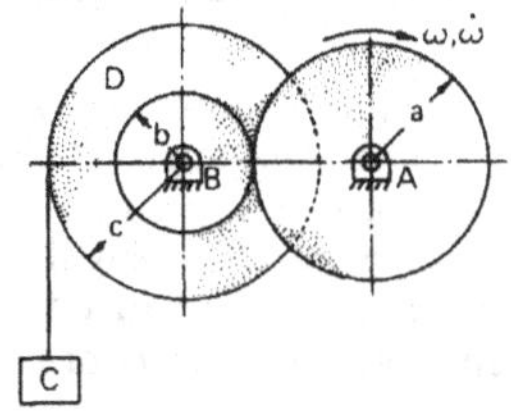

Problem 2.45.

2.46. During an interval of its motion, a gear B of radius a rolls on a horizontal gear rack CD with a constant angular speed $\Omega = 4$ rad/sec. At the moment t_0 shown in the figure, the link OA is parallel to the rack CD. Determine the angular acceleration $\dot{\omega}$ of the link OA at t_0.

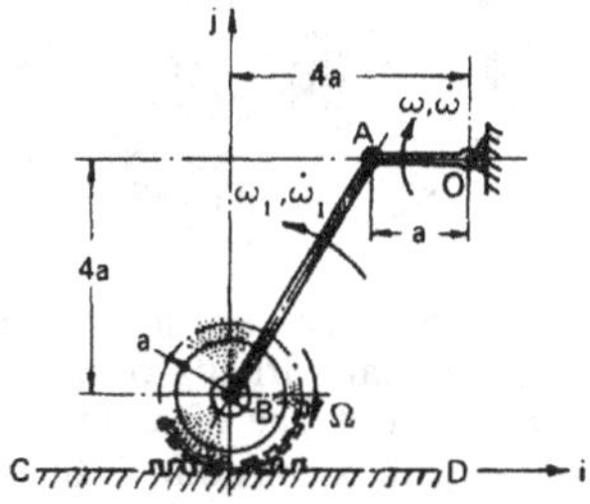

Problem 2.46.

2.47. An inextensible cable is wound on a drum of radius R shown in the figure. The axle O of the drum assembly is guided by a vertical slot so that the pulley of radius r winds up the suspension cable ST, as the end point P descends with speed v and acceleration a. Determine the velocity and acceleration of O; and find the angular velocity and acceleration of the pulley.

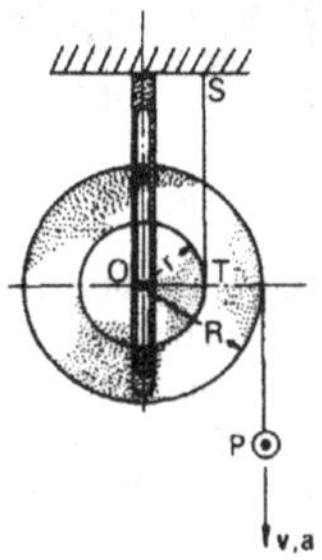

Problem 2.47.

2.48. One end of a rod of length l is pinned at A to a gear G that rolls on its horizontal rack with angular velocity ω and angular acceleration $\dot{\omega}$. The other end of the rod is pinned to a slider B which is constrained to move in a horizontal slot. Determine the velocity and acceleration of B when A is in the position shown.

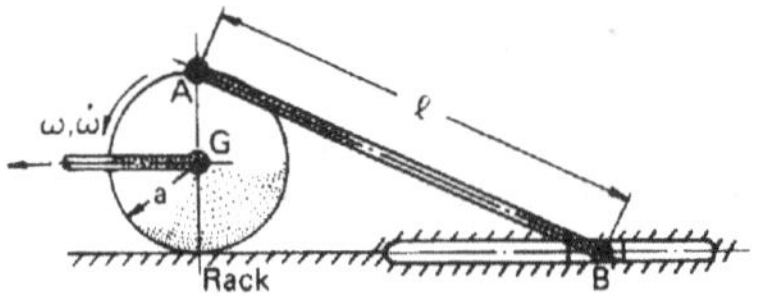

Problem 2.48.

2.49. The large G rotates counterclockwise with a constant angular speed of 4 rad/sec in $\Phi = \{O; \mathbf{i}_k\}$. The right-angled frame AOB is free to turn independently about O, and two planet gears that mesh with G are held in bearings at A and B. At the instant t_0, the frame has a clockwise angular speed of 5 rad/sec, which is decreasing at the rate of 15 rad/sec each second, relative to Φ. Determine relative to Φ the angular velocity $\boldsymbol{\omega}$ and the angular acceleration $\dot{\boldsymbol{\omega}}$ of the planet gears at the moment t_0.

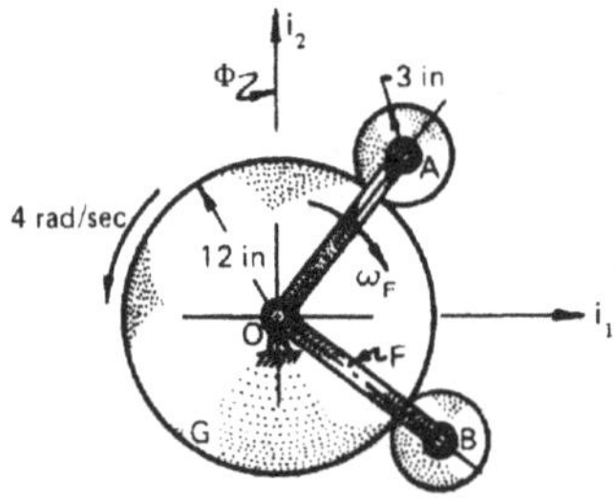

Problem 2.49.

2.50. Elliptic gears sometimes are used in variable speed drives, such as quick return mechanisms. To maintain driving contact without slip, the gears must be identical and each must turn about one of its foci, as shown in the figure. If the gear A rotates counterclockwise with a constant angular speed ω_A, show that the maximum angular speed of the gear B is given by $\omega_B = \omega_A(D/C - 1)$. What is its minimum angular speed?

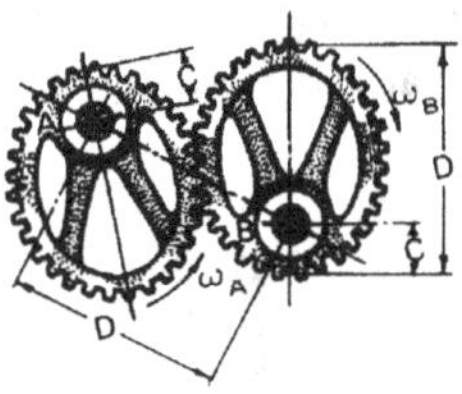

Problem 2.50.

2.51. Two plane, curved rigid bodies have angular speeds ω_A and ω_B about fixed centers at A and B in their plane. Let $\mathbf{x}_A$ and $\mathbf{x}_B$ denote the corresponding position vectors from A and B to the instantaneous point of contact C between the bodies. (a) Show that in driving contact with slip the ratio of the angular speeds is equal to the inverse ratio of the projections of the corresponding position vectors upon the tangent line to the contacting surfaces at C. (b) If the driving contact between the bodies is maintained without slip, show that, in addition to the previous result, the ratio of the angular speeds is equal to the inverse ratio of the projections of the position vectors upon the normal line to the contacting surfaces at C.

2.52. The connecting rod AB of the mechanism shown in the figure is pinned at A to a gear G of radius 8 cm and at B to a slider block that drives the device so that the gear rolls on its horizontal rack. During the interval of interest, the slider block, initially at rest at C, moves with acceleration $\mathbf{a}(B, t) = 18t\mathbf{i}$ cm/sec^2 in $\Phi = \{0; \mathbf{i}_k\}$. Find the angular velocities $\boldsymbol{\omega}_1$ and $\boldsymbol{\omega}_2$ of the rod and the gear, respectively, after 2 sec. What are their angular accelerations at that time?

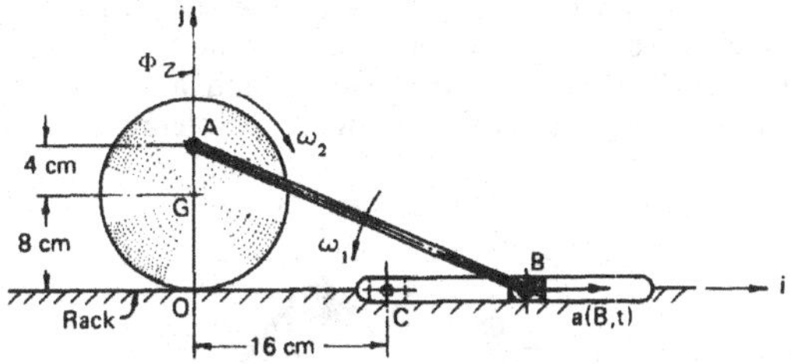

Problem 2.52.

2.53. A rigid rod of length 25 cm, pinned at 0 and B, controls the motion of a slider B through a rack and pinion arrangement shown in the figure. During an interval of interest, the center 0 of the pinion gear G of radius 5 cm is moved with a constant velocity $\mathbf{v}_0 = 40\mathbf{i}$ cm/sec in the fixed frame $\Phi = \{F; \mathbf{i}_k\}$. At the instant shown, the rack gear has a speed of 20 cm/sec which is increasing toward the left at the rate of 15 cm/sec each second. (a) Find the angular velocity and acceleration of G. (b) What is the absolute acceleration of the contact point C on G? (c) Determine the absolute velocity of the slider. (d) Find the angular velocity $\boldsymbol{\Omega}$ of the rod $0B$.

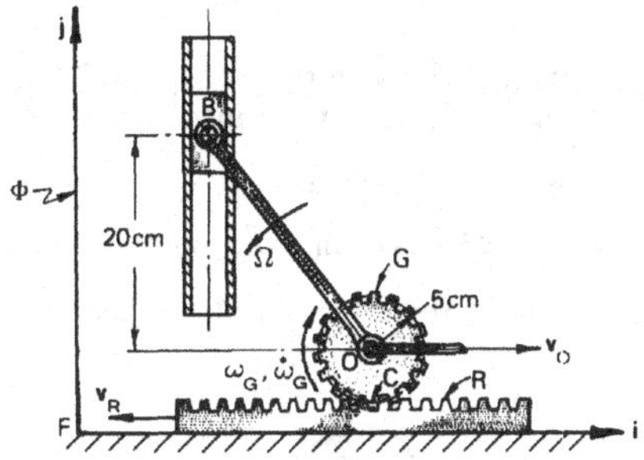

Problem 2.53.

2.54. The slider block B drives the device shown in the figure so that the gear G of radius 6 cm rolls on its horizontal rack. In 2 sec, B moves from its initial rest position at C to the position shown; and during this time, it has the acceleration $\mathbf{a}(B, t) =$

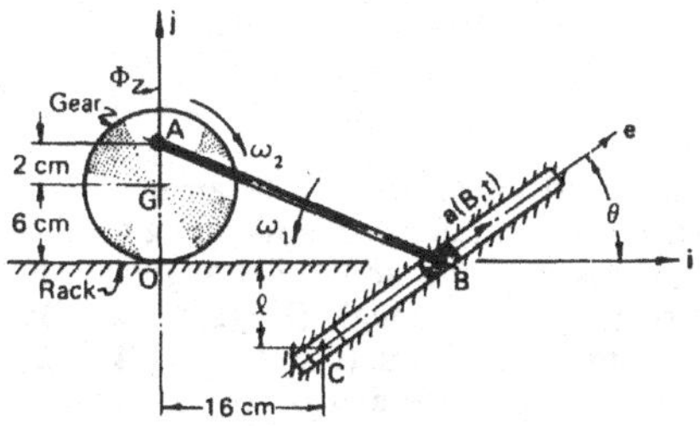

Problem 2.54.

$12t^2\mathbf{e}$ cm/sec^2 in $\Phi = \{O; \mathbf{i}_k\}$. The design specifications for the machine require that the angular speeds of the rod AB and the gear G be the same at the time shown. Find as functions of θ the angular velocities $\boldsymbol{\omega}_1$ and $\boldsymbol{\omega}_2$ of the rod and the gear, respectively, at the time of interest; and determine the values of θ and l needed to meet the design condition. Show how the value of θ may be estimated graphically, and use this trial result to compute its value more precisely.

2.55. The drive wheel D of a mechanism shown in the figure turns at a constant angular rate $\omega_1 = 6$ rad/sec in the ground frame $\Phi = \{F; \mathbf{n}_k\}$. The gear G rolls on the fixed ring gear R. Determine the velocity and acceleration in Φ of the center of G. What are the angular velocity and angular acceleration of G in Φ?

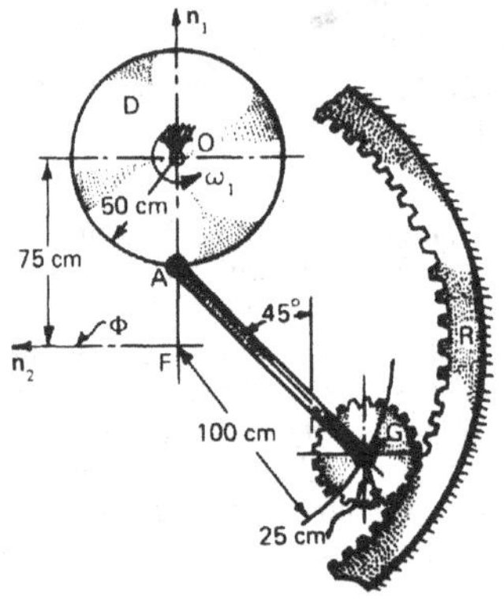

Problem 2.55.

2.56. Two points A and B of a rigid body are 20 cm apart on a line parallel to the vector $\mathbf{u} = 3\mathbf{i} + 4\mathbf{j}$ in $\Phi = \{O; \mathbf{i}_k\}$. At the instant t_0,

$$\mathbf{v}_A = 16\mathbf{i} - 8\mathbf{j} + 5\mathbf{k} \text{ cm/sec}, \qquad \mathbf{v}_B = -16\mathbf{i} + 16\mathbf{j} + 30\mathbf{k} \text{ cm/sec},$$

$$4\mathbf{a}_A = 75\mathbf{i} + 24\mathbf{k} \text{ cm/sec}^2, \quad \text{and} \quad \dot{\boldsymbol{\omega}} = 2\mathbf{i} + 4\mathbf{j} \text{ rad/sec}^2.$$

If the component of $\boldsymbol{\omega}$ in the direction of the line is zero, find the acceleration of B at t_0.

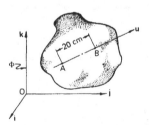

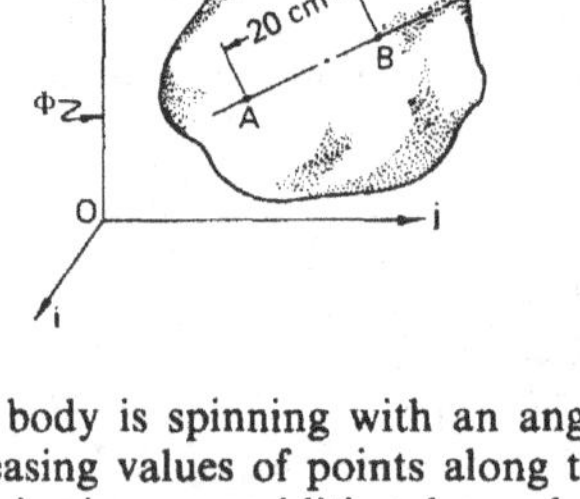

Problem 2.56.

2.57. A rigid body is spinning with an angular speed of 20 rad/sec in the right-hand side of increasing values of points along the line $2x - 4 = 2y - 6 = \sqrt{2}z$ in $\Phi = \{F; \mathbf{i}_k\}$. The body is given an additional angular velocity $\boldsymbol{\omega} = 4\mathbf{i} - 2\mathbf{j} - 5\sqrt{2}\mathbf{k}$ rad/sec about an intersecting line through $Q = (2, 3, 0)$ in Φ. The velocity of the particle at $\mathbf{X} = 10\mathbf{i} + 3\mathbf{j}$ ft from F is $\mathbf{v}_P = 25(\mathbf{i} + \sqrt{2}\mathbf{j}) - 12\mathbf{k}$ ft/sec. Find the velocity of Q, and derive the equation of the new axis of rotation in Φ.

2.58. An inextensible belt CD is used to transmit nonslip rotary motion between two nonintersecting, perpendicular axles of a pulley system shown in the figure. During the start-up period, the pulley A has a constant angular acceleration $\dot{\boldsymbol{\omega}}_A$, and after 10 sec it attains the angular velocity $\boldsymbol{\omega}_A = (1800/\pi)\mathbf{k}$ rpm in $\Phi = \{A; \mathbf{i}_k\}$. (a) Determine for this instant the velocity and acceleration in Φ of the point D of the belt, and the acceleration in Φ of the rim particle P on the pulley B, as shown, in measure units of m/sec^2. (b) More generally, let r_A and r_B denote the respective radii of the pulleys, and derive expressions relating the angular speeds and accelerations of the pulleys for transmission of the motion without slip.

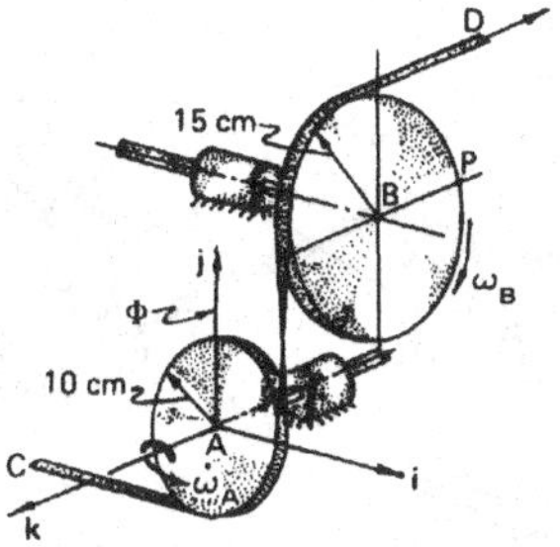

Problem 2.58.

2.59. Two points P and Q of a rigid body at the instant t_0 are located at

$$\mathbf{X}(P, t_0) = 2\mathbf{i} + 3\mathbf{j} + 4\mathbf{k}\ \text{m}, \qquad \mathbf{X}(Q, t_0) = 3\mathbf{i} + 4\mathbf{j} + \mathbf{k}\ \text{m}$$

in $\Phi = \{F; \mathbf{i}_k\}$, and their corresponding velocities are given by $\mathbf{v}(P, t_0) = \mathbf{i} + 4\mathbf{j} + 2\mathbf{k}$ m/sec, $\mathbf{v}(Q, t_0) = 2\mathbf{i} + 3\mathbf{j} + v\mathbf{k}$ m/sec. (a) Determine the unknown component v. (b) The angular velocity of the body has a component in the direction from P to Q equal to $\sqrt{11}$ rad/sec at t_0. Compute the angular velocity of the body at this instant.

2.60. The figure describes a space mechanism having a rigid, triangular crankshaft OAC fixed in vertical bearings at O and C, and rotating with a constant angular velocity $\boldsymbol{\omega}_1 = 10\mathbf{j}$ rad/sec. The connecting rod BD is held at B in a smooth ball socket set in a sleeve bearing that is free to turn on AC, while the end D may rotate in a sleeve bearing on the piston pin. The piston oscillates in its cylinder without rotation of its connecting rod about its axis. At the end of its compression stroke, the piston and its drive linkage are in the configuration shown. (a) Determine at this moment the

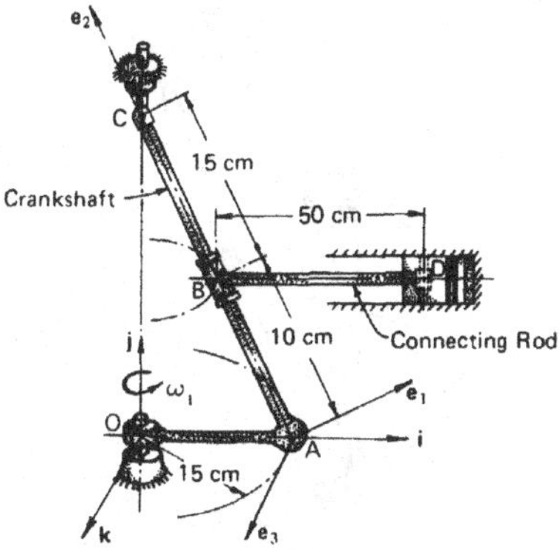

Problem 2.60.

velocities and accelerations of the center points A, B, and D, the angular velocity and acceleration of the connecting rod, and the angular velocity of the crankshaft referred to the imbedded frame $\varphi = \{A; \mathbf{e}_k\}$ indicated. (b) Repeat the analysis for the conditions when the crankshaft has advanced 90° more.

2.61. The crank OA of a four-bar space linkage mechanism has a constant angular speed $\omega = 50$ rad/sec, as illustrated. Find the velocity of the center of the ball joint B, and compute the angular speed Ω of the follower crank CB at the moment shown. Assume that the coupler link AB has no component of angular velocity about its own axis.

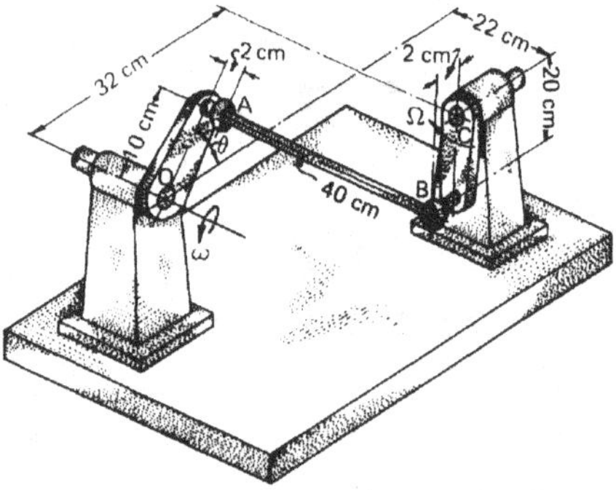

Problem 2.61.

2.62. The crank AB turns about the horizontal axis with a time-varying angular speed $\omega_1(t)$ in $\Phi = \{F; \mathbf{I}_k\}$. The connecting link BC is held in smooth ball joints at its ends, and its motion drives the crank CD to rotate about the vertical axis. Find in Φ the angular velocities and accelerations of the connecting link BC and the crank CD for the configuration shown.

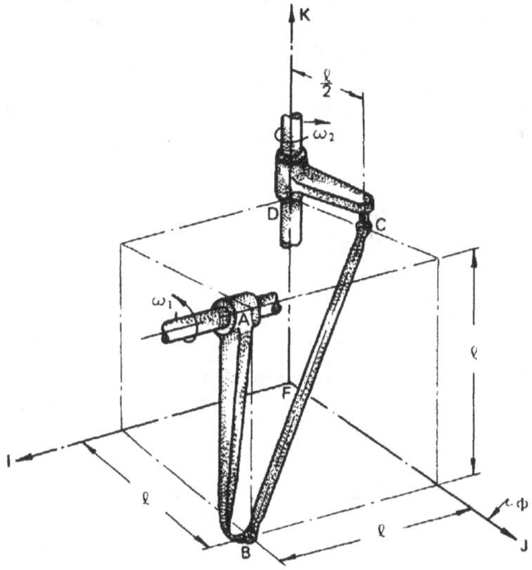

Problem 2.62.

2.63. A gear G of radius r rolls on a fixed circular gear track of radius R, as indicated. Find the velocity and acceleration of the center A of gear G, referred to $\varphi = \{O; \mathbf{t}, \mathbf{n}\}$.

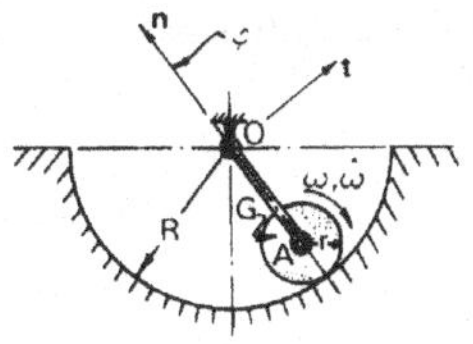

Problem 2.63.

2.64. A thin rigid rod of length l slides in the plane so that its end B has a constant speed $v = 100$ cm/sec along the ground, toward the right. What is the velocity of the point of the rod in contact with the edge A of the supporting wall for the configuration shown?

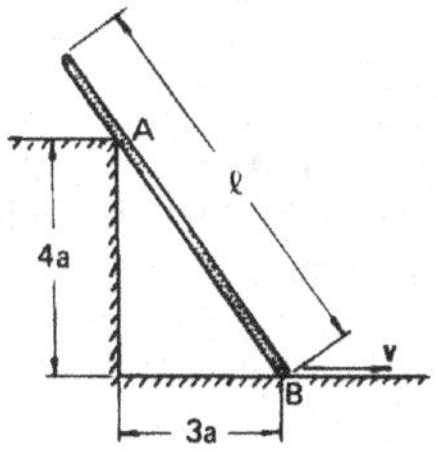

Problem 2.64.

2.65. During a period of its motion, the slider block B has a constant speed of 20 ft/sec directed as shown. A slotted drive link is hinged at B and constrained to slide on a fixed guide pin A. (a) Determine the angular velocity of the drive link AB as a function of the angle θ. (b) Find the velocity and acceleration of the point on the center line of the link instantaneously coincident with the center of the pin when $\theta = 45°$.

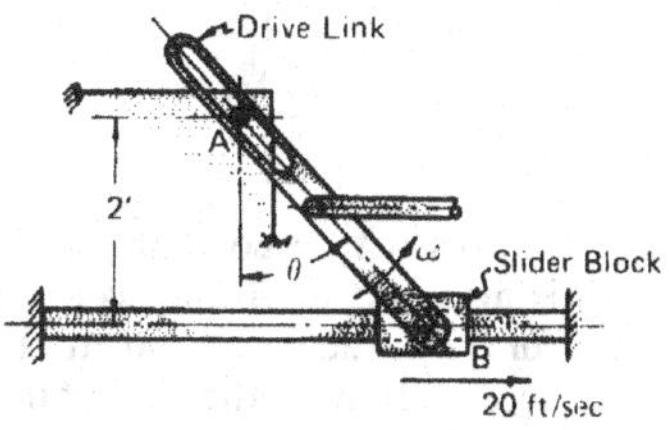

Problem 2.65.

2.66. The end A of a thin rigid bar moves in the plane with a constant speed v along a semicircular wall of radius r, as shown. Find as a function of θ the velocity of the point B of the rod coincident with the edge of the wall, and determine the angular velocity of the rod.

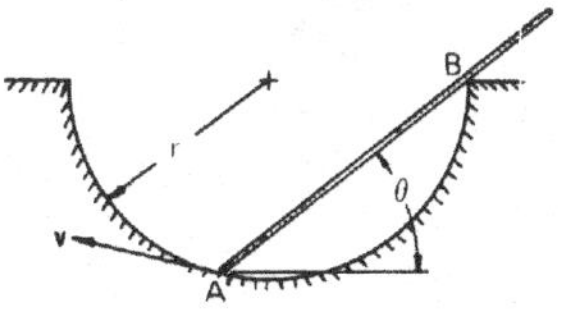

Problem 2.66.

2.67. The figure shows a gear G at the instant t_0 rolling on its horizontal rack so that its center has the velocity $\mathbf{v}_G = 32\mathbf{i}$ cm/sec; the connecting link AB has an angular acceleration $\dot{\omega}_1 = 0.8\mathbf{k}$ rad/sec^2; and the slider block B has acceleration $\mathbf{a}_B = 48\mathbf{i}$ cm/sec^2 in frame $\Phi = \{O; \mathbf{i}_k\}$. Determine the intrinsic velocity and acceleration of the pin center A, and find the radius of curvature of its path at the time t_0. What is the angular acceleration of G at t_0?

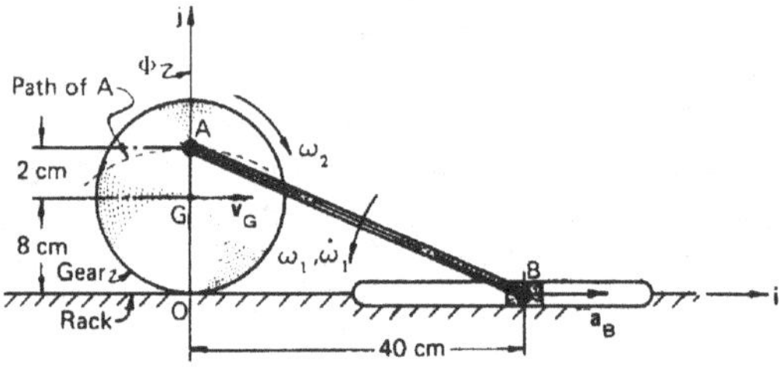

Problem 2.67.

2.68. A gear mechanism consists of a rack R and pinion P of radius $a = 2.5$ cm. The rack is hinged to a slider A that moves in a horizontal slot with constant speed $v_A = 0.25$ m/sec toward the right. Determine for the instant shown the angular velocity and acceleration of the pinion.

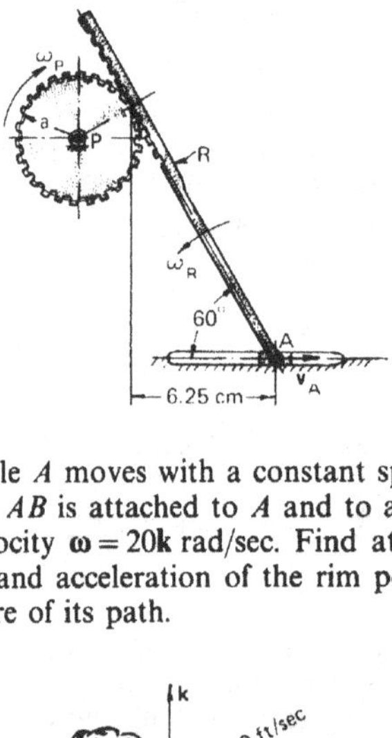

Problem 2.68.

2.69. A vehicle A moves with a constant speed of 40 ft/sec up a plane inclined at 30°. A rigid shaft AB is attached to A and to a drive gear B which turns with a constant angular velocity $\boldsymbol{\omega} = 20\mathbf{k}$ rad/sec. Find at the moment shown in the figure the intrinsic velocity and acceleration of the rim point P on the gear, and determine the radius of curvature of its path.

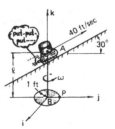

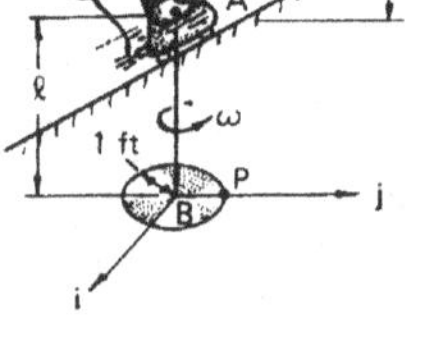

Problem 2.69.

2.70. Prove the invariant projection and parallel axis theorems for rigid body motions. (See Section 2.10.)

2.71. Prove that in a general plane motion of a rigid body there always exist one and only one base point which momentarily is at rest. Therefore, this point, which need not be within the body, is the instantaneous center of rotation for the plane motion. Find its location from an assigned base point. What happens to the instantaneous center of rotation as ω approaches zero?

2.72. In an arbitrary plane motion of a rigid body, let the velocities of two particles O and P in the plane be given at an instant t_0. Show that the intersection point C of lines drawn in the plane perpendicular to $\mathbf{v}_O$ and $\mathbf{v}_P$ at O and P is the instantaneous center of rotation. Describe how you would use this information to find graphically the speed v_Q of any other point Q at r from C at the moment t_0 shown in the figure. What is the angular velocity of the body at t_0?

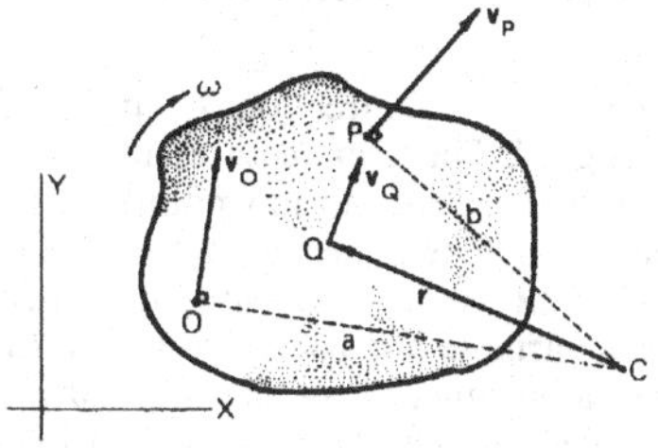

Problem 2.72.

2.73. Apply the results of the last problem to find in terms of assigned quantities the location of the instantaneous center of rotation for each of the following cases:

a. the connecting rod shown in Problem 2.19.
b. the train wheel illustated in Problem 2.36.
c. the coupler BC described in Problem 2.29. Find the angular velocity of BC at t_0.
d. the rod AB in the figure of Problem 2.22. What are the angular velocity of AB and the velocity of A at the position shown?

2.74. Apply the method of instantaneous centers to solve Problem 2.48. Find the instantaneous center both graphically and analytically.

2.75. Construct graphically the instantaneous center of rotation of the coupler bar BC of the four-bar linkage described in Problem 2.31. How many instantaneous centers does a four-bar linkage have? Where are those for Problem 2.31 located? Show that the connecting rod in Problem 2.19 may be viewed as a special case of a four-bar linkage. How many instantaneous centers does it possess? Use the method of instantaneous centers to find the angular speeds of the links BC and CD, and the speed of joint C of the four-bar linkage.

2.76. Locate in terms of assigned quantities the instantaneous center of rotation of the rack shown in Problem 2.68. Determine the angular velocity of the rack at the moment of interest. What is the corresponding angular velocity of the pinion?

2.77. Determine in terms of the assigned quantities the location of the instantaneous center of rotation for the rod described in Problem 2.20. (a) Use the instantaneous center to determine the angular velocity, and find the angular acceleration of the rod as a function of ψ. What is the velocity of point A as a function of ψ? (b) Let

$\Phi = \{F; \mathbf{I}, \mathbf{J}\}$ be a frame fixed in space at the corner of the wall with $\mathbf{J}$ directed vertically; and let $\varphi = \{B; \mathbf{i}, \mathbf{j}\}$ be another frame fixed in the rod with $\mathbf{j}$ directed from B toward A. Derive the standard equations of the locus of instantaneous centers referred to Φ and to φ. These curves are the space centrode and the body centrode for the motion of the rod, respectively. Sketch and label these curves. What is the angular velocity with which the body centrode rolls upon the space centrode for this motion?

2.78. Solve the last problem for the case when the angle $\theta \equiv \angle AFB$ between the wall and the ground may be greater than a right angle. Assume that the rod moves with its ends constrained to the lines FA and FB, and B has the constant velocity $\mathbf{v} = v\mathbf{I}$, as before. What can be said about the case when θ is less than 90°?

2.79. Apply the method of instantaneous centers to determine for the conditions in Problem 2.66 the angular velocity of the rod and its sliding velocity at the supporting edge of the cylindrical wall. Derive the standard equations of the space and body centrodes, and identify them in a sketch for a typical position of the rod. What is the angular speed of the body centrode as it rolls upon the space centrode? (See Problem 2.77.)

2.80. The figure shows the free foci C and D of identical elliptical wheels connected by a link of length l. The gear teeth at the extremities of the major axis suffice to carry the wheels past the dead motion points that occur when the link axis coincides with the line between the fixed axles at A and B. The rotation of wheel A, which has a constant angular velocity $\boldsymbol{\omega}_A$, is transmitted by driving contact without slip to wheel B, which acquires a variable angular velocity $\boldsymbol{\omega}_B$. (a) Use the method of instantaneous centers to find $\boldsymbol{\omega}_B$ as a function of the angle θ and the assigned parameters. Check your result by comparison with the solution to Problem 2.50. (b) What is the angular velocity $\boldsymbol{\omega}_L$ of the link as a function of θ? (c) Determine the angular speed ratios ω_B/ω_A and ω_L/ω_A in terms of the eccentricity defined by $e \equiv d/l$. Hint: It may be helpful to recall the basic property of an ellipse that $|AQ| + |QC| = l$.

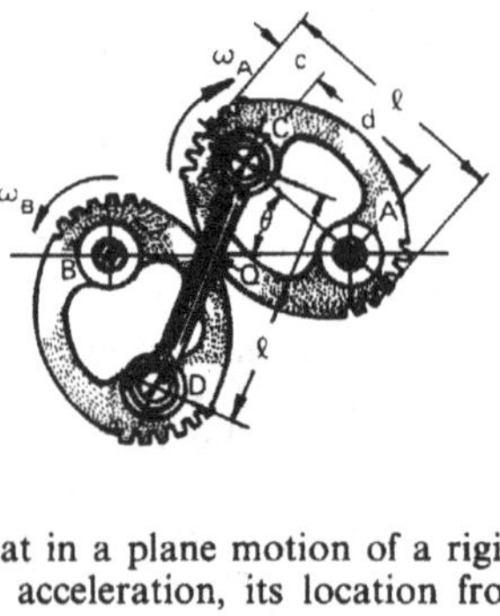

Problem 2.80.

2.81. Prove that in a plane motion of a rigid body there is just one point having zero instantaneous acceleration, its location from the base point O in the plane of motion being given by

$$\mathbf{x} = \frac{\omega^2 \mathbf{a}_O + \dot{\boldsymbol{\omega}} \times \mathbf{a}_O}{\omega^4 + \dot{\omega}^2}.$$

2.82. A wheel of radius r rolls without slipping along a straight horizontal track, its center O having a constant acceleration $\mathbf{a}_O = a\mathbf{i}$ parallel to the track. Find the position relative to O of the point of the wheel that has zero instantaneous acceleration. What is the locus relative to O of all such points? (See Problem 2.81.)

2.83. Let the velocity of a base point O and the angular velocity of the body be assigned so that the velocity of any particle P of a rigid body is determined by (2.115a). And let $\mathbf{v}_n$ and $\mathbf{v}_\alpha$ denote the component vectors of the velocity $\mathbf{v}_O$ normal and parallel, respectively, to the instantaneous axis of rotation defined by $\boldsymbol{\omega}$. Use this decomposition and apply the parallel axis theorem to construct an alternative proof of Chasles' theorem on screw motions. What interpretations may be assigned to screw motions having zero or infinite pitch?

2.84. The angular velocity of the body in the motion described in Problem 2.56 is $\boldsymbol{\omega} = \mathbf{i} - 0.75\mathbf{j} + 2\mathbf{k}$ rad/sec. Find at the moment of interest the instantaneous screw motion of the body, i.e., determine the pitch of the screw and the location of the screw axis from point A.

3

Finite Rigid Body Displacements

3.1. Introduction

Our study of rigid body motion thus far has focused mainly on equations (2.27) and (2.30) for the velocity and acceleration of a body particle. Although they actually represent nothing more than the application of the defining equations (1.8) and (1.9) to the special class of rigid body motions, these rudimentary equations were found to be especially useful because they separate and exhibit clearly the translational and rotational parts of the body's motion in space, and the terms have simple geometrical and physical interpretations. We have seen in several examples that a major advantage of these relations is that the usual differentiation operations have been replaced almost entirely by simple vector algebraic operations. Moreover, as shown in Examples 2.3 and 2.4, equations (2.27) and (2.30) also may be used to obtain information when only the values of the vectors at a particular instant of time and not the vector functions themselves are given in a problem. We shall see in the next chapter that these basic equations play a particularly important role in the description of motion referred to a moving reference frame.

Presently, however, we wish to recall that their derivation evolved from the basic relation (2.7) for the displacement of a rigid body due to a finite rotation about a fixed line, and from the application of primary theorems due to Euler and Chasles for finite rigid body displacements. The proof of Euler's fundamental theorem, Chasles' theorem on screw displacements, and some other results concerning the composition of finite rigid body displacements will be presented in this chapter.

The study of finite rigid body displacements is simplified considerably by application of the elementary properties of matrices and of certain linear transformations called tensors. It is assumed that the reader is familiar with the topics on matrices reviewed in Appendix B. The fundamental properties of (Cartesian) tensors will be introduced following a brief description of the use of index notation and the summation rule. It will be shown later that the rotator is, in fact, a tensor; and we shall learn how the angle and axis of the rotation may be easily found when the components of the rotator are known. The rotator will be linked to another important tensor, called the rotation tensor, that eventually will enable us to relate the rotation of a rigid body about a point to its rotation about a line in the proof of Euler's remarkable theorem. These developments will play a central role in the investigation of the proof of Chasles' notable screw theorem and in the study of the composition of several rotations of a rigid body.

Some of the topics that follow concern subject matter that has grown increasingly important because of intense industrial interest in the design and precise control of finite robot motions and their relation to the study of human body motions, for example. Therefore, the reader may find that understanding the major theorems and learning how to solve some simple problems involving several finite rigid body displacements will prove useful in his future study of topics in other areas, such as biomechanics, robotics, mechanical design analysis, the mechanics of solids and fluids, and applied mathematics. Several examples and applications of the theory will be presented in the text, and numerous additional illustrations are provided for the student in the Problems at the end. We shall begin with some notational matters.

3.2. Index Notation

Index notation will be used in future applications to abbreviate vector components, to identify the elements of matrices, and to represent the components of certain operators called tensors, which will be introduced later. In every case, a principal advantage of the index notation is that it facilitates the writing of repetitious equations in a way that enables one to see at a glance the overall structure of these equations. The three equations

$$a_1 + b_1 = c_1, \qquad a_2 + b_2 = c_2, \qquad a_3 + b_3 = c_3,$$

for example, may be abbreviated by *index notation* to read

$$a_j + b_j = c_j,$$

wherein the *range* of the index $j = 1, 2, 3$. The index j has no special significance—it is only a label that denotes collectively any one of the numbers in its range. Thus, the expression $a_k + b_k = c_k$ for $k = 1, 2, 3$ represents the same set of equations above. On the other hand, without further specification of how the indices are to be chosen from their range, an equation of the sort $a_k + b_i = c_j$ conveys no meaning whatsoever. The same applies to quantities with two or more indices. For instance,

$$A_{ij} + B_{ij} = C_{ij} \quad \text{and} \quad A_{pq} + B_{pq} = C_{pq}$$

represent the same set of nine equations, while $A_{ij} + B_{kp} = C_{pi}$, without further explicit instructions, is meaningless.

3.2.1. The Delta and Permutation Symbols and the Summation Rule

Let us recall that an orthonormal triple of basis vectors $\mathbf{e}_k$ has the following properties [cf. Appendix A, equations (A.4) and (A.10)]:

$$\mathbf{e}_1 \cdot \mathbf{e}_1 = 1, \qquad \mathbf{e}_1 \cdot \mathbf{e}_2 = \mathbf{e}_2 \cdot \mathbf{e}_1 = 0, \ldots, \tag{3.1a}$$

$$\mathbf{e}_1 \times \mathbf{e}_1 = \mathbf{0}, \qquad \mathbf{e}_1 \times \mathbf{e}_2 = -\mathbf{e}_2 \times \mathbf{e}_1 = \mathbf{e}_3, \ldots. \tag{3.1b}$$

These products may be conveniently summarized with the use of the *Kronecker delta* and *permutation* symbols, respectively, defined by

$$\delta_{ij} = \begin{cases} 1 & \text{if } i = j \text{ (no sum)}, \\ 0 & \text{if } i \neq j, \end{cases} \tag{3.2}$$

$$\varepsilon_{ijk} = \begin{cases} +1 & \text{if } ijk = 123, 312, \text{ or } 231, \\ -1 & \text{if } ijk = 321, 132, \text{ or } 213, \\ 0 & \text{otherwise.} \end{cases} \tag{3.3}$$

Expressed in these terms, all 18 relations in (3.1a) and (3.1b) may be written briefly as follows:

$$\mathbf{e}_i \cdot \mathbf{e}_j = \delta_{ij}, \tag{3.4a}$$

$$\mathbf{e}_i \times \mathbf{e}_j = \varepsilon_{ijk} \mathbf{e}_k \tag{3.4b}$$

in which each index i, j, k may have any value in the range 1–3, and in (3.4b) we have introduced the following useful rule.

Summation Rule. *An index that appears twice and only twice in any term is to be summed over its range from* 1 *to* 3, *unless explicitly directed otherwise.*

Let us verify equations (3.4). Upon setting $i = 1$ and $j = 2$ in (3.4a) and using (3.2), we see that $\mathbf{e}_1 \cdot \mathbf{e}_2 = \delta_{12} = 0$. And, similarly, $\mathbf{e}_2 \cdot \mathbf{e}_1 = \delta_{21} = 0$. With $i = j = 1$, we find $\mathbf{e}_1 \cdot \mathbf{e}_1 = \delta_{11} = 1$. These results agree with (3.1a), as required.

Further, putting $i=1$ and $j=2$ in (3.4b), summing over the range of the repeated index k, and using (3.3), we obtain, for example,

$$\mathbf{e}_1 \times \mathbf{e}_2 = \varepsilon_{12k}\mathbf{e}_k = \varepsilon_{121}\mathbf{e}_1 + \varepsilon_{122}\mathbf{e}_2 + \varepsilon_{123}\mathbf{e}_3 = \mathbf{e}_3.$$

Similarly,

$$\mathbf{e}_2 \times \mathbf{e}_1 = \varepsilon_{21k}\mathbf{e}_k = \varepsilon_{213}\mathbf{e}_3 = -\mathbf{e}_3 \quad \text{and} \quad \mathbf{e}_1 \times \mathbf{e}_1 = \varepsilon_{11k}\mathbf{e}_k = 0.$$

It is seen that these relations agree with (3.1b). It is evident from (3.2), (3.3), and the applications in (3.4) that the delta and permutation symbols have the following properties regarding the transposition of their adjacent indices:

$$\delta_{ij} = \delta_{ji}, \tag{3.5a}$$

$$\varepsilon_{ijk} = -\varepsilon_{jik} = \varepsilon_{jki} = -\varepsilon_{kji} = \cdots. \tag{3.5b}$$

These rules are very useful in the manipulation of expressions involving index notation. Their application will be met below.

3.2.2. Some Additional Applications of Index Notation

Any symbol may be used as an index; and, unless explicitly mentioned otherwise, it is understood that the range of all indices is 3. It is especially important to recognize that the particular repeated index used to emphasize summation is unimportant so long as the summation rule is not invalidated. Since the summation index may be replaced by any symbol without altering its meaning, the summation index is called a *dummy index*; it always says: "Sum on me!" An unrepeated index is known as a *free index*; it may be assigned any value in its range.

The application of the index rules is easily demonstrated in the derivation of the well-known formula for the squared magnitude of the vector $\mathbf{v}$. To derive this relation by index notation, we shall first recall that any vector $\mathbf{v}$ may be written as

$$\mathbf{v} = v_k\mathbf{e}_k \tag{3.6a}$$

$$= v_1\mathbf{e}_1 + v_2\mathbf{e}_2 + v_3\mathbf{e}_3 \tag{3.6b}$$

in which

$$v_k = \mathbf{v}\cdot\mathbf{e}_k \qquad \text{for} \quad k = 1, 2, 3 \tag{3.7}$$

are the three scalar components of $\mathbf{v}$ relative to the basis $\mathbf{e}_k$. [Cf. Appendix A, equations (A.1) and (A.5).] Then, we form the scalar product as usual but use the index notation in (3.6a) and recall (3.4a) and (3.2) to deduce

$$\mathbf{v}\cdot\mathbf{v} = v_j\mathbf{e}_j\cdot v_k\mathbf{e}_k = v_jv_k\mathbf{e}_j\cdot\mathbf{e}_k = v_jv_k\delta_{jk} = v_jv_j = v_kv_k. \tag{3.8}$$

Plainly, $v_jv_j = v_1v_1 + v_2v_2 + v_3v_3 = v_1^2 + v_2^2 + v_3^2$. It is clear too that v_kv_k has precisely the same meaning. And so does v_pv_p. Similarly, $\mathbf{v} = v_q\mathbf{e}_q$ is equivalent to (3.6a).

Also with (3.2), we have $\delta_{ij}v_j = \delta_{i1}v_1 + \delta_{i2}v_2 + \delta_{i3}v_3$; hence, for $i = 2$, $\delta_{2j}v_j = v_2$. Indeed, it is evident that in the sum on the index j nothing is obtained until the number j on v_j is the same as the number i on δ_{ij}. That is, in general,

$$\delta_{ij}v_j = v_i. \tag{3.9}$$

This result was used in (3.8) to write $\delta_{jk}v_k = v_j$. Alternatively, $\delta_{jk}v_j = v_k$ also may be used in (3.8). We thus see from these few easy examples that another major advantage of index notation is that it simplifies considerably manipulations of terms in equations.

Finally, it is easy to show with the use of (3.4b) and (3.6a) that the vector product of two vectors $\mathbf{u}$ and $\mathbf{v}$ may be written in the compact index form

$$\mathbf{u} \times \mathbf{v} = \varepsilon_{pqr}u_pv_q\mathbf{e}_r. \tag{3.10}$$

When this formula, with the aid of (3.3), is expanded by the sums indicated, the reader will obtain the familiar result (A.11) given in Appendix A. Notice that in (3.10) and in the first expression in (3.8) care was exercised to avoid repeating the same index more than twice, as this would invalidate the sum rule and create confusion.

Additional applications of the index notation will be encountered below, and others may be found throughout Appendix B. It is expected that the reader is familiar with the elementary matrix operations reviewed there.

3.3. Introduction to Tensors

A tensor may be thought of as a special mathematical machine whose inputs are vectors and whose outputs are other vectors. We may recall, for example, that the rotator $\mathbf{T}$ defined in (2.9) transforms the initial position vectors of particles of a rigid body into their finite displacement vectors by the rule (2.8), namely, $\mathbf{Tx} = \mathbf{d}$; so, on the surface, it appears that the rotator may be a tensor. However, like any machine, our tensor machine functions only in accordance with certain operating instructions that distinguish it from other kinds of mathematical machines. More precisely,* a *tensor* $\mathbf{T}$ is a linear

* In other books, the tensor defined by (3.11) usually is identified as a 2-tensor or a tensor of rank 2 in order to distinguish it from other kinds of multilinear transformations leading to tensors of higher rank. In this ranking of tensors, a scalar is regarded as a tensor of rank 0 and a vector is a tensor of rank 1. However, since there will be no need in this book for tensors of rank greater than 2, the brief though less accurate term "tensor" will be used.

operator that transforms a given vector $\mathbf{u}$ into another vector $\mathbf{Tu}=\mathbf{w}$ such that the following rules hold:

$$\mathbf{T}(\mathbf{u}+\mathbf{v})=\mathbf{Tu}+\mathbf{Tv}, \qquad \mathbf{T}(\lambda\mathbf{u})=\lambda(\mathbf{Tu}) \tag{3.11}$$

where $\mathbf{u}$ and $\mathbf{v}$ are any vectors and λ is any scalar. As indicated above, tensors will be represented by capital letters and printed in bold face type, like vectors. We may recall that the rotator in (2.9) obeys the rule (2.11), which is the combination of the rules (3.11); therefore, we see that the rotator really is a tensor. But more about that later on. For the present, let us focus attention on some simple operating rules that tensors must satisfy and discover some other properties shared by all tensor quantities.

The *zero tensor*, denoted by $\mathbf{0}$, and the *identity tensor*, written as $\mathbf{1}$, are defined by the rules

$$\mathbf{0v}=\mathbf{0}, \tag{3.12a}$$

$$\mathbf{1v}=\mathbf{v}, \tag{3.12b}$$

which hold for *all* vectors $\mathbf{v}$. In other words, the zero tensor transforms every vector into the zero vector; and the identity tensor leaves every vector unchanged.

There are other kinds of tensors that may transform special vectors into the zero vector. Consider, for example, the special tensor $\mathbf{S}$ for which

$$\mathbf{Sv}=\boldsymbol{\Omega}\times\mathbf{v}, \tag{3.13}$$

wherein $\boldsymbol{\Omega}$ is a given vector. This rule plainly satisfies the requirements (3.11), so this special transformation $\mathbf{S}$ really is a tensor. Then, in particular, we may have $\mathbf{Sv}=\mathbf{0}$ for every vector $\mathbf{v}$ parallel to $\boldsymbol{\Omega}$, and no others. So, unlike the rule (3.12), $\mathbf{S}\neq\mathbf{0}$ because this special operator does not transform *every* vector $\mathbf{v}$ into the zero vector.

The *sum* of two tensors and the *scalar multiples* of a tensor are defined by the relations

$$(\mathbf{S}+\mathbf{T})\mathbf{v}=\mathbf{Sv}+\mathbf{Tv}, \qquad (\lambda\mathbf{T})\mathbf{v}=\lambda(\mathbf{Tv}). \tag{3.14}$$

The reader may show that the transformations $\mathbf{S}+\mathbf{T}$ and $\lambda\mathbf{T}$ satisfy the rules (3.11) that qualify them as tensors.

3.3.1. Components of a Tensor

The scalar components of a tensor and the matrix of a tensor will be defined next. Afterwards, some examples that demonstrate the computation of tensor components and the description of their related matrices will be presented.

Index notation is particularly useful in representing the components of tensor quantities. Let $\mathbf{e}_i$ be an orthonormal basis. The set of nine numbers defined by the rule

$$T_{ij} \equiv \mathbf{e}_i \cdot \mathbf{T}\mathbf{e}_j \tag{3.15}$$

are called the (*Cartesian*) *components* of the tensor **T** with respect to the basis $\mathbf{e}_i$. These components may be written in a square matrix

$$[\mathbf{T}] \equiv T = [T_{ij}] = \begin{bmatrix} T_{11} & T_{12} & T_{13} \\ T_{21} & T_{22} & T_{23} \\ T_{31} & T_{32} & T_{33} \end{bmatrix}, \tag{3.16}$$

which is called the *matrix of the tensor* **T** with respect to the basis $\mathbf{e}_i$.

Let us apply (3.15) in a few preliminary examples to derive the components of some tensors. For our first example, we may consider the zero tensor in (3.12). Putting $\mathbf{v} = \mathbf{e}_j$ in (3.12a) and using the definition (3.15), we see that $\mathbf{e}_i \cdot \mathbf{0}\mathbf{e}_j = \mathbf{e}_i \cdot \mathbf{0} = 0$ for every pair of directions $\mathbf{e}_i$; that is, all nine of the components of the zero tensor are zero in every basis, which is to be expected.

The rule (3.12b) yields a more interesting result. In a similar manner, we find with the use of (3.4a)

$$\mathbf{e}_i \cdot \mathbf{1}\mathbf{e}_j = \mathbf{e}_i \cdot \mathbf{e}_j = \delta_{ij}. \tag{3.17}$$

Thus, the nine components of the identity tensor are the same with respect to every orthonormal basis $\mathbf{e}_i$; and they are identified by the Kronecker delta symbol (3.2) whose corresponding matrix is the usual identity matrix I:

$$[\mathbf{1}] \equiv I = [\delta_{ij}] = \begin{bmatrix} 1 & 0 & 0 \\ 0 & 1 & 0 \\ 0 & 0 & 1 \end{bmatrix}. \tag{3.18}$$

For a final introductory example, let us determine the components of the tensor **S** in (3.13). Writing $\mathbf{v} = \mathbf{e}_j$ and using (3.15), (3.4b), and (3.7), we derive

$$\mathbf{e}_i \cdot \mathbf{S}\mathbf{e}_j = \mathbf{e}_i \cdot \boldsymbol{\Omega} \times \mathbf{e}_j = \boldsymbol{\Omega} \cdot \mathbf{e}_j \times \mathbf{e}_i = \varepsilon_{jik} \boldsymbol{\Omega} \cdot \mathbf{e}_k = \varepsilon_{jik} \Omega_k;$$

that is, with (3.3),

$$S_{ij} = -\varepsilon_{ijk} \Omega_k. \tag{3.19}$$

Thus, more explicitly, the matrix of the tensor $\mathbf{S}$ with respect to the basis $\mathbf{e}_i$ is

$$[\mathbf{S}] = [S_{ij}] = \begin{bmatrix} 0 & -\Omega_3 & \Omega_2 \\ \Omega_3 & 0 & -\Omega_1 \\ -\Omega_2 & \Omega_1 & 0 \end{bmatrix}. \tag{3.20}$$

The components of other tensors and their matrices may be found in a similar manner.

3.3.2. The Tensor Product

Let us recall the representation (3.6) for vectors and the manner in which the scalar components with respect to the basis $\mathbf{e}_i$ are defined in (3.7). In view of (3.15), it seems natural that we should seek a similar kind of representation for a tensor in terms of its components and the basis vectors used to define them. As a first step toward accomplishing this, we shall introduce in this section the tensor product of a pair of vectors. In addition, the tensor product will provide another easy operational tool to use with our tensor machine.

If $\mathbf{a}$ and $\mathbf{b}$ are given vectors, their *tensor product* $\mathbf{a} \otimes \mathbf{b}$ may be defined by the requirement that

$$(\mathbf{a} \otimes \mathbf{b})\mathbf{c} = \mathbf{a}(\mathbf{b} \cdot \mathbf{c}) \tag{3.21}$$

hold for *all* vectors $\mathbf{c}$. It is easy to verify that $\mathbf{T} = \mathbf{a} \otimes \mathbf{b}$ obeys (3.11), which qualifies it as a tensor. (See Problem 3.10.) The tensor product is not commutative, because $\mathbf{a}(\mathbf{b} \cdot \mathbf{c}) \neq \mathbf{b}(\mathbf{a} \cdot \mathbf{c})$; hence,

$$\mathbf{a} \otimes \mathbf{b} \neq \mathbf{b} \otimes \mathbf{a}. \tag{3.22}$$

Moreover, (3.21) implies that the tensor product is *associative* with respect to multiplication by a scalar λ and *distributive* with respect to addition of vectors $\mathbf{b}$ and $\mathbf{d}$:

$$\begin{gathered} \lambda(\mathbf{a} \otimes \mathbf{b}) = (\lambda\mathbf{a}) \otimes \mathbf{b} = \mathbf{a} \otimes (\lambda\mathbf{b}), \\ \mathbf{a} \otimes (\mathbf{b} + \mathbf{d}) = \mathbf{a} \otimes \mathbf{b} + \mathbf{a} \otimes \mathbf{d}, \qquad (\mathbf{b} + \mathbf{d}) \otimes \mathbf{a} = \mathbf{b} \otimes \mathbf{a} + \mathbf{d} \otimes \mathbf{a}. \end{gathered} \tag{3.23}$$

Thus, with respect to the basis $\mathbf{e}_k$, these properties reveal that in terms of the vector components, the tensor product may be written

$$\mathbf{a} \otimes \mathbf{b} = a_j \mathbf{e}_j \otimes b_k \mathbf{e}_k = (a_j b_k)(\mathbf{e}_j \otimes \mathbf{e}_k), \tag{3.24}$$

wherein (3.6a) has been recalled. Moreover, it is easy to see by the application of (3.15) in (3.24) and use of (3.7) that $a_j b_k$ are the components of the tensor $\mathbf{a} \otimes \mathbf{b}$; for,

$$\mathbf{e}_j \cdot (\mathbf{a} \otimes \mathbf{b})\, \mathbf{e}_k = \mathbf{e}_j \cdot \mathbf{a}(\mathbf{b} \cdot \mathbf{e}_k) = a_j b_k$$

For the orthonormal basis $\mathbf{e}_k$, (3.21) and (3.4a) yield

$$(\mathbf{e}_i \otimes \mathbf{e}_j)\, \mathbf{e}_k = \mathbf{e}_i(\mathbf{e}_j \cdot \mathbf{e}_k) = \delta_{jk} \mathbf{e}_i. \tag{3.25}$$

This rule often is useful in derivations involving tensor components. We notice, for example, that it yields the relation

$$\mathbf{e}_p \cdot (\mathbf{e}_i \otimes \mathbf{e}_j)\, \mathbf{e}_k = (\mathbf{e}_p \cdot \mathbf{e}_i)(\mathbf{e}_j \cdot \mathbf{e}_k) = \delta_{pi} \delta_{jk}.$$

Example 3.1. The rule (3.21) may be used to express the vector triple product in a tensorial form that is useful in applications. The result is derived from the expansion rule for the triple product of vectors $\mathbf{u}$, $\mathbf{v}$, and $\mathbf{w}$ given in the Appendix A:

$$\mathbf{u} \times (\mathbf{v} \times \mathbf{w}) = (\mathbf{u} \cdot \mathbf{w})\mathbf{v} - (\mathbf{u} \cdot \mathbf{v})\mathbf{w} \qquad [\text{cf. (A.14)}].$$

With the use of (3.12b) and (3.21), it is easily seen that this identity may be expressed in the tensorial form

$$\mathbf{u} \times (\mathbf{v} \times \mathbf{w}) = [(\mathbf{u} \cdot \mathbf{w})\mathbf{1} - \mathbf{w} \otimes \mathbf{u}]\mathbf{v}. \tag{3.26}$$

Hence, the triple product may be viewed as the transformation of a vector $\mathbf{v}$ in accordance with the rule

$$\mathbf{T}\mathbf{v} = -\mathbf{u} \times (\mathbf{w} \times \mathbf{v}) \qquad \text{with} \quad \mathbf{T} \equiv (\mathbf{u} \cdot \mathbf{w})\mathbf{1} - \mathbf{w} \otimes \mathbf{u}. \tag{3.27}$$

3.3.3. The Tensor Representation Theorem

The tensor product $\mathbf{a} \otimes \mathbf{b}$ of two vectors in (3.24) is represented in terms of its nine components $a_j b_k$ and nine tensor products $\mathbf{e}_j \otimes \mathbf{e}_k$ of the corresponding basis vectors. This relation suggests that every tensor $\mathbf{T}$ may have a similar representation. We shall prove this conjecture in the tensor representation theorem below. Also, it will be shown how this representation may be used in manipulation of tensor equations and to express such equations in their component forms with respect to some chosen basis.

Since $\mathbf{e}_i$ is a basis, (3.25) implies that the tensor products $\mathbf{e}_i \otimes \mathbf{e}_j$ form a set of nine *linearly independent* tensor products such that

$$m_{ij} \mathbf{e}_i \otimes \mathbf{e}_j = \mathbf{0} \tag{3.28}$$

may hold if and only if all nine scalars $m_{ij} = 0$. Now, let us consider the set of ten tensors $\mathbf{T}$ and $\mathbf{e}_i \otimes \mathbf{e}_j$. Then there exist ten scalars λ and λ_{ij}, not all zero, such that

$$\lambda \mathbf{T} + \lambda_{ij} \mathbf{e}_i \otimes \mathbf{e}_j = \mathbf{0}.$$

In particular, $\lambda \neq 0$; otherwise, since all of the λ_{ij} are not zero, the foregoing equation would contradict (3.28) for the linear independence of the nine tensor products. Hence, upon writing $T_{ij} \equiv -\lambda_{ij}/\lambda$, we find the important result

$$\mathbf{T} = T_{ij} \mathbf{e}_{ij} \tag{3.29}$$

in which, by definition,

$$\mathbf{e}_{ij} \equiv \mathbf{e}_i \otimes \mathbf{e}_j. \tag{3.30}$$

The set of nine quantities $\mathbf{e}_{ij}$ is called a *tensor basis* associated with the orthonormal basis $\mathbf{e}_i$. It is easy to show that T_{ij} are the components of $\mathbf{T}$ with respect to $\mathbf{e}_i$. Moreover, the representation (3.29) is unique. For, if we suppose that there may be two such representations with respect to the same tensor basis, then their difference would have the form of (3.28). But the linear independence of the tensor basis would then imply that the two sets of components are, in fact, identical. We thus obtain the

Tensor Representation Theorem. *With respect to the orthonormal basis* $\mathbf{e}_i$, *every tensor* $\mathbf{T}$ *has the unique representation* (3.29) *in terms of its components and the linearly independent set of basis tensors* (3.30) *associated with* $\mathbf{e}_i$.

The representation rule (3.29) may now be used to express any tensor in terms of its components in a manner that is similar to that we have grown accustomed to using for vectors. The identity tensor, for example, with components (3.17) in every orthonormal basis, may now be written as

$$\mathbf{1} = \delta_{ij} \mathbf{e}_{ij} = \mathbf{e}_{jj} = \mathbf{e}_j \otimes \mathbf{e}_j. \tag{3.31}$$

Similarly, the tensor whose components are given in (3.19) may be expressed as

$$\mathbf{S} = -\varepsilon_{ijk} \Omega_k \mathbf{e}_{ij}. \tag{3.32}$$

With the aid of these representations, expansion of tensor relations or other manipulations may be carried out in a straightforward manner. For example, the products of these tensors with a given vector $\mathbf{v}$ may be readily derived:

$$\mathbf{1v} = (\mathbf{e}_k \otimes \mathbf{e}_k)\mathbf{v} = \mathbf{e}_k(\mathbf{e}_k \cdot \mathbf{v}) = v_k \mathbf{e}_k = \mathbf{v},$$

$$\mathbf{Sv} = -\varepsilon_{ijk} \Omega_k \mathbf{e}_{ij} \mathbf{v} = \varepsilon_{jik} \Omega_k \mathbf{e}_i(\mathbf{e}_j \cdot \mathbf{v}) = \varepsilon_{kji} \Omega_k v_j \mathbf{e}_i = \mathbf{\Omega} \times \mathbf{v},$$

wherein (3.21), (3.7), the permutation rule (3.3), and (3.10) have been recalled. We recognize, of course, that these are the same as (3.12b) and (3.13).

More importantly, it is readily seen from (3.25) and (3.29) that for any tensor **T**,

$$\mathbf{T}\mathbf{e}_k = T_{ij}(\mathbf{e}_i \otimes \mathbf{e}_j)\,\mathbf{e}_k = T_{ij}\delta_{jk}\mathbf{e}_i = T_{ik}\mathbf{e}_i. \tag{3.33}$$

And further, it is useful to note that for any vector **u**, we may write

$$\mathbf{w} = \mathbf{T}\mathbf{u} = T_{kp}\mathbf{e}_{kp}\mathbf{u} = T_{kp}\mathbf{e}_k(\mathbf{e}_p \cdot \mathbf{u}) = T_{kp}u_p\mathbf{e}_k. \tag{3.34}$$

Hence, the components of the transformed vector are given by

$$w_k = T_{kp}u_p, \tag{3.35}$$

which may be written as the matrix product $w = Tu$. An easy example is provided by (3.9), from which it is seen that $v = Iv$ is the matrix form of (3.12b).

3.3.4. Other Tensor Operations and Rules

Some additional tensor operations will be assembled here, and some principles governing their applications will be derived. Many of the rules are similar to those for matrices, and this will be pointed out along the way. The following topics will discuss the product of tensors, the trace and transpose operations, symmetric and skew tensors, the determinant and the inverse of a tensor, and orthogonal tensors. We shall start with the multiplication of tensors.

3.3.4.1. The Product of Tensors

The *product* $\mathbf{P} = \mathbf{S}\mathbf{T}$ of two tensors **S** and **T** is defined by the condition that

$$(\mathbf{S}\mathbf{T})\mathbf{v} = \mathbf{S}(\mathbf{T}\mathbf{v}) \tag{3.36}$$

hold for *all* vectors **v**. It is easily seen that (3.36) satisfies the primary rules (3.11), and that the components of the product tensor with respect to an orthonormal basis $\mathbf{e}_k$ are given by the product of the matrices of **S** and **T**:

$$P_{jk} = S_{ji}T_{ik}, \qquad [\mathbf{P}] = [\mathbf{S}][\mathbf{T}] \qquad \text{or} \quad P = ST. \tag{3.37}$$

Therefore, the usual rules for matrix multiplication apply to tensors. It is clear that the product of two tensors generally is *not commutative*: $\mathbf{ST} \neq \mathbf{TS}$. (See Problem 3.11.)

3.3.4.2. *The Trace Operation*

The *trace* of a tensor is an operation defined by the rules

$$\operatorname{tr}(\mathbf{S}+\mathbf{T}) = \operatorname{tr}\mathbf{S} + \operatorname{tr}\mathbf{T}, \tag{3.38a}$$

$$\operatorname{tr}(\lambda\mathbf{S}) = \lambda \operatorname{tr}\mathbf{S}, \tag{3.38b}$$

$$\operatorname{tr}(\mathbf{a}\otimes\mathbf{b}) = \mathbf{a}\cdot\mathbf{b}, \tag{3.38c}$$

in which λ is a scalar, $\mathbf{T}$ and $\mathbf{S}$ are tensors, and $\mathbf{a}$ and $\mathbf{b}$ are vectors. It follows easily from (3.38c) that

$$\operatorname{tr}(\lambda\mathbf{a}\otimes\mathbf{b}) = \operatorname{tr}(\mathbf{a}\otimes\lambda\mathbf{b}) = \lambda \operatorname{tr}(\mathbf{a}\otimes\mathbf{b}). \tag{3.39}$$

Introducing the representation of $\mathbf{T}$ in terms of its Cartesian components in (3.29), and applying the rules (3.38) and (3.4a), we obtain

$$\operatorname{tr}(T_{kl}\mathbf{e}_{kl}) = T_{kl} \operatorname{tr}\mathbf{e}_{kl} = T_{kl}\mathbf{e}_k\cdot\mathbf{e}_l = T_{kl}\delta_{kl}.$$

That is,

$$\operatorname{tr}\mathbf{T} = T_{kk}, \tag{3.40}$$

which is the same as the trace of the matrix of $\mathbf{T}$. We recall specifically for the identity tensor (3.31) that $\operatorname{tr}\mathbf{1} = \delta_{kk} = 3$; and it is evident that the usual rules for the trace of a square matrix hold for tensors. For example, the trace of the product of two tensors satisfies the commutative rule

$$\operatorname{tr}(\mathbf{ST}) = \operatorname{tr}(\mathbf{TS}). \tag{3.41}$$

3.3.4.3. *The Transpose of a Tensor*

The transformation $\mathbf{T}^T$ defined by the requirement that

$$\mathbf{T}^T\mathbf{u}\cdot\mathbf{v} = \mathbf{u}\cdot\mathbf{T}\mathbf{v} \tag{3.42}$$

hold for *every pair* of vectors $\mathbf{u}$, $\mathbf{v}$ is readily seen to obey the rules (3.11) qualifying it as a tensor. (See Problem 3.12.) This tensor $\mathbf{T}^T$ is called the *transpose* of the tensor T. The *transposition* operation satisfies the rules

$$\begin{gathered}(\mathbf{S}+\mathbf{T})^T = \mathbf{S}^T + \mathbf{T}^T, \qquad (\lambda\mathbf{S})^T = \lambda\mathbf{S}^T, \\ (\mathbf{a}\otimes\mathbf{b})^T = \mathbf{b}\otimes\mathbf{a}.\end{gathered} \tag{3.43}$$

To confirm the later, let $\mathbf{T} = \mathbf{a}\otimes\mathbf{b}$ in (3.42) and apply (3.21) to deduce

$$(\mathbf{a}\otimes\mathbf{b})^T\mathbf{u}\cdot\mathbf{v} = \mathbf{u}\cdot(\mathbf{a}\otimes\mathbf{b})\mathbf{v} = (\mathbf{u}\cdot\mathbf{a})(\mathbf{b}\cdot\mathbf{v}) = (\mathbf{b}\otimes\mathbf{a})\mathbf{u}\cdot\mathbf{v}.$$

Since this identity must hold for every pair of vectors **u**, **v**, we conclude the last relation in (3.43).

The components of $\mathbf{T}^T$ are easily obtained from (3.42). Let $\mathbf{u} = \mathbf{e}_j$ and $\mathbf{v} = \mathbf{e}_i$, and recall (3.15) to get

$$(\mathbf{T}^T)_{ij} \equiv \mathbf{e}_i \cdot \mathbf{T}^T \mathbf{e}_j = \mathbf{e}_j \cdot \mathbf{T}\mathbf{e}_i = T_{ji}. \tag{3.44}$$

This means that the element in the ith row and jth column of the matrix of the tensor $\mathbf{T}^T$ is equal to the element in the jth row and ith column of the matrix T of the tensor **T**. Hence, T^T denotes the usual transpose of the matrix T, and we write its components as $T^T_{ji} \equiv (T^T)_{ji}$. In fact, all of the familiar rules of transposition of matrices apply to tensors. Specifically,

$$\operatorname{tr} \mathbf{T}^T = \operatorname{tr} \mathbf{T}; \tag{3.45}$$

and the transpose of the product of two tensors **S** and **T** is equal to the reversed product of their transposed tensors:

$$(\mathbf{ST})^T = \mathbf{T}^T \mathbf{S}^T. \tag{3.46}$$

3.3.4.4. Symmetric and Antisymmetric Tensors

A tensor is said to be *symmetric* if $\mathbf{T}^T = \mathbf{T}$. This implies that the components of **T** satisfy $T_{ij} = T_{ji}$. Also, a tensor is said to be *skew* or *antisymmetric* if $\mathbf{T}^T = -\mathbf{T}$. This means that the components of **T** must satisfy $T_{ij} = -T_{ji}$, and hence all the diagonal elements of its matrix T are zero.

The identity tensor in (3.18), or (3.31), is an obvious example of a symmetric tensor; we recall that its components obey the rule (3.5a). The matrix in (3.20) shows at once that the tensor in (3.19), or (3.32), is an antisymmetric tensor. It may be seen with the aid of (3.5b) that $S_{ij} = -S_{ji}$. It should be clear that the rules appropriate to symmetric and skew matrices hold for tensors.

Notice that the tensors defined by

$$\mathbf{T}_S \equiv \tfrac{1}{2}(\mathbf{T} + \mathbf{T}^T), \qquad \mathbf{T}_A \equiv \tfrac{1}{2}(\mathbf{T} - \mathbf{T}^T), \tag{3.47}$$

are symmetric and antisymmetric, respectively; they are known as the *symmetric and antisymmetric parts* of **T**. Hence, every tensor may be decomposed uniquely into the sum of its symmetric and skew parts; for,

$$\mathbf{T} \equiv \mathbf{T}_S + \mathbf{T}_A. \tag{3.48}$$

(See Problem B.7.) The component equations corresponding to (3.47) are written as

$$\mathbf{T}_S = T_{(jk)} \mathbf{e}_{jk}, \qquad \mathbf{T}_A = T_{[jk]} \mathbf{e}_{jk}, \tag{3.49}$$

wherein, by definition,

$$T_{(jk)} \equiv \tfrac{1}{2}(T_{jk} + T_{kj}), \tag{3.50a}$$

$$T_{[jk]} \equiv \tfrac{1}{2}(T_{jk} - T_{kj}). \tag{3.50b}$$

This decomposition is often useful in physical applications because the symmetric and antisymmetric parts of tensors have important and distinct interpretations. In linear elasticity theory, for example, the deformation of a body is described by a symmetric tensor called the strain tensor, but this is accompanied by a local rigid body rotation which is described by an antisymmetric tensor named the rotation tensor. The sum of these tensors is the gradient of the relative displacement vector of neighboring material points, and it characterizes the kinematics of the entire local motion of a deformable solid body. We shall prove later in this chapter that the symmetric and skew parts of the rotator determine the angle and the axis of rotation, respectively, of a rigid body rotation.

The decomposition (3.48) is useful also because symmetric and antisymmetric tensors have special algebraic properties that may be separated in this way. It is easily shown with the use of (3.33) that any tensor satisfies the relations

$$\mathbf{e}_j \times \mathbf{T}\mathbf{e}_j = \mathbf{T}^T\mathbf{e}_j \times \mathbf{e}_j \quad \text{and} \quad \mathbf{e}_j \cdot \mathbf{T}\mathbf{e}_j = \mathbf{T}^T\mathbf{e}_j \cdot \mathbf{e}_j. \tag{3.51}$$

[See (3.42) also.] Hence, it follows readily from (3.47) that

$$\mathbf{e}_j \times \mathbf{T}_S\mathbf{e}_j = \mathbf{0} \quad \text{and} \quad \mathbf{e}_j \cdot \mathbf{T}_A\mathbf{e}_j = 0. \tag{3.52}$$

We note that the latter is simply the trivial identity $\operatorname{tr} \mathbf{T}_A = 0$. Moreover, it may be seen from the foregoing equations and (3.48) that (3.51) is equivalent to

$$\mathbf{e}_j \times \mathbf{T}\mathbf{e}_j = -\mathbf{T}\mathbf{e}_j \times \mathbf{e}_j = \mathbf{e}_j \times \mathbf{T}_A\mathbf{e}_j \tag{3.53}$$

and

$$\mathbf{e}_j \cdot \mathbf{T}\mathbf{e}_j = \mathbf{e}_j \cdot \mathbf{T}_S\mathbf{e}_j = \operatorname{tr} \mathbf{T}_S = \operatorname{tr} \mathbf{T}. \tag{3.54}$$

Another useful application will be illustrated next.

Example 3.2. Use of (3.47) and (3.42) yields the following symmetric and skew parts of the tensor $\mathbf{T}$ in (3.27):

$$\mathbf{T}_S = (\mathbf{u} \cdot \mathbf{w})\mathbf{1} - \tfrac{1}{2}(\mathbf{u} \otimes \mathbf{w} + \mathbf{w} \otimes \mathbf{u}), \tag{3.55a}$$

$$\mathbf{T}_A = \tfrac{1}{2}(\mathbf{u} \otimes \mathbf{w} - \mathbf{w} \otimes \mathbf{u}). \tag{3.55b}$$

It is seen from these relations that $\operatorname{tr}\mathbf{T}_A = 0$, which always is the case for a skew tensor. Thus, with the use of (3.48) and (3.38), we find

$$\operatorname{tr}\mathbf{T} = \operatorname{tr}\mathbf{T}_S = 3(\mathbf{u}\cdot\mathbf{w}) - \mathbf{u}\cdot\mathbf{w} = 2(\mathbf{u}\cdot\mathbf{w}).$$

In the special case when $\mathbf{u} = \mathbf{w}$ is a unit vector, it is seen that $\mathbf{T}_A = \mathbf{0}$ and

$$\mathbf{T} = \mathbf{T}_S = \mathbf{1} - \mathbf{u}\otimes\mathbf{u}, \qquad \operatorname{tr}(\mathbf{u}\otimes\mathbf{u}) = \mathbf{u}\cdot\mathbf{u} = 1, \qquad \operatorname{tr}\mathbf{T}_S = 2. \tag{3.56}$$

Hence, when $\mathbf{u}$ is a unit vector, (3.27) yields the special formula

$$\mathbf{T}\mathbf{v} = -\mathbf{u}\times(\mathbf{u}\times\mathbf{v}) = (\mathbf{1} - \mathbf{u}\otimes\mathbf{u})\mathbf{v}. \qquad \square \tag{3.57}$$

The Vector of a Skew Tensor. The relation (3.19) shows that only skew tensors may have the property (3.13). In fact, since a skew tensor has three nontrivial independent components in three-dimensional space, (3.13) shows that it is always possible to relate to each skew tensor $\mathbf{S}$ a vector $\boldsymbol{\Omega}$ whose scalar components are related to those of $\mathbf{S}$ in accordance with (3.20), namely,

$$\boldsymbol{\Omega} = S_{32}\mathbf{e}_1 + S_{13}\mathbf{e}_2 + S_{21}\mathbf{e}_3 = -S_{23}\mathbf{e}_1 - S_{31}\mathbf{e}_2 - S_{12}\mathbf{e}_3. \tag{3.58}$$

Indeed, it is easy to show that (3.13) always may be solved uniquely for $\boldsymbol{\Omega}$ in terms of $\mathbf{S}$. We set $\mathbf{v} = \mathbf{e}_j$ in (3.13). Then, we form the sum of vector products as indicated below, use the expansion rule (A.14) for the triple product, and recall (3.4a), (3.6),, and (3.7) to derive

$$\mathbf{e}_j\times\mathbf{S}\mathbf{e}_j = \mathbf{e}_j\times(\boldsymbol{\Omega}\times\mathbf{e}_j) = (\mathbf{e}_j\cdot\mathbf{e}_j)\boldsymbol{\Omega} - (\mathbf{e}_j\cdot\boldsymbol{\Omega})\mathbf{e}_j = 2\boldsymbol{\Omega}.$$

The component form of the left-hand side of this equation is derived by application of (3.33) and (3.4b):

$$\mathbf{e}_j\times\mathbf{S}\mathbf{e}_j = \mathbf{e}_j\times S_{pj}\mathbf{e}_p = \varepsilon_{jpk}S_{pj}\mathbf{e}_k = -\varepsilon_{pjk}S_{pj}\mathbf{e}_k.$$

We thus obtain the unique vector associated with the given skew tensor $\mathbf{S}$:

$$\boldsymbol{\Omega} = \tfrac{1}{2}\mathbf{e}_j\times\mathbf{S}\mathbf{e}_j \tag{3.59a}$$

$$= -\tfrac{1}{2}\varepsilon_{pjk}S_{pj}\mathbf{e}_k. \tag{3.59b}$$

This important vector is known as the *vector of the skew tensor*.

Bearing in mind that $\mathbf{S}$ is skew so that $S_{ij} = -S_{ji}$, when (3.59) is written out we shall obtain (3.58). Hence, to each skew tensor $\mathbf{S}$ there corresponds a unique vector $\boldsymbol{\Omega}$ given by (3.58) or (3.59) such that for every vector $\mathbf{v}$, we have

$$\mathbf{S}\mathbf{v} = \boldsymbol{\Omega}\times\mathbf{v}. \tag{3.60}$$

Thus, a skew tensor transforms every vector $\mathbf{v}$ into another vector $\mathbf{S}\mathbf{v}$ which is perpendicular to $\mathbf{v}$ and to the vector $\boldsymbol{\Omega}$ of the skew tensor. And, conversely, for each vector product $\boldsymbol{\Omega}\times\mathbf{v}$ of a given vector $\boldsymbol{\Omega}$ with an arbitrary vector $\mathbf{v}$ there exists a unique tensor

$$\mathbf{S} = -\varepsilon_{ijk}\Omega_k\mathbf{e}_{ij}. \tag{3.61}$$

such that (3.60) holds. The equations (3.13) and (3.32) recalled above have been rewritten as (3.60) and (3.61) for future convenience.

Example 3.3. Find the vector $\boldsymbol{\Omega}$ of the antisymmetric tensor defined in (3.55b).

Solution. The use of (3.24), (3.25), and (3.29) shows tnat the components of $\mathbf{T}_A$ in (3.55) are given by

$$(T_A)_{pj} = \mathbf{e}_p\cdot\mathbf{T}_A\mathbf{e}_j = \tfrac{1}{2}(u_p w_j - w_p u_j).$$

Hence, substitution of this relation into (3.59) and use of (3.5b) yields

$$\boldsymbol{\Omega} = \tfrac{1}{2}\mathbf{e}_j\times\mathbf{T}_A\mathbf{e}_j = -\tfrac{1}{4}\varepsilon_{pjk}(u_p w_j - w_p u_j)\,\mathbf{e}_k = -\tfrac{1}{2}\varepsilon_{pjk}u_p w_j\mathbf{e}_k;$$

that is, with (3.10), we find that

$$\boldsymbol{\Omega} = -\tfrac{1}{2}\mathbf{u}\times\mathbf{w}$$

is the vector of the skew tensor $\mathbf{T}_A = (\mathbf{u}\otimes\mathbf{w} - \mathbf{w}\otimes\mathbf{u})/2$.

As further illustration, let us observe that for an arbitrary vector $\mathbf{v}$, application of (3.60) and (A.14) yields

$$\mathbf{T}_A\mathbf{v} = \boldsymbol{\Omega}\times v = -\tfrac{1}{2}(\mathbf{u}\times\mathbf{w})\times\mathbf{v} = \tfrac{1}{2}[(\mathbf{w}\cdot\mathbf{v})\mathbf{u} - (\mathbf{u}\cdot\mathbf{v})\mathbf{w}];$$

and we recover the original relation in (3.55b):

$$\mathbf{T}_A\mathbf{v} = \tfrac{1}{2}(\mathbf{u}\otimes\mathbf{w} - \mathbf{w}\otimes\mathbf{u})\mathbf{v}.$$

3.3.4.5. The Determinant and the Inverse of a Tensor

The *determinant* det **T** of a tensor **T** is defined as the determinant of its Cartesian component matrix:

$$\det\mathbf{T} = \det[\mathbf{T}] = \det T. \tag{3.62}$$

The elementary rules for determinants are reviewed in Appendix B. In particular, the rule (B.17) holds for tensors:

$$\det(\mathbf{ST}) = (\det\mathbf{S})(\det\mathbf{T}). \tag{3.63}$$

If $\det\mathbf{T}\neq 0$, the tensor is said to be *nonsingular*, hence *invertible*;

otherwise, it is called *singular*. When $\mathbf{T}$ is invertible, there exists a unique tensor $\mathbf{T}^{-1}$, called the *inverse* of $\mathbf{T}$, for which

$$\mathbf{T}\mathbf{T}^{-1} = \mathbf{T}^{-1}\mathbf{T} = \mathbf{1} \tag{3.64}$$

holds. The components of $\mathbf{T}^{-1}$ are the corresponding elements of the inverse of the matrix of $\mathbf{T}$, namely, $T^{-1} = [\mathbf{T}]^{-1}$.

Similarly, other rules appropriate to invertible matrices apply also to tensors. For example, if $\mathbf{S}$ and $\mathbf{T}$ are invertible tensors, their product is invertible also, and

$$(\mathbf{S}\mathbf{T})^{-1} = \mathbf{T}^{-1}\mathbf{S}^{-1}. \tag{3.65}$$

In addition, the transpose of a nonsingular tensor is invertible, hence

$$(\mathbf{T}^{-1})^{T} = (\mathbf{T}^{T})^{-1}. \tag{3.66}$$

3.3.4.6. Orthogonal Tensors

Finally, an *orthogonal tensor* $\mathbf{Q}$ is a nonsingular tensor having the property

$$\mathbf{Q}^{-1} = \mathbf{Q}^{T}. \tag{3.67}$$

Hence, by (3.64) and (3.67), $\mathbf{Q}$ is an orthogonal tensor if and only if

$$\mathbf{Q}\mathbf{Q}^{T} = \mathbf{Q}^{T}\mathbf{Q} = \mathbf{1}. \tag{3.68}$$

Of course, $\det \mathbf{Q} = \pm 1$. When $\det \mathbf{Q} = +1$, $\mathbf{Q}$ is called *proper orthogonal*; otherwise, it is called *improper orthogonal*. Evidently, the rules for orthogonal matrices apply immediately to orthogonal tensors.

This concludes the introduction to tensors. Transformation rules that relate the components of a tensor with respect to different bases will be introduced further on; and other tensor properties will be discussed in connection with the moment of inertia tensor, which will play a fundamental role in our future study of the dynamics of rigid body motions. It may be noted, in addition, that stress, strain, and rate of strain are other examples of tensor quantities that the reader may expect to encounter in the study of solid and fluid mechanics, or more generally, in continuum mechanics. We shall find the foregoing rules useful in the description of finite rigid body rotations, which we shall begin to study next.

3.4. The Rotator

We have learned that the rotator (2.9) is a tensor that transforms the initial position vector $\mathbf{x}$ of a particle P of a rigid body into its finite displacement vector $\mathbf{d}(P)$ due to a rotation through an angle θ in a right-hand sense about a fixed unit direction $\boldsymbol{\alpha}$:

$$\mathbf{d}(P)=\mathbf{T}\mathbf{x}=\sin\theta\,\boldsymbol{\alpha}\times\mathbf{x}+(1-\cos\theta)\boldsymbol{\alpha}\times(\boldsymbol{\alpha}\times\mathbf{x})\qquad [\text{cf. }(2.7)].$$

We are going to show that the vector operator (2.9) may now be replaced by a more useful tensor operator.

Let $\mathbf{S}$ be the skew tensor whose vector $\boldsymbol{\Omega}\equiv\boldsymbol{\alpha}$; and, in (3.57), in write the unit vector $\mathbf{u}\equiv\boldsymbol{\alpha}$. Then application of (3.57) and (3.60) in (2.7) yields

$$\mathbf{d}=\mathbf{T}\mathbf{x}\equiv[\mathbf{S}\sin\theta-(1-\cos\theta)(\mathbf{1}-\boldsymbol{\alpha}\otimes\boldsymbol{\alpha})]\mathbf{x}. \tag{3.69}$$

This identity must hold for all vectors $\mathbf{x}$. It thus follows that the rotator has the unique representation

$$\mathbf{T}=\mathbf{S}\sin\theta+(1-\cos\theta)(\boldsymbol{\alpha}\otimes\boldsymbol{\alpha}-\mathbf{1}) \tag{3.70}$$

in terms of its symmetric and antisymmetric parts given by (3.47):

$$\mathbf{T}_S=(1-\cos\theta)(\boldsymbol{\alpha}\otimes\boldsymbol{\alpha}-\mathbf{1}), \tag{3.71a}$$

$$\mathbf{T}_A=\mathbf{S}\sin\theta, \tag{3.71b}$$

and wherein, by (3.61),

$$\mathbf{S}=S_{ij}\mathbf{e}_{ij}=-\varepsilon_{ijk}\alpha_k\mathbf{e}_{ij}. \tag{3.72}$$

Let the reader show that (3.69) also may be expressed in terms of θ and $\mathbf{S}$ by the relation

$$\mathbf{T}=\mathbf{S}\sin\theta+\mathbf{S}^2(1-\cos\theta),$$

in which $\mathbf{S}^2\equiv\mathbf{S}\mathbf{S}=\boldsymbol{\alpha}\otimes\boldsymbol{\alpha}-\mathbf{1}$. Thus notice that the square of a skew tensor is symmetric.

If values are assigned to θ and $\boldsymbol{\alpha}$, the rotator may be found from (3.70) and expressed in terms of its components with respect to any assigned reference frame $\psi=\{0;\mathbf{e}_k\}$. It is easily shown that (3.15) and (3.70) yield the components

$$T_{ij}=-\varepsilon_{ijk}\alpha_k\sin\theta+(1-\cos\theta)(\alpha_i\alpha_j-\delta_{ij}) \tag{3.73}$$

with respect to frame ψ. The end result may be written as the matrix (3.16) and the matrix product $d=Tx$ may then be used to find in ψ the displacement vector (2.7). The converse problem is more interesting.

Suppose that the rotator is known with respect to a frame ψ. Then (3.70) may be used to find both the angle and the axis of rotation in the terms of the components of $\mathbf{T}$ in ψ. Since $\operatorname{tr} \mathbf{S} = 0$, (3.70) or (3.71a) yields

$$\operatorname{tr} \mathbf{T} = \operatorname{tr} \mathbf{T}_S = 2(\cos\theta - 1). \tag{3.74}$$

[See (3.54) also.] Hence, the angle of rotation is given by

$$\cos\theta = 1 + \tfrac{1}{2} \operatorname{tr} \mathbf{T}. \tag{3.75}$$

The angle of rotation of the body clearly is independent of the reference frame that eventually may be used to determine $\operatorname{tr} \mathbf{T}$ in a particular case. Therefore, $\operatorname{tr} \mathbf{T}$ must have the same value in every reference frame. This important fact will be proved in another way in our later study of the transformation properties of tensors.

In view of (3.53), it is seen that the axial vector of the skew tensor may be found by application of (3.59a) and (3.71b). First, we have

$$\mathbf{e}_j \times \mathbf{T}\mathbf{e}_j = \mathbf{e}_j \times \mathbf{T}_A \mathbf{e}_j = \sin\theta\, \mathbf{e}_j \times \mathbf{S}\mathbf{e}_j = 2\boldsymbol{\alpha} \sin\theta.$$

We may ignore a *trivial rotation* through an angle $\theta = 2n\pi$, $n = 0, 1, 2, \ldots$; and we may assume momentarily that $\theta \neq \pm\pi$. It then follows that the axis of rotation is determined by

$$\boldsymbol{\alpha} = \tfrac{1}{2} \csc\theta\, \mathbf{e}_j \times \mathbf{T}\mathbf{e}_j. \tag{3.76}$$

The case $\theta = \pm\pi$ must be treated separately. We use (3.70) or (3.71) to write

$$\mathbf{T} = \mathbf{T}_S = 2(\boldsymbol{\alpha} \otimes \boldsymbol{\alpha} - \mathbf{1}); \tag{3.77a}$$

hence,

$$\boldsymbol{\alpha} \otimes \boldsymbol{\alpha} = \mathbf{1} + \tfrac{1}{2}\mathbf{T}. \tag{3.77b}$$

The diagonal elements of the corresponding matrices of (3.77b) determine the squares of each of the components of $\boldsymbol{\alpha}$, which also must satisfy $\boldsymbol{\alpha} \cdot \boldsymbol{\alpha} = 1$; and the nondiagonal elements may then be used to fix the signs of the components of $\boldsymbol{\alpha}$. Of course, either $\pm\boldsymbol{\alpha}$ may be used as the solution. In any event, $\boldsymbol{\alpha}$ may be found easily for the special case when $\theta = \pm\pi$.

The reader may show that the component forms of (3.75) and (3.76) are given by

$$\cos\theta = 1 + \tfrac{1}{2} T_{kk} \tag{3.78a}$$

and

$$\alpha_k = -\tfrac{1}{2} \csc\theta\, \varepsilon_{ijk} T_{ij}; \tag{3.78b}$$

and (3.77b) yields

$$\alpha_i \alpha_j = \delta_{ij} + \tfrac{1}{2} T_{ij} \qquad \text{for} \quad \theta = \pm\pi. \tag{3.79}$$

It is useful to observe that when (3.78b) is expanded, the component form of (3.76) may be written as

$$\boldsymbol{\alpha} = -\csc\theta[\mathbf{e}_1 T_{[23]} + \mathbf{e}_2 T_{[31]} + \mathbf{e}_3 T_{[12]}], \tag{3.80}$$

wherein we recall (3.50b). For example, $T_{[23]} \equiv (T_{23} - T_{32})/2$.

Equations (3.75) and (3.76), or (3.80), determine θ and $\boldsymbol{\alpha}$ to within their signs. However, a right-hand rotation through an angle $\hat{\theta} = -\theta$, or equivalently, $2\pi - \theta$, about an axis $\hat{\boldsymbol{\alpha}} = -\boldsymbol{\alpha}$ clearly is equivalent to a right-hand rotation through an angle θ about the axis $\boldsymbol{\alpha}$. In the former case, we visualize in Fig. 3.1 that the rotation through a negative angle about a negative direction is just the mirror reflection of the latter; hence, the pair $\{-\theta, -\boldsymbol{\alpha}\}$ is called the *image rotation* of the pair $\{\theta, \boldsymbol{\alpha}\}$. Since these describe equivalent rotations, it is clear that (3.75) and (3.76), hence (3.80), determine the angle and axis of rotation uniquely. Moreover, it is plain that every rotational displacement may be accomplished by a rotation through an angle which is not greater than a straight angle; so, no generality is sacrificed by our requiring henceforward that $0 \leqslant \theta \leqslant \pi$.

Example 3.4. Find the matrix of the rotator for the rotation of the satellite tracking antenna described in Example 2.1.

Solution. The angle and axis of rotation in $\Phi = \{F; \mathbf{I}_k\}$ are given by

$$\cos\theta = 4/5, \tag{3.81a}$$

$$\sin\theta = 3/5, \tag{3.81b}$$

$$\boldsymbol{\alpha} = \frac{\sqrt{2}}{2}(\mathbf{I} + \mathbf{J}). \tag{3.81c}$$

Hence, the matrix of the components of $\mathbf{T}$ may be obtained by substitution of these values into (3.73). This exercise is left for the reader. We shall consider

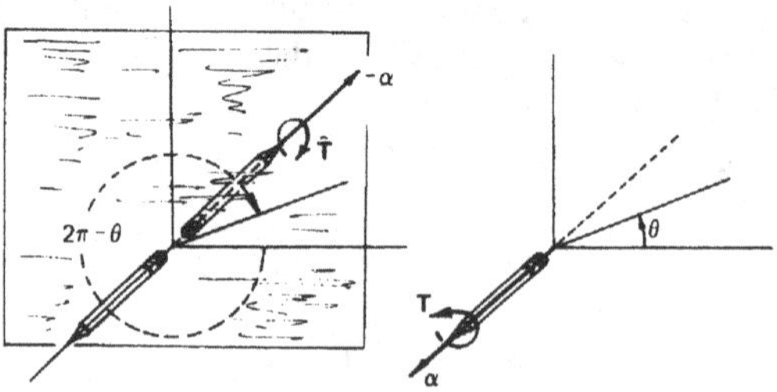

By (3.70), $\hat{\mathbf{T}} \equiv \mathbf{T}(\hat{\theta}, \hat{\alpha}) = \mathbf{T}(-\theta, -\alpha) \overset{\text{or}}{=} \mathbf{T}(2\pi - \theta, \alpha) = \mathbf{T}(\theta, \alpha)$.

Figure 3.1. An equivalent image rotation.

an alternative solution that illustrates some of the tensor relations studied earlier.

Substitution of (3.81a) and (3.81b) into (3.70) provides

$$\mathbf{T} = \tfrac{3}{5}\mathbf{S} + \tfrac{1}{5}(\boldsymbol{\alpha} \otimes \boldsymbol{\alpha} - \mathbf{1}). \tag{3.82}$$

We note that $\{\mathbf{e}_k\} \equiv \{\mathbf{I}_k\}$ so that $\mathbf{e}_{ij} = \mathbf{I}_{ij} = \mathbf{I}_i \otimes \mathbf{I}_j$. Thus, recalling (3.24), we find with (3.81c)

$$\boldsymbol{\alpha} \otimes \boldsymbol{\alpha} = \alpha_i \alpha_j \mathbf{e}_{ij} = \tfrac{1}{2}(\mathbf{I}_{11} + \mathbf{I}_{12} + \mathbf{I}_{21} + \mathbf{I}_{22});$$

(3.31) becomes

$$\mathbf{1} = \mathbf{e}_{kk} = \mathbf{I}_{11} + \mathbf{I}_{22} + \mathbf{I}_{33};$$

and (3.72) yields

$$\mathbf{S} = -\varepsilon_{ijk}\alpha_k \mathbf{e}_{ij} = \frac{\sqrt{2}}{2}(\mathbf{I}_{13} - \mathbf{I}_{31} + \mathbf{I}_{32} - \mathbf{I}_{23}).$$

Then assembling these three expressions in (3.82) yields the result needed. We thus obtain

$$\mathbf{T} = T_{kl}\mathbf{I}_{kl} = \frac{3\sqrt{2}}{10}(\mathbf{I}_{13} - \mathbf{I}_{31} + \mathbf{I}_{32} - \mathbf{I}_{23}) - \frac{1}{10}(\mathbf{I}_{11} + \mathbf{I}_{22} + 2\mathbf{I}_{33} - \mathbf{I}_{12} - \mathbf{I}_{21}).$$

Hence, the matrix of $\mathbf{T}$ in frame Φ may be read from this equation as follows:

$$T = [T_{kl}] = \frac{1}{10}\begin{bmatrix} -1 & 1 & 3\sqrt{2} \\ 1 & -1 & -3\sqrt{2} \\ -3\sqrt{2} & 3\sqrt{2} & -2 \end{bmatrix}. \tag{3.83}$$

Conversely, let us suppose that $\mathbf{T}$ is given by (3.83); and let us verify the original data by application of (3.75) and (3.80). With $\operatorname{tr}\mathbf{T} = -4/10$, we have

$$\cos\theta = 1 + \tfrac{1}{2}\operatorname{tr}\mathbf{T} = 1 - \tfrac{1}{5} = \tfrac{4}{5}.$$

Of course, $\sin\theta = 3/5$; and with (3.83) in (3.80), we find

$$\boldsymbol{\alpha} = -\frac{5}{6}\cdot\frac{1}{10}[\mathbf{I}_1(-3\sqrt{2} - 3\sqrt{2}) + \mathbf{I}_2(-3\sqrt{2} - 3\sqrt{2})] = \frac{\sqrt{2}}{2}(\mathbf{I} + \mathbf{J}).$$

It is seen that the angle and axis of rotation agree with the original conditions in (3.81). □

Example 3.5. An alternative design of the aircraft door panel mechanism illustrated in Problem 2.6 requires that the curved link must execute a finite rotation about a fixed line so that the rotator in frame $\Phi = \{0; \mathbf{i}_k\}$ is given by the matrix

$$T = \begin{bmatrix} -1 & 1 & 0 \\ 0 & -1 & -1 \\ -1 & 0 & -1 \end{bmatrix}. \tag{3.84}$$

The initial position of the guide pin P in the revised design is $\mathbf{x} = 3\mathbf{i} - 5\mathbf{j} + 4\mathbf{k}$ ft from O. Find the angle and axis of the rotation, and determine the displacement of P. What is the terminal location of P in Φ?

Solution. The angle of rotation is obtained by (3.75). With (3.84), we find tr $T = -3$; hence, $\cos\theta = -1/2$, and $\theta = 120°$ or $240°$. However, as mentioned above, nothing is lost by our always choosing the smaller angle; so we shall fix $\theta = 120°$. Then use of (3.84) and $\sin\theta = \sqrt{3}/2$ in (3.80) yields the axis of rotation:

$$\boldsymbol{\alpha} = \frac{1}{\sqrt{3}}(\mathbf{i} + \mathbf{j} - \mathbf{k}). \tag{3.85}$$

Let the reader show that for $\hat{\theta} = 2\pi - \theta = 240°$, $\sin\hat{\theta} = -\sqrt{3}/2$ and the sign of $\boldsymbol{\alpha}$ is reversed, as described earlier.

The displacement of P may be easily computed from the matrix of (2.8): $d = Tx$. For P at $\mathbf{x} = 3\mathbf{i} - 5\mathbf{j} + 4\mathbf{k}$, we have

$$d = [\mathrm{T}][\mathrm{x}] = \begin{bmatrix} -1 & 1 & 0 \\ 0 & -1 & -1 \\ -1 & 0 & -1 \end{bmatrix} \begin{bmatrix} 3 \\ -5 \\ 4 \end{bmatrix} = \begin{bmatrix} -8 \\ 1 \\ -7 \end{bmatrix}.$$

Thus, $\mathbf{d}(P) = -8\mathbf{i} + \mathbf{j} - 7\mathbf{k}$ ft in frame Φ.

The terminal position of P in Φ is given by $\hat{\mathbf{x}}$. We recall $\mathbf{d} = \hat{\mathbf{x}} - \mathbf{x}$, and thereby find

$$\hat{\mathbf{x}}(P) = \mathbf{d} + \mathbf{x} = -5\mathbf{i} - 4\mathbf{j} - 3\mathbf{k} \text{ ft.} \qquad \square$$

In these examples the rotator has been assigned without concern for how it may have been obtained without first having the angle and the axis of rotation. So it is natural to wonder, how may the rotator be derived from other information provided in a problem of the kind illustrated above? If two configurations of the body are known in an assigned frame, we need to learn how we may find a suitable rotator **T** that will take the body from one into the other. Afterwards, we may determine the equivalent angle and axis of rotation required to produce the displacement. Of course, in general, a translation of the body also may be necessary. We shall begin to unravel the answer to this question in the next section.

3.5. The Rotation Tensor

The rotator will now be linked in a simple way to another tensor called the rotation tensor, which is more useful in certain applications, particularly those involving the composition of several rotations about a point, for example. Equations similar to (3.75) and (3.76) will be derived for the rotation tensor; these will determine the angle and the axis of rotation when the rotation tensor is known. And it will be proved that the rotation tensor is an orthogonal tensor. As consequence of this fact, we shall learn eventually that the rotation tensor has a simple geometrical interpretation that will enable us to determine the form of the rotator when only the initial and final orientations of the body are assigned.

To begin, we observe from (2.8) that the terminal position vector of a particle of a rigid body in a rotation about a fixed line may be written as

$$\hat{\mathbf{x}} = \mathbf{R}\mathbf{x}, \tag{3.86}$$

wherein, by definition,

$$\mathbf{R} \equiv \mathbf{T} + \mathbf{1}. \tag{3.87}$$

The tensor $\mathbf{R}$ is named the *rotation tensor.*

The properties of $\mathbf{R}$ are related to those of $\mathbf{T}$ by (3.87). We recall from equation (2.10) that $\mathbf{T}\boldsymbol{\alpha} = \mathbf{0}$ means that points on the axis of rotation suffer no displacement. Application of this result in (3.87) shows that *the rotation tensor preserves the axis of rotation*; that is,

$$\mathbf{R}\boldsymbol{\alpha} = \boldsymbol{\alpha}. \tag{3.88}$$

Moreover, (3.87) also yields $\operatorname{tr}\mathbf{R} = \operatorname{tr}\mathbf{T} + 3$; hence, with (3.74), we obtain

$$\operatorname{tr}\mathbf{R} = 1 + 2\cos\theta. \tag{3.89}$$

Thus, the properties (3.88) and (3.89) together with the constraint $\boldsymbol{\alpha}\cdot\boldsymbol{\alpha} = 1$ determine the axis $\boldsymbol{\alpha}$ and the angle θ of the rotation when $\mathbf{R}$ is prescribed. As before, we may limit $\theta \in [0, \pi]$ with no loss of generality. However, since $\mathbf{R}(-\boldsymbol{\alpha}) = -\boldsymbol{\alpha}$ holds also, (3.88) determines $\boldsymbol{\alpha}$ only to within its sign. This annoying ambiguity may be easily eliminated by application of (3.80). First, we note that $\mathbf{R}_S = \mathbf{T}_S + \mathbf{1}$ and $\mathbf{R}_A = \mathbf{T}_A$ follow from (3.47) and (3.87). Then (3.80) yields

$$\boldsymbol{\alpha} = -\csc\theta[\mathbf{e}_1 R_{[23]} + \mathbf{e}_2 R_{[31]} + \mathbf{e}_3 R_{[12]}]. \tag{3.90}$$

Thus, (3.89) and (3.90) determine θ and $\boldsymbol{\alpha}$ uniquely in terms of the symmetric and skew parts of $\mathbf{R}$. The exceptional case $\theta = \pi$ must be treated separately as described in connection with (3.77b) upon substitution of (3.87).

The rotation tensor has important additional properties that are not shared by the rotator. These are revealed by use of (3.86). Because each of the particles of a rigid body, before and after its rotation, is at its same distance from the point O on the axis of rotation, (3.86) and use of (3.42) show that

$$\mathbf{x} \cdot \mathbf{x} = \hat{\mathbf{x}} \cdot \hat{\mathbf{x}} = \mathbf{R}\mathbf{x} \cdot \mathbf{R}\mathbf{x} = \mathbf{x} \cdot \mathbf{R}^T\mathbf{R}\mathbf{x},$$

that is,

$$\mathbf{x} \cdot (\mathbf{R}^T\mathbf{R} - \mathbf{1})\mathbf{x} = 0.$$

This equation must hold of all particles of the body, i.e., for all $\mathbf{x}$; hence, $(\mathbf{R}^T - \mathbf{1})\mathbf{x} = \mathbf{0}$ must hold for all $\mathbf{x}$. But the only tensor that transforms every vector into the zero vector is the zero tensor; therefore, $\mathbf{R}$ must satisfy $\mathbf{R}^T\mathbf{R} = \mathbf{1}$. It follows by (3.68) that $\mathbf{R}$ is an orthogonal tensor with $\det \mathbf{R} = \pm 1$.

Because a trivial rotation $\mathbf{R} = \mathbf{1}$ must be admissible for every rigid body, only $\det \mathbf{R} = +1$ is allowed. Otherwise, we could choose $\mathbf{R} = -\mathbf{1}$, which is an improper orthogonal tensor. However, in this instance, (3.88) fails to hold, and (3.89) implies that $\cos\theta = -2$, which is impossible. Hence, improper orthogonal tensors do not characterize rigid body rotations. Consequently, every rigid body rotation tensor $\mathbf{R}$ is a proper orthogonal tensor that obeys the rules

$$\mathbf{R}\mathbf{R}^T = \mathbf{R}^T\mathbf{R} = \mathbf{1}, \qquad \det \mathbf{R} = 1. \tag{3.91}$$

Example 3.6. Show that the tensor $\mathbf{R} = R_{ij}\mathbf{e}_{ij}$ having the matrix

$$R = \begin{bmatrix} 1 & 0 & 0 \\ 0 & \cos\theta & -\sin\theta \\ 0 & \sin\theta & \cos\theta \end{bmatrix} \tag{3.92}$$

with respect to the orthonormal basis $\mathbf{e}_k$ is a proper orthogonal tensor. Determine the axis of the rotation it represents.

Solution. It is seen at once that $\det R = \cos^2\theta + \sin^2\theta = 1$; and it is easily shown that

$$R^{-1} = \begin{bmatrix} 1 & 0 & 0 \\ 0 & \cos\theta & \sin\theta \\ 0 & -\sin\theta & \cos\theta \end{bmatrix} = R^T.$$

Therefore, $\mathbf{R}$ is a proper orthogonal tensor. Notice further that $\operatorname{tr} R = 1 + 2\cos\theta$, which agrees with (3.89).

The axis of rotation is found by (3.90); we obtain $\boldsymbol{\alpha} = \mathbf{e}_1$. Thus, (3.92) describes a right-hand rotation through the angle θ about the $\mathbf{e}_1$ direction. Let the reader show that the matrix in Example B.3 in Appendix B is a proper rotation around the $\mathbf{e}_3$ axis; use (3.87) to derive the matrix R for a rotation about the $\mathbf{e}_2$ direction; and compare the forms of the three matrices mentioned. In addition, apply (3.88) to find the axis of the rotation (3.92).

3.6. Change of Basis and Transformation Laws

A vector $\mathbf{v}$ will have different scalar components when referred to different bases. But the vector itself is unchanged; it remains the same geometrical object regardless of the basis used to represent it. Thus, for the bases $\mathbf{e}_j$ and $\mathbf{e}'_k$ relative to which $\mathbf{v}$ has the respective components v_j and v'_k, we may write

$$\mathbf{v} = v_j\mathbf{e}_j = v'_k\mathbf{e}'_k. \tag{3.93}$$

The same thing applies to tensors. A tensor will have different components when referred to different bases, but the tensor itself is unchanged. A tensor is an invariant entity. Thus, a tensor $\mathbf{T}$ having components T_{ij} and T'_{kl} with respect to the corresponding bases $\mathbf{e}_{ij}$ and $\mathbf{e}'_{kl}$ may be written as

$$\mathbf{T} = T_{ij}\mathbf{e}_{ij} = T'_{kl}\mathbf{e}'_{kl}. \tag{3.94}$$

The fundamental rules that relate the different components of vectors and tensors are called *transformation laws*. We shall see in later chapters that these laws play an important role in rigid body dynamics. To derive the transformation laws, we shall need to introduce relations connecting the different bases.

3.6.1. A Change of Basis

Let $\{\mathbf{e}_p\}$ and $\{\mathbf{e}'_q\}$ be two sets of orthonormal basis vectors so that

$$\mathbf{e}_i \cdot \mathbf{e}_j = \delta_{ij} \tag{3.95a}$$

and

$$\mathbf{e}'_k \cdot \mathbf{e}'_l = \delta_{kl}. \tag{3.95b}$$

We recall that each of these six sets of conditions specifies the unit magnitudes and mutual orthogonality of the basis vectors in each set. In addition, we may form the dot product

$$\mathbf{e}_i' \cdot \mathbf{e}_j = \cos\langle \mathbf{e}_i', \mathbf{e}_j \rangle \equiv A_{ij}, \tag{3.96}$$

in which $\langle \mathbf{e}_i', \mathbf{e}_j \rangle$ denotes the angle between the unit vector $\mathbf{e}_i'$ and the unit vector $\mathbf{e}_j$. Hence, the A_{ij} in (3.96) define the nine direction cosines of the three vectors $\{\mathbf{e}_i'\}$ relative to the basis set $\{\mathbf{e}_j\}$. It is important to notice that their definition requires that we write the first index of A_{ij} as the prime label. Of course, it is certainly true that

$$\mathbf{e}_i' \cdot \mathbf{e}_j = \cos\langle \mathbf{e}_i', \mathbf{e}_j \rangle = \cos\langle \mathbf{e}_j, \mathbf{e}_i' \rangle = \mathbf{e}_j \cdot \mathbf{e}_i'; \tag{3.97}$$

however, we caution that $A_{ij} \neq A_{ji}$. For example, it is clear that $\cos\langle \mathbf{e}_1', \mathbf{e}_2 \rangle = \cos\langle \mathbf{e}_2, \mathbf{e}_i' \rangle$; but $A_{12} \neq A_{21}$, for plainly,

$$A_{12} = \cos\langle \mathbf{e}_1', \mathbf{e}_2 \rangle \neq \cos\langle \mathbf{e}_2', \mathbf{e}_1 \rangle = A_{21}.$$

Thus, we shall agree in (3.96) that A_{ij} *always is read as the cosine of the angle between* $\mathbf{e}_i'$ *and* $\mathbf{e}_j$. Moreover, let us agree to extend the index rules to terms enclosed in the cuneiform brackets.

We recall from (3.6) and (3.7) that every vector $\mathbf{v}$ may be written in the form

$$\mathbf{v} = v_p \hat{\mathbf{e}}_p = (\mathbf{v} \cdot \hat{\mathbf{e}}_p)\, \hat{\mathbf{e}}_p \tag{3.98}$$

with respect to the basis $\hat{\mathbf{e}}_p$. In particular, we may identify $\mathbf{v} = \mathbf{e}_i'$ and $\hat{\mathbf{e}}_j = \mathbf{e}_j$, and apply (3.96) to obtain

$$\mathbf{e}_i' = (\mathbf{e}_i' \cdot \mathbf{e}_j)\, \mathbf{e}_j = A_{ij} \mathbf{e}_j. \tag{3.99}$$

And similarly, reversing the roles of the primed and unprimed sets and using (3.97) while retaining the definition (3.96), we have also

$$\mathbf{e}_j = (\mathbf{e}_j \cdot \mathbf{e}_i')\, \mathbf{e}_i' = A_{ij} \mathbf{e}_i'. \tag{3.100}$$

Either of the transformation rules (3.99) and (3.100) is called *a change of basis.*

The same thing may be seen in more geometrical terms by our recalling that any unit vector $\mathbf{n}$, say, may be expressed in terms of the three cosines of the direction angles α_j that it makes with the familiar orthonormal basis directions $\mathbf{i}_j$ at a point Q, as shown in Fig. 3.2a. Namely, $\mathbf{n} = \cos\alpha_j\, \mathbf{i}_j$. If we write $\cos\langle \mathbf{n}, \mathbf{i}_j \rangle$ for the cosine of the angle α_j between $\mathbf{n}$ and the $\mathbf{i}_j$ direction and agree, as mentioned above, to extend the summation rule to terms in the cuneiform brackets, then $\mathbf{n}$ may be written more conveniently as $\mathbf{n} =$

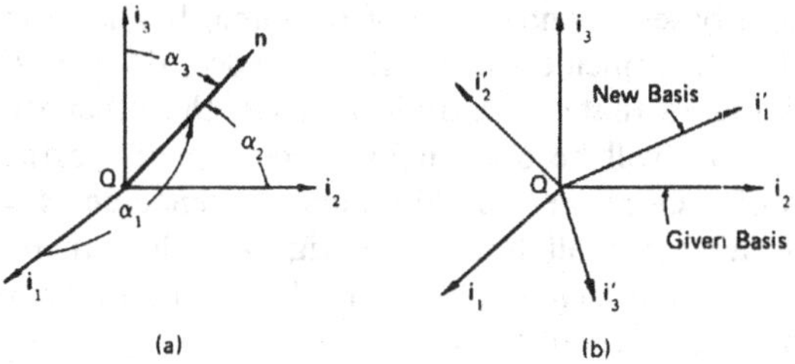

Figure 3.2. Direction angles of a unit vector **n**, and a change of basis.

$\cos\langle \mathbf{n}, \mathbf{i}_j \rangle \mathbf{i}_j$. Now let $\{\mathbf{i}'_p\}$ be another orthonormal basis, as shown in Fig. 3.2b. Then, for example, with $\mathbf{n} = \mathbf{i}'_1$, we shall have $\mathbf{i}'_1 = \cos\langle \mathbf{i}'_1, \mathbf{i}_j \rangle \mathbf{i}_j$. Similar equations may be written for $\mathbf{i}'_2$ and $\mathbf{i}'_3$. Thus, in general, we shall have $\mathbf{i}'_i = \cos\langle \mathbf{i}'_i, \mathbf{i}_j \rangle \mathbf{i}_j = A_{ij}\mathbf{i}_j$, which, with (3.96), is the same as (3.99) applied to the usual Cartesian bases.

Naturally, the change of basis (3.99) induces a change of tensor product basis so that, for example,

$$\mathbf{e}'_{ij} = \mathbf{e}'_i \otimes \mathbf{e}'_j = A_{ik}\mathbf{e}_k \otimes A_{jl}\mathbf{e}_l = A_{ik}A_{jl}(\mathbf{e}_k \otimes \mathbf{e}_l).$$

A similar relation is induced by use of (3.100). We thus obtain for *a change of tensor product basis*

$$\mathbf{e}'_{ij} = A_{ik}A_{jl}\mathbf{e}_{kl}, \qquad \mathbf{e}_{kl} = A_{ik}A_{jl}\mathbf{e}'_{ij}. \tag{3.101}$$

3.6.1.1. *The Basis Transformation Matrix*

The matrix $A = [A_{pq}] = [\cos\langle \mathbf{e}'_p, \mathbf{e}_q \rangle]$ of direction cosines for the change of basis (3.99) or (3.100) is named the *basis transformation matrix.* To discover its fundamental properties, we use the change of bases in (3.99) and (3.100) to form the scalar products in (3.95). For example, with (3.99) and (3.95), we derive

$$\delta_{ip} = \mathbf{e}'_i \cdot \mathbf{e}'_p = A_{ij}\mathbf{e}_j \cdot A_{pk}\mathbf{e}_k = A_{ij}A_{pk}\mathbf{e}_j \cdot \mathbf{e}_k = A_{ij}A_{pk}\delta_{jk},$$

that is,

$$A_{ik}A_{pk} = \delta_{ip} \quad \text{or} \quad AA^T = I. \tag{3.102a}$$

And, similarly, with (3.100), we find

$$A_{ki}A_{kp} = \delta_{ip} \quad \text{or} \quad A^TA = I. \tag{3.102b}$$

The properties (3.102) show that A is an orthogonal matrix. Of course,

$\det A = \pm 1$. If the bases $\mathbf{e}'_i$ and $\mathbf{e}_j$ are of the same hand, their mutual directions can be made to coincide so that $A = I$; hence, $\det A = +1$. If the bases have opposite hand so that one is right-handed, the other left-handed, then one direction $\mathbf{e}_1$, say, will be oppositely directed to its partner $\mathbf{e}'_1$ when the other two pairs of vectors are mutually aligned, hence $\det A = -1$. We have assumed in this text that all frames are right-handed; therefore, for every change of basis, we shall have $\det A = +1$. Hence, the transformation matrix A is a proper orthogonal matrix.

Another interesting result may be obtained by forming from (3.99) the tensor product

$$\mathbf{Q} \equiv \mathbf{e}'_j \otimes \mathbf{e}_j = A_{jk}\mathbf{e}_k \otimes \mathbf{e}_j = A^T_{kj}\mathbf{e}_k \otimes \mathbf{e}_j. \qquad (3.103a)$$

or from (3.100)

$$\mathbf{Q} \equiv \mathbf{e}'_k \otimes \mathbf{e}_k = \mathbf{e}'_k \otimes A_{jk}\mathbf{e}'_j = A^T_{kj}\mathbf{e}'_k \otimes \mathbf{e}'_j. \qquad (3.103b)$$

Hence, $\mathbf{Q}$ is an orthogonal tensor whose matrix with respect to either basis is the transpose of A: $Q = Q' = A^T$. Indeed, the orthogonality property of $\mathbf{Q}$ may be demonstrated independently of the foregoing properties of A. It is easily shown from the product rule (3.36) that the tensor product obeys the rule

$$(\mathbf{a} \otimes \mathbf{b})(\mathbf{c} \otimes \mathbf{d}) = (\mathbf{b} \cdot \mathbf{c})(\mathbf{a} \otimes \mathbf{d}). \qquad (3.104)$$

The proof is left for the student in Problem 3.13. Thus, with the help of (3.104), and recalling (3.43), (3.95a), and (3.31), we derive from (3.103)

$$\mathbf{Q}\mathbf{Q}^T = (\mathbf{e}'_j \otimes \mathbf{e}_j)(\mathbf{e}_q \otimes \mathbf{e}'_q) = (\mathbf{e}_j \cdot \mathbf{e}_q)(\mathbf{e}'_j \otimes \mathbf{e}'_q) = \mathbf{e}'_j \otimes \mathbf{e}'_j = \mathbf{1}.$$

The reader may show similarly that $\mathbf{Q}^T\mathbf{Q} = \mathbf{1}$. Therefore, in accordance with (3.68), the tensor $\mathbf{Q}$ defined by (3.103a), with respect to the two sets of orthonormal bases indicated, is an orthogonal tensor.

Final, let us consider the product

$$\mathbf{Q}\mathbf{e}_k = (\mathbf{e}'_p \otimes \mathbf{e}_p)\,\mathbf{e}_k = \delta_{pk}\mathbf{e}'_p = \mathbf{e}'_k.$$

We thus learn that the tensor $\mathbf{Q}$ defined by (3.103) is the orthogonal tensor that transforms the basis vector $\mathbf{e}_k$ into the corresponding basis vector $\mathbf{e}'_k$; and hence the change of basis (3.99) and (3.100) may be written as

$$\mathbf{e}'_k = \mathbf{Q}\mathbf{e}_k = A_{kl}\mathbf{e}_l, \qquad (3.105a)$$

$$\mathbf{e}_l = \mathbf{Q}^T\mathbf{e}'_l = A_{kl}\mathbf{e}'_k. \qquad (3.105b)$$

The tensor $\mathbf{Q}$ is named the *basis transformation tensor*. Its matrix with respect to either of the orthonormal bases in (3.103) is the transpose of the basis transformation matrix defined in (3.96), hence

$$Q = A^T, \quad \text{namely,} \quad Q_{jk} = A^T_{jk} \equiv A_{kj} = \cos\langle \mathbf{e}_j, \mathbf{e}'_k \rangle \tag{3.106}$$

may be read as the array of nine direction cosines of the triple $\mathbf{e}_j$ relative to the basis $\mathbf{e}'_k$, the first index of Q_{jk} being the unprimed label.

3.6.2. Transformation Laws for Vectors and Tensors

The rules that relate the different components of vectors and tensors may now be readily derived by application of the change of basis to the representation equations (3.93) and (3.94). We shall begin with the vector equation (3.93) and by substitution of (3.100) obtain

$$\mathbf{v} = v_j A_{kj} \mathbf{e}'_k = v'_k \mathbf{e}'_k.$$

Since $\mathbf{e}'_k$ is a basis, this equality implies that

$$v'_k = A_{kj} v_j \quad \text{or} \quad v' = Av = Q^T v, \tag{3.107a}$$

where the second expression is the matrix form of the first and (3.106) is recalled. Similarly, the inverse of this formula may be found by use of (3.99) in (3.93); or it may be gotten directly from (3.107a) by use of (3.102). In any event, we find

$$v_j = A_{kj} v'_k \quad \text{or} \quad v = A^T v' = Qv'. \tag{3.107b}$$

The rule in either of the alternative forms of (3.107) is known as the (*Cartesian*) *vector component transformation law*.

A similar procedure that uses (3.101) in (3.94) yields the following variants of the (*Cartesian*) *tensor component transformation law* expressed in both index and matrix form:

$$T'_{ij} = A_{ik} A_{jl} T_{kl} \quad \text{or} \quad T' = ATA^T = Q^T T Q; \tag{3.108a}$$

$$T_{kl} = A_{ik} A_{jl} T'_{ij} \quad \text{or} \quad T = A^T T' A = QTQ^T. \tag{3.108b}$$

Herein we remember (3.106). Notice that (3.108b) also may be derived from (3.108a) by use of the orthogonality rules (3.102) for A.

A change of basis in the plane is easy to visualize. An example that illustrates a plane change of basis and the use of the vector transformation law will be given next. An application of the tensor transformation law will be demonstrated in the following subsection.

Example 3.7. Let $\mathbf{v} = 3\mathbf{i}_1 + 4\mathbf{i}_2$ in the familiar $\mathbf{i}_k$ basis. Find the components of $\mathbf{v}$ referred to a basis $\mathbf{i}'_k$ obtained by a counterclockwise rotation of the first set through 45° about their mutual axis $\mathbf{i}_3 = \mathbf{i}'_3$ perpendicular to the plane, as shown in Fig. 3.3.

Solution. Since this problem is so easily visualized, let us gather some confidence by first deriving the result geometrically, and afterwards confirm the result by application of the tools introduced above. It is seen from Fig. 3.3 that the projections of the vector $\mathbf{v}$ upon the $\mathbf{i}'_k$ directions in the plane are given by

$$v'_1 = 3\cos 45^\circ + 4\sin 45^\circ = \frac{7\sqrt{2}}{2}, \qquad v'_2 = 4\cos 45^\circ - 3\sin 45^\circ = \frac{\sqrt{2}}{2};$$

hence, referred to $\mathbf{i}'_k$, our original vector becomes

$$\mathbf{v} = 3\mathbf{i}_1 + 4\mathbf{i}_2 \tag{3.109a}$$

$$= \frac{\sqrt{2}}{2}(7\mathbf{i}'_1 + \mathbf{i}'_2). \tag{3.109b}$$

This is to be compared with (3.93). The given components in (3.109a) and the transformed components in (3.109b) correspond to the v_k and the v'_k components of $\mathbf{v}$ in (3.93).

Now let us repeat the work by use of the vector transformation law (3.107a). We start with the easy construction of the basis transformation matrix defined by (3.96):

$$A = \begin{bmatrix} A_{11} & A_{12} & A_{13} \\ A_{21} & A_{22} & A_{23} \\ A_{31} & A_{32} & A_{33} \end{bmatrix} = \begin{bmatrix} \cos\langle \mathbf{i}'_1, \mathbf{i}_1 \rangle & \cos\langle \mathbf{i}'_1, \mathbf{i}_2 \rangle & \cos\langle \mathbf{i}'_1, \mathbf{i}_3 \rangle \\ \cos\langle \mathbf{i}'_2, \mathbf{i}_1 \rangle & \cos\langle \mathbf{i}'_2, \mathbf{i}_2 \rangle & \cos\langle \mathbf{i}'_2, \mathbf{i}_3 \rangle \\ \cos\langle \mathbf{i}'_3, \mathbf{i}_1 \rangle & \cos\langle \mathbf{i}'_3, \mathbf{i}_2 \rangle & \cos\langle \mathbf{i}'_3, \mathbf{i}_3 \rangle \end{bmatrix}.$$

The nine direction cosines are obtained by aid of Fig. 3.3; we find easily for the prescribed rotation

$$A = \begin{bmatrix} \cos 45^\circ & \cos 45^\circ & 0 \\ -\sin 45^\circ & \cos 45^\circ & 0 \\ 0 & 0 & 1 \end{bmatrix} = \begin{bmatrix} \sqrt{2}/2 & \sqrt{2}/2 & 0 \\ -\sqrt{2}/2 & \sqrt{2}/2 & 0 \\ 0 & 0 & 1 \end{bmatrix}. \tag{3.110}$$

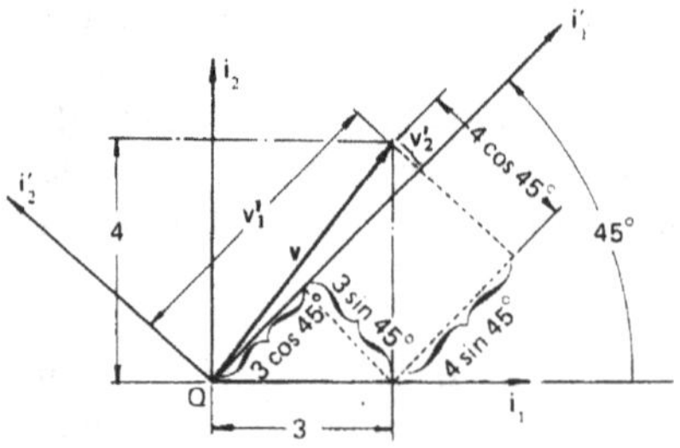

Figure 3.3. Geometrical construction of a particular change of basis.

Thus, with the given column matrix $v=(v_k)=(3, 4, 0)$ and (3.110), the vector transformation law (3.107a) yields

$$v' = Av = \begin{bmatrix} \sqrt{2}/2 & \sqrt{2}/2 & 0 \\ -\sqrt{2}/2 & \sqrt{2}/2 & 0 \\ 0 & 0 & 1 \end{bmatrix} \begin{bmatrix} 3 \\ 4 \\ 0 \end{bmatrix} = \begin{bmatrix} 7\sqrt{2}/2 \\ \sqrt{2}/2 \\ 0 \end{bmatrix}, \qquad (3.111)$$

which is the column matrix of components v'_k in $\mathbf{i}'_k$. This delivers the result (3.109b) derived before. The new method appears longer, whereas actually, after one masters understanding of the ingredients needed to do the calculation, the entire story is presented in only one line by (3.111). Moreover, it often is easier than having to perceive the geometry associated with the individual component construction. To see this more graphically, let us work the converse problem.

Suppose that we are given (3.109b), and we wish to find v_k for the same conditions. With A given in (3.110), we find by (3.107b)

$$v = A^T v' = \begin{bmatrix} \sqrt{2}/2 & -\sqrt{2}/2 & 0 \\ \sqrt{2}/2 & \sqrt{2}/2 & 0 \\ 0 & 0 & 1 \end{bmatrix} \begin{bmatrix} 7\sqrt{2}/2 \\ \sqrt{2}/2 \\ 0 \end{bmatrix} = \begin{bmatrix} 3 \\ 4 \\ 0 \end{bmatrix}; \qquad (3.112)$$

and this yields the desired result (3.109a).

3.6.2.1. Invariant Properties of Tensors

We know from their definitions in (3.12) that the zero and the identity tensors have the same components with respect to every Cartesian reference system. Nevertheless, these facts also are evident from (3.108). Indeed, if $T=0$, (3.108a) shows that $T'=0$ also; and for $T=I$, (3.108a) together with (3.102a) confirms that $T'=I$ too. Otherwise, in general, the components of a tensor will change under a change of basis, and (3.108) serves to determine the transformed tensor components from the assigned set when the basis transformation matrix is given. On the other hand, however, the tensor transformation law (3.108) also implies certain invariant properties that *all* tensors possess.

The angle of rotation of a rigid body, we recall, is a physically invariant quantity that has nothing to do with the choice of reference frame that we may use in the description of the rotator or the rotation tensor. Hence, it was concluded earlier on the basis of (3.74) and (3.89) that this physical invariance implied that tr $\mathbf{T}$ and tr $\mathbf{R}$ also had to be the same for every reference system. In fact, this converse result may now be proved for every tensor.

We recall the rule (3.41) for matrices of tensors; and we show by (3.108a) and (3.102b) that

$$\operatorname{tr} T' = \operatorname{tr}(ATA^T) = \operatorname{tr}(A^TAT) = \operatorname{tr}(IT) = \operatorname{tr} T.$$

Hence, the trace of a tensor is invariant under a change of basis. And, similarly, the determinant rule (3.63) and the orthogonality property of A yields the invariant property

$$\det T' = \det(ATA^T) = (\det A)(\det T)(\det A^T) = \det T.$$

Thus, *the trace and the determinant of a tensor are invariant under a change of basis.* The three invariants

$$I_1 \equiv \operatorname{tr} \mathbf{T}, \qquad I_2 \equiv \tfrac{1}{2}[I_1^2 - \operatorname{tr}(\mathbf{T}^2)], \qquad I_3 \equiv \det \mathbf{T}, \tag{3.113}$$

are known as the *principal invariants* of the tensor **T**. These invariants play a central role in the mechanics of materials.

The rotator and the rotation tensor are examples of physical tensor quantities whose components must obey the transformation law (3.108) under a change of basis. Thus, if either **T** or **R** is assigned, its components in any other basis may be determined. Since tr **T**, hence tr **R**, has the same value in every reference system, the angle of rotation is unchanged; only the description of the axis of rotation is affected by the change of basis. The transformed axis of rotation may be found from the vector transformation law (3.107); and the transformed components of **T**, hence also those of **R**, may be computed from (3.70), or (3.87), in the usual way. However, this approach usually is somewhat more tedious than the application of the tensor transformation law and the subsequent calculation of the axis from (3.80) or (3.90). But this is more a matter of judgement or personal preference, and either method will yield the desired result. A typical calculation is illustrated below.

Example 3.8. The rotator for the aircraft door panel mechanism described in Example 3.5 is given by (3.84) in frame $\Phi = \{O; \mathbf{i}_k\}$. Find the rotator referred to a frame $\Phi' = \{O; \mathbf{i}'_k\}$ resulting from the change of basis provided by (3.110). Determine the angle and axis of rotation in Φ'.

Solution. The components of **T** in Φ' may be found from the tensor transformation law. Substitution of (3.110) and (3.84) into (3.108a) and factorization of common elements to simplify the matrices yields

$$T' = ATA^T = \frac{1}{2}\begin{bmatrix} 1 & 1 & 0 \\ -1 & 1 & 0 \\ 0 & 0 & \sqrt{2} \end{bmatrix}\begin{bmatrix} -1 & 1 & 0 \\ 0 & -1 & -1 \\ -1 & 0 & -1 \end{bmatrix}\begin{bmatrix} 1 & -1 & 0 \\ 1 & 1 & 0 \\ 0 & 0 & \sqrt{2} \end{bmatrix}.$$

Hence, the matrix of **T** in Φ' is

$$T' = \frac{1}{2}\begin{bmatrix} -1 & 1 & -\sqrt{2} \\ -1 & -3 & -\sqrt{2} \\ -\sqrt{2} & \sqrt{2} & -2 \end{bmatrix}. \tag{3.114}$$

We observe that $\text{tr } T' = -3$ is the same as $\text{tr } T$ in Example 3.5. Hence, the angle of rotation is the same; namely, $\theta = 120°$. The axis of rotation in Φ' is provided by (3.80). With $\sin\theta = \sqrt{3}/2$ and (3.114), we get

$$\boldsymbol{\alpha} = \frac{1}{\sqrt{3}}(\sqrt{2}\mathbf{i}'_1 - \mathbf{i}'_3). \tag{3.115}$$

To check this answer, let us apply the vector transformation law to $\boldsymbol{\alpha}$ in Φ. With (3.110) and (3.85) in (3.107a), we have, with factorization,

$$\alpha' = A\alpha = \frac{\sqrt{2}}{2\sqrt{3}}\begin{bmatrix} 1 & 1 & 0 \\ -1 & 1 & 0 \\ 0 & 0 & \sqrt{2} \end{bmatrix}\begin{bmatrix} 1 \\ 1 \\ -1 \end{bmatrix} = \frac{1}{\sqrt{3}}\begin{bmatrix} \sqrt{2} \\ 0 \\ -1 \end{bmatrix},$$

which agrees with (3.115). These values together with $\theta = 120°$ may be used in (3.70) to find **T** referred to Φ'. This calculation, which returns us to (3.114), is left for the student.

3.7. Rotation about a Fixed Point

The general finite displacement of the particles of a rigid body is described by (2.12). Let us assume that the base point O is fixed so that $\mathbf{b} = \mathbf{0}$. Then the displacement of any other point P is given by

$$\mathbf{d}(P) = \hat{\mathbf{x}} - \mathbf{x}, \tag{3.116}$$

in which we recall that $\hat{\mathbf{x}}$ and $\mathbf{x}$ are the final and the initial position vectors of P from O referred to an assigned spatial frame φ, say. Suppose there exists another point $Q \neq O$ for which $\mathbf{d}(Q) = \mathbf{0}$ also. Then because the body is rigid, no particle on the line OQ can be farther from O and Q after the displacement than it was initially; hence $\mathbf{d} = \mathbf{0}$ for every point on the line OQ. In this case, the displacement may be effected by a rotation (2.7) about the fixed line OQ. In general, however, (3.116) describes a rotation about a fixed point, and we do not yet know for this case if such a line OQ exists. Therefore, let us assume for the moment that there are no points P besides the base point O for which

(3.116) vanishes. Then, as defined before, (3.116) is a rotation about the fixed point O.

Now let us recall that a change of basis by a proper orthogonal transformation is a rigid body rotation about a fixed point, namely, the origin of the reference frame used initially. Moreover, we known that a rigid body rotation about a fixed line also may be described by a proper orthogonal tensor. Our future objective is to determine how these two ideas may be related to one another. A relation that describes an arbitrary rigid body rotation about a fixed base point in terms of the nine direction cosines relating the orientation of the body reference frame in its initial and its final configurations will be derived next. The connection of the result with a rotation about a fixed line will be studied in the following section.

To start with, we recall that $\hat{\mathbf{x}}$ is the displaced position vector of P in the spatial frame $\varphi = \{O; \mathbf{i}_k\}$, and $\mathbf{x}$ is the position vector of P in the body frame $\varphi' = \{O; \mathbf{i}'_k\}$, which was coincident with φ initially, as described in Fig. 2.6. Thus, when viewed in φ alone, we envision the *same particle* P identified by two vectors $\mathbf{x} = x_k \mathbf{i}_k$ and $\hat{\mathbf{x}} = \hat{x}_k \mathbf{i}_k$ separated by an angle ψ, as shown in Fig. 3.4. This is the usual representation considered in (3.116) and in all of our earlier equations for the displacement vector in an assigned spatial frame φ. Hence, as usual, (3.116) may be written in the following familiar component form in φ:

$$\mathbf{d}(P) = \hat{\mathbf{x}} - \mathbf{x} = (\hat{x}_k - x_k)\,\mathbf{i}_k. \tag{3.117}$$

However, the same situation may be viewed differently. In the body frame φ' in Fig. 2.6, the point P has always the same position vector $\mathbf{p}$, say. Therefore, it appears always to an observer in φ' that P has the coordinate components $\{p'_k\} \equiv \{x_k\} = (x_1, x_2, x_3)$ so that $\mathbf{p} = p'_k \mathbf{i}'_k = x_k \mathbf{i}'_k$ referred to frame φ'. Of course, these also are the initial coordinates of P in the spatial frame φ, because the frames coincides initially. But the *same vector* $\mathbf{p}$ after the

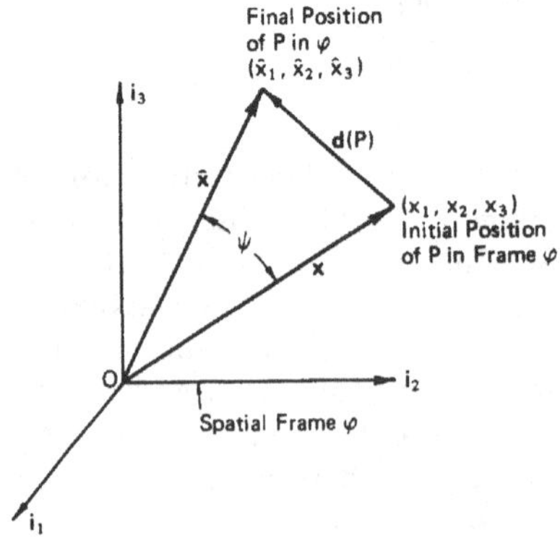

Figure 3.4. Displacement of a particle P viewed in the spatial frame φ.

displacement, though unchanged in the body frame φ', is identified differently by another observer in the spatial frame φ as the vector $\hat{\mathbf{x}}$ having coordinate components $\{p_k\} = \{\hat{x}_k\} = (\hat{x}_1, \hat{x}_2, \hat{x}_3)$. Hence, referred to φ, the vector $\mathbf{p} = p_k \mathbf{i}_k = \hat{x}_k \mathbf{i}_k$. Notice that the numbers $\hat{x}_k$ and x_k are the same as those in (3.117). A plane example in Fig. 3.5 shows the relative rotation of the body frame φ' through an angle θ about the $\mathbf{i}_1 = \mathbf{i}'_1$ axis in the spatial frame φ and the single position vector of the point P from O identified as $\mathbf{x}$ by the observer in φ' and as $\hat{\mathbf{x}}$ by the observer in φ. We may ignore the observers, and note that in any case we may write

$$\mathbf{p}(P) = p_k \mathbf{i}_k = \hat{x}_k \mathbf{i}_k = p'_k \mathbf{i}'_k = x_k \mathbf{i}'_k, \tag{3.118}$$

because the same vector always may be referred to any two bases in just this way in accordance with (3.93). This arrangement is special, because, by construction, the component numbers in (3.118) are the same as those in (3.117). Therefore, any relation connecting these components that derives from (3.118) holds also for the components in (3.117). Thus, in this sense, the two points of view are equivalent.

In the present viewpoint, the two sets of components for the same vector $\mathbf{p}$ in (3.118) may be related through the orientation angles between the two frames by a change of basis applied to φ'. Since $\hat{\mathbf{x}} = \hat{x}_k \mathbf{i}_k = \mathbf{p}$ and $\mathbf{x} = x_k \mathbf{i}_k$ referred to the spatial frame φ, substitution of the change of basis (3.105a) into the last term in (3.118) yields

$$\hat{\mathbf{x}} = \hat{x}_k \mathbf{i}_k = x_k \mathbf{Q} \mathbf{i}_k = \mathbf{Q}(x_k \mathbf{i}_k) = \mathbf{Q}\mathbf{x},$$

or alternatively,

$$\hat{\mathbf{x}} = \hat{x}_k \mathbf{i}_k = x_k A_{kl} \mathbf{i}_l = A_{kl} x_k \mathbf{i}_l.$$

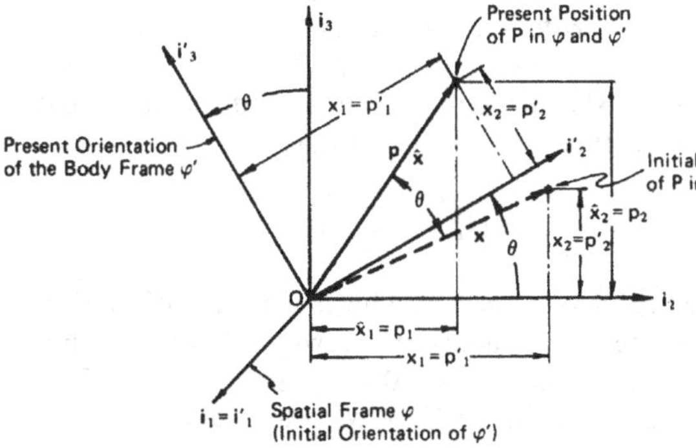

Figure 3.5. The position vector **p** representing both **x** and $\hat{\mathbf{x}}$, which are viewed in separate frames.

Therefore, the terminal location in φ of a particle of a rigid body due to a finite rotation about a fixed point is given by

$$\hat{\mathbf{x}} = \mathbf{Q}\mathbf{x} \tag{3.119a}$$

or

$$\hat{\mathbf{x}} = \hat{x}_k \mathbf{i}_k = A_{kl} x_k \mathbf{i}_l \tag{3.119b}$$

in terms of the nine direction cosines relating the orientation of the body reference frame in its initial and final configurations in φ. This is the result that we set out to obtain here. The corresponding displacement (3.117) may be written as

$$\mathbf{d}(P) = \hat{\mathbf{x}} - \mathbf{x} = (\mathbf{Q} - \mathbf{1})\mathbf{x} = (A_{kl} - \delta_{kl})\, x_k \mathbf{i}_l, \tag{3.120}$$

in φ. We recall from (3.106) that $\mathbf{Q}$ is the orthogonal basis transformation tensor whose matrix is $A^T = [\cos\langle \mathbf{i}'_k, \mathbf{i}_l \rangle]^T$.

3.8. Euler's Theorem

It was assumed in the last section that there were no points P besides the base point O for which (3.116) vanishes. This led to the displacement equation (3.120) for a rotation about the fixed point O. We are going to prove that, in fact, this assumption is false. Hence, when a rigid body is turned about a fixed point, there is always a material line through this point whose particles suffer no total displacement, hence the *same* displacement may be produced by a rotation around this line. This is the substance of Euler's theorem.

Euler's Theorem. *An arbitrary displacement of a rigid body with a fixed point O is equivalent to a rotation about a line through O.*

Proof. If there exists through O an imbedded line that suffers no resultant displacement, then (3.120) shows that

$$(\mathbf{Q} - \mathbf{1})\mathbf{x} = \mathbf{0} \tag{3.121}$$

must hold for all places $\mathbf{x}$ on that line. This is a system of three homogeneous linear equations in the components of $\mathbf{x}$; with (3.106), its matrix form is $(A^T - I)x = 0$. The trivial solution $\mathbf{x} = \mathbf{0}$ corresponds to the fixed base point at O. For nontrivial solutions of (3.121), we must have

$$\det(\mathbf{Q} - \mathbf{1}) = \det(A^T - I) = 0. \tag{3.122}$$

Since A is a proper orthogonal matrix, application of (3.102) and the elementary rules for determinants yield

$$\det(A^T - I) = \det(A^T - I)^T = \det[A(I - A^T)] = (\det A)\det(I - A^T).$$

However, this determinant being of third order, it follows that

$$\det(A^T - I) = \det(I - A^T) = -\det(A^T - I),$$

which clearly satisfies the criterion (3.122) for existence of an equivalent fixed axis of rotation; and the theorem is proved.

Therefore, besides $\mathbf{x} = \mathbf{0}$, other solutions of (3.121) exist. Indeed, if the vector $\mathbf{x}^*$ is such a solution, so also is $\mathbf{x} = k\mathbf{x}^*$ for $-\infty < k < \infty$. We recognize this as the vector equation of a straight line through O. This line is the axis of rotation. This remarkable result shows that no matter how the body may have been brought from its initial configuration into its final configuration, there always is a unique axis about which the body may be turned to move every particle of the body from its initial place to its final place in φ. Let the reader show that for nontrivial rotations the axis of rotation is indeed unique.

Comparison of (3.86) with (3.119) reveals the identity

$$\mathbf{R} = \mathbf{Q}, \tag{3.123a}$$

hence,

$$R = A^T = [\cos\langle \mathbf{i}'_j, \mathbf{i}_k \rangle]^T, \tag{3.123b}$$

by which the rotation about a fixed point may be reduced to a rotation about a fixed line, in accordance with Euler's theorem. The angle θ of the rotation is determined by (3.89), in which $\operatorname{tr} \mathbf{R} = \operatorname{tr} A^T$. And the unique axis of rotation in the spatial frame φ is found by aid of (3.90), in which $R_A = A_A^T = -A_A$. Of course, the case $\theta = \pi$ must be treated separately as described in (3.77), or it may be found by use of (3.88). The equivalent rotation about a fixed line characterized by (3.89) and (3.90) is called the *(equivalent) Euler rotation.*

The result (3.123) also may be cast in terms of the rotator. We find with (3.87), (3.122), and (3.123) the relations

$$\mathbf{T} = \mathbf{Q} - \mathbf{1}, \tag{3.124a}$$

$$T = A^T - I, \tag{3.124b}$$

$$\det \mathbf{T} = 0. \tag{3.124c}$$

Therefore, as remarked in Chapter 2, it follows from Euler's theorem and (2.12) that the general displacement of a rigid body is equivalent to a parallel translation **b** of the base point O together with a displacement $\mathbf{Tx}$ due to a rotation about a line through O:

$$\mathbf{d}(P) = \mathbf{b} + \mathbf{Tx} \qquad \text{[cf. (2.13)]}.$$

Another useful form of (2.13) may be obtained by aid of (3.87). Recalling the description of the vectors defined in Fig. 2.6, the reader may show that (2.13) yields

$$\hat{\mathbf{X}}(P) = \hat{\mathbf{B}}(O) + \mathbf{Rx}(P), \tag{3.125}$$

wherein $\hat{\mathbf{X}}(P)$ and $\hat{\mathbf{B}}(O)$ are the terminal position vectors of P and O in the spatial frame $\Phi = \{F; \mathbf{I}_k\}$.

Euler's theorem has shown that the components of the rotation tensor, hence also those of the rotator, may be found by construction of the basis transformation matrix A in (3.123b). Thus, the angle and the axis of the equivalent Euler rotation may be obtained from (3.89) and (3.90) when the initial and final orientations of the body reference frame are known. This will be illustrated next.

Example 3.9. A certain mechanism is to be designed to rotate an antenna panel of a spacecraft about a point O so that the side facing the $\mathbf{i}_1$ direction in the initial configuration ultimately must face the initial $\mathbf{i}_3$ direction in the terminal configuration shown in Fig. 3.6. Determine the equivalent Euler rotation required for the design.

Solution. The spatial frame $\varphi = \{O; \mathbf{i}_j\}$ is chosen to coincide with the body frame $\varphi' = \{O; \mathbf{i}'_k\}$ in its initial orientation. The orientation of the body frame in the terminal configuration of the panel is shown in Fig. 3.6. The diagram is used with (3.123b) to construct the rotation matrix

$$R = A^T = \begin{bmatrix} 0 & 0 & 1 \\ 0 & -1 & 0 \\ 1 & 0 & 0 \end{bmatrix}. \tag{3.126}$$

Thus, $\operatorname{tr} R = -1$ and (3.89) yields $\cos\theta = -1$. Hence, $\theta = \pi$ is the equivalent angle of rotation.

The axis of rotation cannot be determined by (3.90) for this exceptional case; rather (3.77b) or (3.88) must be applied directly. Use of (3.87) in (3.77) yields

$$\boldsymbol{\alpha} \otimes \boldsymbol{\alpha} = \tfrac{1}{2}(\mathbf{R} + \mathbf{1}). \tag{3.127}$$

With (3.126), the diagonal components in (3.127) yield the three equations

$$\alpha_1^2 = 1/2, \qquad \alpha_2^2 = 0, \qquad \alpha_3^2 = 1/2; \tag{3.128}$$

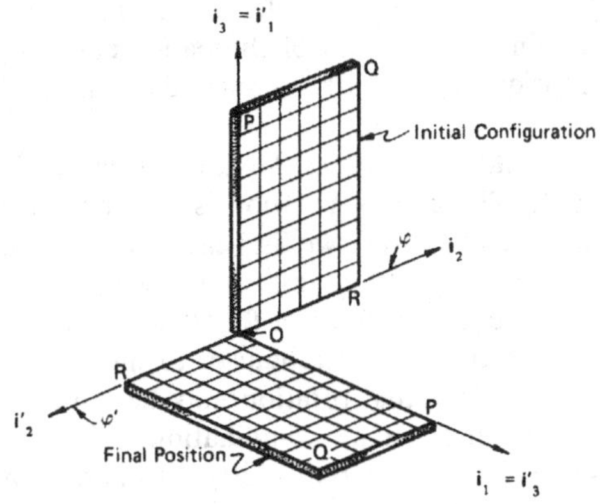

Figure 3.6. Finite rotation of an antenna panel.

and the single nonzero, nondiagonal element $\alpha_1\alpha_3 = 1/2$ shows that α_1 and α_3 have the same sign. We thus conclude with (3.128) that the equivalent Euler rotation consists of a right-hand rotation of $\theta = 180°$ about either of the directions given by

$$\boldsymbol{\alpha} = \pm\frac{\sqrt{2}}{2}(\mathbf{i}_1 + \mathbf{i}_3) \tag{3.129}$$

in the spatial frame φ. This axis lies in the plane of $\mathbf{i}_1$ and $\mathbf{i}_3$ at 45° from the $\mathbf{i}_1$ axis.

It may be helpful to show that the results flow also from (3.88). Use of (3.126) gives

$$R\alpha = \begin{bmatrix} 0 & 0 & 1 \\ 0 & -1 & 0 \\ 1 & 0 & 0 \end{bmatrix}\begin{bmatrix} \alpha_1 \\ \alpha_2 \\ \alpha_3 \end{bmatrix} = \begin{bmatrix} \alpha_3 \\ -\alpha_2 \\ \alpha_1 \end{bmatrix} = \alpha = \begin{bmatrix} \alpha_1 \\ \alpha_2 \\ \alpha_3 \end{bmatrix}.$$

Therefore, $\alpha_1 = \alpha_3$ and $\alpha_2 = 0$. Since $\boldsymbol{\alpha}$ is a unit vector, we have also $\alpha_1^2 + \alpha_2^2 + \alpha_3^2 = 2\alpha_1^2 = 1$; that is, $\alpha_1 = \alpha_3 = \pm\sqrt{2}/2$. Hence, the axis of the Euler rotation is given by (3.129), as before. □

Example 3.10. In another rotational maneuver of the panel mechanism from its initial configuration shown in Fig. 3.6, the rotation matrix about the point O is given by

$$R = \begin{bmatrix} 1/2 & 0 & -\sqrt{3}/2 \\ 0 & 1 & 0 \\ \sqrt{3}/2 & 0 & 1/2 \end{bmatrix}. \tag{3.130}$$

Find the equivalent Euler angle and axis of the rotation through O. Sketch the final orientation of the panel, and use the diagram to confirm the basis transformation matrix associated with (3.130).

Solution. With tr $R=2$, (3.89) yields $\theta=\cos^{-1}(1/2)=60°$ for the rotation angle. The axis of rotation is determined by (3.90). Thus, with $\sin\theta=\sqrt{3}/2$ and use of (3.130), we find easily $\boldsymbol{\alpha}=-\mathbf{i}_2$. Of course, a rotation of 300° about the axis $\mathbf{i}_2$ is the same; but, recognizing this trivial equivalence, we have agreed earlier to restrict $\theta \in [0, \pi]$. It must be born in mind, however, that in the design of the mechanism to move the panel to the terminal state found above, the actual rotation may be done in an infinite variety of ways. There are infinitely many combinations of rotations whose resultant leads to the same rotation matrix (3.130); but there is only one Euler rotation to which all are equivalent.

The Euler rotation consists of turning the body through 60° about the axis $\boldsymbol{\alpha}=-\mathbf{i}_2$ in the conventional right-hand sense. Therefore, the antenna panel has the final orientation sketched in Fig. 3.7. Thus, referring to the figure and recalling (3.123b), we find the transposed basis transformation matrix

$$A^T=\begin{bmatrix}\cos 60° & 0 & \cos 150° \\ 0 & 1 & 0 \\ \cos 30° & 0 & \cos 60°\end{bmatrix}=\begin{bmatrix}1/2 & 0 & -\sqrt{3}/2 \\ 0 & 1 & 0 \\ \sqrt{3}/2 & 0 & 1/2\end{bmatrix},$$

which is seen to be the same as R in (3.130). □

It is important to realize that there are infinitely many lines through the fixed base point O about which the body may be turned so as to move any *single* given particle P from its initial place to its final position that resulted from a previous arbitrary rotation about O. Euler's rotation is remarkable because, among these infinity of rotations, only Euler's rotation will move *every* particle of a rigid body, that is the entire body, from its initial con-

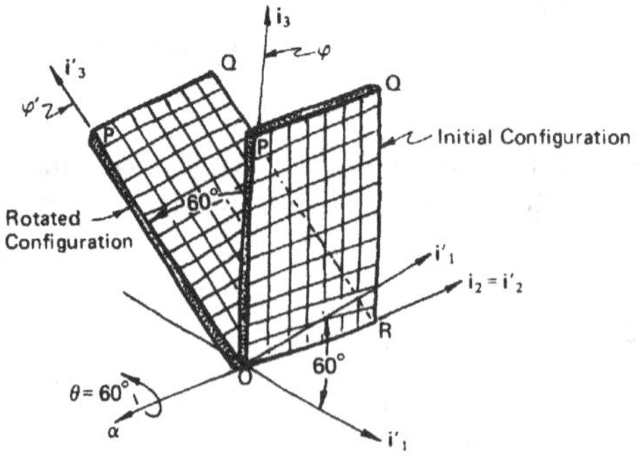

Figure 3.7. Sketch of the panel orientation in φ after a 60° rotation about $\boldsymbol{\alpha}=-\mathbf{i}_2$.

figuration in φ into its final configuration in φ, however that terminal state actually may have been achieved. This distinction may be readily visualized in Fig. 3.4. Therein, we recall that the rotation about the fixed point O carried the particle P at $\mathbf{x}$ in φ into the place $\hat{\mathbf{x}}$ in φ, the angle between these vectors being ψ, say. We thus visualize that the same displacement of the *single* particle P may be effected by a rotation through the angle ψ about a fixed axis

$$\boldsymbol{\alpha} = (\mathbf{x} \times \hat{\mathbf{x}})/|\mathbf{x} \times \hat{\mathbf{x}}|, \tag{3.131}$$

which is perpendicular to the plane of $\mathbf{x}$ and $\hat{\mathbf{x}}$. We note that this may be rewritten as

$$\mathbf{x} \times \hat{\mathbf{x}} = |\mathbf{x}|^2 \sin \psi \boldsymbol{\alpha}. \tag{3.132}$$

But this *simple rotation* generally is *not* the equivalent Euler rotation, because it does not move *every* particle of the body into its terminal state resulting from the given rotation about the fixed point O. This will be illustrated in a numerical example below. The reader may show that (3.132) actually is equivalent to (2.7) and, in fact, may be used to derive it.

Example 3.11. The initial and final position vectors of the antenna horn described in Example 2.1 in a rotation about the fixed point F are given as

$$\mathbf{x}(H) = \frac{3\sqrt{2}}{2}(\mathbf{I} + \mathbf{J}) + 6\mathbf{K}\ \text{m}, \qquad \hat{\mathbf{x}}(H) = \frac{3\sqrt{2}}{10}(11\mathbf{I} - \mathbf{J}) + \frac{24}{5}\mathbf{K}\ \text{m}$$

in the spatial frame $\Phi = \{F; \mathbf{I}_k\}$. Find the angle and axis of the simple rotation described in (3.132) which transforms $\mathbf{x}$ into $\hat{\mathbf{x}}$ by a rotation in their plane, and thereby show that this rotation is not the unique Euler rotation.

Solution. It is seen from the data that $|\mathbf{x}| = |\hat{\mathbf{x}}| = 3\sqrt{5}$ m, and we compute

$$\mathbf{x} \times \hat{\mathbf{x}} = \tfrac{9}{5}(9\sqrt{2}\mathbf{I} + 7\sqrt{2}\mathbf{J} - 8\mathbf{K})\ \text{m}^2.$$

Hence, $|\mathbf{x} \times \hat{\mathbf{x}}| = 162/5\ \text{m}^2$. The angle of the plane rotation is given by

$$\sin \psi = \frac{|\mathbf{x} \times \hat{\mathbf{x}}|}{|\mathbf{x}|^2} = 18/25;$$

and gathering these values in (3.131), we find the axis of the simple rotation that moves $\mathbf{x}$ into $\hat{\mathbf{x}}$, as assigned:

$$\boldsymbol{\alpha} = \tfrac{1}{18}(9\sqrt{2}\mathbf{I} + 7\sqrt{2}\mathbf{J} - 8\mathbf{K}).$$

It is evident that the axis and angle of rotation found here are not the same as those assigned in Example 2.1, wherein the fixed axis of rotation clearly is the Euler axis about which the entire body is turned. Hence, the displacement (2.7) for a rigid body rotation about a fixed line is not automatically equivalent to a rotation about the fixed point shown in Fig. 3.4. For any given change of basis, i.e., for any pair of orientations of the body reference frame in a pure rotation about a fixed point, the angle and the axis of the equivalent rotation about a fixed line generally cannot be found in this way. Indeed, this illustration shows clearly the importance of Euler's marvelous theorem.

3.9. Fundamental Invariant Property of the Rotator

The order of the motions of translation and rotation in the general displacement (2.13) obviously may be reversed, and they may occur simultaneously. It is also clear that the choice of base point being arbitrary, a given displacement may be constructed in an infinite variety of ways. Therefore, this raises the question of whether or not the rotator may be affected when the base point is shifted arbitrarily to another point in the body. The question is settled by the following theorem.

Rotator Invariance Theorem. *For a given displacement of a rigid body, the rotator is independent of the choice of base point; consequently, the axes of rotation corresponding to all base points are parallel and the angles of rotation about them are equal.*

Proof. Let P be any particle of the rigid body $\mathscr{B}$; and, for the same displacement of $\mathscr{B}$, let us assume that $\mathbf{T}$ and $\mathbf{T}^*$ are distinct rotators of $\mathscr{B}$ about lines through two base points O and O^* whose displacements are $\mathbf{b}$ and $\mathbf{b}^*$, respectively, as diagrammed in Fig. 3.8. Since the displacement of any particle of the rigid body, by definition, is the same for every choice of base point used

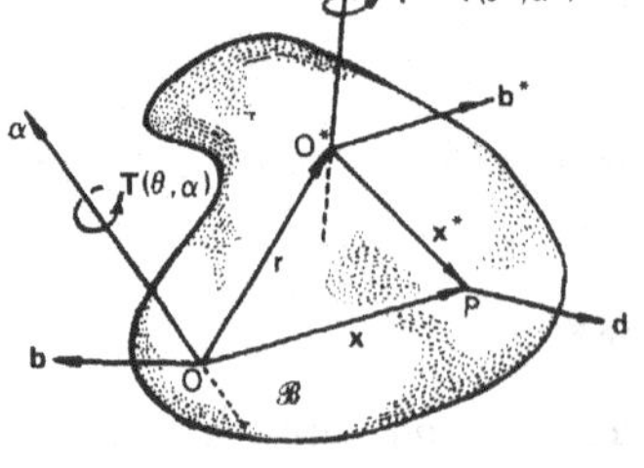

Figure 3.8. A change of base point in a general displacement of a rigid body.

in the decomposition described by (2.13), with the vectors defined in Fig. 3.8, we may write for both O and O^*

$$\begin{aligned}\mathbf{d}(P) &= \mathbf{b} + \mathbf{T}\mathbf{x} = \mathbf{b}^* + \mathbf{T}^*\mathbf{x}^* \\ &= (\mathbf{b} + \mathbf{T}\mathbf{r}) + \mathbf{T}\mathbf{x}^*,\end{aligned} \tag{3.133}$$

wherein $\mathbf{x} = \mathbf{r} + \mathbf{x}^*$ has been used. But with O as the base point, it is seen that the term in parentheses is equal to $\mathbf{b}^*$. Therefore, $(\mathbf{T}^* - \mathbf{T})\,\mathbf{x}^* = \mathbf{0}$. Because P is an arbitrary particle, this equation must hold for all $\mathbf{x}^*$; hence, $(\mathbf{T}^* - \mathbf{T})$ must be the zero tensor. It follows that the rotator is invariant with respect to the choice of base point: $\mathbf{T}^* \equiv \mathbf{T}(\theta^*, \boldsymbol{\alpha}^*) = \mathbf{T}(\theta, \boldsymbol{\alpha})$.

Moreover, it is now evident from (3.74) and either (3.76) or (3.80) that the rotation angles θ^* and θ about the axes $\boldsymbol{\alpha}^*$ and $\boldsymbol{\alpha}$, respectively, may differ by at most a trivial rotation or an image rotation, neither of which is of special consequence. Therefore, $\boldsymbol{\alpha}^* = \boldsymbol{\alpha}$ and $\theta^* = \theta$. Thus, in a given rigid body displacement, the axes of rotation corresponding to all base points are parallel and the angles of rotation about them are the same. This completes the proof of the invariance theorem. In sum, it shows that *the Euler rotation in a given rigid body displacement is invariant with respect to the choice of base point.*

Use of the rotator invariance theorem in (3.133) yields

$$\mathbf{d}(P) = \mathbf{b} + \mathbf{T}\mathbf{x} \tag{3.134a}$$

$$= \mathbf{b}^* + \mathbf{T}\mathbf{x}^*. \tag{3.134b}$$

That is to say, *the finite displacement of a particle of a rigid body due to a translation* $\mathbf{b}$ *and a rotation* $\mathbf{T}$ *about a base point* O *is equivalent to an identical rotation* $\mathbf{T}$ *about any other base point* O^* *together with a new translation* $\mathbf{b}^*$ *given by*

$$\mathbf{b}^* = \mathbf{b} + \mathbf{T}\mathbf{r}, \tag{3.135}$$

in which $\mathbf{r} = \mathbf{x} - \mathbf{x}^*$ *is the position vector of* O^* *from* O. (See Fig. 3.9.)

A change of base point, therefore, plainly results in a different trans-

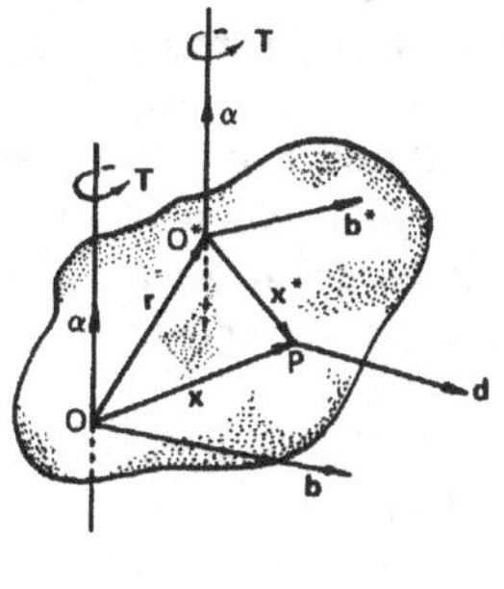

Figure 3.9. The Euler rotation for a given displacement is the same for every choice of base point.

lational displacement given by (3.135) for the new base point. Consequently, for each choice of base point the translational displacement of a rigid body will vary, but the Euler rotation will not.

3.10. The Parallel Axis Theorem

The foregoing results enable one to find both the translational and rotational parts of a rigid body displacement corresponding to any proposed base point when these displacements are known for an assigned base point. An easy result that derives from (3.134) and the fact that $\mathbf{Tx}\cdot\mathbf{a}=\mathbf{0}$ for all particles is summarized by the Invariant Projection Theorem.

Invariant Projection Theorem. *The projections on the Euler axis of rotation of the displacements of all points of a rigid body are equal; that is, for all particles P*

$$\mathbf{d}(P)\cdot\mathbf{a}=\mathbf{b}\cdot\mathbf{a}. \tag{3.136}$$

In particular, for a rotation about a fixed point (or line), we have $\mathbf{b}=\mathbf{0}$; and, trivially, (3.136) verifies that the displacement of every material point is perpendicular to the Euler axis of rotation. In this case, (3.134) and (3.135) show that the displacement of the particle P is given by

$$\mathbf{d}(P)=\mathbf{Tx}=\mathbf{b}^*+\mathbf{Tx}^* \tag{3.137a}$$

with

$$\mathbf{b}^*=\mathbf{Tr} \tag{3.137b}$$

and

$$\mathbf{b}^*\cdot\mathbf{a}=0, \tag{3.137c}$$

where $\mathbf{x}$ and $\mathbf{x}^*$ are the position vectors of P from O and O^*, respectively, and $\mathbf{r}=\mathbf{x}-\mathbf{x}^*$ is the position vector of O^* from O. (See Fig. 3.9 with $\mathbf{b}=\mathbf{0}$.) The result (3.137) thus yields the following theorem.

Parallel Axis Theorem. *The displacement of a particle of a rigid body due to a pure rotation about any line is equivalent to a displacement due to an equal rotation about a parallel line together with a translation perpendicular to that line, and conversely.*

A simple application that demonstrates the parallel axis theorem will be studied next.

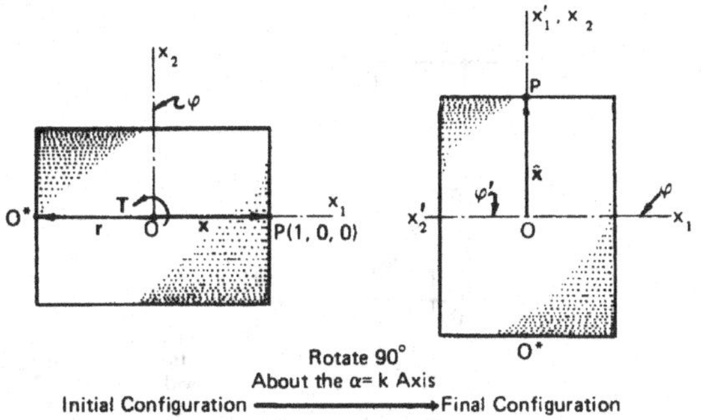

Figure 3.10. Rigid body displacement in a pure 90° rotation about O.

Example 3.12. A particle P of a rectangular plate is located initially at $\mathbf{x}(P) = 1\mathbf{i}$ ft from the base point O in the spatial frame φ. The plate is turned 90° counterclockwise about the axis $\boldsymbol{\alpha} = \mathbf{k}$ so that the final position vector of P referred to φ is $\hat{\mathbf{x}}(P) = 1\mathbf{j}$ ft, as shown in Fig. 3.10. Thus, the displacement of P referred to φ due to the pure rotation about O is given easily by

$$\mathbf{d}(P) = \hat{\mathbf{x}}(P) - \mathbf{x}(P) = (-\mathbf{i} + \mathbf{j})\text{ ft}. \tag{3.138}$$

Of course, the same thing may be computed from (3.137a) for a pure rotation about the point O. With $\mathbf{d} = \mathbf{T}\mathbf{x} = (\mathbf{R} - \mathbf{1})\mathbf{x}$, we have

$$d(P) = Tx(P) = \begin{bmatrix} -1 & -1 & 0 \\ 1 & -1 & 0 \\ 0 & 0 & 0 \end{bmatrix} \begin{bmatrix} 1 \\ 0 \\ 0 \end{bmatrix} = \begin{bmatrix} -1 \\ 1 \\ 0 \end{bmatrix} \text{ft},$$

which agrees with (3.138).

Now let us consider the same plate to be rotated 90° about a parallel axis at the new base point O^* initially at $\mathbf{r} = -\mathbf{i}$ ft from O in φ, as shown in Fig. 3.10. The initial position vector of P from O^* is $\mathbf{x}^*(P) = \mathbf{x}(P) - \mathbf{r} = 2\mathbf{i}$ ft. The parallel axis theorem states that the same displacement given in (3.138) for a pure rotation about O can be accomplished by a translation $\mathbf{b}^*$ perpendicular to the axis $\boldsymbol{\alpha} = \mathbf{k}$ together with the same rotation about a parallel axis through O^*. Since $\mathbf{T}$ is the same as before, the rotational part of the displacement of P about O^* is given as

$$Tx^*(P) = \begin{bmatrix} -1 & -1 & 0 \\ 1 & -1 & 0 \\ 0 & 0 & 0 \end{bmatrix} \begin{bmatrix} 2 \\ 0 \\ 0 \end{bmatrix} = \begin{bmatrix} -2 \\ 2 \\ 0 \end{bmatrix}. \tag{3.139}$$

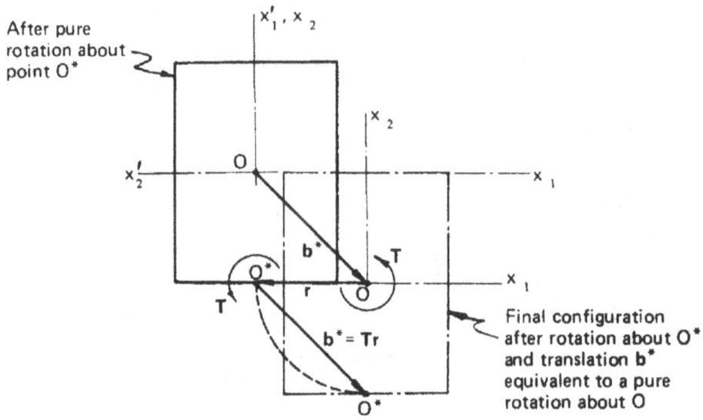

Figure 3.11. The displacement due to a pure rotation about an axis at O is equal to the displacement due to the same rotation about a parallel axis at O^* and a translation $\mathbf{b}^*$ perpendicular to the axis of rotation.

Thus, use of (3.138) and (3.139) in (3.137) yields $\mathbf{b}^* = \mathbf{d}(P) - \mathbf{T}\mathbf{x}^*(P) = (-\mathbf{i}+\mathbf{j}) - (-2\mathbf{i}+2\mathbf{j}) = (\mathbf{i}-\mathbf{j})$ ft, which is certainly perpendicular to $\boldsymbol{\alpha}$. This is the required translational displacement of O^* in φ. The reader may confirm that the translation $\mathbf{b}^*$ obtained from (3.137b) yields the same result. The displacement is illustrated in Fig. 3.11. Notice that $\mathbf{b}^*$ is simply the chord displacement of O^* on the circle of radius OO^* due to a pure rotation about O.

3.10.1. The Center of Rotation

A displacement of a rigid body $\mathscr{B}$ each of whose particles is displaced parallel to a given plane is called a *plane displacement* of $\mathscr{B}$. Any plane cross section of $\mathscr{B}$ parallel to the assigned plane, which is named the *displacement plane*, may be chosen as the plane for discussion; it requires no special distinction. It follows that the axis of rotation must be perpendicular to the displacement plane. If $\mathbf{T} = \mathbf{0}$, the plane displacement is a pure translation with $\mathbf{d}(P) = \mathbf{b}$ for all particles P. If there exists a base point O for which $\mathbf{b} = \mathbf{0}$, the plane displacement is a pure rotation about an axis at O.

It is not difficult to prove that in an *unconstrained* displacement of a rigid body in space, there is, in general, no point whose displacement is zero. (See Problem 3.36.) On the other hand, it can be shown that *for a plane displacement of a rigid body, there always exists one and only one point whose displacement is zero.* This is a consequence of the parallel axis theorem applied to the case in which every particle suffers a plane displacement consisting of a parallel translation and a rotation about an axis normal to the plane. Hence, the entire motion, by the parallel axis theorem, is equivalent to a pure rotation about a parallel axis at a point O^*, say, situated in the displacement plane. The point O^* whose displacement is zero is known as the *center of* *rotation.* In general, this unique base point will not be within the body, unless, of course, it happens to be a fixed material point about which the body is rotating normal to the displacement plane.

It is easy to see geometrically that because a particle P in the displacement plane must be equidistant from the center of rotation, the point O^* must lie on the perpendicular bisector of the displacement vector $\mathbf{d}(P)$ of the particle. But the same thing is true for every particle situated in the displacement plane; hence, the center of rotation may be constructed graphically by locating the point of intersection of the normal bisectors of the displacement vectors of two particles in the displacement plane. Of course, it may happen in some problems that the center of rotation can not be accurately located in this way. Therefore, we are led to question how the center of rotation may be computed. For the plane displacement, this is an easy problem in vector geometry, which is left as an exercise for the reader. (See Problem 3.37.) The parallel axis theorem and existence of a center of rotation in a plane displacement lead to some useful graphical applications which may be found in most standard texts on kinematics of machines. A few example are included in the exercises at the end of the chapter. (See Problems 3.38–3.40.) Some additional theoretical results will be presented next.

3.11. Chasles' Screw Displacement Theorem

The choice of base point to be used in (2.13) is totally arbitrary, so any convenient choice is admissible. In fact, we recall again that the base point need not be a material point of the body; so every conceivable point of space is a potential candidate for use as a base point. Because of this arbitrariness in the selection of a base point, the displacement of a rigid body particle may be described in an infinite variety of ways. Therefore, we are led to question if there may be any particularly special choices of base point for which the displacement of the particles of a rigid body may be most simply, perhaps uniquely, described. We are going to show with the aid of the parallel axis theorem that the answer is provided by the following remarkable theorem due to Chasles:

The Screw Displacement Theorem. *Any displacement of a rigid body is reducible to a unique screw displacement consisting of the Euler rotation about an axis and a translation along that axis.*

Proof. Let the displacement of a base point O and the Euler rotation of the rigid body be assigned; then the general displacement of any particle P at $\mathbf{x}$ from O is provided by (3.134a). We wish to prove that there exists a base point O^*, say, whose displacement $\mathbf{b}^*$ is parallel to the axis of the Euler rotation; hence, with O^* as the base point, the result follows. Therefore, we shall need to determine the new parallel translation $\mathbf{b}^*$ and the location $\mathbf{r}$ of O^* from O.

We may always write $\mathbf{b}=\mathbf{b}_n+\mathbf{b}_\alpha$ in terms of the component vectors $\mathbf{b}_n$ and $\mathbf{b}_\alpha$ normal and parallel, respectively, to the axis of rotation, as shown in Fig. 3.12. Hence, (3.134a) may be written

$$\mathbf{d}(P)=\mathbf{b}_\alpha+[\mathbf{b}_n+\mathbf{Tx}(P)]. \tag{3.140}$$

However, in accordance with the parallel axis theorem, the displacement terms consisting of a rotation about a line and a translation perpendicular to the line is equivalent to a pure rotation about a parallel axis through a base point at O^*, say. Thus, the term in the brackets in (3.140) may be replaced by the pure Euler rotation $\mathbf{Tx}^*(P)$ to yield

$$\mathbf{d}(P)=\mathbf{b}_\alpha+\mathbf{Tx}^*(P) \tag{3.141a}$$

with

$$\mathbf{b}_n=-\mathbf{Tr}, \tag{3.141b}$$

where $\mathbf{r}\equiv\mathbf{x}-\mathbf{x}^*$ is the position vector of O^* from O, as shown in Fig. 3.12. But (3.141a) states that the assigned displacement $\mathbf{d}(P)$ is equal to the same Euler rotation about an axis at O^* and a new translation $\mathbf{b}^*\equiv\mathbf{b}_\alpha$ parallel to that axis, and Chasles' theorem follows.

This displacement (3.141) is recognized as a typical screw displacement, from which the theorem derives its name. The axis of the Euler rotation at O^* is called the *screw axis.** The *pitch of the screw*, defined by $p\equiv(\mathbf{b}\cdot\boldsymbol{\alpha})/\theta$, is identified as the ratio of the screw translational displacement to the angle of rotation. Thus, in terms of the familiar screw displacement of a nut, the rectilinear distance along the screw axis through which the nut advances when turned through a given angle is simply the product of the pitch of the screw and the circular measure of its angle of rotation. Finally, we see from (3.136) that *the pitch of the screw is invariant with respect to the choice of base point.* It is apparent that zero pitch means pure rotation, while infinite pitch is a pure translation.

The system of three algebraic equations (3.141b), in which the normal component of the assigned base point translation is given by

$$\mathbf{b}_n=\mathbf{b}-\mathbf{b}_\alpha \tag{3.142a}$$

with

$$\mathbf{b}_\alpha=(\mathbf{b}\cdot\boldsymbol{\alpha})\boldsymbol{\alpha}\equiv p\theta\boldsymbol{\alpha}, \tag{3.142b}$$

* In older works, the screw axis sometimes is called the central axis, and Chasles' theorem is known as the central axis theorem. The screw theorem forms the foundation upon which the classical kinematical theory of screws was built. See the treatise by Ball referenced at the end of Chapter 2.

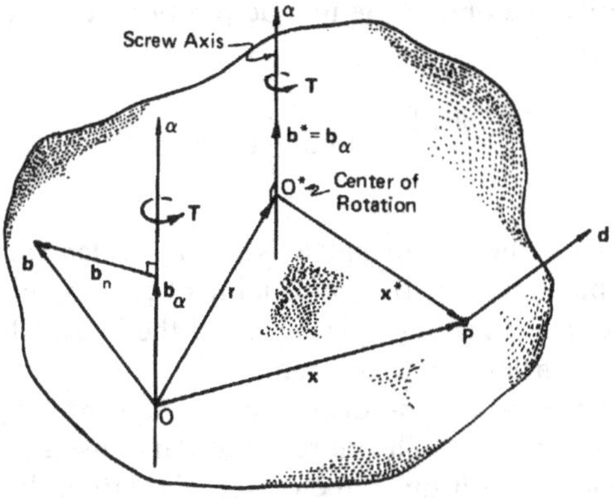

Figure 3.12. Schematic for Chasles' screw displacement theorem.

determines the position vector $\mathbf{r}$ of the new base point O^*. However, since $\det \mathbf{T} = 0$ in (3.124c), the system (3.141b) does not determine $\mathbf{r}$ uniquely; for, if $\mathbf{r}$ is a solution of (3.141b), so is $\mathbf{x}' = \mathbf{r} + k\boldsymbol{\alpha}$ for all values of the constant k. We recognize this relation as the vector equation of the screw axis through O^*; it implies that no special point on the new axis of rotation is distinguished. Since any point on the new axis may be used, without loss of generality, we may choose O^* so that $\mathbf{r}$ is the shortest vector from O to the new parallel axis. Then, in addition to (3.141b), we have the condition

$$\mathbf{r} \cdot \boldsymbol{\alpha} = 0. \tag{3.143}$$

The point O^* located in this way will be called the *center of rotation*. Of course, since $\mathbf{r}$ will vary with the choice of base point, the location of the center of rotation with respect to the base point used will vary along the axis of rotation. Thus, for an assigned base point and with (3.141b) and (3.143), the location of the center of rotation O^* yielding the screw displacement (3.141a) may be uniquely determined, as will be shown next.

With the aid of (3.70) and (3.143), it is seen that (3.141b) may be written as

$$\mathbf{T}\mathbf{r} = \mathbf{S}\mathbf{r} \sin\theta - (1 - \cos\theta)\mathbf{r} = -\mathbf{b}_n. \tag{3.144}$$

We recall next that $\boldsymbol{\alpha}$ is the vector of the skew tensor $\mathbf{S}$ so that $\mathbf{S}\mathbf{r} = \boldsymbol{\alpha} \times \mathbf{r}$. And, with (3.143), it may be seen that $\boldsymbol{\alpha} \times \mathbf{S}\mathbf{r} = -\mathbf{r}$. Upon forming the cross product of (3.144) with $\boldsymbol{\alpha}$, we find

$$\mathbf{r} \sin\theta + (1 - \cos\theta)\, \mathbf{S}\mathbf{r} = \boldsymbol{\alpha} \times \mathbf{b}_n. \tag{3.145}$$

Thus, for nontrivial values of θ, (3.144) may be solved for the vector $\mathbf{S}\mathbf{r}$; and when the result is substituted into (3.145) and reduced by use of a trigonometric identity, we obtain the unique position vector of the base point O^* on the screw axis:

$$\mathbf{r} = \frac{1}{2}\left[\mathbf{b}_n + \cot\frac{\theta}{2}\,\boldsymbol{\alpha}\times\mathbf{b}_n\right], \tag{3.146}$$

in which θ and $\boldsymbol{\alpha}$ are the angle and axis of the Euler rotation and $\mathbf{b}_n$ in (3.142a) is the displacement of the base point O perpendicular to the Euler axis determined for the assigned displacement of the body. This completes the technical details in Chasles' screw theorem.

If the displacement is a plane displacement, then $\mathbf{b}^* = \mathbf{b}_\alpha = \mathbf{0}$, and $\mathbf{b}_n = \mathbf{b}$ lies in the displacement plane. Therefore, in a plane displacement, the center of rotation O^* coincides with the plane center of rotation described earlier in Section 3.10.1, and (3.146) has the easy geometrical description shown in Problem 3.37. Thus, the interpretation of the center of rotation in the general case is similar, but an additional translation along the screw axis is superimposed on the rotation. And, as in the plane case, it is evident that in general the screw axis will not be within the body.

The foregoing method of reduction of any given displacement of a rigid body to a screw displacement is unique. It is clearly impossible to have distinct screw axes through separate base points. For any nontrivial rotation about either axis, the particles situated upon the other line will not be restored to their original line after the rotation. This observation may be proved analytically as follows.

Suppose that there are two screw axes $\boldsymbol{\alpha}'$ and $\boldsymbol{\alpha}^*$ through noncoaxial points O' and O^* such that (3.141a) holds for the given displacement $\mathbf{d}(P)$. Invariance of the rotator with respect to the choice of base point implies at once that the axes must be parallel; hence, $\boldsymbol{\alpha}' = \boldsymbol{\alpha}^* \equiv \boldsymbol{\alpha}$, say. Therefore, for distinct screw displacements with respect to the noncoaxial base points,

$$\mathbf{d}(P) = \mathbf{b}'_\alpha + \mathbf{T}\mathbf{x}'(P) = \mathbf{b}^*_\alpha + \mathbf{T}\mathbf{x}^*(P),$$

wherein $\mathbf{x}'(P)$ and $\mathbf{x}^*(P)$ denote the position vectors of P from O' and O^*, respectively. Also, $\mathbf{b}'_\alpha$ and $\mathbf{b}^*_\alpha$ are the corresponding screw translations along the separate screw axes at O' and O^*. But invariance of the pitch implies that $\mathbf{b}^*_\alpha = \mathbf{b}'_\alpha$; and it follows from the previous equation that $\mathbf{T}\mathbf{r} = \mathbf{0}$, where $\mathbf{r} = \mathbf{x}' - \mathbf{x}^*$. Since $\mathbf{T} \neq \mathbf{0}$, this means that $\mathbf{r}$ must be parallel to $\boldsymbol{\alpha}$; hence, contrary to the hypothesis, O' and O^* lie on the same axis. Therefore, we may conclude the following uniqueness theorem: *There is at most one screw displacement by which a given displacement of a rigid body may be accomplished.* Thus, among all possible displacements of a rigid body, Chasles' unique screw displacement has unparalleled simplicity.

In the next section, it will be shown that successive displacements of a body may be represented by successive screws whose composition is an equivalent unique screw displacement. The following example will illustrate the essential aspects of Chasles' notable screw theorem.

Example 3.13. A rigid body suffers a displacement described by a rotation

$$R=\frac{\sqrt{2}}{4}\begin{bmatrix} 1+\sqrt{3} & 1-\sqrt{3} & 0 \\ -1+\sqrt{3} & 1+\sqrt{3} & 0 \\ 0 & 0 & 2\sqrt{2} \end{bmatrix}$$

about a point at (3, −2, −4) cm in the fixed frame $\Phi=\{F;\mathbf{I}_k\}$, together with a certain translation. Another particle P, initially at $\mathbf{X}=\mathbf{I}-2\mathbf{J}-5\,\mathbf{K}$ cm in Φ, has been moved to the place $\hat{\mathbf{X}}=4\mathbf{I}+6\mathbf{J}-\mathbf{K}$ cm. Find the equivalent screw displacement of the body. What is the axial advance of the body in one revolution about the screw axis?

Solution. Any base point whose displacement is known may be used to determine the screw translation and the location of the screw axis in Φ. Since the rotation is independent of the choice of base point, the point $P\equiv O$ is a convenient choice for which the displacement is given by

$$\mathbf{b}=\hat{\mathbf{X}}-\mathbf{X}=3\mathbf{I}+8\mathbf{J}+4\mathbf{K}\text{ cm}.$$

It is easily seen from (3.90) that the Euler axis of the rotation is $\boldsymbol{\alpha}=\mathbf{K}$. Hence, with $\mathbf{b}$ above, (3.142b) yields the screw translation

$$\mathbf{b}_\alpha=4\boldsymbol{\alpha}=4\mathbf{K}\text{ cm},$$

which is the same for all points of the body; and (3.142a) yields the normal translation of the base point O:

$$\mathbf{b}_n=3\mathbf{I}+8\mathbf{J}\text{ cm}.$$

The angle of the equivalent rotation is found by (3.89). Thereby, we obtain from R above

$$\theta=\cos^{-1}[\sqrt{2}(1+\sqrt{3})/4]=15^\circ=\pi/12\text{ rad}.$$

The screw axis passes through the base point O^* located at $\mathbf{r}$ from the base point O, in accordance with (3.146). Use of the foregoing angle and normal translation vector yields

$$\cot\frac{\theta}{2}=\frac{\sin\theta}{1-\cos\theta}=\frac{2(2-\sqrt{3})^{1/2}}{4-\sqrt{2}-\sqrt{6}},$$

and

$$\boldsymbol{\alpha} \times \mathbf{b}_n = -8\mathbf{I} + 3\mathbf{J} \text{ cm}.$$

Substitution of these calculations into (3.146) gives

$$\mathbf{r} = \left[\frac{3}{2} - \frac{8(2-\sqrt{3})^{1/2}}{4-\sqrt{2}-\sqrt{6}}\right]\mathbf{I} + \left[4 + \frac{3(2-\sqrt{3})^{1/2}}{4-\sqrt{2}-\sqrt{6}}\right]\mathbf{J} \text{ cm}$$
$$= -28.884\mathbf{I} + 15.394\mathbf{J} \text{ cm}.$$

Therefore, the screw axis passes through the center of rotation O^* whose position vector $\mathbf{B}^*$ from the origin in Φ is given by

$$\mathbf{B}^* \equiv \mathbf{r} + \mathbf{X} = -27.883\mathbf{I} + 13.394\mathbf{J} - 5\mathbf{K} \text{ cm}.$$

We thus find that the given displacement may be reduced to a unique screw displacement consisting of a rotation of 15° about an axis $\boldsymbol{\alpha} = \mathbf{K}$ through the point at $\mathbf{B}^*$ together with a pure translation of the body through 4 cm along the axis. The pitch of the screw, defined in (3.142), is $p = 48/\pi$ cm/rad. Thus, in one revolution about the screw axis, the body will advance a distance of $2\pi p = 96$ cm along the axis. (See Problems 3.41 and 3.42.)

3.12. Composition of Finite Rotations

It was shown in Section 2.7 that successive finite rotations of a rigid body about concurrent axes are neither additive nor commutative. Therefore, as illustrated in Fig. 2.7, the displacement of a rigid body generally will depend upon the order in which the rotations are performed. We saw in a few earlier examples that when the successive rotations are easy to visualize, their composition may be readily written down by use of the direction cosines between the body imbedded axes in their terminal state and those of the spatial set with which they were coincident initially. Needless to say, it is not always easy to perceive the successive and the resultant effects of several complex rotations, so it will be useful to derive the rule for their composition.

Let us begin by recalling (3.119a) and (3.123a) for a rotation about a fixed point O. After a rotation $\mathbf{R}$ in the spatial frame $\varphi = \{O; \mathbf{i}_k\}$, the position vector $\hat{\mathbf{x}}$ of the particle P which initially was at the place $\mathbf{x}$ will be given by (3.86):

$$\hat{\mathbf{x}} = \mathbf{R}\mathbf{x} \qquad [\text{cf. } (3.86)].$$

The specific form of the matrix of $\mathbf{R}$ in φ may be found as described earlier in terms of the direction angles of the imbedded body axes; or, for a rotation about any axis $\boldsymbol{\alpha}$ in φ, the matrix R may be determined from its relation to the rotator in (3.87), whichever method may be the more appropriate.

Let us consider N successive finite rotations $\mathbf{R}_1, \mathbf{R}_2, \ldots, \mathbf{R}_N$ about N concurrent axes through O, all referred to φ; and let $\mathbf{x} \equiv \hat{\mathbf{x}}_0, \hat{\mathbf{x}}_1, \hat{\mathbf{x}}_2, \ldots, \hat{\mathbf{x}}_{N-1}, \hat{\mathbf{x}}_N \equiv \mathbf{x}^*$ denote the corresponding successive position vectors of the particle P which initially is at $\mathbf{x}$ from O, all referred to φ. Then, by (3.86), we have N equations $\hat{\mathbf{x}}_1 = \mathbf{R}_1 \hat{\mathbf{x}}_0$, $\hat{\mathbf{x}}_2 = \mathbf{R}_2 \hat{\mathbf{x}}_1, \ldots$; or, more briefly, $\hat{\mathbf{x}}_k = \mathbf{R}_k \hat{\mathbf{x}}_{k-1}$, $k = 1, 2, \ldots, N$ (no sum on k). Hence, successive elimination of $\hat{\mathbf{x}}_{k-1}$ from these equations yields the final position vector $\mathbf{x}^*$ of P in terms of its initial place $\mathbf{x}$; namely,

$$\mathbf{x}^*(P) = \mathbf{R}^* \mathbf{x}(P) \tag{3.147a}$$

with

$$\mathbf{R}^* \equiv \mathbf{R}_N \mathbf{R}_{N-1} \cdots \mathbf{R}_2 \mathbf{R}_1. \tag{3.147b}$$

It is easy to show that $\mathbf{R}^*$ satisfies (3.91) and thus represents the equivalent Euler rotation about O. The resultant angle of rotation θ^* and the resultant axis $\boldsymbol{\alpha}^*$ may now be found in the usual way from (3.89) and (3.90) applied to $\mathbf{R}^*$. The resultant axis of rotation generally will depend on the order of the rotations in (3.147b), but it follows from the rule (3.41) that θ^* is independent of that order. That is, for rotations $\mathbf{R}_1$ and $\mathbf{R}_2$, in general $\mathbf{R}_2 \mathbf{R}_1 \neq \mathbf{R}_1 \mathbf{R}_2$; but it is true always that

$$\operatorname{tr} \mathbf{R}^* = 1 + 2\cos\theta^* = \operatorname{tr}(\mathbf{R}_1 \mathbf{R}_2) = \operatorname{tr}(\mathbf{R}_2 \mathbf{R}_1). \tag{3.148}$$

Of course, when the base point O has the displacement $\mathbf{b}$ in $\Phi = \{F; \mathbf{I}_k\}$, which is parallel to φ, the resultant displacement of P in Φ may be written in terms of the resultant rotator as follows:

$$\mathbf{d}^*(P) = \mathbf{b} + \mathbf{T}^* \mathbf{x}(P) \qquad \text{with} \quad \mathbf{T}^* = \mathbf{R}^* - \mathbf{1}. \tag{3.149}$$

It is important to recall that tensors have different components with respect to different bases; therefore, it is essential to bear in mind in (3.147b) the reference bases to which the rotation tensors are referred. If all of the rotations are referred to the same frame $\psi = \{O; \mathbf{e}_k\}$, say, then the resultant rotation matrix R^* of (3.147b) is obtained from the product of the matrices of the tensors formed in the order indicated in (3.147b), and R^* also will be referred to the frame ψ. In general, however, if different frames are used to follow the successive rotations, (3.147b) does not always reduce simply to the product of matrices of the tensors in the order indicated there. To see the difference, we shall illustrate each case in its turn.

Let us consider first two rigid body rotations $\mathbf{R}_1 : \mathbf{e}_k \to \mathbf{e}_k^1$ and $\mathbf{R}_2 : \mathbf{e}_k \to \mathbf{e}_k^2$

with respect to the *same* basis $\mathbf{e}_k$. Then (3.103) and (3.123) yield the representations

$$\mathbf{R}_1 = \mathbf{e}_k^1 \otimes \mathbf{e}_k = R_{pk}^1 \mathbf{e}_p \otimes \mathbf{e}_k \quad \text{and} \quad \mathbf{R}_2 = \mathbf{e}_q^2 \otimes \mathbf{e}_q = R_{rq}^2 \mathbf{e}_r \otimes \mathbf{e}_q,$$

in which $R_{pk}^n = A_{kp}^n \equiv \cos\langle \mathbf{e}_k^n, \mathbf{e}_p \rangle$ for $n = 1, 2$ (that is, $R_n = A_n^T$) are assumed to be known. The resultant rotation is formed by the product of these tensors in accordance with (3.147b). Recalling (3.104) and the orthogonality relations (3.95), we find

$$\mathbf{R}_2 \mathbf{R}_1 = R_{rp}^2 R_{pk}^1 \mathbf{e}_r \otimes \mathbf{e}_k = \mathbf{R}^*,$$

which yields the following matrix equation for the resultant rotation:

$$R_{rk}^* = R_{rp}^2 R_{pk}^1, \qquad \text{that is,} \quad R^* = R_2 R_1.$$

We thus find that the resultant rotation is the product of the matrices of the rotation tensors, with respect to the basis $\mathbf{e}_k$, formed in the order indicated in (3.147b); and R^* is referred to the same basis.

Now let us suppose that to find the components of the second rotation it proves more convenient to use a *different* basis such that $\mathbf{R}_2$: $\mathbf{e}_k^1 \to \mathbf{e}_k^2$, say. This is a rotation with respect to the basis $\mathbf{e}_k^1$ that was reached after the first rotation considered before. Thus, (3.103) and (3.123) yield the representations

$$\mathbf{R}_1 = \mathbf{e}_k^1 \otimes \mathbf{e}_k = R_{pk}^1 \mathbf{e}_p \otimes \mathbf{e}_k \quad \text{and} \quad \mathbf{R}_2 = \mathbf{e}_q^2 \otimes \mathbf{e}_q^1 = R_{rq}^2 \mathbf{e}_r^1 \otimes \mathbf{e}_q^1,$$

in which $R_{pk}^2 = A_{kp}^2 \equiv \cos\langle \mathbf{e}_k^2, \mathbf{e}_p^1 \rangle$ so that $R_2 = A_2^T$, and R_{pk}^1 is the same as previously. These arrays are assumed to be known from the geometry. The resultant rotation is formed by the product of these tensors in accordance with (3.147b). Substituting the change of basis $\mathbf{e}_k^1 = R_{pk}^1 \mathbf{e}_p$ into the last term above and recalling (3.104) and the orthogonality property (3.91) applied to R_1, we obtain

$$\mathbf{R}^* = \mathbf{R}_2 \mathbf{R}_1 = R_{rq}^2 R_{sr}^1 R_{mq}^1 (\mathbf{e}_s \otimes \mathbf{e}_m)\, R_{pk}^1 (\mathbf{e}_p \otimes \mathbf{e}_k) = R_{sr}^1 R_{rq}^2 \mathbf{e}_s \otimes \mathbf{e}_q = R_{sq}^* \mathbf{e}_{sq}.$$

This yields the following matrix equation for the resultant rotation:

$$R_{sq}^* = R_{sr}^1 R_{rq}^2, \qquad \text{that is,} \quad R^* = R_1 R_2.$$

Thus, when the resultant rotation is referred to the initial or the terminal basis, but the successive rotations are performed with respect to different bases for computational convenience, say, we see that the order of the product of the matrices may not be the same as the order of the product of their tensors in the fundamental equation (3.147b). In fact, in this example, the order is reversed. (See Problem 3.49 also.)

Although the construction of the basis transformation matrices for successive rotations is straightforward, the last example shows that the method sometimes may become awkward when the axes of rotation are not the axes of one of the primary reference frames being used. In such cases, it is often simpler to apply the formula (3.87) for the rotation tensor in terms of the angle and axis of the assigned rotation in an appropriate frame $\psi = \{O; \mathbf{e}_k\}$. We combine (3.70) and (3.87) to obtain

$$\mathbf{R} = \mathbf{1} + \mathbf{S}\sin\theta + (1 - \cos\theta)(\boldsymbol{\alpha}\otimes\boldsymbol{\alpha} - \mathbf{1}), \tag{3.150}$$

wherein, for the reader's convenience, we recall (3.72) and (3.24):

$$\mathbf{S} = -\varepsilon_{ijk}\alpha_k \mathbf{e}_{ij} \tag{3.151a}$$

and

$$\boldsymbol{\alpha}\otimes\boldsymbol{\alpha} = \alpha_i\alpha_j\mathbf{e}_{ij} \tag{3.151b}$$

referred to ψ. For each assigned axis and angle of rotation, $\mathbf{R}$ may be computed from (3.150); and the resultant rotation may be found by the matrix product of the successive rotation matrices in the form provided in (3.147b). The angle and the axis of the resultant Euler rotation may then be computed in the frame ψ by aid of (3.89) and (3.90), as usual. It is useful to observe in calculations that the matrix of $\mathbf{S}$ is skew and that of $\boldsymbol{\alpha}\otimes\boldsymbol{\alpha}$ is symmetric.

Example 3.14. The solar panel of a spacecraft receives three rotations about axes $\boldsymbol{\alpha}_k$ in the spatial frame $\psi = \{O; \mathbf{e}_k\}$, as shown in Fig. 3.13. The first rotation is 90° about the panel axis $\boldsymbol{\alpha}_1$; the second is 90° about the satellite body axis $\boldsymbol{\alpha}_2$; and the last is a 180° turn of the satellite about the line $\boldsymbol{\alpha}_3$. Find the angle and axis of the equivalent Euler rotation. What is the final orientation in ψ of the satellite body axis?

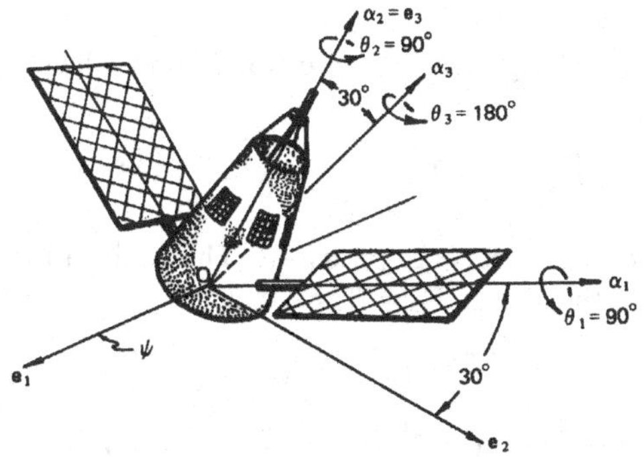

Figure 3.13. Finite rotations of the solar panel of a spacecraft.

Solution. The equation for the kth rotation $\mathbf{R}_k$ through the angle θ_k about the axis $\boldsymbol{\alpha}_k$ may be obtained from (3.150):

$$\mathbf{R}_k = \mathbf{1} + \mathbf{S}_k \sin\theta_k + (1 - \cos\theta_k)(\boldsymbol{\alpha}_k \otimes \boldsymbol{\alpha}_k - \mathbf{1}), \tag{3.152}$$

without sum on k. Thus, with $\theta_1 = \pi/2$, the first rotation is given by $\mathbf{R}_1 = \mathbf{S}_1 + \boldsymbol{\alpha}_1 \otimes \boldsymbol{\alpha}_1$, wherein the axis is obtained from the geometry in Fig. 3.13:

$$\boldsymbol{\alpha}_1 = \frac{\sqrt{3}}{2}\mathbf{e}_2 + \frac{1}{2}\mathbf{e}_3$$

in ψ. Then use of this result in (3.151) yields the matrices

$$[\mathbf{S}_1] = \begin{bmatrix} 0 & -1/2 & \sqrt{3}/2 \\ 1/2 & 0 & 0 \\ -\sqrt{3}/2 & 0 & 0 \end{bmatrix}, \qquad [\boldsymbol{\alpha}_1 \otimes \boldsymbol{\alpha}_1] = \begin{bmatrix} 0 & 0 & 0 \\ 0 & 3/4 & \sqrt{3}/4 \\ 0 & \sqrt{3}/4 & 1/4 \end{bmatrix}$$

referred to ψ. It follows from the formula given earlier that the matrix in ψ of the rotation tensor $\mathbf{R}_1 = R^1_{pq}\mathbf{e}_{pq}$ is given by

$$R_1 = \begin{bmatrix} 0 & -1/2 & \sqrt{3}/2 \\ 1/2 & 3/4 & \sqrt{3}/4 \\ -\sqrt{3}/2 & \sqrt{3}/4 & 1/4 \end{bmatrix}.$$

The second rotation is easily obtained by construction of a basis transformation array, as described in Example 3.6, or by use of the same formula given above with $\theta_2 = \pi/2$ and $\boldsymbol{\alpha}_2 = \mathbf{e}_3$. The reader will find the tensor $\mathbf{R}_2 = R^2_{ij}\mathbf{e}_{ij}$ whose matrix in ψ is

$$R_2 = \begin{bmatrix} 0 & -1 & 0 \\ 1 & 0 & 0 \\ 0 & 0 & 1 \end{bmatrix}.$$

The final rotation is gotten by use of (3.152), in which

$$\theta_3 = \pi \quad \text{and} \quad \boldsymbol{\alpha}_3 = -\frac{1}{2}\mathbf{e}_1 + \frac{\sqrt{3}}{2}\mathbf{e}_3.$$

We shall need to compute $\mathbf{R}_3 = 2\boldsymbol{\alpha}_3 \otimes \boldsymbol{\alpha}_3 - \mathbf{1}$. Thus, with (3.151b), we find the following matrix of the tensor $\mathbf{R}_3 = R^3_{rs}\mathbf{e}_{rs}$ in ψ:

$$R_3 = \begin{bmatrix} -1/2 & 0 & -\sqrt{3}/2 \\ 0 & -1 & 0 \\ -\sqrt{3}/2 & 0 & 1/2 \end{bmatrix}.$$

Since all three rotation tensors are referred to the same frame, the resultant Euler rotation in frame ψ is obtained from the matrix product $R^* = R_3 R_2 R_1$ in accordance with (3.147b). Execution of the products yields

$$R^* = \begin{bmatrix} 1 & 0 & 0 \\ 0 & 1/2 & -\sqrt{3}/2 \\ 0 & \sqrt{3}/2 & 1/2 \end{bmatrix}.$$

The angle of rotation is given by (3.89) in which $\operatorname{tr} \mathbf{R}^* = 2$; we find $\theta^* = 60°$. Finally, use of θ^* and R^* in (3.90) yields $\boldsymbol{\alpha}^* = \mathbf{e}_1$ for the axis of the resultant rotation of the solar panel in ψ.

The satellite body axis is rotated in ψ due to the rotation R_3 alone. Its final orientation $\boldsymbol{\alpha}_2'$ may be found from (3.105a) or (3.86), namely, $\boldsymbol{\alpha}_2' = \mathbf{R}_3 \boldsymbol{\alpha}_2 = \mathbf{R}_3 \mathbf{e}_3$. Thus,

$$[\boldsymbol{\alpha}_2'] = \begin{bmatrix} -1/2 & 0 & -\sqrt{3}/2 \\ 0 & -1 & 0 \\ -\sqrt{3}/2 & 0 & 1/2 \end{bmatrix} \begin{bmatrix} 0 \\ 0 \\ 1 \end{bmatrix} = \begin{bmatrix} -\sqrt{3}/2 \\ 0 \\ 1/2 \end{bmatrix},$$

that is, $\boldsymbol{\alpha}_2' = -\sqrt{3}/2\mathbf{e}_1 + 1/2\mathbf{e}_3$ in ψ.

3.12.1. Euler Angles and Rotations

A rigid body having one point fixed has three degrees of freedom that may be described by three independent rotation angles. There is, of course, no unique choice for these angles, so the reader may expect to encounter different descriptions in the literature. The most widely used description is the set known as the Euler angles. There is, nonetheless, no standard sequence for execution of the corresponding rotations, hence different formulas for the composition of rotations described by the same set of Euler angles also occur in the literature. In addition, the definition of the transformation matrix A used in other texts sometimes is the transpose of the definition employed here. Therefore, the reader is cautioned to check the definition of the Euler angles, the sequence of the Euler rotations, and the description of the basis transformation array, when consulting other works. Herein, we shall adopt a sequence that seems to have wide appeal.

The *Euler angles* are a set of three independent parameters that specify the orientation of a rigid body in the spatial frame φ; they relate the orientation of the body frame φ' to φ. The set of Euler angles will be introduced below. Three corresponding consecutive rotation matrices will be expressed in terms of these angles, and their composition representing the equivalent Euler rotation of the body in φ will be derived.

We shall begin the sequence with the body frame $\varphi' = \{O; \mathbf{i}'_k\}$ coincident with the spatial frame $\varphi = \{O; \mathbf{i}_k\}$, as usual; and let the body receive three successive proper rotations about certain body axes as described next and illustrated in Fig. 3.14. The first rotation is through an angle ϕ about the axis $\mathbf{k}' = \mathbf{k}$ in φ so that $\mathbf{R}_1: \mathbf{i}_k \to \mathbf{i}^1_k$, which identifies the body frame in its first displaced position. The next rotation is through an angle θ about the body axis $\mathbf{i}' \equiv \mathbf{i}^1_1$ in φ so that $\mathbf{R}_2: \mathbf{i}^1_k \to \mathbf{i}^2_k$, the body frame in its second displaced position. Finally, the body is turned through an angle ψ about the body axis $\mathbf{k}' \equiv \mathbf{i}^2_3$ so that $\mathbf{R}_3: \mathbf{i}^2_k \to \mathbf{i}'_k$, the body frame in its terminal position in φ. The three angles $\{\phi, \theta, \psi\}$ define the set of Euler angles.

We apply (3.123) to obtain for the first rotation the matrix $R_1 = A_1^T = [\cos\langle \mathbf{i}^1_j, \mathbf{i}_k \rangle]^T$. We thus find

$$R_1 = \begin{bmatrix} \cos\phi & -\sin\phi & 0 \\ \sin\phi & \cos\phi & 0 \\ 0 & 0 & 1 \end{bmatrix}. \tag{3.153}$$

Similarly, the second and third rotation matrices are given by $R_2 = A_2^T = [\cos\langle \mathbf{i}^2_k, \mathbf{i}^1_l \rangle]^T$ and $R_3 = A_3^T = [\cos\langle \mathbf{i}'_m, \mathbf{i}^2_n \rangle]^T$. Hence,

$$R_2 = \begin{bmatrix} 1 & 0 & 0 \\ 0 & \cos\theta & -\sin\theta \\ 0 & \sin\theta & \cos\theta \end{bmatrix}, \qquad R_3 = \begin{bmatrix} \cos\psi & -\sin\psi & 0 \\ \sin\psi & \cos\psi & 0 \\ 0 & 0 & 1 \end{bmatrix}. \tag{3.154}$$

Since all three rotations are with respect to different bases following the successive positions of the body frame, in accordance with the discussion of this case in the last section, the composition rule (3.147b) yields the resultant rotation formula $R^* = R_1 R_2 R_3$. Executing the indicated matrix products of (3.153) and (3.154), we derive the resultant Euler rotation

$$R^* = \begin{bmatrix} c\phi\, c\psi - s\phi\, c\theta\, s\psi & -c\phi\, s\psi - s\phi\, c\theta\, c\psi & s\phi\, s\theta \\ s\phi\, c\psi + c\phi\, c\theta\, s\psi & -s\phi\, s\psi + c\phi\, c\theta\, c\psi & -c\phi\, s\theta \\ s\theta\, s\psi & s\theta\, c\psi & c\theta \end{bmatrix}, \tag{3.155}$$

wherein, for brevity, c and s denote cos and sin. The corresponding transformation matrix is $A^* = [\cos\langle \mathbf{i}'_j, \mathbf{i}_k \rangle] = R^{*T}$; hence, (3.155) are the components of the resultant rotation tensor $\mathbf{R}^*$ with respect to either basis of φ or φ', as described in (3.103). Equations (3.106) and (3.123) show that $A^* = R^{*T}$ is the transformation from the space frame to the body frame, whereas $A^{*T} = R^*$ is the inverse transformation from the body frame to the space frame. Therefore,

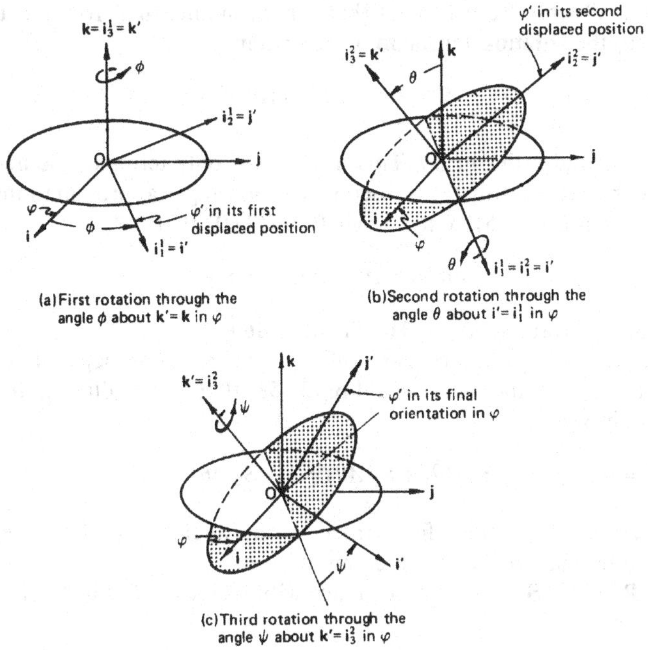

Figure 3.14. Euler angles and rotations.

the components x_k in φ and x'_k in φ' of the position vector $\mathbf{x}$ of a particle P, for example, are related by

$$x' = A^* x \quad \text{or} \quad x = A^{*T} x' \tag{3.156}$$

under the change of basis

$$\mathbf{i}'_p = \mathbf{R}^* \mathbf{i}_p = A^*_{pq} \mathbf{i}_q \quad \text{or} \quad \mathbf{i}_q = \mathbf{R}^{*T} \mathbf{i}'_q = A^*_{pq} \mathbf{i}'_p. \tag{3.157}$$

3.12.2. Review of the Composition of Infinitesimal Rotations

It was shown in Section 2.7 that consecutive infinitesimal rotations are vectors; and their composition, therefore, is both additive and commutative. We are now able to show that the same result may be derived from (3.147b). It will be shown that the resultant of two infinitesimal rotations $\mathbf{R}_1$ and $\mathbf{R}_2$ is indeed commutative, that is, $\mathbf{R}^* = \mathbf{R}_2 \mathbf{R}_1 = \mathbf{R}_1 \mathbf{R}_2$; and the result will be reduced to the additive rule (2.22) derived earlier.

We first observe from (3.150) that for an infinitesimal rotation through an angle $\Delta\theta$, the rotation tensor may be written as

$$\mathbf{R} = \mathbf{1} + \mathbf{S}\,\Delta\theta + \mathbf{0}(\Delta\theta)^2 \tag{3.158}$$

to terms of second order in $\Delta\theta$. This means that only terms of the first order in $\Delta\theta$ need be retained in the subsequent small angle approximations. Also, for the skew tensor (3.151a), it follows from (3.60) that

$$\Delta\theta\,\mathbf{S}\mathbf{v} = \Delta\theta\,\boldsymbol{\alpha} \times \mathbf{v} \equiv \Delta\boldsymbol{\theta} \times \mathbf{v} \tag{3.159}$$

holds for an arbitrary vector $\mathbf{v}$. Herein $\Delta\boldsymbol{\theta} = \Delta\theta\boldsymbol{\alpha}$.

Let us consider two successive infinitesimal rotations $\mathbf{R}_1$ and $\mathbf{R}_2$ about concurrent axes. We find with the aid of (3.158) that the resultant infinitesimal rotation is given by

$$\mathbf{R}^* = \mathbf{R}_2\mathbf{R}_1 = \mathbf{1} + \mathbf{S}_1\,\Delta\theta_1 + \mathbf{S}_2\,\Delta\theta_2 = \mathbf{1} + \mathbf{S}_2\,\Delta\theta_2 + \mathbf{S}_1\,\Delta\theta_1 = \mathbf{R}_1\mathbf{R}_2,$$

in which terms larger than first order in the angles have been neglected. Hence, infinitesimal rotations are commutative.

Since $\mathbf{R}^* = \mathbf{1} + \mathbf{S}^*\,\Delta\theta^*$ in accordance with (3.158), it follows from the last equation and (3.159) that

$$[\Delta\boldsymbol{\theta}^* - (\Delta\boldsymbol{\theta}_1 + \Delta\boldsymbol{\theta}_2)] \times \mathbf{v} = \mathbf{0}$$

must hold for every vector $\mathbf{v}$. Thus, the equivalent infinitesimal rotation satisfies the additive rule derived in (2.22), namely,

$$\Delta\boldsymbol{\theta}^* = \Delta\boldsymbol{\theta}_1 + \Delta\boldsymbol{\theta}_2 = \Delta\boldsymbol{\theta}_2 + \Delta\boldsymbol{\theta}_1 \qquad [\text{cf. } (2.22)].$$

This completes the demonstration.

3.12.3. Composition of Rotations about Nonintersecting Axes

In two consecutive, arbitrary displacements of a rigid body, the translational displacements are independent of the rotations. The rotations may be compounded separately and their contribution to the displacement may then be added to the sum of the translations to obtain the total displacement of any particle of the body. The case when the axes of rotation are concurrent has been studied earlier. In this section, we shall examine the situation in which the body receives successive rotations about nonintersecting axes. The special case of reversed rotations about parallel axes also is described; and it will be shown how any given displacement may be represented as the sum of rotational displacements about nonintersecting axes.

We shall begin with two consecutive pure rotations about nonintersecting axes. The displacement of a particle P in a pure rotation $\mathbf{T}_1$ about an axis $\boldsymbol{\alpha}_1$ is given by $\mathbf{d}_1 = \mathbf{x}_1 - \mathbf{x} = \mathbf{T}_1\mathbf{x}$, where $\mathbf{x}$ and $\mathbf{x}_1$ are the initial and final positions of P from the base point O, as shown in Fig. 3.15a. The subsequent displacement of P in another pure rotation $\mathbf{T}_2$ about a nonintersecting axis $\boldsymbol{\alpha}_2$ through the base point O' is given by $\mathbf{d}_2 = \mathbf{T}_2\mathbf{x}_1'$, where $\mathbf{x}_1'$ is the location of P from O'. But the same displacement, by the parallel axis theorem, may be accomplished by the same rotation $\mathbf{T}_2$ about a parallel axis through O together with a translation $\mathbf{b}_O$ perpendicular to $\boldsymbol{\alpha}_2$. Therefore, as described in Fig. 3.15b, $\mathbf{d}_2 = \mathbf{x}_2 - \mathbf{x}_1 = \mathbf{T}_2\mathbf{x}_1' = \mathbf{b}_O + \mathbf{T}_2\mathbf{x}_1$, wherein $\mathbf{x}_2$ is the ultimate position vector of P from O. With (3.87), we have $\mathbf{x}_1 = \mathbf{R}_1\mathbf{x}$ and $\mathbf{x}_2 = \mathbf{b}_O + \mathbf{R}_2\mathbf{x}_1$; hence, $\mathbf{x}_2 = \mathbf{b}_O + \mathbf{R}^*\mathbf{x}$, where $\mathbf{R}^* = \mathbf{T}^* + \mathbf{1} = \mathbf{R}_2\mathbf{R}_1$ in terms of the rotation tensors. Therefore, the resultant displacement due to the rotations alone is given by

$$\mathbf{d}(P) = \mathbf{x}_2 - \mathbf{x} = \mathbf{b}_O + \mathbf{T}^*\mathbf{x} \qquad \text{with} \quad \mathbf{b}_O \cdot \boldsymbol{\alpha}_2 = 0. \tag{3.160}$$

In words, *the displacement resulting from the composition of two consecutive pure rotations about nonintersecting axes is equivalent to a rotation about a line through the first base point O together with a parallel translation which is equal to a pure rotation of O about the axis at O′.*

If $\mathbf{b}_1$ and $\mathbf{b}_2$ denote the translational parts of two displacements whose rotations are characterized above, then, with (3.160), the resultant displacement will be given by

$$\mathbf{d}^*(P) = \mathbf{b}^* + \mathbf{T}^*\mathbf{x}, \tag{3.161}$$

where $\mathbf{b}^* = \mathbf{b}_O + \mathbf{b}_1 + \mathbf{b}_2$ is the resultant translational displacement of the base point O. For the special case when the axes intersect in O, we have $\mathbf{b}_O = \mathbf{0}$.

If the body is returned to its initial configuration after two successive rotations about concurrent axes, then by (3.87) we have $\mathbf{T}^* = \mathbf{R}^* - \mathbf{1} =$

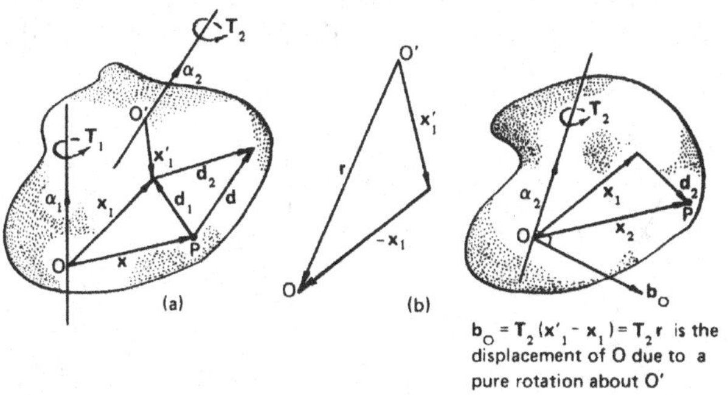

Figure 3.15. The composition of consecutive rotations about nonintersecting lines in (a) is equal to a rotation about $\boldsymbol{\alpha}_2$ and a normal translation $\mathbf{b}_O$ in (b).

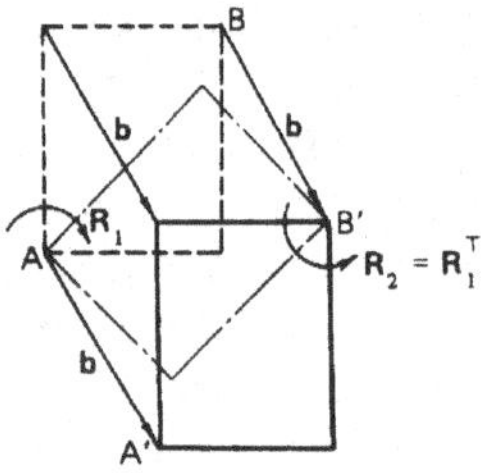

Figure 3.16. A rotation about any point followed by a reversed rotation about another point is a "walking" translation.

$\mathbf{R}_2\mathbf{R}_1 - \mathbf{1} = \mathbf{0}$; hence, $\mathbf{R}_2 = \mathbf{R}_1^T$. In this case, $\mathbf{R}_2$ is *equivalent* to a rotation which is the reverse of $\mathbf{R}_1$; and from (3.160), we have the following conclusion: *A rotation about a line followed by a reversed rotation about a parallel line is equivalent to a parallel translation perpendicular to the axis of rotation. Moreover, the translation is equal to the displacement resulting from the first rotation of the base point on the second axis.*

A rotation $\mathbf{R}_1$ about a point A followed by a reversed rotation $\mathbf{R}_2 = \mathbf{R}_1^T$ about another point B is shown in Fig. 3.16. The displacement of B after the first rotation is $\mathbf{b}$. Since the second axis is through the displaced point denoted as B', the reversed rotation produces no motion of B'; and the resultant displacement is a pure translation in which all points of the body experience the same translational displacement $\mathbf{d}(P) = \mathbf{b}_O = \mathbf{b}$ in accordance with (3.160). Thus, reversed rotations about distinct parallel lines will produce a displacement that is similar to a common walking motion.

Although the composition of rotations is not additive, under appropriate conditions, any given displacement may be decomposed into a sum of rotational displacements about certain lines which generally do not intersect. To see this, let us suppose in Fig. 3.17a that for a given displacement (3.161) the resultant rotation about the base point O is decomposed into two consecutive rotations $\mathbf{R}_1$ and $\mathbf{R}_2$ about concurrent axes $\boldsymbol{\alpha}_1$ and $\boldsymbol{\alpha}_2$, such that $\boldsymbol{\alpha}_1$ is any given direction and $\boldsymbol{\alpha}_2$ is perpendicular to the translation vector $\mathbf{b}^*$. Since $\mathbf{T}^* = \mathbf{T}_1 + \mathbf{T}_2\mathbf{R}_1$, (3.161) may be written as $\mathbf{d}^* = \mathbf{T}_1\mathbf{x} + \mathbf{b}^* + \mathbf{T}_2\bar{\mathbf{x}}$, where $\bar{\mathbf{x}} = \mathbf{R}_1\mathbf{x}$ is the position vector from O to the final position P' of the particle P

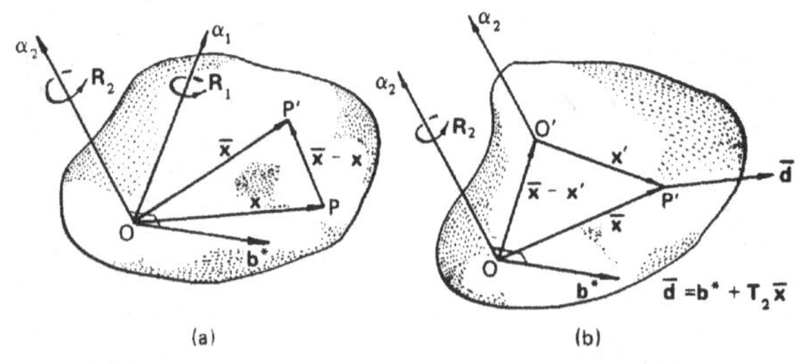

Figure 3.17. Schematic of consecutive rotational displacements about nonintersecting axes.

after the first rotation alone. Since $\boldsymbol{\alpha}_2 \cdot \mathbf{b}^* = 0$, we may apply the parallel axis theorem in Fig. 3.17b to find a parallel line through another base point O' such that $\bar{\mathbf{d}} = \mathbf{b}^* + \mathbf{T}_2\bar{\mathbf{x}} = \mathbf{T}_2\mathbf{x}'$ is a pure rotation about O', where $\mathbf{x}'$ is the location of P' from O'. Thus, as shown in Fig. 3.18, the given displacement $\mathbf{d}^*(P)$ is the sum of two pure rotational displacements $\mathbf{d}_1 = \mathbf{T}_1\mathbf{x}$ and $\mathbf{d}_2 = \mathbf{T}_2\mathbf{x}'$ about O and O', so that

$$\mathbf{d}^*(P) = \mathbf{T}_1\mathbf{x} + \mathbf{T}_2\mathbf{x}'. \tag{3.162}$$

Hence, any given rigid body displacement (3.161) *may be represented by the sum of two consecutive pure rotational displacements about, in general, nonintersecting axes* $\boldsymbol{\alpha}_1$ *and* $\boldsymbol{\alpha}_2$, *where* $\boldsymbol{\alpha}_1$ *is any assigned direction and* $\boldsymbol{\alpha}_2$ *is perpendicular to the parallel translation vector of the assigned displacement.*

3.12.4. Composition of Screw Displacements

We have learned that every rigid body displacement may be reduced to a unique screw displacement. It seems natural, therefore, that successive screws about nonintersecting axes ought to be reducible to another unique screw displacement. The composition of these screw displacements will be studied next.

For two consecutive screw displacements characterized by (3.141) and generally having nonintersecting axes $\boldsymbol{\alpha}_1$ and $\boldsymbol{\alpha}_2$, we have $\mathbf{b}^* = \mathbf{b}_{\alpha 1} + \mathbf{b}_{\alpha 2} + \mathbf{b}_O$ in (3.161), with $\mathbf{b}_O \cdot \boldsymbol{\alpha}_2 = 0$. As in the proof of Chasles' theorem, $\mathbf{b}^*$ may be expressed in terms of its vector components parallel and normal to the equivalent screw axis $\boldsymbol{\alpha}^*$, namely, $\mathbf{b}^* = \mathbf{b}^*_{\alpha*} + \mathbf{b}^*_{n*}$; and the parallel axis theorem may be applied to find a line parallel to $\boldsymbol{\alpha}^*$ such that $\mathbf{T}^*\mathbf{x}^* = \mathbf{b}^*_{n*} + \mathbf{T}^*\mathbf{x}$, where $\mathbf{x}^*$ is the initial location of P from a point O^* on the resultant screw axis. With (3.161), we thus reach the unique resultant screw displacement

$$\mathbf{d}^*(P) = \mathbf{b}^*_{\alpha*} + \mathbf{T}^*\mathbf{x}^*. \tag{3.163}$$

Here we have $\mathbf{b}^*_{\alpha*} = (\mathbf{b}^* \cdot \boldsymbol{\alpha}^*)\,\boldsymbol{\alpha}^* = p^*\theta^*\boldsymbol{\alpha}^*$ for the screw translation and the pitch. In the special case when the component screws are simple rotations about a point, we recover (3.141) from (3.163). We summarize: *Any dis-*

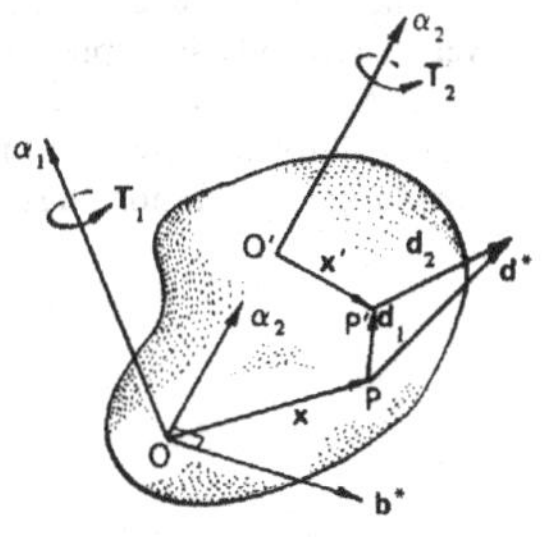

Figure 3.18. A general displacement viewed as consecutive rotations about nonintersecting axes.

placement of a rigid body, however complex, is reducible to a unique screw displacement.

This completes our study of finite rigid body displacements. In the next chapter, we shall return to the basic equations for the velocity and acceleration of a rigid body particle in their important application to the study of motion referred to a moving reference frame. The equations developed in this chapter for a change of basis will be useful in some applications encountered there. And late in Chapter 4, we shall learn how our fundamental equation for finite displacements of a rigid body may be used in an elegant derivation of the relative motion relation introduced and applied earlier throughout that chapter. Otherwise, the reader who in a first reading may have omitted detailed study of the present work will find no serious difficulty when these topics are interlaced with some of the examples further on. The use of tensors in the study of rigid body dynamics in later chapters, however, may require review of some of the topics on tensors covered here.

References

1. Beatty, M., Vector representation of rigid body rotation, *American Journal of Physics* **31**, 134–135 (1963). Kinematics of finite rigid body displacements, *Ibid.* **34**, 949–954 (1966). These papers contain the main theorems on finite rigid body displacements presented in this chapter. A coordinate free vector construction of the rotation tensor and numerous additional references are given in another paper: Vector analysis of rigid rotations, *Journal of Applied Mechanics* **44**, 501–502 (1977).
2. Coleman, B. D., Markovitz, H., and Noll, W., *Viscometric Flows of Non-Newtonian Fluids. Theory and Experiments*, Springer-Verlag, New York, 1966. A brief introduction to tensors is contained in the Appendix on Mathematical Concepts.
3. Long, R., *Engineering Science Mechanics*, Prentice-Hall, Englewood Cliffs, New Jersey, 1963. Index notation and the elements of Cartesian tensors are introduced and applied in Chapter 1 in the construction of the component equation (3.73). See also Chapter 6 on rigid body motions.
4. Paul, B., On the composition of finite rotations, *American Mathematical Monthly* **70**, 859–862 (1963). The author essentially applies (2.7) twice and cleverly manipulates the resultant equations to derive by vector methods the axis and angle of the equivalent Euler rotation for the composition of two finite rotations.
5. Room, T. G., The composition of rotations in Euclidean three space, *American Mathematical Monthly* **59**, 688–692 (1952). The composition of finite rotations is studied by a combination of vector methods and spherical trigonometry.
6. Wilson, E. B., *Vector Analysis—The Lectures by J. W. Gibbs*, Dover, New York, 1960. The theory of dyadics (tensors) is applied in Chapter 6. Some interesting results on finite rotations cast in dated notation may be found there.

Problems

3.1. Write out each of the following expressions in which the range of the indices, except as noted, is 3:

(a) v_2 if $v_i = \varepsilon_{ijk}\omega_j x_k$;
(b) T_{12} if $T_{ij} = A_{ki}A_{ej}T'_{ke}$;
(c) a_3 if $a_s = A_{sj}b_j + B_{sk}c_k$, $j = 1, 2$.

3.2. Translate the following sets of equations into index notation, and specify the range of all indices:

(a) $u_1 = v_1^2 w_1 + v_2^2 w_1 + v_3^2 w_1$;
$u_2 = v_1^2 w_2 + v_2^2 w_2 + v_3^2 w_2$;
$u_3 = v_1^2 w_3 + v_2^2 w_3 + v_3^2 w_3$;

(b) $v_1 = A_{111} + A_{122} + A_{133}$;
$v_2 = A_{211} + A_{222} + A_{233}$;
$v_3 = A_{311} + A_{322} + A_{333}$;

(c) $A_{11} = B_{11}C_{11} + B_{12}C_{21}$;
$A_{12} = B_{11}C_{12} + B_{12}C_{22}$;
$A_{21} = B_{21}C_{11} + B_{22}C_{21}$;
$A_{22} = B_{21}C_{12} + B_{22}C_{22}$;

(d) $v_1^2 = x_1^2 - 2x_1 y_1 + y_1^2$;
$v_1 v_2 = x_1 x_2 - x_1 y_2 - y_1 x_2 + y_1 y_2$;
$v_2 v_1 = x_2 x_1 - x_2 y_1 - y_2 x_1 + y_2 y_1$;
$v_2^2 = x_2^2 - 2x_2 y_2 + y_2^2$.

3.3. Substitute the index relations as described below. How many terms would occur in the result, if you had to write it out? Write out the one having the smaller number of terms.

(a) $a_i = M_{ij}b_j$ into $c_j = N_{jk}a_k$.
(b) $u_i = B_{ij}v_j$ and $C_{ij} = p_i q_j$ into $w_k = C_{kl}u_l$.
(c) $u_i = A_{ik}v_k$ into $\lambda = u_k v_k$.
(d) $v_i = B_{ij}A_{jkk}$ into $\lambda = v_i C_{jij}$.
(e) $A_{ij} = B_{ik}C_{kj}$ into $\lambda = A_{mk}C_{mk}$.

3.4. (a) How many terms would you have in the equation for C_{1212}, if you were to write it out from the formula below?

$$C_{ijkl} = A_{pi}A_{qj}A_{rk}A_{sl}C'_{pqrs}.$$

(b) Show that $(A_{ijk} + A_{jki} + A_{jik})x_i x_j x_k = 3A_{ijk}x_i x_j x_k$. How many terms would you have in the right-hand term, if you were to write it out?

3.5. Use index notation to simplify the following expressions:

(a) $\varepsilon_{ijk}\delta_{ip}\delta_{jq} = ?$
(b) $A_{ip}\delta_{pq}\delta_{ir}\delta_{rk} = ?$
(c) $\{16\varepsilon_{ijk}\beta_j + 9(\beta_i\beta_k - \delta_{ik})\}\,\beta_i\beta_k = ?$ where $\boldsymbol{\beta}\cdot\boldsymbol{\beta} = 1$.

3.6. If P is a point with coordinates p_i in frame $\Phi = \{O; \mathbf{i}_k\}$, show that

$$p = p_i \cos\alpha_i,$$

where p is the length of the segment $\overline{OP}$ and α_i are its direction angles in Φ.

3.7. If $\overline{OP}$ and $\overline{OQ}$ are line segments with direction angles α_i and β_i, respectively, in frame $\varphi = \{O; \mathbf{i}_k\}$, prove that

$$\cos\psi = \cos\alpha_i \cos\beta_i$$

where ψ is the angle between $\overline{OP}$ and $\overline{OQ}$.

3.8. (a) Derive (3.10), and show by its expansion that it yields the familiar result (A.11) in Appendix A. (b) Verify for a few sets of indices the following ε–δ identity

$$\varepsilon_{ijk}\varepsilon_{pqk} = \delta_{ip}\delta_{jq} - \delta_{iq}\delta_{jp},$$

and use this result together with (3.10) to derive the expansion formula (A.14) for the vector triple product. (c) Find identities for $\varepsilon_{pqr}\varepsilon_{kqr}$ and $\varepsilon_{klm}\varepsilon_{klm}$.

3.9. (a) Use index notation to show that in an unconstrained spatial motion of a rigid body there is in general no point having zero velocity, but that in general there exists one point, and only one point, with instantaneous zero acceleration. As indicated in Problem 2.71, however, under special constraints there may be one or more points with zero velocity. (See also Problems 2.72, 2.81, and 2.82). (b) Find in the frame $\varphi = \{O; \mathbf{i}_k\}$ the instantaneous location of the unique point of zero acceleration at the instant when $\mathbf{a}_O = a_O\mathbf{i}$, $\boldsymbol{\omega} = \omega\mathbf{k}$ and $\dot{\boldsymbol{\omega}} = \dot{\omega}_k\mathbf{i}_k$, referred to φ.

3.10. Prove that the tensor product $\mathbf{a} \otimes \mathbf{b}$ of vectors $\mathbf{a}$ and $\mathbf{b}$ is in fact a tensor. Derive the relations (3.23).

3.11. Prove that the product of two tensors $\mathbf{S}$ and $\mathbf{T}$ defined by the rule (3.36) is indeed a tensor; and show that its components relative to an orthonormal basis $\mathbf{e}_k$ are determined by the product of the matrices of these tensors. Further, let $\mathbf{U}$ be another tensor and λ a scalar. Establish that the following rules hold for the products of tensors:

(a) $(\mathbf{ST})\mathbf{U} = \mathbf{S}(\mathbf{TU})$;
(b) $\lambda(\mathbf{ST}) = (\lambda\mathbf{S})\mathbf{T} = \mathbf{S}(\lambda\mathbf{T})$;
(c) $\mathbf{S}(\mathbf{T} + \mathbf{U}) = \mathbf{ST} + \mathbf{SU}$;
(d) $(\mathbf{S} + \mathbf{T})\mathbf{U} = \mathbf{SU} + \mathbf{TU}$;
(e) $\mathbf{1T} = \mathbf{T1} = \mathbf{T}$.

3.12. The transpose $\mathbf{T}^T$ of the tensor $\mathbf{T}$ is defined by (3.42). Prove that $\mathbf{T}^T$ satisfies (3.11) and hence really is a tensor; and show that transposition obeys the rules (3.43) and (3.46).

3.13. Let $\mathbf{p}$, $\mathbf{q}$, $\mathbf{u}$, and $\mathbf{v}$ be vectors. Show that

$$(\mathbf{p} \otimes \mathbf{q})(\mathbf{u} \otimes \mathbf{v}) = (\mathbf{q} \cdot \mathbf{u})(\mathbf{p} \otimes \mathbf{v}).$$

Hence, show also that for an orthonormal basis $\mathbf{e}_k$,

$$(\mathbf{e}_i \otimes \mathbf{e}_j)(\mathbf{e}_k \otimes \mathbf{e}_l) = \delta_{jk}\mathbf{e}_i \otimes \mathbf{e}_l.$$

Use the last relation to derive the component representation for the product $\mathbf{ST}$ of the tensors $\mathbf{S}$ and $\mathbf{T}$.

3.14. The inner (scalar) product of tensors $\mathbf{S}$ and $\mathbf{T}$ is defined by $\mathbf{S} \cdot \mathbf{T} \equiv \operatorname{tr}(\mathbf{S}^T\mathbf{T}) = \operatorname{tr}(\mathbf{ST}^T)$. Show that for vectors $\mathbf{p}$, $\mathbf{q}$, $\mathbf{u}$, and $\mathbf{v}$

$$(\mathbf{p} \otimes \mathbf{q}) \cdot (\mathbf{u} \otimes \mathbf{v}) = (\mathbf{p} \cdot \mathbf{u})(\mathbf{q} \cdot \mathbf{v});$$

hence, also for the orthonormal basis $\mathbf{e}_k$,

$$(\mathbf{e}_i \otimes \mathbf{e}_j) \cdot (\mathbf{e}_k \otimes \mathbf{e}_l) = \delta_{ik}\delta_{jl}.$$

Apply the last relation to show that $\mathbf{S} \cdot \mathbf{T} = S_{kl}T_{kl}$. Thus, in particular, show that the magnitude of a tensor, defined by $|\mathbf{T}| = (\mathbf{T} \cdot \mathbf{T})^{1/2}$, is simply the square root of the sum of squares of the components of $\mathbf{T}$. Find the magnitude of the identity tensor using only index notation. Determine the magnitude of an arbitrary orthogonal tensor $\mathbf{Q}$.

3.15. Derive equations (3.51)–(3.54).

3.16. Use (3.69) to determine the vector $\mathbf{T}\mathbf{e}_j$ as a function of θ and $\boldsymbol{\alpha}$. Apply the result to derive (3.73) and (3.75).

3.17. Derive Rodrigues' formula (circa 1840) for the displacement of a particle of a rigid body in a rotation about a fixed line, namely,

$$\mathbf{d}(P) = \tan(\theta/2)\boldsymbol{\alpha} \times (\hat{\mathbf{x}} + \mathbf{x}),$$

in which $\hat{\mathbf{x}}$ denotes the terminal position of P from the origin on the axis of rotation and the other terms are the same as those in (2.7).

3.18. The function $\tan(\theta/2)$ enters Rodrigues' formula in the previous problem. Show that in terms of the symmetric and skew parts of the rotator $\mathbf{T}$

$$\tan\frac{\theta}{2} = \frac{[-2\,\mathrm{tr}(\mathbf{T}_A^2)]^{1/2}}{4 + \mathrm{tr}\,\mathbf{T}_S} = \frac{2[T_{12}^2 + T_{23}^2 + T_{31}^2]^{1/2}}{4 + T_{11} + T_{22} + T_{33}}.$$

Note that the square of a tensor is defined by $\mathbf{T}^2 = \mathbf{T}\mathbf{T}$.

3.19. An electric motor must be arranged to turn a flood control valve through an angle $\theta < \pi$ in a right-hand sense about a fixed directed line through the origin and the point $P = (1, 2, 0)$ ft in frame $\Phi = \{O; \mathbf{I}_k\}$. An engineer has shown that any design is acceptable for which $T_{11} = -2/5$, $T_{22} = -1/10$, and $T_{33} = -1/2$ are satisfied in Φ. Determine the angle and axis of rotation, and compute the remaining six components of T in Φ.

3.20. Find the matrix of the rotator for the rotation of the control link described in Problem 2.6. Use the result, conversely, to compute the angle and the axis of the rotation. Compute by matrix methods the displacement of the pin P.

3.21. A cardboard box is initially positioned with its centroid C at the place $(4, 6, 7)$ ft in frame $\Phi = \{O; \mathbf{I}_k\}$ fixed in a packaging machine. The empty carton must be transported by a rotation about a line through the point $P = (2, 0, -1)$ ft in Φ to a new configuration where the box is filled and sealed. The rotator required for the operation is specified by

$$T = \frac{1}{9}\begin{bmatrix} 4\sqrt{3} - 8 & -(1 + \sqrt{3}) & 5 - \sqrt{3} \\ 5 - \sqrt{3} & \frac{5}{2}(\sqrt{3} - 2) & \frac{1}{2}(5 - 4\sqrt{3}) \\ -(1 + \sqrt{3}) & \frac{1}{2}(11 - 4\sqrt{3}) & \frac{5}{2}(\sqrt{3} - 2) \end{bmatrix}.$$

Determine the rotation angle, and find the equation in Φ of the line of rotation. What is the displacement of C? Find the location from C of the point T which initially was at the top of the container at the place $(4, 6, 9)$ ft in Φ.

3.22. An alternative design for the door panel mechanism described in Problem 2.6 requires the rotation

$$R=\begin{bmatrix} 1 & 0 & 0 \\ 0 & 1/2 & -\sqrt{3}/2 \\ 0 & \sqrt{3}/2 & 1/2 \end{bmatrix}$$

The guide pin P initially is at the place $(1, 1, 1)$ ft from O in Φ. Determine the angle of the rotation and axis of the shaft OA. Find the displacement of P.

3.23. The matrix of a rigid body rotation tensor is given by

$$R=\frac{\sqrt{2}}{10}\begin{bmatrix} 3 & 5 & 4 \\ -3 & 5 & -4 \\ 4\sqrt{2} & 0 & -3\sqrt{2} \end{bmatrix}$$

referred to $\varphi = \{O; \mathbf{i}_k\}$. (a) Find the angle and the fixed axis of rotation in φ. What is the terminal location and the displacement in φ of the particle P initially at $\mathbf{x} = 4\mathbf{i} - \sqrt{2}\mathbf{j} - 3\mathbf{k}$ ft from O? (b) Find the matrix of the rotator and use it to compute the displacement of P. Compare the result with the previous solution.

3.24. A rigid body is rotated about a line through the point C at $(1, 3, 5)$ ft in $\Phi = \{O; \mathbf{I}_k\}$. The rotation tensor in Φ is

$$R=\begin{bmatrix} \sqrt{2}/2 & \sqrt{2}/4 & \sqrt{6}/4 \\ 0 & \sqrt{3}/2 & -1/2 \\ -\sqrt{2}/2 & \sqrt{2}/4 & \sqrt{6}/4 \end{bmatrix}.$$

Determine the final position in Φ of the particle P whose place initially was $\mathbf{X} = \sqrt{2}\mathbf{I}_1 + (2\sqrt{3/3})\,\mathbf{I}_2 + 2\mathbf{I}_3$ ft. What is the displacement of P? Find the axis and the angle of rotation of the body in Φ.

3.25. A rigid body is rotated about a line through the point A at $(2, 5, -7)$ m in $\Phi = \{F; \mathbf{I}_k\}$. The rotation matrix in the frame $\Phi' = \{A; \mathbf{I}_k\}$ is given by

$$R=\begin{bmatrix} 0 & -\sqrt{2}/2 & -\sqrt{2}/2 \\ 1 & 0 & 0 \\ 0 & -\sqrt{2}/2 & \sqrt{2}/2 \end{bmatrix}.$$

(a) What is the final position relative to Φ' of the particle P initially at $\mathbf{X} = 2\mathbf{I}_1 + \mathbf{I}_2 - \mathbf{I}_3$ m in Φ? (b) What is the final location of P in Φ? (c) Determine the displacement of P referred to Φ'. (d) Find the angle and the axis of rotation referred to Φ'. What is the axis when referred to Φ?

3.26. The Cartesian coordinates of a point P in $\varphi = \{O; \mathbf{i}_k\}$ are given as $(3\sqrt{2}, -2\sqrt{2}, 5)$. Find the coordinates of the same point in $\varphi' = \{O; \mathbf{i}'_k\}$, if φ' has the orientation relative to φ shown in the figure.

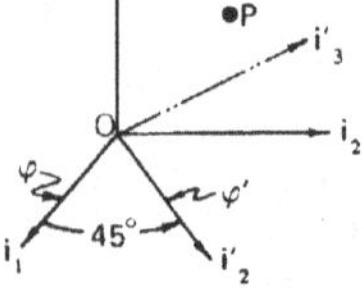

Problem 3.26.

3.27. A change of basis is described by the basis transformation tensor

$$\mathbf{Q} = \tfrac{1}{2}(-\mathbf{e}_{11} + \mathbf{e}_{22} + 2\mathbf{e}_{33}) - \sqrt{3}\,\mathbf{e}_{(12)}$$

referred to the initial frame $\psi = \{O; \mathbf{e}_k\}$. Is this a proper or an improper orthogonal transformation? Find the basis transformation matrix, and sketch the orientation of the transformed frame $\psi' = \{O; \mathbf{e}'_k\}$. What can be said about the nature of frame ψ'?

3.28. The rotator (3.83) describes the finite rotation of the satellite tracking antenna shown for Example 2.1. Construct the basis transformation matrix for a change of frame from $\Phi = \{F; \mathbf{I}_k\}$ into $\Phi' = \{F; \mathbf{I}'_k\}$ produced by a right-hand rotation through 45° about the $\mathbf{K} = \mathbf{K}'$ axis. (a) Use the tensor transformation law to find the matrix T' of $\mathbf{T}$ referred to Φ'; and determine from the result the angle and axis of the rotation relative to Φ'. Compute $\det \mathbf{T}$ in Φ and Φ'. (b) Apply the vector transformation law to the axis in (3.81c), and compare the result with that found previously. Do these values agree physically with the description in Fig. 2.5?

3.29. A hot, rectangular steel ingot must be transported by a machine to a new location and orientation. A point on the top of the ingot is located at $A = (20, 10, 20)$ ft, and the ingot has its center at $C = (20, 10, 10)$ ft in its initial configuration relative to a fixed reference frame Φ. In its final position the center is at $O' = (40, 30, 30)$ ft and the ingot has experienced a right-hand rotation of 90° about $\mathbf{i}_1$ followed by a similar rotation of 180° about $\mathbf{i}_3$ of the parallel spatial frame $\varphi = \{O'; \mathbf{i}_k\}$. Find the equivalent Euler rotation of the ingot. What is the total displacement in Φ of the point A?

3.30. Show that the relation (3.132) for a simple rotation is equivalent to (2.7). Begin with (3.132) and derive by vector methods the finite displacement equation

$$\mathbf{d}(P) = \sin\psi\, \boldsymbol{\alpha} \times \boldsymbol{\xi} + (1 - \cos\psi)\boldsymbol{\alpha} \times (\boldsymbol{\alpha} \times \boldsymbol{\xi}),$$

in which $\boldsymbol{\xi}$ is the position vector of P from an arbitrary point on the axis of rotation.

3.31. Compute the angle and the axis of the simple rotation of the guide pin P whose initial and final position vectors $\mathbf{x}$ and $\hat{\mathbf{x}}$ are given in Example 3.5. How do these values compare with those found in the example for the Euler rotation about the fixed line?

3.32. A rigid body is rotated through 45° about a fixed line $x = 2y = 3z$ in the right-hand sense of increasing values of the coordinates along the line. Determine the displacement of the particle P at the place $(-1, 2, 0)$ m in $\varphi = \{O; \mathbf{i}_k\}$. Find the simple rotation that yields the same displacement of P. Explain the unusual nature of the result.

3.33. A rigid body particle B, which initially is at the place $(1, 2, 3)$ ft in $\Phi = \{F; \mathbf{I}_k\}$, moves to $(3, 2, 1)$ ft at O' in Φ. The body also performs rotations about O' that are equivalent to

$$R = \begin{bmatrix} 0 & 1 & 0 \\ 0 & 0 & -1 \\ -1 & 0 & 0 \end{bmatrix}.$$

The initial location of another particle P is $(4, 5, 6)$ ft in Φ. Determine the final position of P relative to the parallel spatial frame $\varphi = \{O'; \mathbf{i}_k\}$ and relative to Φ. What is the projection on the axis of rotation of the resultant displacement of a body point Q whose initial place in Φ is $(2, 4, 6)$ ft? What is it for P?

3.34. An equilateral triangular box of side 3 ft and depth 1 ft has the initial orientation shown at A. The box is displaced to the configuration B such that the face OPQ, initially in the xy plane of $\varphi = \{O; \mathbf{i}_k\}$, is parallel to the yz plane of φ at B. (a) Find the equivalent Euler rotation of the box. (b) What are the displacement and terminal position vector in $\Phi = \{F; \mathbf{I}_k\}$ of the point Q? (c) Determine the position vector of P at B in φ; and find the displacement of P in Φ.

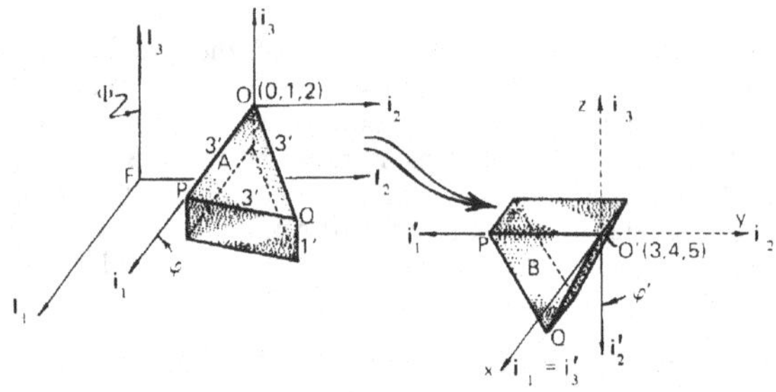

Problem 3.34.

3.35. An envelope must be transported to a new location and orientation suitable for zip code reading in a mail sorting machine. The center O of the envelope, initially at $\mathbf{B} = 3\mathbf{I} + 5\mathbf{J} + 8\mathbf{K}$ cm, is moved to $\hat{\mathbf{B}} = 8\mathbf{I} + 10\mathbf{J} + 13\mathbf{K}$ cm in the machine frame $\Phi = \{F; \mathbf{I}_k\}$, and the envelope has undergone right-hand rotations of 45° about the $\mathbf{i}_2$ axis followed by 90° about the $\mathbf{i}_3$ axis of the spatial frame $\varphi = \{O; \mathbf{i}_k\}$ parallel to Φ. The design of a suction device to move the envelope requires that the equivalent angle and axis of rotation through O be provided. Determine the data required, and furnish the final location and total displacement in Φ of the point P which was at the place $\mathbf{X} = 3\mathbf{I} + 5\mathbf{J} + 13\mathbf{K}$ cm initially.

3.36. Recall the rotator properties (3.124). Prove that for an arbitrary unconstrained displacement of a rigid body in space, there exists, in general, no point whose displacement may be zero. See also the next problem.

3.37. In every plane displacement of a rigid body there is always one and only one point whose displacement is zero. This point is the unique center of rotation for the plane displacement of the entire body. Thus, consider the right triangle whose base is one-half of the displacement vector $\mathbf{b}$ of any assigned base point O, and whose hypotenuse is the position vector $\mathbf{r}$ of the center of rotation O^* from O, as shown. The angle, of course, is one-half of the angle θ of the rotation of the body about the axis $\boldsymbol{\alpha}$ normal to the displacement plane, and $\mathbf{b} \cdot \boldsymbol{\alpha} = 0$. Apply this geometry to prove by a vector algebraic construction that the unique location of the center of rotation in the plane displacement is given by

$$\mathbf{r} = \frac{1}{2}\left[\mathbf{b} + \frac{\sin\theta}{1-\cos\theta}\,\boldsymbol{\alpha} \times \mathbf{b}\right].$$

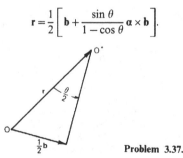

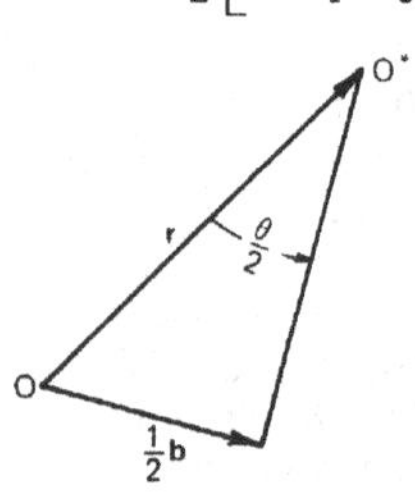

Problem 3.37.

3.38. The rectangular plate shown in the figure undergoes a right-hand rotation of 45° about a normal axis through A, followed by a reversed rotation about a parallel line through D. Determine the parallel translation by which the same resultant displacement may be effected. Verify the result graphically.

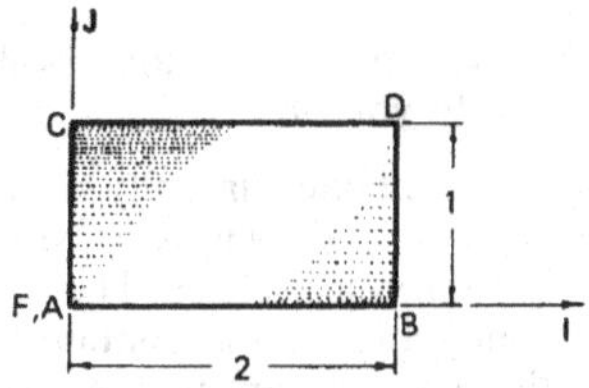

Problem 3.38.

3.39. The plate shown in its initial configuration in the previous problem undergoes a displacement such that the final position of A in $\Phi = \{F; \mathbf{I}_k\}$ is $(3, -2, 0)$ ft, and it also suffers a right-hand rotation of 30° about an axis through A normal to its plane. Compute the location of the center of rotation. Verify the result by a graphical construction.

3.40. A bottling machine is to be designed to operate continuously as follows. Two empty bottles are received by a carrier at A shown in the diagram. The filling operation is to begin after the carrier has turned 90° counterclockwise to the position at $B = (4, -2, 0)$ ft in $\Phi = \{A; \mathbf{I}_k\}$. The capping operation starts at C where the carrier has been rotated an additional 90°, and the bottles subsequently are removed from the machine when the carrier is at D and rotated 90° more. The design is to be based upon a single continuous circular motion of the carrier. Compute the location of the center O of rotation, and thus determine the radius r of the carrier table. Find the locations C and D of the carrier for the capping operation and the removal of the bottles from the machine. Show in a carefully drawn sketch how this information may be obtained graphically.

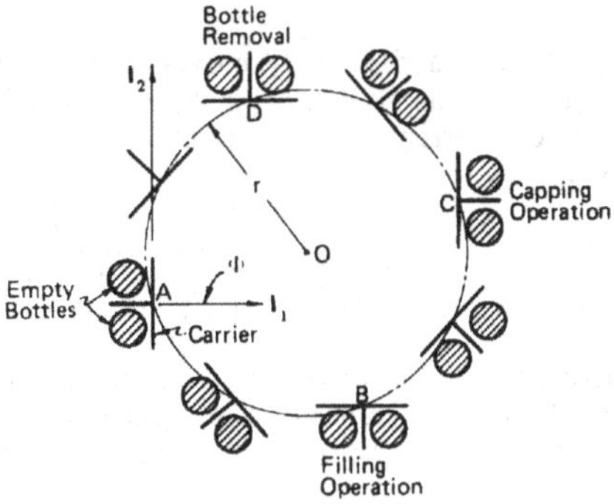

Problem 3.40.

3.41. Let the rotation matrix be defined by the transpose of the matrix given in Example 3.13, while all else remains the same. Rework the example, and describe the screw displacement. What differences do you find? Is this displacement the same as a reversed screw?

3.42. Rework the Example 3.13 with the point at $(3, -2, -4)$ cm as the base point O, while all else remains the same. Are the results the same? Find the position vector of the new center of rotation from the center of rotation O^* found in the example.

3.43. Reduce the displacement described in Problem 3.33 to a screw displacement. With B as the base point, determine the center of rotation and the pitch of the screw. Find the screw translation and its magnitude. What is the axis and the rotation of the screw? Solve this problem two ways: (i) by application of the linear system (3.141b), and (ii) by use of (3.146).

3.44. Repeat the work in the previous problem with the particle Q as the base point. What differences do you observe?

3.45. A rectangular container, initially situated with its center C at $(1, 2, 3)$ ft in the fixed frame $\Phi = \{F; \mathbf{I}_k\}$, as shown, is to be transported in a packing machine to a new configuration determined as follows. The center C must be moved to $(3, 4, 5)$ ft in Φ; and in its final configuration, the container must be rotated about a fixed axis through C. The rotation is equivalent to consecutive right-hand rotations of 90° about each of the $\mathbf{i}_2$ and the $\mathbf{i}_3$ axes of the spatial frame $\varphi = \{C; \mathbf{i}_k\}$, in their turn. Determine the angle and axis of the equivalent rotation required, and find the displacement of a point T whose initial location at the top of the container was $(1, 2, 4)$ ft in Φ. Locate the center of rotation of the equivalent screw displacement with respect to C, and characterize the screw.

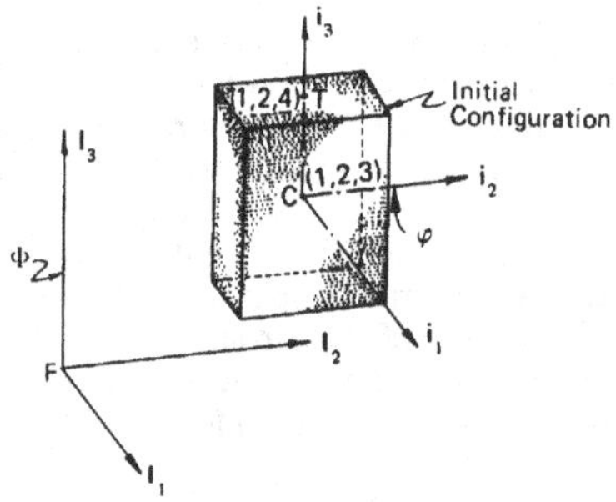

Problem 3.45.

3.46. A point A of a rigid body moves from its initial place at the origin to the place $(1, 2, 3)$ m in frame $\Phi = \{F; \mathbf{I}_k\}$. Another particle P is initially at the place $(1, 4, 6)$ m in Φ. Concurrently, the body executes certain rotations relative to Φ that are equivalent to

$$R = \frac{1}{3}\begin{bmatrix} 2 & 1 & 2 \\ -2 & 2 & 1 \\ -1 & -2 & 2 \end{bmatrix}.$$

Find the equivalent screw displacement, i.e., find the screw axis and angle of rotation, the magnitude of the screw translation, the pitch of the screw, and the location in Φ of the center of rotation with respect to A. What is the terminal location of P in Φ?

3.47. A rigid body executes three successive right-hand turns about the $\mathbf{i}_k$ axes of the spatial frame $\varphi = \{O; \mathbf{i}_k\}$. The first rotation is 45° about $\mathbf{i}$, the second is 90° about $\mathbf{j}$, and the last is 180° about $\mathbf{k}$. Construct the three rotation matrices in frame φ, and find the equivalent Euler rotation. Check your result by geometrical construction of the basis transformation matrix relating the imbedded body frame φ' to φ.

3.48. Determine by matrix multiplication the resultant Euler rotation equivalent to consecutive 90° right-hand rotations of a rigid body about the $\mathbf{i}$ and $\mathbf{k}$ axes of the spatial frame $\varphi = \{O; \mathbf{i}_j\}$. Find the Euler axis and angle of rotation. Apply the same rotations in reverse order, and repeat the previous part. Are the results the same?

3.49. A rigid body undergoes three consecutive proper rotations about axes of the body frame $\varphi' = \{O; \mathbf{e}'_k\}$, which initially is coincident with the spatial frame $\varphi = \{O; \mathbf{e}_k\}$, such that $\mathbf{R}_1\colon \mathbf{e}_k \to \mathbf{e}^1_k$, $\mathbf{R}_2\colon \mathbf{e}^1_k \to \mathbf{e}^2_k$, and $\mathbf{R}_3\colon \mathbf{e}^2_k \to \mathbf{e}^3_k \equiv \mathbf{e}'_k$, where $\mathbf{e}^n_k$, $n = 1, 2, 3$, denotes the successive body bases. Find the resultant rotation matrix referred to the body frame φ' in its terminal configuration in φ. Retain the usual definition of the basis transformation matrix.

3.50. The three rotations of the body about successive positions of the body axes as described in the previous problem are given as follows: $\mathbf{R}_1$ is 30° about $\mathbf{e}'_1$, which initially is coincident with $\mathbf{e}_1$; $\mathbf{R}_2$ is 90° about the $\mathbf{e}'_3$ body axis in its new orientation in φ; and $\mathbf{R}_3$ is 90° about the $\mathbf{e}'_1$ body axis in its new orientation resulting from all earlier rotations of the body in φ. What are the resultant angle and axis of rotation of the body referred to the body frame φ'? Check your result by constructing the resultant basis transformation matrix relating φ' to φ.

3.51. A rigid body executes three consecutive proper rotations of 90° about each of the three axes of the spatial frame $\varphi = \{F; \mathbf{i}_p\}$, each in its turn. Determine the resultant rotation by two methods: (a) Construct a diagram relating the initial and final orientations of the body frame, and find the basis transformation $A^*\colon \mathbf{i}_p \to \mathbf{i}'_p$. (b) Construct the rotation matrices R_1, R_2, and R_3 with respect to φ, and find the resultant rotation matrix R^* in φ. Are the results from (a) and (b) the same? Compute the resultant angle and axis of the rotation; and draw a sketch showing φ, φ', θ^*, and $\boldsymbol{\alpha}^*$.

3.52. Repeat the previous problem for consecutive 90° proper rotations about the axes of the body frame $\varphi' = \{F; \mathbf{i}'_p\}$, each in its turn. The matrices in part (b) should be referred to φ'. Are the results the same as those found in the last problem? (See Problem 3.49.)

3.53. A rigid body suffers three successive proper rotations as follows: (i) 45° about the $\mathbf{i}_1$ axis of the spatial frame $\varphi = \{F; \mathbf{i}_p\}$ followed by (ii) a 45° turn about the $\mathbf{i}'_2$ axis of the body frame $\varphi' = \{F; \mathbf{i}'_j\}$ in its new position, and (iii) a 90° rotation about the $\mathbf{i}_3$ axis of φ. Find the resultant Euler rotation with respect to φ.

3.54. The solar panel of a spacecraft suffers two rotations about axes $\boldsymbol{\alpha}_k$ in the spatial frame $\varphi = \{O; \mathbf{e}_k\}$, as indicated. The first rotation is 90° about the panel axis $\boldsymbol{\alpha}_1$, and the second is a 180° turn of the satellite about the axis $\boldsymbol{\alpha}_2$. Both axes are in the 23-plane. Describe the resultant rotation of the panel in φ; and find the final orientation in φ of the body frame $\varphi' = \{O; \mathbf{e}'_k\}$, which initially was coincident with φ. Sketch the final orientation of φ' in φ.

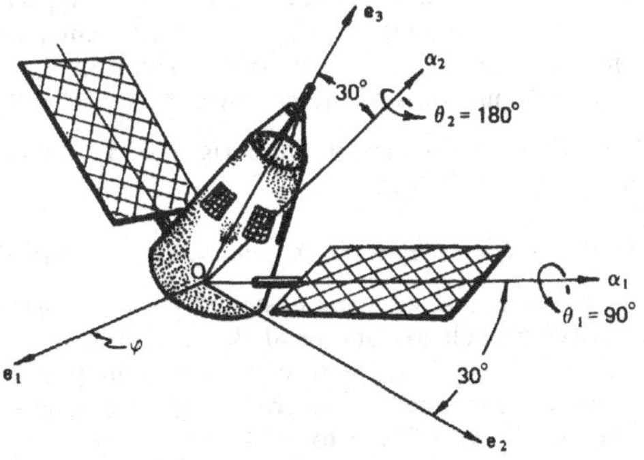

Problem 3.54.

3.55. The attitude of a spacecraft is adjusted by consecutive rotations produced by jet thrust nozzles. The first rotation is through a right angle about the axis $\boldsymbol{\alpha}_1$ in the vertical plane of a reference frame $\varphi = \{O; \mathbf{i}_k\}$ oriented with respect to the earth; and the next rotation is through a straight angle about the axis $\boldsymbol{\alpha}_2$ in the horizontal plane of frame φ, as illustrated. Find the resultant axis and angle of rotation of the spacecraft.

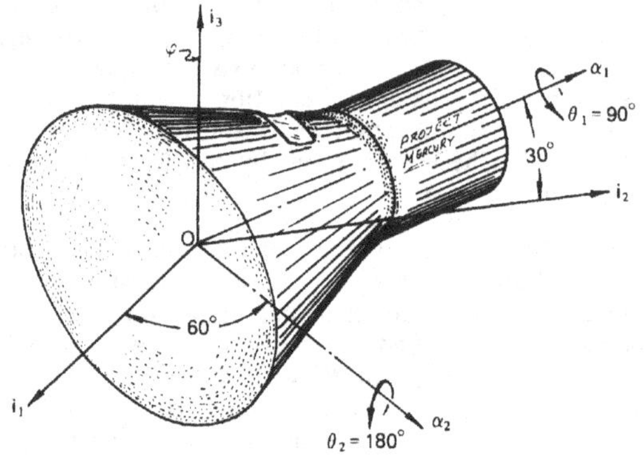

Problem 3.55.

3.56. Three consecutive, right-hand rotations about the $\mathbf{i}_k$ axes of a fixed spatial frame $\varphi = \{F; \mathbf{i}_k\}$ are required to bring a rigid body into its final orientation in φ. The body is held rigidly by a robotic device that controls its rotation. The first rotation is through an angle ϕ about $\mathbf{i}_3$; the second is through an angle θ about $\mathbf{i}_1$; and the third is through an angle ψ about $\mathbf{i}_2$. (a) Find the resultant rotation matrix. (b) If $\phi = 45°$, $\theta = 30°$, and $\psi = 60°$, determine the angle of the Euler rotation and the direction angles of the resultant axis in φ. Illustrate the results in a sketch.

3.57. The machine that controls the robot in the previous problem executes a further reversed resultant rotation of the rigid body/robot assembly about an axis through the particle P at $\mathbf{x}(P) = \mathbf{i}_2 + 2\mathbf{i}_3$ units in φ. Find the magnitude and the direction in φ of the displacement of P, and describe the displacement of the body. Sketch the results.

3.58. Apply (3.152) in the spatial frame φ to derive the matrices $\mathbf{R}_k$ for the three consecutive rotations through the Euler angles $\{\phi, \theta, \psi\}$ about axes $\boldsymbol{\alpha}_k = \{\mathbf{k}, \mathbf{i}_1^1, \mathbf{k}'\}$, as described in the text, all referred to φ. Show that the resultant rotation determined by (3.147b) is given by the matrix product $R^* = R_3 R_2 R_1$ with respect to φ; and show that the product yields the result (3.155) derived differently in the text. It may be helpful to notice from (3.147a) that each new axis of rotation may be computed from its former direction by multiplication of the rotation; hence, $\mathbf{i}_1^1 = \mathbf{R}_1 \mathbf{i}$ yields $\mathbf{i}_1^1$ in φ, for example.

3.59. Review the theory of finite rigid body displacements by completing your part in the following dialogue.

A Dialogue Concerning the Kinematics of a Rigid Body.

A certain machine must be designed to transport sealed rectangular boxes from one conveyor belt to another at the rate of 20 boxes per minute. You have observed that in accordance with certain design specifications the change of location may be accomplished in several simple steps; and with the aid of the diagram shown below, you begin to explain the procedure to your supervisor, who fre-

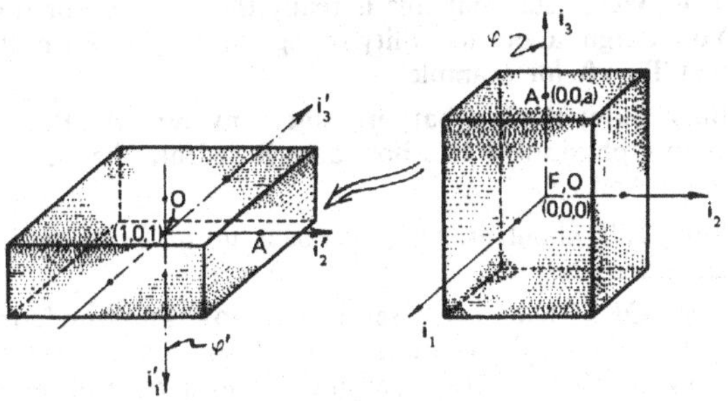

Problem 3.59.

quently interrupts your explanation for more information. Your conversation begins.

You: Let's start with a reference frame $\varphi = \{F; \mathbf{i}_k\}$ fixed in space, and call this the spatial frame. And let's take another frame $\varphi' = \{O; \mathbf{i}'_k\}$ imbedded in the box with the origin O at its centroid and initially coincident with the spatial frame at F. We'll call it the body frame. The displacement described in the specs can be accomplished like this: first rotate the box 90° counterclockwise about the $\mathbf{i}_2$ axis of φ, then 90° about the $\mathbf{i}_3$ direction in the same way. Now translate the box a suitable unit distance along the $\mathbf{i}_1$ axis, and, finally, move it the same distance along the $\mathbf{i}_3$ direction. This brings the box to the desired location.

Boss: Is it possible, since the angles are the same, to do the rotations in reverse order and have the same result?

You:

Boss: Really? Why does that happen?

You:

(*) Boss: Of course! I should have remembered that. Can you show me how you might replace these two rotations by a single rotation about O? I'd like to know the equivalent angle and axis required.

You: Sure; that's not hard. We can use the resultant rotation tensor $\mathbf{R}^*$ to calculate the rotation angle from the relation ____________. Then the axis is obtained from the formula ____________. Actually, the total displacement is related to the rotation tensor through the rotator equation

$$\mathbf{T}^* = \mathbf{S}^* \sin\theta^* + (1 - \cos\theta^*)(\boldsymbol{\alpha}^* \otimes \boldsymbol{\alpha}^* - \mathbf{1})$$

(**) where θ^* is the equivalent... .

Boss: Whoa! Let's put on the brakes so I can catch up. I got lost in the stars. I'm familiar with vectors and matrices, and I know a little about tensors; but I've never seen these tensor formulas before. Outline for me how you get this stuff and explain your notation so I can get a general picture of what this is all about. Start with your ...ah...ah... rotator.

You:

Boss: That's much clearer! I notice that the equation you got a moment ago for the particle displacement enables us to write your $\mathbf{T}$ as a vector operator:

$$\mathbf{T} = \sin\theta\, \boldsymbol{\alpha} \times [\ \ ] + (1 - \cos\theta)\boldsymbol{\alpha} \times (\boldsymbol{\alpha} \times [\ \ ]).$$

It occurred to me that this might be useful.

You: Well, yeah, but this is really the same tensor thing I wrote down before. You're right about its utility though; from your form of **T** we can see straight off that $\mathbf{T}\boldsymbol{\alpha} = \mathbf{0}$, for example.

Boss: Sure; that's what prompted my remark. But what does it mean, you know... physically? And how does **R** get into the picture? And what about those stars? And... .

You: Talk about brakes! Let's back up to **T**. Physically, the equation says that ________________.

Boss: OK. That makes sense. But now I'm wondering about your choice of origin on the axis of rotation. Intuitively, I'd say it doesn't matter. But can you show me that the displacement due to a rotation about the fixed axis doesn't depend on your choice of reference point on it? It just seems so obvious.

You: It really is. You can see it from ________________.

Boss: Well, this is all very interesting; but we seem to have drifted from our main problem. When I interrupted before (see ** above), you were telling me that you needed something more than **T*** to get the answer. And you still haven't explained the stars.

You: The stars only serve to distinguish the resultant rotation terms from the others. And it's **T*** that we want to find; but, so far, we haven't related it to the data based on the design specs. To do this, we need to relate the different frames.

Boss: The different frames?

You: Yeah. You'll remember that we started out with two reference frames; one fixed in space, the other fixed in the body. We haven't used these before. The displacements I described at the beginning were with respect to the spatial frame. Notice that after a displacement of the body, a particle has a new spatial location, but its location with respect to the body frame is unchanged. We can use this simple fact to relate a rigid body rotation about a point to the rotator for the fixed axis that we discussed before.

Boss: I vaguely recall my old math prof, Dr. Whitmore, a tiny fellow who the students dubbed Wee Willy Whitmore, mentioned this in connection with an important theorem due to...ah...ah... It's on the tip of my tongue—oh, what is his name? He was a prolific, 18th century mathematician that Wee Willy just loved to talk about.

You: You probably mean ________________.

Boss: That's it! He corresponded with another guy named Bernoulli about bending of slender beams, and in the process of studying this problem developed special topics in applied mathematics; and he created some fundamental principles of mechanics of solids and fluids. In fact, he developed a great deal of the stuff that engineers use today. I guess I'm beginning to sound like his press agent Willy. Anyway, I don't remember his theorem on rigid rotations, so perhaps you could refresh my memory.

You:

Boss: That really is a remarkable result. But I still don't see its specific practical value here.

You: Let me show you the proof, then I'm sure you'll see how this theorem leads to information relating the rotation tensor **R*** to our problem data.

Boss: OK. But just sketch the major ideas. I'm anxious to get the answer to our problem before going to the management meeting this morning.

You:

Boss: You were right. Now that we have this result, I do see the connection between the resultant angle and axis of rotation and the other matrices you've described. How did you know that the trace of neither the rotator nor the rotation tensor depends on your choice of reference frame?

You:

Boss: I'm convinced. Let's put this to work to get the answer to our problem (see * above).

You: I'll compute the resultant rotation you asked about earlier. I can now relate the two frames to find ____________; and then use the formulas ____________, which I wrote down before, to get the equivalent axis and angle you wanted: ____________.

Boss: That was slick. Of course, you realize that you haven't accounted for the unit translations.

You: That's the easiest part. From that important rotation theorem I proved before, we know that the most general displacement of a rigid body is equivalent to ____________. So, on this basis, I can give you the total displacement of any point of the body.

Boss: Then show me the general result; and afterwards demonstrate it for the centroid and the point A indicated in your sketch.

You:

Boss: I'm impressed with your thorough understanding of finite rigid body displacements. I haven't looked at this stuff for years. Your tensor algebraic methods actually are easier to follow than the longhand algebraic and geometrical proofs that Wee Willy was so fond of, so this has been an enlightening review for me. I can see from your analysis how the displacements of all points of the body can be easily found. You have a good start on this problem.

You: Start?

Boss: Yes. I'd like you to consider the possibility of producing the same displacement of the body by other means that might allow the design group to come up with the simplest possible mechanism for our purpose. For the design specs you used, what would be the simplest possible translation and rotation that could be applied? And how would it be executed?

I've got to leave for the management meeting; but if you need a few ideas to clarify this, you might consult Mr. Chasles, who I understand has had considerable experience with problems of this type. You might also look into the possible velocities and accelerations that may be involved. As you know, the device must handle 20 boxes per minute; but the design group would like us to look at the possibility of improving the rate for other potential applications. I'll check back with you later. You're doing a good job.

You:

4

Motion Referred to a Moving Reference Frame and Relative Motion

4.1. Introduction

Thus far in our studies, the motion of a material point has been referred mainly to an assigned reference frame that often we thought of as being "fixed in space." We have seen, however, that it is sometimes more convenient to refer the motion to a reference frame which itself is in arbitrary motion relative to some (possibly moving) assigned frame. To trace the motion of a long-range ballistic missile or space rocket, for example, it is essential that the motion of the earth be taken into account. In such cases it becomes necessary to refer the body's motion to a moving reference system imbedded in the earth; and, in this instance, a reference frame fixed in the distant "fixed stars" may be chosen as a suitable assigned reference frame with respect to which the earth's motion is known.

Sometimes several reference frames in motion one relative to the other are encountered in a problem. The motion of a maneuverable target object in retreat from a pursuing rocket fired from an aircraft clearly is perceived differently by the pilot in the aircraft reference frame and by the rocket guidance control system in the rocket reference frame. In this instance, the motion of the target object is referred to two frames, and it would be of interest to know how the target's motions as seen by the two "observers" are related.

It is an easy geometrical problem to refer a vector, such as velocity or acceleration, to any desired frame, since this means only that the vector is represented in terms of the basis that defines the frame. This was demonstrated earlier in the easy Example 3.7 and in the Example 2.10 in which the velocity was referred to an imbedded reference frame. The intrinsic representation of a motion, studied in Chapter 1, is another case where the velocity and acceleration of a particle were referred to a moving frame. In general, if any vector associated with the motion of a particle is known in one reference frame, then we may determine by orthogonal projections its components along lines parallel to the basis of any frame that bears a known orientation with the assigned one, whether the frames are moving or not. This kind of construction may be used to relate the motion of a material point as seen by two observers associated with these frames; and this procedure also will enable us to determine the influence of the motion of the reference frame on the motion of a particle apparent to a moving observer. In addition, we shall learn the extent of the error committed when the effect of the motion of a moving frame, such as the earth, is ignored.

Equations relating the angular velocities and angular accelerations of several reference bodies will be derived in this chapter. Generalized formulas that relate the velocities and accelerations of a particle in two reference systems will be constructed and applied to a variety of problems. These general equations will include most of our earlier results. Some useful examples of motion referred to special moving frames that are convenient for problems where cylindrical or spherical coordinates are appropriate also will be developed. Special topics that use matrix and tensor methods described in Chapter 3 will be presented at the end. But, first, we must learn how to compute the rate of change of a vector which is referred to a moving frame. Let us begin with an example.

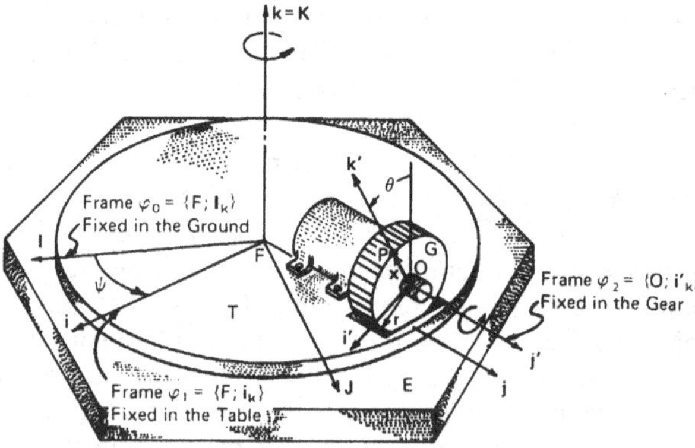

Figure 4.1. A motion $\mathbf{x}(P, t)$ of a gear particle P referred to the moving frames φ_1 and φ_2, and to the fixed ground frame φ_0.

4.2. An Introductory Example

A particular reference frame usually is chosen to simplify the form or the derivation of equations and to ease calculations. A vector **v**, for example, may be a constant vector in one reference frame, and a function of one or more variables in another. Therefore, the derivatives of the same vector referred to different reference frames will be different. This will be illustrated in the following example.

The Fig. 4.1 of the motor on a rotating table shows three reference frames: the frame $\varphi_0 = \{F; \mathbf{I}_k\}$ is fixed in the ground E; frame $\varphi_1 = \{F; \mathbf{i}_k\}$ is in the table T; and frame $\varphi_2 = \{O; \mathbf{i}'_k\}$ is imbedded in the gear G of radius r. A point P on the rim of G has a position vector $\mathbf{x}$ from its center. Of course, in each of these frames the same vector $\mathbf{x}$ has a unique representation $\mathbf{x} = x_k \mathbf{e}_k$ in which the scalar components x_k describe the behavior of $\mathbf{x}$ when referred to the reference frame with basis vectors $\mathbf{e}_k$. But in another frame, i.e., when $\mathbf{x}$ is referred to a different basis, the components will be different. In the present case, it is easily seen from the geometry in Fig. 4.1 that

$$\mathbf{x} = r\mathbf{k}' \text{ is a constant vector in } \varphi_2; \tag{4.1a}$$

$$\mathbf{x} = r(\sin\theta\,\mathbf{i} + \cos\theta\,\mathbf{k}) = \mathbf{x}(\theta) \text{ in } \varphi_1; \tag{4.1b}$$

$$\mathbf{x} = r\sin\theta(\cos\psi\,\mathbf{I} + \sin\psi\,\mathbf{J}) + r\cos\theta\,\mathbf{K} = \mathbf{x}(\theta, \psi) \text{ in } \varphi_0. \tag{4.1c}$$

We thus see that the position vector $\mathbf{x}$ is a function of ψ and θ in φ_0; it depends on θ in frame φ_1; and in φ_2, it is a constant vector.

It follows from (4.1) that the rate of change of $\mathbf{x}$ with respect to θ is given by

$$\frac{\partial\mathbf{x}}{\partial\theta} = \begin{cases} \mathbf{0} \text{ in } \varphi_2; \\ r(\cos\theta\,\mathbf{i} - \sin\theta\,\mathbf{k}) \text{ in } \varphi_1; \\ r\cos\theta(\cos\psi\,\mathbf{I} + \sin\psi\,\mathbf{J}) - r\sin\theta\,\mathbf{K} \text{ in } \varphi_0; \end{cases} \tag{4.2}$$

and its change with respect to ψ is

$$\frac{\partial\mathbf{x}}{\partial\psi} = \begin{cases} \mathbf{0} \text{ in } \varphi_2; \\ \mathbf{0} \text{ in } \varphi_1; \\ r\sin\theta(-\sin\psi\,\mathbf{I} + \cos\psi\,\mathbf{J}) \text{ in } \varphi_0. \end{cases} \tag{4.3}$$

Thus, assuming that the time t is the same for all observers, with the use of (4.2) and (4.3) in which $\theta = \theta(t)$ and $\psi = \psi(t)$, we obtain the following time rates of change of $\mathbf{x}$ in the frames indicated:

$$\frac{d\mathbf{x}}{dt} = \mathbf{0} \text{ in } \varphi_2, \tag{4.4a}$$

$$\frac{d\mathbf{x}}{dt}=\frac{\partial\mathbf{x}}{\partial\theta}\dot{\theta}=r\dot{\theta}(\cos\theta\,\mathbf{i}-\sin\theta\,\mathbf{k})\text{ in }\varphi_1, \tag{4.4b}$$

$$\frac{d\mathbf{x}}{dt}=\frac{\partial\mathbf{x}}{\partial\theta}\dot{\theta}+\frac{\partial\mathbf{x}}{\partial\psi}\dot{\psi}=r\dot{\theta}\cos\theta(\cos\psi\,\mathbf{I}+\sin\psi\,\mathbf{J})-r\dot{\theta}\sin\theta\,\mathbf{K}$$
$$+\,r\dot{\psi}\sin\theta(-\sin\psi\,\mathbf{I}+\cos\psi\,\mathbf{J})\text{ in }\varphi_0. \tag{4.4c}$$

Notice that the constant vector $\mathbf{k}'$ seen by an observer fixed in the gear frame φ_2 is perceived as a time-varying vector

$$\mathbf{k}'(t)=\sin\theta(t)\,\mathbf{i}+\cos\theta(t)\,\mathbf{k} \tag{4.5}$$

by an observer in the table frame φ_1, because the angle $\theta=\theta(t)$ that $\mathbf{k}'$ makes with directions $\mathbf{i}$ and $\mathbf{k}$ fixed in frame φ_1 changes with time. Consequently, $d\mathbf{k}'/dt=\mathbf{0}$ for the observer in φ_2, whereas for the observer in φ_1

$$\frac{d\mathbf{k}'(t)}{dt}=\dot{\theta}(\cos\theta\,\mathbf{i}-\sin\theta\,\mathbf{k}), \tag{4.6}$$

in which $\dot{\theta}$ is the angular speed of the gear relative to the table. We thus see that (4.4b) also may be derived from (4.1a) directly. The reader may show that the last of (4.4) may be obtained in a similar manner from (4.1b). It will be seen that the time derivatives of $\mathbf{i}(t)$ and $\mathbf{k}(t)$ will involve the angular speed of the table relative to the ground. The general manner in which the angular velocity of a moving reference frame affects the time derivative of any vector referred to that frame will be studied in the next section.

The increasing complexity of the expressions in (4.1)–(4.4) underscores the simplicity which seems to be achieved when the position vector for the motion is referred to a moving reference frame, the simplest relations being those obtained for the frame φ_2 imbedded in the gear. We shall see more of this as we move ahead. Of course, the three time derivatives (4.4) of the same vector referred to different frames have distinct meanings, and for this reason it is useful to introduce special notations for clarity. This will be done in the next section, where we shall derive the simple, general formula for the derivative of any vector referred to a moving frame.

4.3. Derivative of a Vector Referred to a Moving Reference Frame

Any reference frame having a motion relative to a preferred, assigned frame is called a *moving frame*. Thus, in this sense, the *preferred frame* often is thought of as being fixed in space. However, it is not necessary to impose the

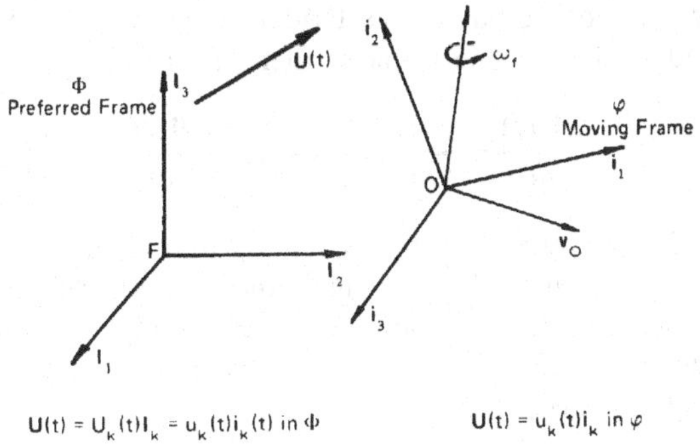

Figure 4.2. A vector $\mathbf{U}(t)$ referred to a preferred and to a moving frame.

condition that the preferred frame be at absolute rest in space. In fact, the term "absolute rest" is without meaning in our universe; so let us agree that *the term "absolute" is to be applied only in reference to quantities referred to the preferred frame, whether it be fixed in space or not.* The terms "preferred" and "moving" will be used only in the simple relative sense described above; and, whenever it may be convenient, we may reverse our choice of labels for the two frames and achieve parallel results. We shall assume that all observers employ the same time reference, i.e., *all observers use the same standard clock.*

Let the preferred frame be denoted by $\Phi = \{F; \mathbf{I}_k\}$ and the moving frame by $\varphi = \{O; \mathbf{i}_k\}$. Suppose that the frame φ has angular velocity $\boldsymbol{\omega}_f$ relative to the frame Φ and translational velocity $\mathbf{v}_O$, as shown in Fig. 4.2. Let $\mathbf{U}(t)$ be any vector-valued function of the time t. The scalar components of $\mathbf{U}(t)$ are $U_k(t)$ when referred to the preferred frame Φ and $u_k(t)$ when referred to the moving frame φ at time t. Thus,* $U(t) = U_k(t)\,\mathbf{I}_k$ is the representation at time t of the vector $\mathbf{U}(t)$ as seen by the observer F fixed in frame Φ and referred to that frame, whereas the relation $\mathbf{U}(t) = u_k(t)\,\mathbf{i}_k(t)$ is the representation of $\mathbf{U}(t)$ as seen by the *same* observer F but referred to the moving frame. Of course, the basis vectors $\mathbf{i}_k$ have always the direction of the moving coordinate axes, so that apparent to an observer O in frame φ, the vector $\mathbf{U}(t)$ at time t has the same representation $\mathbf{U} = u_k(t)\,\mathbf{i}_k$; however, as indicated, the moving observer perceives no change in $\mathbf{i}_k$ with time t. Consequently, the rate of change of $\mathbf{U}(t)$ will appear differently to the two observers. At time $t + \Delta t$, the preferred observer F sees a change in $u_k(t)$ and a change in $\mathbf{i}_k(t)$, but only the change in $u_k(t)$ is apparent to the moving observer at O in φ. We wish to relate the rates of change of the vector $\mathbf{U}(t)$ as seen by these two observers.

Consider the vector $\mathbf{U}(t) = u_k(t)\,\mathbf{i}_k(t)$ as seen by the preferred observer F

* The summation rule introduced in Section 3.2.1 is used here.

and referred to the moving basis $\mathbf{i}_k(t)$. Evidently, the rate of change of $\mathbf{U}(t)$ in Φ when referred to the moving frame is then given by

$$\frac{d\mathbf{U}(t)}{dt} = \frac{du_k(t)}{dt}\,\mathbf{i}_k(t) + u_k(t)\frac{d\mathbf{i}_k(t)}{dt}. \tag{4.7}$$

To determine $d\mathbf{i}_k(t)/dt$, we recall that $\mathbf{i}_k(t)$ is a unit vector along the kth axis of the moving frame. Since the moving frame is rigid, we can treat the end point Q of the vector $\mathbf{i}_k(t)$ as a point of a rigid body, as shown in Fig. 4.3. Then the velocity of Q relative to O in the frame Φ is given by (2.29), in which we put $\mathbf{x}(t) = \mathbf{i}_k(t)$; hence,

$$\frac{d\mathbf{i}_k(t)}{dt} = \boldsymbol{\omega}_f \times \mathbf{i}_k. \tag{4.8}$$

Upon substituting (4.8) into (4.7), we get in Φ

$$\frac{d\mathbf{U}(t)}{dt} = \frac{du_k(t)}{dt}\,\mathbf{i}_k(t) + \boldsymbol{\omega}_f \times u_k(t)\,\mathbf{i}_k(t). \tag{4.9}$$

Finally, let us introduce a special partial differentiation symbol $\delta/\delta t$ defined by

$$\frac{\delta\mathbf{U}(t)}{\delta t} \equiv \frac{du_k(t)}{dt}\,\mathbf{i}_k(t). \tag{4.10}$$

That is, $\delta\mathbf{U}(t)/\delta t$ is the rate of change of the vector $\mathbf{U}(t)$ as though the basis vectors $\mathbf{i}_k(t)$ were fixed. Therefore, it represents the rate of change of $\mathbf{U}(t)$ apparent to an observer O in frame φ at time t. Putting (4.10) into (4.9) and writing $\dot{\mathbf{U}}(t) \equiv d\mathbf{U}(t)/dt$, we see that the *total (or absolute) time rate of change of the vector* $\mathbf{U}(t)$ as seen by the preferred observer, but referred to the moving frame, is given by

$$\dot{\mathbf{U}}(t) = \frac{\delta\mathbf{U}(t)}{\delta t} + \boldsymbol{\omega}_f(t) \times \mathbf{U}(t). \tag{4.11}$$

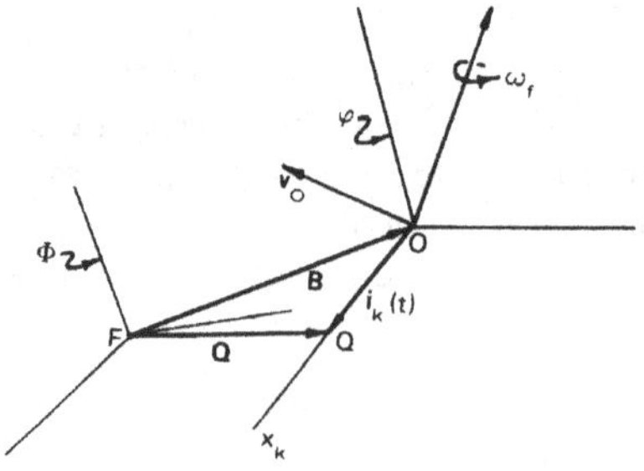

Figure 4.3. Schema for the calculation of $d\mathbf{i}_k(t)/dt$ in frame Φ.

We note that when the vector $\mathbf{U}(t)$ is fixed in the moving frame, the moving observer perceives no change in $\mathbf{U}(t)$; hence, $\delta\mathbf{U}/\delta t=\mathbf{0}$ and

$$\dot{\mathbf{U}}(t)=\boldsymbol{\omega}_f(t)\times\mathbf{U}(t). \tag{4.12}$$

This result extends (4.8) to any vector fixed in the moving frame.

The vector $\delta\mathbf{U}(t)/\delta t$, the time rate of change of $\mathbf{U}(t)$ in φ, is called the *apparent time rate of change* of $\mathbf{U}(t)$; and $\boldsymbol{\omega}_f\times U$, which is due to the relative rotation of the moving frame, is called the *convective time rate of change* of $\mathbf{U}(t)$. Notice that the translational velocity of frame φ has no influence on the rate of change of $\mathbf{U}(t)$, for if there were no rotation of the moving frame, the basis vectors $\mathbf{i}_k$ would have constant directions in frame Φ; therefore, they would be the same for both observers. We may thus summarize the result (4.11) as follows: *The absolute time rate of change $d\mathbf{U}(t)/dt$ of a vector $\mathbf{U}(t)$ in the preferred frame is equal to the sum of the apparent time rate of change $\delta\mathbf{U}(t)/\delta t$ in the moving frame and the convective time rate of change $\boldsymbol{\omega}_f\times\mathbf{U}(t)$.*

The formula (4.11) applies to any vector $\mathbf{U}(t)$ that is referred to a moving reference frame having angular velocity $\boldsymbol{\omega}_f$; consequently, we see that the differential operator d/dt when operating on any vector referred to a moving frame must be replaced by the following linear differential vector operator:

$$\frac{d}{dt}[\ \ ]=\frac{\delta}{\delta t}[\ \ ]+\boldsymbol{\omega}_f\times[\ \ ]. \tag{4.13}$$

The following important relation derives at once from (4.11) or (4.13):

$$\dot{\boldsymbol{\omega}}_f(t)\equiv\frac{d}{dt}\boldsymbol{\omega}_f(t)=\frac{\delta}{\delta t}\boldsymbol{\omega}_f(t). \tag{4.14}$$

That is, *the absolute and apparent time rates of change of the angular velocity $\boldsymbol{\omega}_f$ of the moving frame are the same.* This means only that if first we write $\boldsymbol{\omega}_f$ in the frame Φ, differentiate the result keeping the directions $\mathbf{I}_k$ fixed, and afterwards change the basis from $\mathbf{I}_k$ to $\mathbf{i}_k$, the derivative $d\boldsymbol{\omega}_f/dt$ would be the same as the result $\delta\boldsymbol{\omega}_f/\delta t$ obtained by first referring the vector $\boldsymbol{\omega}_f$ to the moving frame directions $\mathbf{i}_k$, and then differentiating $\boldsymbol{\omega}_f$ while keeping the basis vectors $\mathbf{i}_k$ fixed.

The fundamental equation (4.11) will be applied in three examples that follow. In the first two examples, the time derivative of the angular velocity of the arm of a robotic device and of the position vector of a point on the arm, both referred to a moving reference frame, will be determined. Time derivatives of the same vectors referred to a fixed reference frame also will be computed for comparison. The third example concerns evaluation of the first and second time derivatives of an angular velocity vector for a gear driven by a motor on a rotating platform. These examples will demonstrate the simplicity achieved by use of the basic relation (4.11); and they will prepare the way for the future derivation of generalized equations for the total angular acceleration of one of several connected rigid bodies, and for the velocity and acceleration of a particle viewed from two reference frames in relative motion.

Example 4.1. At the instant t_0, the telescopic arm OA of a robot shown in Fig. 4.4 is being lowered with an angular speed $\dot{\beta} = 0.2$ rad/sec, which is increasing at the rate of 0.3 rad/sec each second, relative to the swivel yoke. The yoke has a constant angular speed $\dot{\alpha} = 0.5$ rad/sec about a fixed vertical axis $\mathbf{K}$ in a frame $\Phi = \{O; \mathbf{I}_k\}$ fixed in the machine foundation. Let $\boldsymbol{\omega}_2$ denote the angular velocity of the arm relative to the yoke, whose angular velocity relative to the machine is denoted by $\boldsymbol{\omega}_1$, as indicated in Fig. 4.4. Find the time rate of change in Φ of the vector $\boldsymbol{\omega}_2$ referred to a reference frame $\varphi = \{O; \mathbf{i}_k\}$ imbedded in the yoke so that $\mathbf{k} = \mathbf{K}$.

Solution. It is evident from Fig. 4.4 that the angular velocity $\boldsymbol{\omega}_2$ relative to the yoke frame φ has a simple representation when referred to φ. But this frame is moving with angular velocity $\boldsymbol{\omega}_1 = 0.5\mathbf{k}$ rad/sec relative to the machine frame Φ. Thus, the time derivative in Φ of the vector $\boldsymbol{\omega}_2$, which is referred to a frame moving with angular velocity

$$\boldsymbol{\omega}_f \equiv \boldsymbol{\omega}_1 = \dot{\alpha}\mathbf{k} = 0.5\mathbf{k} \text{ rad/sec} \tag{4.15a}$$

relative to Φ, may be determined easily by (4.11):

$$\dot{\boldsymbol{\omega}}_2 = \frac{\delta \boldsymbol{\omega}_2}{\delta t} + \boldsymbol{\omega}_f \times \boldsymbol{\omega}_2, \tag{4.15b}$$

in which the definition of the δ-derivative in (4.10) is to be recalled. At the moment of interest, the angular velocity and the angular acceleration of OA relative to φ are, respectively,

$$\boldsymbol{\omega}_2 = \dot{\beta}\mathbf{j} = 0.2\mathbf{j} \text{ rad/sec}, \qquad \frac{\delta \boldsymbol{\omega}_2}{\delta t} = \ddot{\beta}\mathbf{j} = 0.3\mathbf{j} \text{ rad/sec}^2 \tag{4.15c}$$

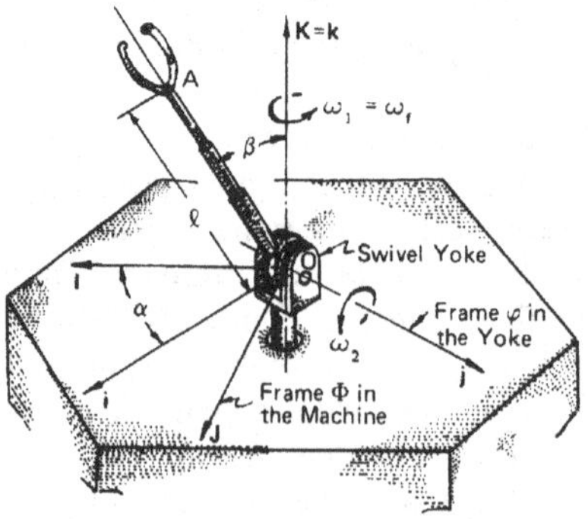

Figure 4.4. A robotistic device with an angular velocity $\boldsymbol{\omega}_2$ referred to a rotating frame φ fixed in the supporting yoke.

referred to φ. Hence, with (4.15a) and the first equation in (4.15c), we have

$$\boldsymbol{\omega}_f \times \boldsymbol{\omega}_2 = 0.5\mathbf{k} \times 0.2\mathbf{j} = -0.1\mathbf{i}\ \text{rad/sec}^2. \tag{4.15d}$$

Substitution of the second equation in (4.15c) and (4.15d) into (4.15b) gives the solution at the instant t_0:

$$\dot{\boldsymbol{\omega}}_2 = 0.3\mathbf{j} - 0.1\mathbf{i}\ \text{rad/sec}^2. \tag{4.15e}$$

This is the time rate of change in the fixed frame Φ of the angular velocity $\boldsymbol{\omega}_2$, but referred to frame φ, i.e., expressed in terms of the vector basis of the moving frame φ.

This completes the solution of the problem. It may prove helpful, however, to examine the familiar method in which we first write $\boldsymbol{\omega}_2$ in the basis of the preferred frame Φ, differentiate the result as usual, and then, by a change of basis, transform the derived formula back to φ for comparison with (4.15e).

In this approach, we must first express the basis vectors $\mathbf{i}_k$ of frame φ in terms of the basis vectors $\mathbf{I}_k$ of frame Φ:

$$\mathbf{i} = \cos\alpha\,\mathbf{I} + \sin\alpha\,\mathbf{J}, \qquad \mathbf{j} = -\sin\alpha\,\mathbf{I} + \cos\alpha\,\mathbf{J}, \qquad \mathbf{k} = \mathbf{K}. \tag{4.16a}$$

In the present case, only the second of (4.16a) is needed. Then the first equation in (4.15c) becomes

$$\boldsymbol{\omega}_2 = \dot{\beta}(-\sin\alpha\,\mathbf{I} + \cos\alpha\,\mathbf{J}). \tag{4.16b}$$

This is the angular velocity of the arm OA relative to the yoke, as before, but now referred to frame Φ. Differentiation of (4.16b) in frame Φ yields

$$\dot{\boldsymbol{\omega}}_2 = -(\ddot{\beta}\sin\alpha + \dot{\alpha}\dot{\beta}\cos\alpha)\,\mathbf{I} + (\ddot{\beta}\cos\alpha - \dot{\alpha}\dot{\beta}\sin\alpha)\,\mathbf{J}. \tag{4.16c}$$

This is the angular acceleration of the arm OA relative to φ but referred to Φ.

But the same vector may be referred to any frame whatever. Therefore, substitution into (4.16c) of the change of basis from $\mathbf{I}_k$ into $\mathbf{i}_k$, which is left for the reader, delivers the same physical result as (4.16c) but referred to φ again. We find

$$\dot{\boldsymbol{\omega}}_2 = -\dot{\alpha}\dot{\beta}\mathbf{i} + \ddot{\beta}\mathbf{j}. \tag{4.16d}$$

Indeed, it will be seen that this formula follows more easily from (4.15b) upon substitution of the first expressions in (4.15c) and (4.15a). Naturally, use here of the assigned numerical values for $\dot{\alpha}$, $\dot{\beta}$, and $\ddot{\beta}$ in (4.16d) will yield the same instantaneous result (4.15e) derived algebraically from (4.15b). □

In the first solution based upon (4.15b), the usual differentiation operations were not needed. In the second approach, however, it was necessary to write $\boldsymbol{\omega}_2$ in terms of the variables α and β and execute the differentiation of the vector function (4.16b). Notice further that some messy geometry was avoided by use of the fundamental rule (4.11). It is seen that the geometrical considerations and differentiation operations essential in the familiar approach are reduced almost entirely to simpler vector algebraic operations through use of the general relation (4.11). As a consequence, this basic equation may be applied directly in problems where only the instantaneous values of the various quantities are assigned. The simplicity of the general formula (4.16d) referred to the moving frame compared with the corresponding relation (4.16c) referred to the fixed frame certainly is evident.

Example 4.2. Let us suppose that during the motion of the robot described in the last example, the length of the telescopic arm is a computer-controlled function of time denoted by $l(t)$ in Fig. 4.4. (a) What is the time rate of change in φ of the position vector of A referred to the moving yoke frame at an arbitrary time? (b) What is its time derivative in Φ?

Solution. (a) The position vector of A on the telescopic arm in Fig. 4.4 is given in frame φ by

$$\mathbf{x}(A, t) = l(t)[\sin \beta(t)\, \mathbf{i} + \cos \beta(t)\, \mathbf{k}]. \tag{4.17a}$$

The time derivative of this vector in the frame φ, in which the basis vectors $\mathbf{i}_k$ are fixed, is determined by use of (4.10); we get

$$\left.\frac{d\mathbf{x}(A, t)}{dt}\right|_{\varphi} \equiv \frac{\delta \mathbf{x}(A, t)}{\delta t} = (\dot{l} \sin \beta + l\dot{\beta} \cos \beta)\, \mathbf{i} + (\dot{l} \cos \beta - l\dot{\beta} \sin \beta)\, \mathbf{k}. \tag{4.17b}$$

We shall see later that this is the velocity of A as seen by an observer situated in the yoke frame φ and referred to the same frame; it is the velocity of A relative to φ.

(b) The time derivative in Φ of the position vector in (4.17a), which is a vector referred to the moving yoke frame, is determined by use of (4.11); we have

$$\frac{d\mathbf{x}(A, t)}{dt} = \frac{\delta x(A, t)}{\delta t} + \boldsymbol{\omega}_f \times \mathbf{x}(A, t). \tag{4.17c}$$

The first term on the right-hand side is given by (4.17b). The constant angular velocity of the moving frame is $\boldsymbol{\omega}_f \equiv \boldsymbol{\omega}_1 = 0.5\mathbf{k}$ rad/sec; thus, with (4.17a), we obtain $\boldsymbol{\omega}_f \times \mathbf{x} = (l/2) \sin \beta \, \mathbf{j}$. Using this result and (4.17b) in (4.17c), we find the solution

$$\frac{d\mathbf{x}(A, t)}{dt} = (\dot{l} \sin \beta + l\dot{\beta} \cos \beta) \, \mathbf{i} + \frac{l}{2} \sin \beta \, \mathbf{j} + (\dot{l} \cos \beta - l\dot{\beta} \sin \beta) \, \mathbf{k}. \tag{4.17d}$$

This is the velocity of A as seen by an observer situated in the machine foundation Φ but referred to the moving frame φ for convenience; it is the (absolute) velocity of A relative to Φ.

The physical ideas illustrated here will be expanded further on. In preparation for our future studies, the student should find it helpful to determine the derivative of (4.17b) in the frames φ and Φ. □

Example 4.3. An electric motor M shown in Fig. 4.5 is attached to a platform that rotates with a constant angular speed of 10 rev/sec about a vertical axis. The motor drives a gear G at a constant angular speed $\omega_1 = 300$ rev/min relative to the platform. (a) Find the time derivative of the angular velocity vector $\boldsymbol{\omega}_1$ in the fixed spatial frame Φ. (b) What is $\ddot{\boldsymbol{\omega}}_1$ in Φ? (c) What are these derivatives in the moving frame φ fixed in the platform? Refer all vectors to the frame φ shown in Fig. 4.5.

Solution. (a) The moving frame $\varphi = \{O; \mathbf{i}_k\}$ is fixed in the platform with $\mathbf{j}$ directed along the center line of the motor axle. Thus, with modified units and referred to the moving frame φ, $\boldsymbol{\omega}_f = 20\pi\mathbf{k}$ rad/sec is the constant angular velocity of the platform frame φ relative to Φ, and $\boldsymbol{\omega}_1 = 10\pi\mathbf{j}$ rad/sec is the angular velocity of the gear relative to the platform. It is clear from (4.10) that in the rotating φ-frame $\delta\boldsymbol{\omega}_1/\delta t = \mathbf{0}$; and it follows from the general rule (4.11) or by (4.12) that

$$\dot{\boldsymbol{\omega}}_1 = \frac{\delta \boldsymbol{\omega}_1}{\delta t} + \boldsymbol{\omega}_f \times \boldsymbol{\omega}_1 = -200\pi^2 \, \mathbf{i} \text{ rad/sec}^2. \tag{4.18a}$$

This is the time derivative of $\boldsymbol{\omega}_1$ in Φ, but referred to the moving frame. Because $\boldsymbol{\omega}_1$ is a constant vector in φ, clearly, the change in $\boldsymbol{\omega}_1$ seen by the

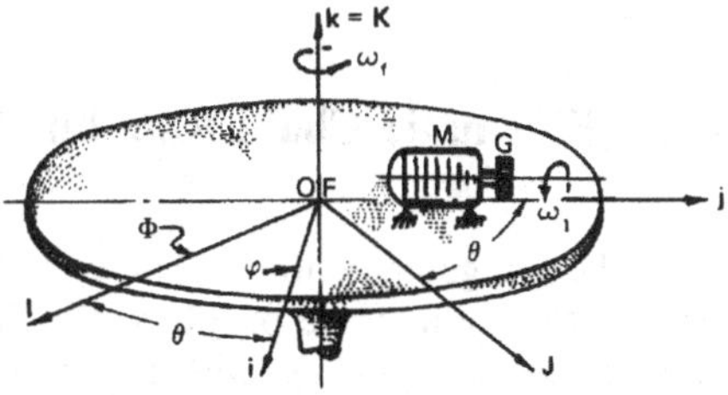

Figure 4.5. A constant vector $\boldsymbol{\omega}_1$ in the rotating platform frame φ has a nonzero time derivative in the fixed frame Φ due to the rotation of φ relative to Φ.

observer in Φ is due solely to the change in its direction due to the rotation of φ, which is in the direction of $-\mathbf{i}$ indicated in (4.18a).

(b) To find $\ddot{\boldsymbol{\omega}}_1$, we observe from (4.18a) that $\dot{\boldsymbol{\omega}}_1$ is a constant vector in the platform frame φ; and hence $\delta\dot{\boldsymbol{\omega}}_1/\delta t = \mathbf{0}$. This is the time derivative of $\dot{\boldsymbol{\omega}}_1$ in φ. Then, with (4.11), the time derivative of $\dot{\boldsymbol{\omega}}_1$ in Φ is

$$\ddot{\boldsymbol{\omega}}_1 = \frac{\delta\dot{\boldsymbol{\omega}}_1}{\delta t} + \boldsymbol{\omega}_f \times \dot{\boldsymbol{\omega}}_1 = -4000\pi^3\mathbf{j}\ \text{rad/sec}^3. \tag{4.18b}$$

Clearly, the corresponding derivative in φ is $\delta^2\boldsymbol{\omega}_1/\delta t^2 = \mathbf{0}$.

(c) Because $\boldsymbol{\omega}_1$ is a constant vector in φ, all of its derivatives vanish in φ: $\delta^n\boldsymbol{\omega}_1/\delta t^n = \mathbf{0}$, as we saw above for $n = 1, 2$. This completes the problem solution.

We have determined in (4.18a) the angular acceleration $\dot{\boldsymbol{\omega}}_1$ in frame Φ when the angular velocity vector $\boldsymbol{\omega}_1$ is referred to the reference frame φ which is turning with angular velocity $\boldsymbol{\omega}_f$ in Φ. However, as shown in Example 4.1, we may obtain the same results by referring $\boldsymbol{\omega}_1$ to the preferred frame Φ. First, we must write the moving basis $\mathbf{i}_k$ in terms of the fixed basis $\mathbf{I}_k$. From the geometry of Fig. 4.5, we have

$$\mathbf{i} = \cos\theta\,\mathbf{I} + \sin\theta\,\boldsymbol{J}, \qquad \mathbf{j} = \cos\theta\,\mathbf{J} - \sin\theta\,\mathbf{I}, \qquad \mathbf{k} = \mathbf{K}. \tag{4.18c}$$

Thus, referred to Φ, we have the general formula

$$\boldsymbol{\omega}_1 = 10\pi\mathbf{j} = 10\pi(\cos\theta\,\mathbf{J} - \sin\theta\,\mathbf{I})\ \text{rad/sec}; \tag{4.18d}$$

and with $\dot{\theta} = |\boldsymbol{\omega}_f| = 20\pi$ rad/sec, we obtain

$$\dot{\boldsymbol{\omega}}_1 = -10\pi\dot{\theta}(\sin\theta\,\mathbf{J} + \cos\theta\,\mathbf{I}) = -200\pi^2\mathbf{i}\ \text{rad/sec}^2, \tag{4.18e}$$

in which the change of basis back to φ is left for the reader. This is the same as (4.18a). Construction of $\ddot{\boldsymbol{\omega}}_1$ is left to the student. This method illustrates again that an important advantage of the procedure developed in this chapter of referring a vector to a moving frame is that unnecessarily complicated geometrical considerations may be avoided, particularly in three-dimensional problems, and the calculations are reduced in large measure to easy vector algebraic computations.

4.4. Kinematic Chain Rule for Angular Velocity Vectors

The foregoing examples show that several angular velocity vectors may arise naturally in some problems. In this section, we shall develop a rule for their composition. Thus, with this objective in mind, let us write $\boldsymbol{\omega}_{10}$ for the angular velocity of the frame φ relative to frame Φ. The subscripts 1 and 0 denote the frames φ and Φ, respectively, and $\boldsymbol{\omega}_{10}$ is read as the angular velocity of frame 1 relative to frame 0. Now, as far as the moving observer is concerned, frame 1 serves as his preferred frame, while frame 0 seems to be moving with angular velocity $\boldsymbol{\omega}_{01}$. Intuitively, we should expect that these simple relative angular velocities are related by $\boldsymbol{\omega}_{10} = -\boldsymbol{\omega}_{01}$. This will be proved below.

4.4.1. The Simple Relative Angular Velocity Rule

It proves convenient in the following construction to append to the derivatives of $\mathbf{U}(t)$ subscripts 1 and 0 to emphasize the frame wherein the rate is computed. Moreover, we note that for any frame f the following notations are equivalent:

$$\dot{\mathbf{U}}_f \equiv \left.\frac{d\mathbf{U}}{dt}\right|_f \equiv \left.\frac{\delta\mathbf{U}}{\delta t}\right|_f; \tag{4.19}$$

that is, *the time rate of change of* $\mathbf{U}$ *in any frame* f *is always obtained by differentiation of* $\mathbf{U}$ *in* f *with the basis vectors in* f *kept fixed,* as emphasized in (4.10). This was seen earlier for the derivatives in (4.4). Equation (4.4a), for example, is the derivative of the position vector $\mathbf{x}$ in (4.1a) with the basis vectors $\mathbf{i}'_k$ in φ_2 kept fixed; hence, either of the representations $\dot{\mathbf{x}}_{\varphi_2} = d\mathbf{x}/dt|_{\varphi_2} = \delta\mathbf{x}/\delta t|_{\varphi_2}$ expresses more precisely the content of (4.4a). The formula (4.4b) is the derivative of the same vector referred to φ_1 in which the basis vectors $\mathbf{i}_k$ are fixed; thus, any one of the forms $\dot{\mathbf{x}}_{\varphi_1} = d\mathbf{x}/dt|_{\varphi_1} = \delta\mathbf{x}/\delta t|_{\varphi_1}$ identifies better the result (4.4b) for $\mathbf{x}$ given by (4.1b) in frame φ_1. And similar notation clarifies (4.4c) as the derivative of (4.1c) for $\mathbf{x}$ referred to frame φ_0 wherein the basis $\mathbf{I}_k$ is fixed. Clearly, then, the three derivatives (4.4) have different meanings which the special notation defined in (4.19) helps to clarify, especially when these different derivatives of the same vector occur in the same or related equations. Indeed, with the notation of (4.19), our basic relation (4.11) may be written as

$$\dot{\mathbf{U}}_0 = \dot{\mathbf{U}}_1 + \boldsymbol{\omega}_{10} \times \mathbf{U}. \tag{4.20a}$$

In words, *the (absolute) time derivative of* $\mathbf{U}$ *in frame* 0 *is equal to the sum of the (apparent) time derivative of* $\mathbf{U}$ *in frame* 1 *and its convective time rate of change due to the angular velocity of frame* 1 *relative to frame* 0.

Because our choice of preferred frame is arbitrary, we may reverse our previous choice in (4.11) and thus obtain for the same arbitrary vector $\mathbf{U}(t)$ the parallel relation

$$\dot{\mathbf{U}}_1 = \dot{\mathbf{U}}_0 + \boldsymbol{\omega}_{01} \times \mathbf{U}. \tag{4.20b}$$

Addition of the last two equations yields

$$(\boldsymbol{\omega}_{10}+\boldsymbol{\omega}_{01})\times\mathbf{U}=\mathbf{0}. \tag{4.20c}$$

Since $\mathbf{U}$ is arbitrary, (4.20c) implies that $\boldsymbol{\omega}_{10}+\boldsymbol{\omega}_{01}=\mathbf{0}$. We thus obtain the following *simple relative angular velocity rule*:

$$\boldsymbol{\omega}_{10}=-\boldsymbol{\omega}_{01}. \tag{4.20d}$$

Hence, the angular velocity of frame 1 *relative to frame* 0 *is equal and oppositely directed to the angular velocity of frame* 0 *relative to frame* 1.

4.4.2. The Composition Rule for Several Angular Velocity Vectors

The special notation introduced above may be easily extended to account for any number of moving reference frames; and the set of equations (4.20) may be expressed in a more general form. This will be done next; and the main result will be used to derive the general composition rule for several angular velocity vectors.

Let us consider n reference frames φ_i, $i=1, 2..., n$, each with angular velocity $\boldsymbol{\omega}_{i0}$ relative to a preferred frame $\Phi\equiv\varphi_0$; and, henceforward, let us agree for convenience to refer to the frame φ_k simply as the frame k. Then $\boldsymbol{\omega}_{ij}$ denotes the angular velocity of frame i relative to frame j; and the obvious extension of the notation introduced in (4.19) and (4.20) for any vector $\mathbf{U}$ leads to the general formula*

$$\dot{\mathbf{U}}_j=\dot{\mathbf{U}}_i+\boldsymbol{\omega}_{ij}\times\mathbf{U}. \tag{4.21}$$

That is, *the time rate of change of a vector* $\mathbf{U}$ *in frame* j *is equal to its time rate of change in frame* i *plus the convective time rate of change of* $\mathbf{U}$ *due to the angular velocity of frame* i *relative to frame* j.

By changing the frame labels in (4.21), we may write

$$\dot{\mathbf{U}}_k=\dot{\mathbf{U}}_i+\boldsymbol{\omega}_{ik}\times\mathbf{U}=\dot{\mathbf{U}}_j+\boldsymbol{\omega}_{jk}\times\mathbf{U}, \qquad i, j, k=0, 1, 2,..., n.$$

Therefore, $(\boldsymbol{\omega}_{ik}-\boldsymbol{\omega}_{ij}-\boldsymbol{\omega}_{jk})\times\mathbf{U}=\mathbf{0}$ follows upon substitution of (4.21) into the third term. Because $\mathbf{U}$ is arbitrary, this delivers the following *general composition rule for angular velocity vectors*:

$$\boldsymbol{\omega}_{ik}=\boldsymbol{\omega}_{ij}+\boldsymbol{\omega}_{jk}, \qquad i, j, k=0, 1, 2,..., n. \tag{4.22}$$

Two easy results derive from (4.21) and (4.22). First, it follows easily from (4.21) that $\boldsymbol{\omega}_{jj}=\mathbf{0}$ for all $j=0, 1, 2,..., n$. Plainly, *the frame* j *has no angular velocity relative to itself.* As a consequence, the replacement of the index k by i in (4.22) reveals that *the angular velocity of frame* i *relative to*

* The summation rule for repeated indices is suspended for all chain rule relations.

frame j is equal to the opposite of the angular velocity of frame j relative to frame i:

$$\boldsymbol{\omega}_{ij} = -\boldsymbol{\omega}_{ji}. \tag{4.23}$$

Finally, (4.22) may be used to establish the following *kinematic chain rule for angular velocity vectors*:

$$\boldsymbol{\omega}_{n0} = \boldsymbol{\omega}_{n,n-1} + \boldsymbol{\omega}_{n-1,n-2} + \cdots + \boldsymbol{\omega}_{21} + \boldsymbol{\omega}_{10}. \tag{4.24}$$

To derive this useful result, we put $j = i - 1$, $k = 0$ in (4.22) to obtain the relation

$$\boldsymbol{\omega}_{i0} = \boldsymbol{\omega}_{i,i-1} + \boldsymbol{\omega}_{i-1,0}, \qquad i = 1, 2, \ldots, n. \tag{4.25a}$$

It thus follows by (4.25) that

$$\begin{aligned}
\boldsymbol{\omega}_{n0} &= \boldsymbol{\omega}_{n,n-1} + \boldsymbol{\omega}_{n-1,0};\\
\boldsymbol{\omega}_{n-1,0} &= \boldsymbol{\omega}_{n-1,n-2} + \boldsymbol{\omega}_{n-2,0};\\
&\vdots\\
\boldsymbol{\omega}_{30} &= \boldsymbol{\omega}_{32} + \boldsymbol{\omega}_{20};\\
\boldsymbol{\omega}_{20} &= \boldsymbol{\omega}_{21} + \boldsymbol{\omega}_{10}.
\end{aligned}$$

Upon substituting the second of these into the first, and so on, we reach the important result stated in (4.24). In particular, for the special case $n = 3$, we see in the construction above and from (4.24) that

$$\boldsymbol{\omega}_{30} = \boldsymbol{\omega}_{32} + \boldsymbol{\omega}_{20} = \boldsymbol{\omega}_{32} + \boldsymbol{\omega}_{21} + \boldsymbol{\omega}_{10}. \tag{4.25b}$$

The composition rule (4.22) plainly represents the generalized kinematic chain rule for angular velocity vectors. Some applications of these important results are presented below.

Example 4.4. Recall the earlier Example 4.3 of a motor-driven gear mounted on a rotating platform shown in Fig. 4.5. Therein, the constant

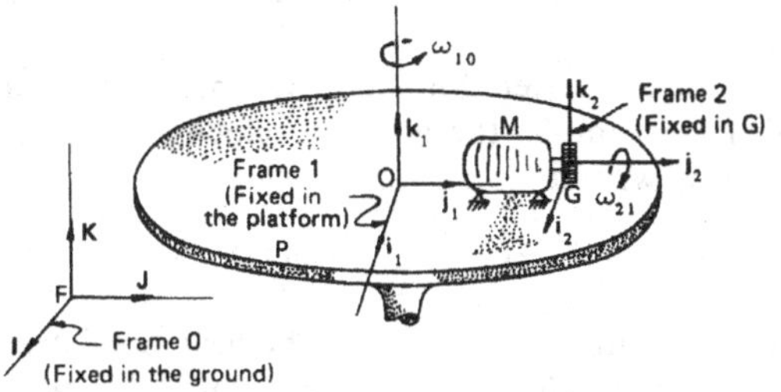

Figure 4.6. Application of multiple reference frames to a motor-driven gear mounted on a spinning platform.

angular velocity of the gear relative to the platform is given as $\boldsymbol{\omega}_1 = 10\pi\mathbf{j}_2$ rad/sec; and $\boldsymbol{\omega}_f = 20\pi\boldsymbol{k}_1$ rad/sec is the constant angular velocity of the platform relative to the ground. Determine the total angular velocity of the gear relative to the ground, but referred to a frame fixed in the platform. Label and identify carefully all references frames used.

Solution. The problem statement suggests that three imbedded reference frames are relevant. The frames may be labeled in any convenient manner. We shall use the consecutive numerical scheme shown in Fig. 4.6. Thus, frame $2 = \{G; \mathbf{i}_2, \mathbf{j}_2, \mathbf{k}_2\}$ is fixed in the gear G; frame $1 = \{O; \mathbf{i}_1, \mathbf{j}_1, \mathbf{k}_1\}$ is imbedded in the platform P; and frame $0 = \{F; \mathbf{I}, \mathbf{J}, \mathbf{K}\}$ is in the ground F.

Recalling the assigned data, we identify $\boldsymbol{\omega}_{21} = \boldsymbol{\omega}_1 = 10\pi\mathbf{j}_2$ rad/sec as the constant angular velocity of frame 2 (the gear) relative to frame 1 (the platform), and $\boldsymbol{\omega}_{10} = \boldsymbol{\omega}_f = 20\pi\mathbf{k}_1$ rad/sec as the constant angular velocity of frame 1 relative to frame 0 (the ground). We note also that $\mathbf{j}_2 = \mathbf{j}_1$. Therefore, the total angular velocity of the gear frame 2 relative to the ground frame 0, but referred to the platform frame 1, is given by the kinematic chain rule (4.24) for the case $n = 2$:

$$\boldsymbol{\omega}_{20} = \boldsymbol{\omega}_{21} + \boldsymbol{\omega}_{10} = 10\pi\mathbf{j}_1 + 20\pi\mathbf{k}_1 \text{ rad/sec}. \tag{4.26a}$$

The choice of labels to be assigned to the various frames is unimportant; but their use must be consistent with the composition rules derived above. In the present problem, for example, the gear frame may be labeled as frame 5, and the platform frame as frame 3. Then if the ground frame is identified as frame 4, the total angular velocity of frame 5 relative to frame 4, but referred to the platform frame 3, is given by suitable use of the general kinematic chain rule (4.22):

$$\boldsymbol{\omega}_{54} = \boldsymbol{\omega}_{53} + \boldsymbol{\omega}_{34} = 10\pi\mathbf{j}_1 + 20\pi\mathbf{k}_1 \text{ rad/sec}, \tag{4.26b}$$

wherein we have identified $\boldsymbol{\omega}_{53} = \boldsymbol{\omega}_1$ as the angular velocity of the gear frame 5 relative to the platform frame 3 and $\boldsymbol{\omega}_{34} = \boldsymbol{\omega}_f$ as the angular velocity of the platform frame 3 relative to the ground frame 4. Of course, the result agrees with (4.26a).

Sometimes the use of letters for the subscripts in the chain rule (4.22) or (4.24) is helpful. For example, the angular velocity of the gear G relative to the ground F may be written as

$$\boldsymbol{\omega}_{GF} = \boldsymbol{\omega}_{GP} + \boldsymbol{\omega}_{PF}, \tag{4.26c}$$

where $\boldsymbol{\omega}_{GP}$ denotes the angular velocity of the gear G relative to the platform P and $\boldsymbol{\omega}_{PF}$ is the angular velocity of the platform P relative to the ground F.

Any one of the foregoing frame labeling schemes may be used. However, use of consecutive numerical labels that reflect the chain of angular velocities involved in the problem usually is more convenient, especially for the computation of the total angular acceleration studied later. In any case, *care must be exercised to name and identify the various frames or the bodies in which the frames are imbedded.* □

Example 4.5. Recall the data for the moment of interest t_0 described in Example 4.1, and suppose further that the claw attached to the telescopic arm of the robot turns about the arm axis with an angular velocity $\boldsymbol{\omega}_3 = 0.1\boldsymbol{\gamma}$ rad/sec relative to the arm, as indicated. The data are shown in Fig. 4.7, in which four appropriate reference frames also are defined. Find for the instant t_0 the total angular velocity of the claw in the machine frame 0, but referred to frame 1 fixed in the yoke.

Solution. Let us write $\boldsymbol{\omega}_{32} = 0.1\boldsymbol{\gamma}$ rad/sec for the angular velocity of the claw frame $3 = \{A; \boldsymbol{\gamma}, \mathbf{e}, \mathbf{f}\}$ relative to the arm frame $2 = \{O; \boldsymbol{\gamma}, \mathbf{i}', \mathbf{j}\}$ whose angular velocity relative to the yoke frame $1 = \{O; \mathbf{i}, \mathbf{j}, \mathbf{k}\}$ is written as $\boldsymbol{\omega}_{21} = \boldsymbol{\omega}_2 = 0.2\mathbf{j}$ rad/sec. Let $\boldsymbol{\omega}_{10} = \boldsymbol{\omega}_1 = 0.5\mathbf{k}$ rad/sec denote the angular velocity of the yoke frame 1 relative to the preferred, machine frame $0 = \{O; \mathbf{I}, \mathbf{J}, \mathbf{K}\}$. Then the total angular velocity of the claw frame 3 relative to the machine frame 0 is given by use of the chain rule (4.25b):

$$\boldsymbol{\omega}_{30} = \boldsymbol{\omega}_{32} + \boldsymbol{\omega}_{21} + \boldsymbol{\omega}_{10} = 0.1\boldsymbol{\gamma} + 0.2\mathbf{j} + 0.5\mathbf{k} \text{ rad/sec.} \tag{4.27a}$$

But this is not yet referred to frame 1; it remains to refer $\boldsymbol{\gamma}$ to frame 1. The geometry provides

$$\boldsymbol{\gamma} = \sin\beta_0 \mathbf{i} + \cos\beta_0 \mathbf{k} \qquad \text{with} \quad \beta_0 \equiv \beta(t_0). \tag{4.27b}$$

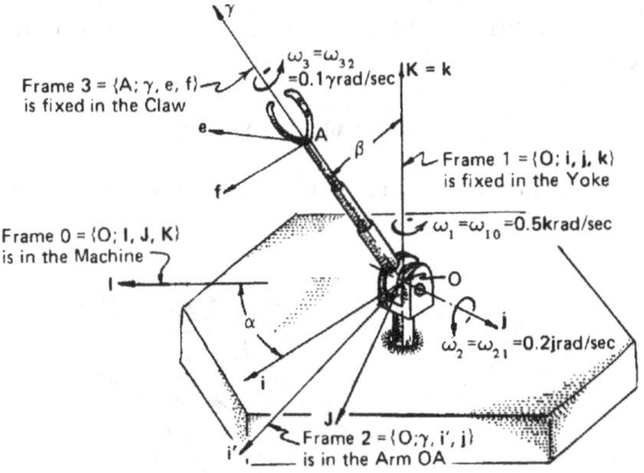

Figure 4.7. Multiple reference frames applied to the complex rotations of a robot.

Then use of (4.27b) in (4.27a) delivers

$$\boldsymbol{\omega}_{30} = 0.1 \sin \beta_0 \mathbf{i} + 0.2\mathbf{j} + (0.5 + 0.1 \cos \beta_0)\, \mathbf{k} \text{ rad/sec}, \tag{4.27c}$$

in which β_0 is the angular placement of the arm at the time t_0 shown in Fig. 4.7. The result (4.27c) is the total angular velocity of the claw frame 3 relative to the machine frame 0 and referred to the yoke frame 1, at the instant of interest.

4.4.2.1. Application to a Universal Joint Mechanism

The universal joint illustrated in Fig. 4.8 is a mechanism used to connect rotating shafts that intersect in a constant angle ϕ. Each connecting shaft terminates in a U-shaped yoke. The yokes are connected by a rigid cross link, the ends of which are set in bearings in the yokes at A, B, C, and D. When the drive yoke turns as shown in Fig. 4.8, the cross link must rotate relative to the yoke about its axle AB. The motion of the cross link about the axle CD and relative to the follower yoke is similar. We are going to show by use of the general chain rule (4.22) that even if the angular speed ω_1 of the drive shaft is constant, the angular speed ω_2 of the follower shaft will not be uniform. We seek the ratio ω_2/ω_1 of the angular speeds and its maximum and minimum values. The variation in the angular speed ratio with the angle of rotation of the drive yoke for various shaft angles will be described graphically at the end.

Let $\boldsymbol{\omega}_{10} \equiv \boldsymbol{\omega}_1$ and $\boldsymbol{\omega}_{20} \equiv \boldsymbol{\omega}_2$ denote the respective angular velocities of the drive shaft and follower shaft, and write $\boldsymbol{\omega}_{30}$ for the unknown angular velocity of the cross link, all relative to a preferred frame $\varphi_0 = \{F; \mathbf{i}_k\}$. These vectors are identified in Fig. 4.9 as the angular velocities in φ_0 of three reference frames: $\varphi_1 = \{O; \boldsymbol{\sigma}_1, \boldsymbol{\gamma}_1, \boldsymbol{\mu}_1\}$ fixed in the drive yoke; $\varphi_2 = \{O; \boldsymbol{\sigma}_2, \boldsymbol{\gamma}_2, \boldsymbol{\mu}_2\}$ imbedded in the follower yoke; and $\varphi_3 = \{O; \boldsymbol{\gamma}_1, \boldsymbol{\gamma}_2, \boldsymbol{\gamma}_3\}$ attached to the rigid cross link. The kinematic chain rule (4.22) thus yields two relations connecting these vectors:

$$\boldsymbol{\omega}_{30} = \boldsymbol{\omega}_{31} + \boldsymbol{\omega}_{10} = \boldsymbol{\omega}_{32} + \boldsymbol{\omega}_{20}, \tag{4.28a}$$

wherein $\boldsymbol{\omega}_{31}$ is the angular velocity of the cross link (frame 3) about the axle

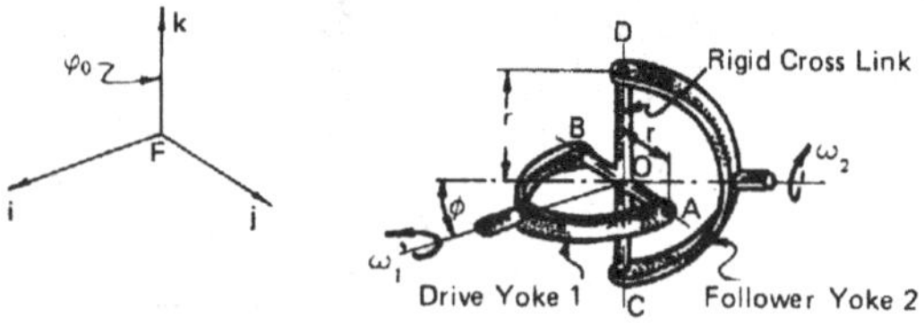

Figure 4.8. A typical universal joint mechanism.

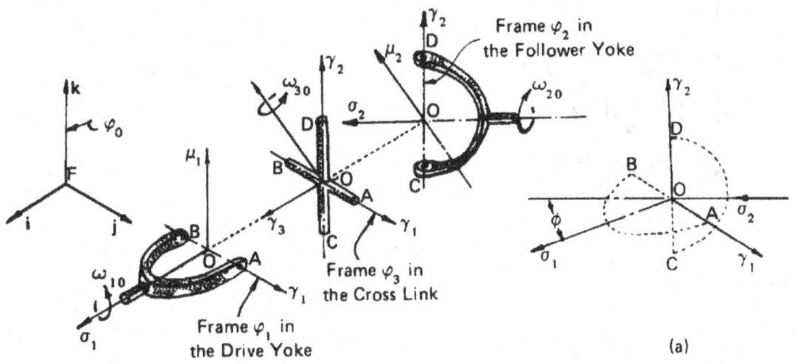

Figure 4.9. Exploded view of the universal joint showing the three imbedded frames.

AB relative to the drive yoke (frame 1), and $\boldsymbol{\omega}_{32}$ has a corresponding meaning.

We note in Figs. 4.9 and 4.9a that the unit vectors $\boldsymbol{\gamma}_1$ and $\boldsymbol{\gamma}_2$ are along the arms of the cross link, and the unit vectors $\boldsymbol{\sigma}_1$ and $\boldsymbol{\sigma}_2$ are along the shafts. Hence, (4.28a) may be written as

$$\omega_{31}\boldsymbol{\gamma}_1 + \omega_{10}\boldsymbol{\sigma}_1 = \omega_{32}\boldsymbol{\gamma}_2 + \omega_{20}\boldsymbol{\sigma}_2. \tag{4.28b}$$

To eliminate ω_{31} and ω_{32}, we form the scalar product of (4.28b) with the unit vector $\boldsymbol{\gamma}_3 = \boldsymbol{\gamma}_1 \times \boldsymbol{\gamma}_2$, and thereby obtain the desired formula for the ratio of the angular speeds:

$$\omega_2/\omega_1 \equiv \omega_{20}/\omega_{10} = (\boldsymbol{\gamma}_1 \times \boldsymbol{\gamma}_2 \cdot \boldsymbol{\sigma}_1)/(\boldsymbol{\gamma}_1 \times \boldsymbol{\gamma}_2 \cdot \boldsymbol{\sigma}_2). \tag{4.28c}$$

This completes our application of the kinematic chain rule for angular velocity vectors. The rest of the analysis concerns the interpretation of (4.28c) in terms of the shaft angle and the angle of rotation of the drive yoke.

Since $\boldsymbol{\gamma}_2$ is perpendicular to both $\boldsymbol{\gamma}_1$ and $\boldsymbol{\sigma}_2$, we may write $\boldsymbol{\gamma}_2 = \alpha\boldsymbol{\sigma}_2 \times \boldsymbol{\gamma}_1$, where α is an unknown scalar. Therefore, with $\boldsymbol{\gamma}_1 \times \boldsymbol{\gamma}_2 = \boldsymbol{\gamma}_1 \times (\alpha\boldsymbol{\sigma}_2 \times \boldsymbol{\gamma}_1) = \alpha[\boldsymbol{\sigma}_2 - (\boldsymbol{\gamma}_1 \cdot \boldsymbol{\sigma}_2)\boldsymbol{\gamma}_1]$, and noting also that $\boldsymbol{\gamma}_1 \cdot \boldsymbol{\sigma}_1 = 0$, we obtain from (4.28c)

$$\omega_2/\omega_1 = \boldsymbol{\sigma}_1 \cdot \boldsymbol{\sigma}_2/[1 - (\boldsymbol{\gamma}_1 \cdot \boldsymbol{\sigma}_2)^2] = \cos\phi/[1 - (\boldsymbol{\gamma}_1 \cdot \boldsymbol{\sigma}_2)^2] \tag{4.28d}$$

where ϕ is the shaft angle shown in Fig. 4.9a.

Let θ denote the angular placement of the director $\boldsymbol{\gamma}_1$ of the cross link arm AB, which is just the rotation angle of the drive yoke; and chose φ_0 so that its xy plane contains $\boldsymbol{\sigma}_1$ and $\boldsymbol{\sigma}_2$ with $\boldsymbol{\sigma}_1 = \mathbf{i}$, as shown in Fig. 4.10. Then the figure geometry gives

$$\boldsymbol{\gamma}_1 = \cos\theta\,\mathbf{j} + \sin\theta\,\mathbf{k}, \qquad \boldsymbol{\sigma}_2 = \cos\phi\,\mathbf{i} - \sin\phi\,\mathbf{j}. \tag{4.28e}$$

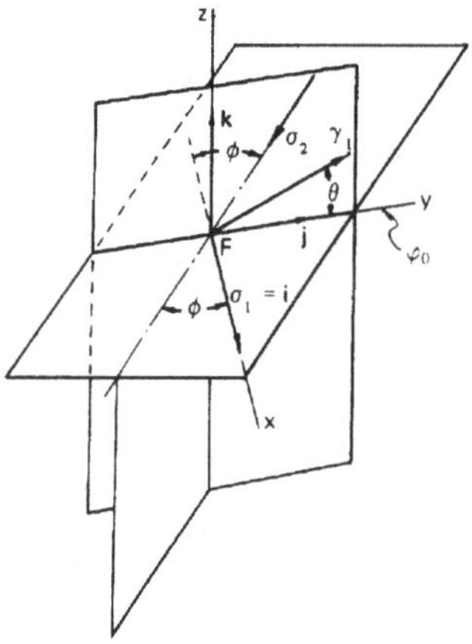

Figure 4.10. Auxiliary geometry describing the angle of rotation $\theta(t)$ of the drive yoke.

Thus, finally, use of (4.28e) in the second equality of (4.28d) yields the angular speed ratio

$$\omega_2/\omega_1 = \cos\phi/(1 - \sin^2\phi\cos^2\theta). \tag{4.28f}$$

This formula shows that although both shafts must complete one revolution in the same time, the ratio of their angular speeds varies with the angle of rotation $\theta(t)$ of the driver and is a function also of the shaft angle ϕ. Thus, even if the angular speed ω_1 of the drive shaft is constant, the angular speed ω_2 of follower shaft will not be uniform, except for $\phi = 0$.

The denominator in (4.28f) achieves its smallest value $\cos^2\phi$ at $\theta = (0, \pi)$ and its greatest value 1 at $\theta = (\pi/2, 3\pi/2)$; therefore, the angular velocity ratio has the maximum and minimum values

$$\begin{aligned} &\max(\omega_2/\omega_1) = 1/\cos\phi \qquad \text{at } \theta = 0, \pi; \\ &\min(\omega_2/\omega_1) = \cos\phi \qquad \text{at } \theta = \pi/2, 3\pi/2. \end{aligned} \tag{4.28g}$$

Thus, ω_2/ω_1 attains its greatest value when the cross link is in the position shown in Fig. 4.8, and its least value occurs after a further 90° rotation of the drive shaft.

A polar graph of the ratio (4.28f) is shown in Fig. 4.11 for a quarter revolution of the drive yoke and for various shaft angles. Notice the considerable variation in the angular speed ratio as the shaft angle ϕ is increased. Usually, however, the universal joint is used in circumstances where the shaft angle is small so that ω_2 is very nearly equal to ω_1. This is evident in Fig. 4.11 for the shaft angle $\phi = 10°$, for example.

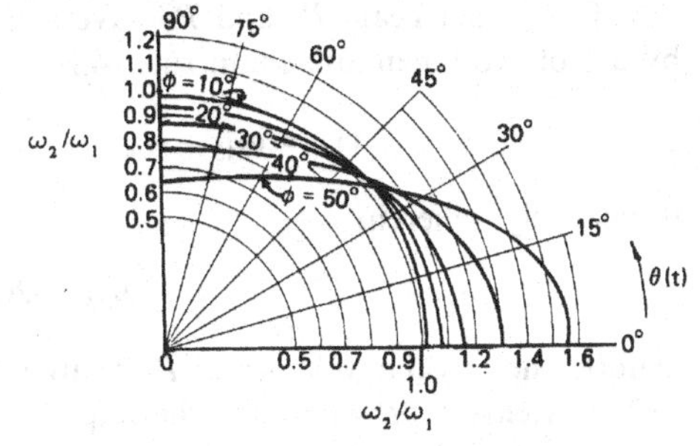

Figure 4.11. Polar plot of the angular speed ratio ω_2/ω_1 as a function of the angle of rotation $\theta(t)$ of the drive shaft for $0 \leq \theta \leq \pi/2$ and for various values of the angle ϕ between the shafts.

4.4.2.2. Application to a Planetary Gear Train

A planetary gear train shown in Fig. 4.12 consists of two bevel gears P_1 and P_2 having perpendicular axles set in bearings connected at A and B to a curved link. The link S is called the sun and P_1 and P_2, with pitch angles θ_1 and θ_2, are named planets. The link turns with angular velocity $\boldsymbol{\omega}_S$ relative to a preferred frame $\varphi_0 = \{F; \mathbf{I}, \mathbf{J}, \mathbf{K}\}$ fixed in the machine foundation. The planet gear P_1, whose axis is fixed in φ_0, has angular velocity $\boldsymbol{\omega}$ relative to the frame $\varphi_3 = \{O; \mathbf{i}, \mathbf{j}, \mathbf{k}\}$ fixed in S. The rotation of P_2 is controlled by the rotations of P_1 and S. We wish to determine the absolute angular velocities of P_1 and P_2 and the angular velocity of P_2 relative to φ_3, all referred to φ_3. Use of the general chain rule (4.22) will be illustrated.

Let $\varphi_1 = \{A; \mathbf{a}, \mathbf{b}, \mathbf{k}\}$ and $\varphi_2 = \{B; \mathbf{i}, \mathbf{m}, \mathbf{n}\}$ define frames imbedded in P_1 and P_2, respectively; and write $\boldsymbol{\omega}_{30} \equiv \boldsymbol{\omega}_S$ for the absolute angular velocity of the link in φ_0, and $\boldsymbol{\omega}_{13} \equiv \boldsymbol{\omega}$ for the angular velocity of P_1 relative to S. Then

$$\boldsymbol{\omega}_{30} \equiv \boldsymbol{\omega}_S = \omega_S \mathbf{k}, \qquad \boldsymbol{\omega}_{13} \equiv \boldsymbol{\omega} = \omega \mathbf{k}. \tag{4.29a}$$

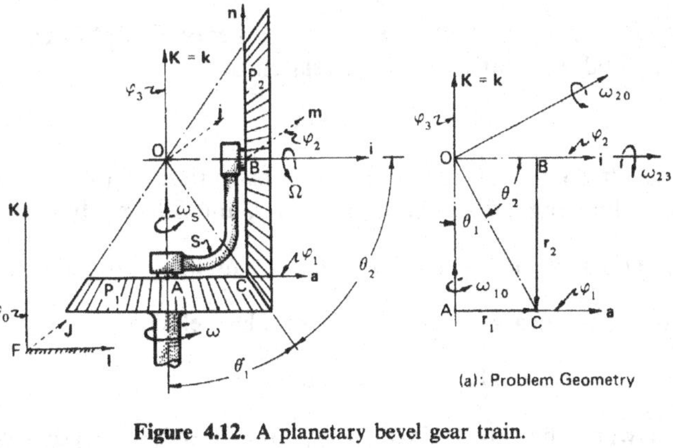

Figure 4.12. A planetary bevel gear train.

The planet gears P_1 and P_2 have absolute angular velocities determined by use of the kinematic chain rule (4.22):

$$\boldsymbol{\omega}_{10} = \boldsymbol{\omega}_{13} + \boldsymbol{\omega}_{30}, \qquad \boldsymbol{\omega}_{20} = \boldsymbol{\omega}_{23} + \boldsymbol{\omega}_{30}, \tag{4.29b}$$

respectively, wherein

$$\boldsymbol{\omega}_{23} = \Omega \mathbf{i}, \tag{4.29c}$$

denotes the angular velocity of P_2 relative to S. Use of (4.29a) and (4.29c) in (4.29b) yields expressions for the angular velocities of P_1 and P_2 relative to φ_0, but referred to φ_3:

$$\boldsymbol{\omega}_{10} = (\omega + \omega_S)\,\mathbf{k}, \qquad \boldsymbol{\omega}_{20} = \Omega \mathbf{i} + \omega_S \mathbf{k} \tag{4.29d}$$

This completes our immediate use of the chain rules (4.29b). To complete the solution, it remains to find Ω.

To relate the angular speeds of the gears, we use (2.27) and observe that the velocity in φ_0 of the point of rolling contact at C in Fig. 4.12a must satisfy the equations

$$\mathbf{v}_C = \mathbf{v}_A + \boldsymbol{\omega}_{10} \times \mathbf{r}_1 = \mathbf{v}_B + \boldsymbol{\omega}_{20} \times \mathbf{r}_2, \tag{4.29e}$$

in which $\mathbf{r}_1 = r_1 \mathbf{a}$ and $\mathbf{r}_2 = -r_2 \mathbf{k}$. Since the axis OA is fixed in φ_0, $\mathbf{v}_O = \mathbf{v}_A = \mathbf{0}$ in φ_0; and because B belongs also to S, we have $\mathbf{v}_B = \mathbf{v}_O + \boldsymbol{\omega}_{30} \times \mathbf{r}_1$ in φ_0. Thus, with the aid of these relations and the first equation in (4.29b), (4.29e) may be written as

$$\boldsymbol{\omega}_{13} \times \mathbf{r}_1 = \boldsymbol{\omega}_{20} \times \mathbf{r}_2, \tag{4.29f}$$

which yields the single component equation

$$\omega r_1 = \Omega r_2. \tag{4.29g}$$

It is clear in Fig. 4.12a that $r_1/r_2 = \tan\theta_1$; hence, by (4.29g), the angular speeds of P_2 and P_1 relative to S are related by

$$\Omega = \omega \tan\theta_1. \tag{4.29h}$$

This result completes the solution. Substitution of (4.29h) into (4.29d) gives the absolute angular velocities of P_1, P_2, and S in φ_0, but referred to φ_3:

$$\boldsymbol{\omega}_{10} = (\omega + \omega_S)\,\mathbf{k}, \qquad \boldsymbol{\omega}_{20} = \omega \tan\theta_1\, \mathbf{i} + \omega_S \mathbf{k}, \qquad \boldsymbol{\omega}_{30} = \omega_S \mathbf{k}. \tag{4.29i}$$

Relative to φ_3, the gears and link have angular velocities

$$\boldsymbol{\omega}_{13} = \omega \mathbf{k}, \qquad \boldsymbol{\omega}_{23} = \omega \tan\theta_1\, \mathbf{i}, \qquad \boldsymbol{\omega}_{33} = \mathbf{0}. \tag{4.29j}$$

Notice that when the link is fixed in φ_0, $\boldsymbol{\omega}_S = \mathbf{0}$ and the second equation in

(4.29i) yields the familiar bevel gear formula (2.85). Also, if P_1 is fixed to S, $\omega = 0$ and the entire assembly turns with angular velocity $\boldsymbol{\omega}_S$. Of course, if P_1 is fixed in φ_0, then $\omega = -\omega_S$ and the rotation of P_2 is determined by the speed of S.

For a numerical example, let $\theta_1 = 60°$, $\omega = 20$ rad/sec, and $\omega_S = 40$ rad/sec. Then, we find from (4.29j) the angular velocities of the gears relative to the link:

$$\boldsymbol{\omega}_{13} = 20\mathbf{k} \text{ rad/sec}, \qquad \boldsymbol{\omega}_{23} = 20\sqrt{3}\,\mathbf{i} \text{ rad/sec}.$$

Of course, the link has no angular velocity relative to φ_3. The absolute angular velocities of the gears and the link relative to the machine φ_0 come from (4.29i):

$$\boldsymbol{\omega}_{10} = 60\mathbf{k} \text{ rad/sec}, \qquad \boldsymbol{\omega}_{20} = 20(\sqrt{3}\,\mathbf{i} + 2\mathbf{k}) \text{ rad/sec},$$
$$\boldsymbol{\omega}_{30} = 40\mathbf{k} \text{ rad/sec}.$$

Therefore, the absolute angular speeds are found to be

$$|\boldsymbol{\omega}_{10}| = 60 \text{ rad/sec}, \qquad |\boldsymbol{\omega}_{20}| = 20\sqrt{7} \text{ rad/sec},$$
$$|\boldsymbol{\omega}_{30}| = 40 \text{ rad/sec}.$$

This ends our study of the kinematic chain rules for angular velocity vectors. Other examples will be encountered ahead; but there is need first to develop some additional tools to handle the composition of angular acceleration vectors.

4.5. The Composition Rule for Several Angular Accelerations

The rules for mutual relative angular accelerations and for the composition of multiple angular acceleration vectors will be derived below. The latter requires use of the kinematic chain rule for angular velocity vectors. Because rotational effects are introduced by each member of the chain, it turns out that angular acceleration vectors do not obey the simple additive rule found for angular velocity vectors. It will be necessary in the derivation to exercise care in execution of the time derivatives of the chain equations (4.22) and (4.24); hence, some special notation similar to (4.19) will be required for this purpose.

It will prove helpful in the subsequent construction to observe the following equivalent notations:

$$\dot{\boldsymbol{\omega}}_{ij} \equiv \dot{\boldsymbol{\omega}}_{ij}\big|_j \equiv \left.\frac{d\boldsymbol{\omega}_{ij}}{dt}\right|_j \equiv \frac{\delta\boldsymbol{\omega}_{ij}}{\delta t}. \tag{4.30}$$

All are derivatives of $\boldsymbol{\omega}_{ij}$ with respect to time in the frame indicated by the second index j. Thus, $\dot{\boldsymbol{\omega}}_{ij}$ is the time derivative in frame j of the angular velocity vector of frame i relative to frame j, and referred to frame j; hence, it defines the angular acceleration of frame i relative to frame j as seen by the observer in frame j. The different forms (4.30) are used as a matter of convenience to express clearly the ideas that follow. However, after the main results are obtained, typically only the first and last of (4.30) are used.

4.5.1. The Simple Relative Angular Acceleration Rule

To begin, let $\mathbf{U} = \boldsymbol{\omega}_{ij}$ in (4.21). Then, with (4.30), we have

$$\dot{\boldsymbol{\omega}}_{ij}|_j = \dot{\boldsymbol{\omega}}_{ij}|_i, \tag{4.31}$$

which is a restatement of (4.14): *The time derivative of $\boldsymbol{\omega}_{ij}$ in frame j is equal to its time derivative in frame i.*

Let us next recall the simple relative angular velocity relation:

$$\boldsymbol{\omega}_{ij} = -\boldsymbol{\omega}_{ji}. \qquad \text{Ref. (4.23)}$$

We anticipate, of course, that a similar rule should hold for mutual angular accelerations. Indeed, differentiation of this result in frame i and use of (4.31) and (4.30) yields the *simple relative angular acceleration rule*:

$$\dot{\boldsymbol{\omega}}_{ij} = -\dot{\boldsymbol{\omega}}_{ji}. \tag{4.32}$$

That is, *the angular acceleration of frame i relative to frame j (computed in frame j) is equal and oppositely directed to the angular acceleration of frame j relative to frame i (computed in frame i).*

It is evident from (4.32) that *no reference frame may have an angular acceleration relative to itself*: $\dot{\boldsymbol{\omega}}_{jj} = \mathbf{0}$.

4.5.2. The Composition of Multiple Angular Acceleration Vectors

We shall derive first the fundamental composition rule based on the derivative of the kinematic chain formula (4.24). Afterwards, the general composition rule will be presented. We thus begin with the differentiation in frame 0 of the chain rule (4.25a); this gives

$$\dot{\boldsymbol{\omega}}_{i0}|_0 = \dot{\boldsymbol{\omega}}_{i,i-1}|_0 + \dot{\boldsymbol{\omega}}_{i-1,0}|_0. \tag{4.33}$$

We must recall that $\boldsymbol{\omega}_{i,i-1}$ is a vector referred to the moving frame $i-1$ which has the absolute angular velocity $\boldsymbol{\omega}_{i-1,0}$ in frame 0. Then application of (4.21) to the first term on the right-hand side of (4.33) and use of (4.30) gives the following formula for the absolute angular acceleration of frame i relative to frame 0:

$$\dot{\boldsymbol{\omega}}_{i0} = \left[\frac{\delta \boldsymbol{\omega}_{i,i-1}}{\delta t} + \boldsymbol{\omega}_{i-1,0} \times \boldsymbol{\omega}_{i,i-1}\right] + \dot{\boldsymbol{\omega}}_{i-1,0}. \tag{4.34}$$

Since $\delta \boldsymbol{\omega}_{i,i-1}/\delta t = \dot{\boldsymbol{\omega}}_{i,i-1}|_{i-1}$, with the notation of (4.30), it follows by (4.34) that in general $\dot{\boldsymbol{\omega}}_{i0} \neq \dot{\boldsymbol{\omega}}_{i,i-1} + \dot{\boldsymbol{\omega}}_{i-1,0}$. Therefore, (4.34) shows that the simple additive chain rule found for angular velocity vectors does not hold for the composition of angular acceleration vectors because additional convective acceleration terms are introduced by elements of the angular velocity chain. The simple connection of (4.34) with the chain rule (4.25a) is apparent.

With $i = 1, 2, \ldots, n$ in (4.34), we may obtain the following useful and easily remembered *fundamental composition rule for the absolute angular acceleration of frame n in frame* 0:

$$\begin{aligned} \dot{\boldsymbol{\omega}}_{n0} &= \left[\frac{\delta \boldsymbol{\omega}_{n,n-1}}{\delta t} + \boldsymbol{\omega}_{n-1,0} \times \boldsymbol{\omega}_{n,n-1}\right] \\ &\quad + \left[\frac{\delta \boldsymbol{\omega}_{n-1,n-2}}{\delta t} + \boldsymbol{\omega}_{n-2,0} \times \boldsymbol{\omega}_{n-1,n-2}\right] \\ &\quad + \cdots + \left[\frac{\delta \boldsymbol{\omega}_{21}}{\delta t} + \boldsymbol{\omega}_{10} \times \boldsymbol{\omega}_{21}\right] + \dot{\boldsymbol{\omega}}_{10}. \end{aligned} \tag{4.35}$$

The connection of (4.35) with the kinematic chain rule (4.24) surely is evident.

For example, application of the composition rule (4.35) to the chain $\boldsymbol{\omega}_{30} = \boldsymbol{\omega}_{32} + \boldsymbol{\omega}_{21} + \boldsymbol{\omega}_{10}$ gives the absolute angular acceleration of frame 3 relative to frame 0:

$$\dot{\boldsymbol{\omega}}_{30} = \left[\frac{\delta \boldsymbol{\omega}_{32}}{\delta t} + \boldsymbol{\omega}_{20} \times \boldsymbol{\omega}_{32}\right] + \left[\frac{\delta \boldsymbol{\omega}_{21}}{\delta t} + \boldsymbol{\omega}_{10} \times \boldsymbol{\omega}_{21}\right] + \frac{\delta \boldsymbol{\omega}_{10}}{\delta t}. \tag{4.36}$$

In this relation, $\delta \boldsymbol{\omega}_{32}/\delta t$ is the angular acceleration of frame 3 relative to frame 2 referred to (i.e., as seen in) frame 2, and $\boldsymbol{\omega}_{20} \times \boldsymbol{\omega}_{32}$ is the convective rate of change of $\boldsymbol{\omega}_{32}$ due to the angular velocity $\boldsymbol{\omega}_{20}$ of frame 2 in frame 0, which is the part of the total angular acceleration of frame 3 relative to frame 2 that the moving observer does not see. The second bracketed term in (4.36) may be interpreted similarly. Of course, the last term in (4.36) is the absolute angular acceleration of frame 1 relative to frame 0. Although each δ-derivative may be written in the superimposed dot notation described in (4.30), the notation used in (4.36) is recommended. The form of the derivative (4.36) is not unique. The total angular velocity $\boldsymbol{\omega}_{30}$ also may be written as shown in (4.28a), hence other forms of (4.36) are possible. Let the reader show that the derivatives obtained from (4.28a) are equivalent to those in (4.36).

The rule (4.34) may be readily generalized by differentiation of (4.22) in frame k and use of (4.21). We thereby obtain the following *composition rule for several angular acceleration vectors*:

$$\dot{\boldsymbol{\omega}}_{ik} = \left[\frac{\delta\boldsymbol{\omega}_{ij}}{\delta t} + \boldsymbol{\omega}_{jk} \times \boldsymbol{\omega}_{ij}\right] + \dot{\boldsymbol{\omega}}_{jk}. \tag{4.37}$$

Finally and more generally, it may be seen from (4.22) and (4.21) that the time derivative of $\boldsymbol{\omega}_{ik}$ in any frame p is determined by

$$\dot{\boldsymbol{\omega}}_{ik}\,|_p = \dot{\boldsymbol{\omega}}_{ij}\,|_p + \dot{\boldsymbol{\omega}}_{jk}\,|_p = \frac{\delta\boldsymbol{\omega}_{ij}}{\delta t} + \boldsymbol{\omega}_{jp} \times \boldsymbol{\omega}_{ij} + \frac{\delta\boldsymbol{\omega}_{jk}}{\delta t} + \boldsymbol{\omega}_{kp} \times \boldsymbol{\omega}_{jk}. \tag{4.38}$$

Clearly, when $p = k$, (4.38) yields (4.37). Henceforward, our use of the special notation will be restricted mainly to the first and last terms in (4.30), as appears in (4.34)–(4.37). Some examples will be presented next.

Example 4.6. Let us return to Example 4.4 of the motor on the rotating platform shown in Fig. 4.6; and recall that the total angular velocity $\boldsymbol{\omega}_{20}$ of the gear relative to the ground is given by (4.26a):

$$\boldsymbol{\omega}_{20} = \boldsymbol{\omega}_{21} + \boldsymbol{\omega}_{10} = 10\pi\mathbf{j}_1 + 20\pi\mathbf{k}_1 \text{ rad/sec.} \qquad \text{Ref. (4.26a)}$$

Note that $\mathbf{j}_1$ belongs to the platform frame 1, and $\mathbf{k}_1$ is fixed both in frame 1 and in the ground frame 0. Find the absolute angular acceleration of the gear referred to the platform frame. What is its time derivative in frame 0?

Solution. Application of the composition rule (4.35) to the first equation in (4.26a) gives the total angular acceleration of the gear relative to the ground observer:

$$\dot{\boldsymbol{\omega}}_{20} = \frac{\delta\boldsymbol{\omega}_{21}}{\delta t} + \boldsymbol{\omega}_{10} \times \boldsymbol{\omega}_{21} + \frac{\delta\boldsymbol{\omega}_{10}}{\delta t}. \tag{4.39a}$$

Since the gear frame 2 spins with constant angular velocity $\boldsymbol{\omega}_{21} = 10\pi\mathbf{j}_1$ rad/sec in frame 1, and the platform turns with constant angular velocity $\boldsymbol{\omega}_{10} = 20\pi\mathbf{k}_1$ rad/sec in frame 0, the first and last terms on the right in (4.39a) are zero. Therefore, with $\mathbf{i} = \mathbf{i}_1$ in the platform frame, we find

$$\dot{\boldsymbol{\omega}}_{20} = \boldsymbol{\omega}_{10} \times \boldsymbol{\omega}_{21} = -200\pi^2\mathbf{i} \text{ rad/sec}^2. \tag{4.39b}$$

This is the absolute angular acceleration of the gear referred to frame 1. The result agrees with (4.18a) derived earlier.

Because the first equation in (4.39b) holds for all times, its derivative in frame 0 is given by

$$\ddot{\boldsymbol{\omega}}_{20} = \dot{\boldsymbol{\omega}}_{10} \times \boldsymbol{\omega}_{21} + \boldsymbol{\omega}_{10} \times \left[\frac{\delta \boldsymbol{\omega}_{21}}{\delta t} + \boldsymbol{\omega}_{10} \times \boldsymbol{\omega}_{21} \right]$$

$$= \boldsymbol{\omega}_{10} \times (\boldsymbol{\omega}_{10} \times \boldsymbol{\omega}_{21}) = -4000\pi^3 \mathbf{j} \text{ rad/sec}^3, \tag{4.39c}$$

in which (4.20a) was used with $\mathbf{U} = \boldsymbol{\omega}_{21}$ and we have put $\mathbf{j} = \mathbf{j}_1$ to reach identity with our earlier result (4.18b). Note also that $\mathbf{i}$ being fixed in frame 1, $d\mathbf{i}/dt = \boldsymbol{\omega}_{10} \times \mathbf{i}$, and hence (4.39c) may be derived from (4.39b) somewhat differently, as described earlier in (4.8). □

Example 4.7. Suppose that at the instant t_0 the gear in Fig. 4.6 has an angular speed of 3 rad/sec and an angular acceleration of 5 rad/sec^2 relative to the platform, which has an angular speed of 7 rad/sec and is decelerating at the rate of 3 rad/sec^2 relative to the ground. What is the angular acceleration of the gear relative to the ground at the time t_0, but referred to the platform frame?

Solution. We are given

$$\boldsymbol{\omega}_{21} = 3\mathbf{j} \text{ rad/sec}, \qquad \boldsymbol{\omega}_{10} = 7\mathbf{k} \text{ rad/sec},$$

$$\frac{\delta \boldsymbol{\omega}_{21}}{\delta t} = 5\mathbf{j} \text{ rad/sec}^2, \qquad \dot{\boldsymbol{\omega}}_{10} = -3\mathbf{k} \text{ rad/sec}^2,$$

wherein we have put $\mathbf{j} = \mathbf{j}_2$ and $\mathbf{k} = \mathbf{k}_1$. All vectors are referred to the platform frame. Thus, application of the kinematic chain and composition rules (4.24) and (4.35), respectively, gives

$$\boldsymbol{\omega}_{20} = \boldsymbol{\omega}_{21} + \boldsymbol{\omega}_{10} = 3\mathbf{j} + 7\mathbf{k} \text{ rad/sec},$$

$$\dot{\boldsymbol{\omega}}_{20} = \left[\frac{\delta \boldsymbol{\omega}_{21}}{\delta t} + \boldsymbol{\omega}_{10} \times \boldsymbol{\omega}_{21} \right] + \dot{\boldsymbol{\omega}}_{10} = 5\mathbf{j} - 21\mathbf{i} - 3\mathbf{k} \text{ rad/sec}^2.$$

This is the angular acceleration of the gear relative to the ground, but referred to the platform frame.

4.5.2.1. Application to Robotics

Modern manufacturing systems frequently use sophisticated electomechanical robots to execute repetitive and routine production operations like soldering, spray painting, materials handling, and machine tool changing. The robot's movements are largely controlled by computer electronics to perform a variety of tasks, but its function is primarily mechanical. In fact, many industrial robots are simply programmable, mechanical manipulators designed to reorient and reposition various objects such as materials, tools, and parts, in an assembly or mass production operation. Discrete movements of the manipulator between fixed end point positions or continuous motions along precisely determined trajectories according to assigned sequences are recorded in the robot's memory and may be executed upon command.

The simple robotic device described in our earlier examples is a model of a typical mechanical manipulator whose general motion may be rather complex. To investigate its motions and the forces and torques needed to effect them, the angular rates of rotation are important, particularly in high-speed operations. This study would involve the use of the composition rules for angular velocity and angular acceleration vectors. The following example will demonstrate their application in the general description of the rotational rates of the manipulator claw of the robot. The frames and angular velocity vectors are shown in Fig. 4.7, but now we shall ignore the special numerical values assigned before. We want to find the total angular velocity and angular acceleration of the claw in the machine frame, but referred to the yoke frame.

Let $\dot{\psi}$ be the angular speed of the claw relative to the telescopic arm; write $\dot{\beta}$ for the angular speed of the arm relative to the yoke, and let $\dot{\alpha}$ denote the angular speed of the yoke relative to the machine frame. Then the corresponding relative angular velocity vectors indicated in Fig. 4.7 are given by

$$\boldsymbol{\omega}_{32} = \dot{\psi}\boldsymbol{\gamma}, \qquad \boldsymbol{\omega}_{21} = \dot{\beta}\mathbf{j}, \qquad \boldsymbol{\omega}_{10} = \dot{\alpha}\mathbf{k}. \tag{4.40a}$$

Therefore, with the aid of $\boldsymbol{\gamma} = \sin\beta\,\mathbf{i} + \cos\beta\,\mathbf{k}$ in the first equation in (4.40a), the total angular velocity of the claw in the machine frame and referred to the yoke frame is given by the kinematic chain rule (4.25b). We thereby obtain

$$\boldsymbol{\omega}_{30} = \dot{\psi}\sin\beta\,\mathbf{i} + \dot{\beta}\mathbf{j} + (\dot{\psi}\cos\beta + \dot{\alpha})\,\mathbf{k}. \tag{4.40b}$$

The corresponding total angular acceleration may be found from (4.36). We observe in (4.40a) that $\boldsymbol{\gamma}$ is fixed in frame 2; $\mathbf{j}$ is in frame 1; and $\mathbf{k}$ is in frame 0. Then the relative angular acceleration vectors derive from (4.40a):

$$\frac{\delta\boldsymbol{\omega}_{32}}{\delta t} = \ddot{\psi}\boldsymbol{\gamma}, \qquad \frac{\delta\boldsymbol{\omega}_{21}}{\delta t} = \ddot{\beta}\mathbf{j}, \qquad \frac{\delta\boldsymbol{\omega}_{10}}{\delta t} = \ddot{\alpha}\mathbf{k}; \tag{4.40c}$$

and the convective terms in (4.36) are given by

$$\begin{aligned}\boldsymbol{\omega}_{20} \times \boldsymbol{\omega}_{32} &= (\dot{\beta}\mathbf{j} + \dot{\alpha}\mathbf{k}) \times \dot{\psi}\boldsymbol{\gamma} = \dot{\beta}\dot{\psi}\mathbf{i}' + \dot{\alpha}\dot{\psi}\sin\beta\,\mathbf{j},\\ \boldsymbol{\omega}_{10} \times \boldsymbol{\omega}_{21} &= \dot{\alpha}\mathbf{k} \times \dot{\beta}\mathbf{j} = -\dot{\alpha}\dot{\beta}\mathbf{i}.\end{aligned} \tag{4.40d}$$

Finally, use of $\mathbf{i}' = \cos\beta\,\mathbf{i} - \sin\beta\,\mathbf{k}$ in the first equation in (4.40d) and collection of (4.40c) and (4.40d) into (4.36) delivers the absolute angular acceleration of the claw manipulator as seen by the observer in the machine frame but, for convenience, referred to the yoke frame 1:

$$\dot{\boldsymbol{\omega}}_{30} = (\ddot{\psi}\sin\beta + \dot{\beta}\dot{\psi}\cos\beta - \dot{\beta}\dot{\alpha})\,\mathbf{i} + (\ddot{\beta} + \dot{\alpha}\dot{\psi}\sin\beta)\,\mathbf{j} + (\ddot{\alpha} + \ddot{\psi}\cos\beta - \dot{\beta}\dot{\psi}\sin\beta)\,\mathbf{k}. \tag{4.40e}$$

It is an exercise for the reader to confirm this result by differentiation of (4.40b). What are the absolute angular velocity and acceleration of the claw referred to the arm frame? Consider what one would do to refer the results to frame 3 in the claw itself.

4.5.2.2. *Application to a Planetary Gear Train*

Let us recall the planetary, bevel gear train shown in Fig. 4.12. Consider the case when the link S has a constant angular velocity $\boldsymbol{\omega}_{30} = \boldsymbol{\omega}_S$ relative to the machine frame φ_0, and the planet gear P_1 has the constant angular velocity $\boldsymbol{\omega}_{13} = \boldsymbol{\omega}$ relative to the sun frame φ_3. The general composition rules (4.22) and (4.37) will be applied to determine the angular acceleration of the planet gear P_2 in φ_0, but referred to φ_3. An earlier basic procedure also will be reviewed.

With the aid of the composition rules (4.22) and (4.37), we have the following relations for the total angular velocity and angular acceleration of the planet gear P_2 in φ_0:

$$\boldsymbol{\omega}_{20} = \boldsymbol{\omega}_{23} + \boldsymbol{\omega}_{30}, \tag{4.41a}$$

$$\dot{\boldsymbol{\omega}}_{20} = \frac{\delta\boldsymbol{\omega}_{23}}{\delta t} + \boldsymbol{\omega}_{30} \times \boldsymbol{\omega}_{23} + \dot{\boldsymbol{\omega}}_{30}. \tag{4.41b}$$

Since $\boldsymbol{\omega}_{30} = \omega_S\mathbf{k}$ is a constant vector in φ_0, $\dot{\boldsymbol{\omega}}_{30} = \mathbf{0}$. We recall from the second equation in (4.29j) that $\boldsymbol{\omega}_{23} = \omega\tan\theta_1\,\mathbf{i}$ is constant in frame φ_3, so $\delta\boldsymbol{\omega}_{23}/\delta t = \mathbf{0}$; and with the aid of the aforementioned relations, we find $\boldsymbol{\omega}_{30} \times \boldsymbol{\omega}_{23} = \omega\omega_S\tan\theta_1\,\mathbf{j}$. Therefore, the angular acceleration of P_2 in φ_0 but referred to φ_3 is

$$\dot{\boldsymbol{\omega}}_{20} = \omega\omega_S\tan\theta_1\,\mathbf{j}. \tag{4.41c}$$

This is the absolute angular acceleration of P_2; it arises from the change in the direction $\mathbf{i}$ of its axis of rotation in the second equation in (4.29i). And since the second equality of (4.29i) is valid for all times t, we may differentiate it directly to obtain $\dot{\boldsymbol{\omega}}_{20} = \omega\tan\theta_1\,d\mathbf{i}/dt$. Then (4.8) yields $d\mathbf{i}/dt = \boldsymbol{\omega}_{30} \times \mathbf{i} = \omega_S\mathbf{j}$, which leads to the same result derived in (4.41c). Similarly,

$\ddot{\boldsymbol{\omega}}_{20}$ also varies due to the change in the direction $\mathbf{j}$: $d\mathbf{j}/dt = \boldsymbol{\omega}_{30} \times \mathbf{j} = \omega_S \mathbf{k} \times \mathbf{j} = -\omega_S \mathbf{i}$. Thus, from (4.41c) follows

$$\ddot{\boldsymbol{\omega}}_{20} = -\omega\omega_S^2 \tan\theta_1 \, \mathbf{i}. \tag{4.41d}$$

Higher derivatives of this kind seldom arise in applications. We may notice in passing that because $\boldsymbol{\omega}_{10} = (\omega + \omega_S)\,\mathbf{k}$ is constant in φ_0, the absolute angular acceleration $\dot{\boldsymbol{\omega}}_{10}$ of P_1 is zero.

4.5.2.3. Application to Gyroscopic Rotations

Rapidly rotating parts of propulsion machinery, such as the flywheel of an automobile engine, the steam turbine of a ship, and the fan jet or propellor of an aircraft, can have significant mechanical effects on the structural integrity and on the control of vehicles propelled by such machinery. Powerful forces and torques may be induced by any tendency to alter the spatial direction of the axis of rotation of the spinning body. The effect, for example, of the whirling action of an aircraft propellor or jet during a simple turn in the flight plane of the aircraft is to cause the nose to rise or to fall. This tendency of the aircraft to pitch forward or backward may be compensated by adjustment of the control flaps, but this causes large stresses to be induced on the structural framework. Analysis of these so-called gyroscopic effects requires understanding of the rotational motion of the spinning device in a rotating reference frame. The following example illustrates the application of the composition rules (4.24) and (4.35) to the description of a spinning disk supported by a shaft fixed in an aircraft executing simultaneous turn and roll maneuvers.

A rotating disk D shown in Fig. 4.13 is driven at a variable angular speed $\dot{\phi}$ through a shaft BD supported in bearings in an aircraft. Gyroscopic effects arise when the aircraft executes a turn, roll, or dive maneuver. Let us consider the case when the aircraft turns with angular speed $\dot{\psi}$ about a vertical axis fixed in space and simultaneously rolls with angular speed $\dot{\theta}$ about its central axis BP. We wish to find the angular velocity and angular acceleration of D relative to the ground frame, but not necessarily referred to it. Afterwards, we shall evaluate these quantities at the initial instant when $\psi = \theta = 0$.

We begin as usual by defining some appropriate reference frames. The frame $0 = \{G; \mathbf{i}, \mathbf{j}, \mathbf{k}\}$ denotes the ground frame; and frame $1 = \{B; \mathbf{a}, \mathbf{b}, \mathbf{k}\}$, which is fixed to the vertical plane ABD as shown in Fig. 4.13, rotates with angular speed $\dot{\psi}$ about the vertical axis $\mathbf{k}$ of frame 0. Hence, this frame follows the aircraft turn so that the aircraft has only a pure roll relative to it. Frame $2 = \{B; \mathbf{b}, \mathbf{c}, \mathbf{d}\}$ is fixed in the aircraft with $\mathbf{d}$ directed along the shaft BD and with $\mathbf{b}$ in the direction of the aircraft axis normal to the plane ABD; and frame $3 = \{D; \mathbf{d}, \mathbf{e}, \mathbf{f}\}$ is fixed in the disk at D. Thus, $\boldsymbol{\omega}_{32} = \dot{\phi}\mathbf{d}$ is the angular velocity of the disk frame 3 relative to the aircraft frame 2. The air-

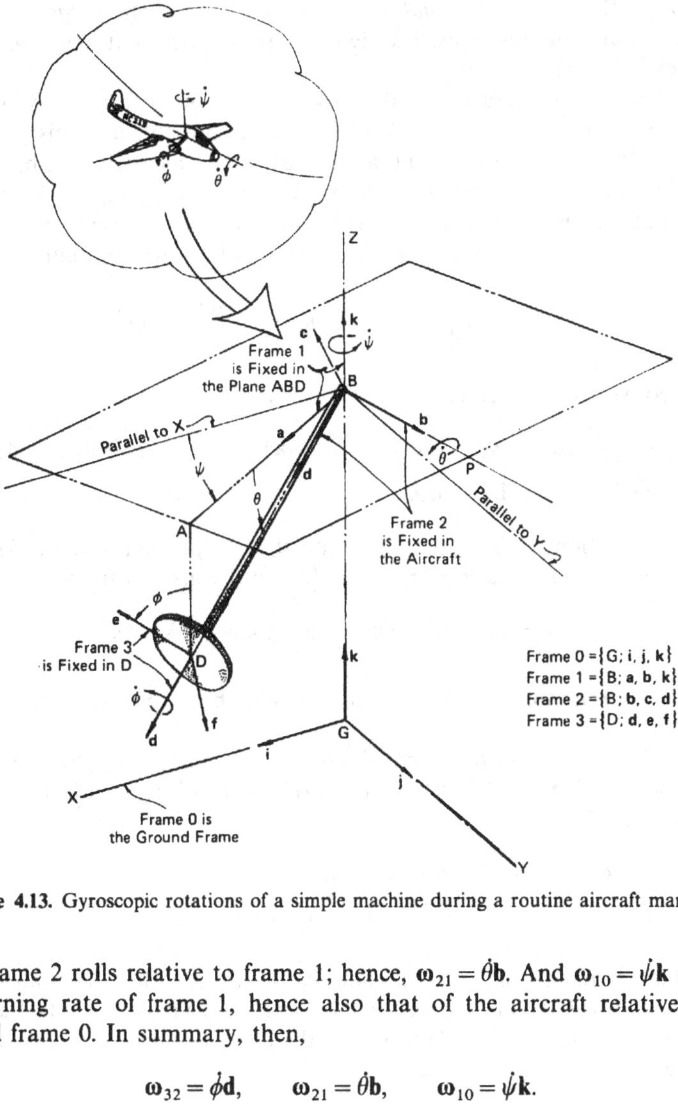

Figure 4.13. Gyroscopic rotations of a simple machine during a routine aircraft maneuver.

craft frame 2 rolls relative to frame 1; hence, $\boldsymbol{\omega}_{21} = \dot{\theta}\mathbf{b}$. And $\boldsymbol{\omega}_{10} = \dot{\psi}\mathbf{k}$ denotes the turning rate of frame 1, hence also that of the aircraft relative to the ground frame 0. In summary, then,

$$\boldsymbol{\omega}_{32} = \dot{\phi}\mathbf{d}, \qquad \boldsymbol{\omega}_{21} = \dot{\theta}\mathbf{b}, \qquad \boldsymbol{\omega}_{10} = \dot{\psi}\mathbf{k}. \tag{4.42a}$$

Therefore, the total angular velocity of the disk frame 3 relative to the ground frame 0 is given by the kinematic chain rule (4.24) for the case $n = 3$. Hence, by (4.2b)

$$\boldsymbol{\omega}_{30} = \boldsymbol{\omega}_{32} + \boldsymbol{\omega}_{21} + \boldsymbol{\omega}_{10} = \dot{\phi}\mathbf{d} + \dot{\theta}\mathbf{b} + \dot{\psi}\mathbf{k}. \tag{4.42b}$$

It is a straightforward geometrical problem to relate the vectors **d** and **b** to the ground frame; however, we shall omit this step. The rate $\dot{\phi}$ is called the *spin* speed; $\dot{\psi}$ is titled the *precessional* speed; and $\dot{\theta}$ is named the *nutational* speed. These terms are commonly used to describe gyroscopic rotations of any spinning body, such as D.

The angular acceleration of the disk in the ground frame is derived by application of the composition rule (4.35). In the present case, this yields the formula (4.36) in which we recall that $\delta\boldsymbol{\omega}_{32}/\delta t$ is the time derivative of the vector $\boldsymbol{\omega}_{32}$ keeping the basis vectors in frame 2 fixed; $\delta\boldsymbol{\omega}_{21}/\delta t$ requires in differentiation that the basis of frame 1 be fixed; and $\delta\boldsymbol{\omega}_{10}/\delta t = \dot{\boldsymbol{\omega}}_{10}$ in the fixed ground frame, as usual. Thus, with the aid of (4.42a), we determine

$$\frac{\delta\boldsymbol{\omega}_{32}}{\delta t} = \ddot{\phi}\mathbf{d}, \qquad \frac{\delta\boldsymbol{\omega}_{21}}{\delta t} = \ddot{\theta}\mathbf{b}, \qquad \frac{\delta\boldsymbol{\omega}_{10}}{\delta t} = \ddot{\psi}\mathbf{k};$$

and the convective acceleration terms are given by

$$\boldsymbol{\omega}_{20} \times \boldsymbol{\omega}_{32} = (\boldsymbol{\omega}_{21} + \boldsymbol{\omega}_{10}) \times \boldsymbol{\omega}_{32} = (\dot{\theta}\mathbf{b} + \dot{\psi}\mathbf{k}) \times \dot{\phi}\mathbf{d} = -\dot{\theta}\dot{\phi}\mathbf{c} + \dot{\psi}\dot{\phi}(\mathbf{k} \times \mathbf{d}),$$
$$\boldsymbol{\omega}_{10} \times \boldsymbol{\omega}_{21} = \dot{\psi}\mathbf{k} \times \dot{\theta}\mathbf{b} = -\dot{\psi}\dot{\theta}\mathbf{a}.$$

Hence, substitution of these quantities into the composition rule (4.36) yields the absolute angular acceleration of the disk in the ground frame:

$$\dot{\boldsymbol{\omega}}_{30} = \ddot{\phi}\mathbf{d} + \ddot{\theta}\mathbf{b} - \dot{\theta}\dot{\phi}\mathbf{c} - \dot{\psi}\dot{\theta}\mathbf{a} + \ddot{\psi}\mathbf{k} + \dot{\psi}\dot{\phi}(\mathbf{k} \times \mathbf{d}). \tag{4.42c}$$

It remains to write this in terms of some specified basis, but we shall omit this step.

At the initial instant when $\theta = \psi = 0$, we see easily that $\mathbf{a} = \mathbf{d} = \mathbf{i}$, $\mathbf{b} = \mathbf{k} \times \mathbf{d} = \mathbf{j}$, and $\mathbf{c} = \mathbf{k}$. Therefore, (4.42b) and (4.42c) provide the desired initial values

$$\begin{aligned} \boldsymbol{\omega}_{30}|_{t=0} &= \dot{\phi}\mathbf{i} + \dot{\theta}\mathbf{j} + \dot{\psi}\mathbf{k}\,|_{t=0}, \\ \dot{\boldsymbol{\omega}}_{30}|_{t=0} &= (\ddot{\phi} - \dot{\psi}\dot{\theta})\,\mathbf{i} + (\ddot{\theta} + \dot{\psi}\dot{\phi})\,\mathbf{j} + (\ddot{\psi} - \dot{\theta}\dot{\phi})\,\mathbf{k}\,|_{t=0}, \end{aligned} \tag{4.42d}$$

which are referred to the ground frame 0. Of course, when the spin is constant, $\ddot{\phi} = 0$ in (4.42d). Let the reader show that (4.36) may be applied directly to the first equation in (4.42d) to derive the same instantaneous result given by the second equation in (4.42d).

4.6. Velocity and Acceleration Referred to a Moving Frame

The fundamental rules (4.10) and (4.11) were applied in Example 4.2 to determine in a moving reference frame and in a preferred frame the time derivative of a position vector which was referred to the moving frame. The example revealed the relationship of the velocities of the same particle as seen by the observers in these frames. In this section, generalized formulas that relate the velocities and accelerations of a particle in two relatively moving reference frames will be derived and interpreted in appropriate physical terms. These basic formulas may be used to compute the velocity and acceleration of any material point whatever. In fact, it will be seen that they also include our earlier equations for the velocity and acceleration of a point of a rigid body.

4.6.1. The Velocity Equation

Consider a particle P at $\mathbf{X}(t)$ from F in the preferred frame Φ and at $\mathbf{x}(t)$ from O in the moving frame φ, and let $\mathbf{B}(t)$ denote the position vector of O from F, as shown in Fig. 4.14. Then $\mathbf{X}=\mathbf{B}+\mathbf{x}$. Differentiating this equation with respect to time in Φ and introducing $\mathbf{v}_P \equiv \mathbf{v}(P, t)=\dot{\mathbf{X}}$ and $\mathbf{v}_O \equiv \mathbf{v}(O, t)=\dot{\mathbf{B}}$ for the respective velocities of P and O in Φ, we have

$$\mathbf{v}_P=\mathbf{v}_O+\dot{\mathbf{x}}. \tag{4.43}$$

The preferred observer, in order to relate his observations to those of the moving observer, chooses to refer $\mathbf{x}(t)$ to frame φ so that

$$\mathbf{x}(t)=x_k(t)\,\mathbf{i}_k(t). \tag{4.44}$$

Thus, because $\mathbf{x}(t)$ is referred to a moving frame which is turning with angular

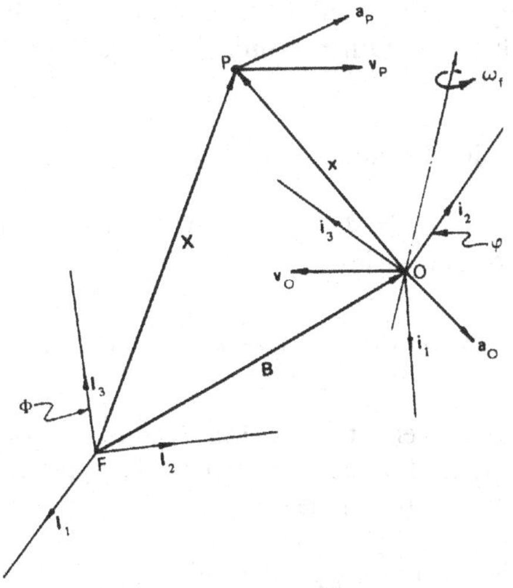

Figure 4.14. Schema for the description of the motion of a particle referred to a moving frame.

velocity $\boldsymbol{\omega}_f$ relative to Φ, the rule (4.11) must be applied to (4.44) to determine the time derivative of $\mathbf{x}(t)$ in Φ. We thereby obtain

$$\dot{\mathbf{x}}(t) = \frac{\delta \mathbf{x}(t)}{\delta t} + \boldsymbol{\omega}_f(t) \times \mathbf{x}(t). \tag{4.45}$$

Finally, use of (4.45) in (4.43) yields the velocity of P in Φ:

$$\mathbf{v}_P = \frac{\delta \mathbf{x}}{\delta t} + \mathbf{v}_O + \boldsymbol{\omega}_f \times \mathbf{x}. \tag{4.46}$$

The velocity of P in Φ is named the *absolute velocity* of P to distinguish it from the other velocity terms in (4.46). We remember that the moving observer also may assign to $\mathbf{x}(t)$ the same representation (4.44) in which the basis $\mathbf{i}_k$ is considered independent of t; and in this sense the preferred representation (4.44) may be regarded as the position vector of P in frame φ at the instant t. Hence, the apparent rate of change $\delta\mathbf{x}/\delta t$ in (4.46) is called the *relative velocity* of P; it is the velocity of P as seen by the moving observer at time t. Further, we recall from (2.27) that the sum $\mathbf{v}_O + \boldsymbol{\omega}_f \times \mathbf{x}$ is the *rigid body velocity* of the particle P; that is, in (4.46), this is the velocity that the particle P would have if it were fixed in the moving frame. Indeed, if P is fixed in φ, then the angular velocities of the frame φ and of the body containing P are the same and $\delta\mathbf{x}/\delta t = \mathbf{0}$; hence, (4.46) reduces to (2.27). As a consequence, the absolute velocity sometimes is called the *total velocity* to coincide with our earlier usage. In summary, therefore, the velocity equation (4.46) states that the *absolute velocity of a particle P is equal to the relative velocity of P in φ plus the rigid body velocity of P in Φ.*

The reader may find it helpful to review our earlier Example 4.2 in light of the general result (4.46). Other examples will be presented further on.

4.6.2. The Acceleration Equation

The acceleration of P in Φ is obtained by differentiation of (4.46) with respect to time in Φ. This yields

$$\mathbf{a}_P \equiv \dot{\mathbf{v}}_P = \dot{\overline{\frac{\delta \mathbf{x}}{\delta t}}} + \mathbf{a}_O + \boldsymbol{\omega}_f \times \dot{\mathbf{x}} + \dot{\boldsymbol{\omega}}_f \times \mathbf{x}, \tag{4.47}$$

wherein $\mathbf{a}_O \equiv \dot{\mathbf{v}}_O$ is the acceleration of O in Φ and, from (4.14), $\dot{\boldsymbol{\omega}}_f = \delta\boldsymbol{\omega}_f/\delta t$ is the angular acceleration of frame φ relative to Φ. Application of the operator (4.13) to the first term in (4.47) yields

$$\dot{\overline{\frac{\delta \mathbf{x}}{\delta t}}} = \frac{\delta}{\delta t}\left[\frac{\delta \mathbf{x}}{\delta t}\right] + \boldsymbol{\omega}_f \times \left[\frac{\delta \mathbf{x}}{\delta t}\right] = \frac{\delta^2 \mathbf{x}}{\delta t^2} + \boldsymbol{\omega}_f \times \frac{\delta \mathbf{x}}{\delta t};$$

and with the aid of (4.45), the third term on the right in (4.47) becomes

$$\boldsymbol{\omega}_f \times \dot{\mathbf{x}} = \boldsymbol{\omega}_f \times \frac{\delta \mathbf{x}}{\delta t} + \boldsymbol{\omega}_f \times (\boldsymbol{\omega}_f \times \mathbf{x}).$$

Substitution of the last two relations into (4.47) delivers the formula for the acceleration of P in Φ:

$$\mathbf{a}_P = \frac{\delta^2 \mathbf{x}}{\delta t^2} + \mathbf{a}_O + \boldsymbol{\omega}_f \times (\boldsymbol{\omega}_f \times \mathbf{x}) + \dot{\boldsymbol{\omega}}_f \times \mathbf{x} + 2\boldsymbol{\omega}_f \times \frac{\delta \mathbf{x}}{\delta t}. \tag{4.48}$$

The acceleration of P in Φ is named the *absolute acceleration* of P to distinguish it from the other acceleration terms in (4.48). The apparent rate of change of the relative velocity, namely, $\delta^2\mathbf{x}/\delta t^2$ in (4.48), is called the *relative acceleration* of P; it is the acceleration of P apparent to the moving observer in φ. Moreover, we observe from (2.30) that $\mathbf{a}_O + \boldsymbol{\omega}_f \times (\boldsymbol{\omega}_f \times \mathbf{x}) + \dot{\boldsymbol{\omega}}_f \times \mathbf{x}$ is the *rigid body acceleration* of P; it is the acceleration that the particle would have if it were fixed in the moving frame. Clearly, if, in fact, P was fixed in φ, then $\delta\mathbf{x}/\delta t = \mathbf{0}$, $\delta^2\mathbf{x}/\delta t^2 = \mathbf{0}$ and the angular velocities and accelerations of the frame φ and of the body containing P are the same; hence, (4.48) reduces to (2.30). Therefore, to coincide with our earlier usage, the absolute acceleration sometimes is referred to as the *total acceleration.* Finally, the acceleration $2\boldsymbol{\omega}_f \times \delta\mathbf{x}/\delta t$ is titled the *Coriolis acceleration.* With the foregoing lexicon of terms, the content of (4.48) may be summarized as follows: *The absolute acceleration of a particle P is equal to the sum of the relative acceleration of P in φ, the rigid body acceleration of P in Φ and the Coriolis acceleration of P in Φ.*

We have found that when the particle is fixed in the moving frame, (4.46) and (4.48) reduce to the basic equations (2.27) and (2.30) for the velocity and acceleration of a particle of a rigid body. Further, if the "moving" frame suffers no motion whatever in Φ, then $\mathbf{v}_O = \mathbf{0}$, $\mathbf{a}_O = \mathbf{0}$, $\boldsymbol{\omega}_f = \mathbf{0}$, and (4.46) and (4.48) reduce again to the basic definitions (1.8) and (1.10) for the velocity and acceleration of a particle in the "fixed" frame φ. Thus, the fundamental equations for the velocity and acceleration of a particle whose motion is referred to a moving frame contain most of our earlier basic equations. The results (4.46) and (4.48) will provide a clear picture later of corrections that must be made by a moving observer to account for the motion of his frame in the solution of dynamical problems.

Applications of (4.46) and (4.48) are diverse. They may be applied to study the motion of a discrete particle, such as an electron, or the motion of a particle of any deformable or rigid body whatever. The remainder of this chapter is devoted to the study of several examples. We shall begin with the simplest case when the frame rotation vanishes.

4.7. Simple Relative Motion

The important case when $\boldsymbol{\omega}_f = \mathbf{0}$ reduces (4.46) and (4.48) to the following *equations for simple relative motion*:

$$\mathbf{v}_P = \frac{\delta \mathbf{x}}{\delta t} + \mathbf{v}_O, \tag{4.49a}$$

$$\mathbf{a}_P = \frac{\delta^2 \mathbf{x}}{\delta t^2} + \mathbf{a}_O. \tag{4.49b}$$

These relations may be rewritten in the form of simple kinematic chain rules.

To effect the transformation, let us begin by replacing $\mathbf{v}_P$ and $\mathbf{v}_O$ by $\mathbf{v}_{PF}$ and $\mathbf{v}_{OF}$, respectively. These are read as the velocities of P and O relative to point F in Φ. We observe also that (4.49a) may be rewritten as $\delta\mathbf{x}/\delta t = \mathbf{v}_P - \mathbf{v}_O \equiv \mathbf{v}_{PO}$, which is simply the velocity of P relative to point O in φ. Similarly, $\mathbf{a}_{PF}$ and $\mathbf{a}_{OF}$ are the accelerations of points P and O relative to point F, respectively; and $\delta^2\mathbf{x}/\delta t^2 = \mathbf{a}_P - \mathbf{a}_O \equiv \mathbf{a}_{PO}$ is the acceleration of P relative to point O. Thus, the transformation of (4.49) to these terms yields the following elementary chain rules for simple relative motion:

$$\mathbf{v}_{PF} = \mathbf{v}_{PO} + \mathbf{v}_{OF}, \qquad \mathbf{a}_{PF} = \mathbf{a}_{PO} + \mathbf{a}_{OF}. \tag{4.50}$$

It must be remembered that (4.50) may be used only when there is no relative rotation of the frames. Of course, in a specific problem, any convenient arrangement of letters may be used to tag the points; and certainly numerical labels also may be introduced.

The rules (4.50) may be readily extended to any number of purely translating frames. For the triple of points i, j, and k, say, (4.50) yields the *generalized kinematic chain rule for simple relative motion*:

$$\mathbf{v}_{ik} = \mathbf{v}_{ij} + \mathbf{v}_{jk}, \qquad \mathbf{a}_{ik} = \mathbf{a}_{ij} + \mathbf{a}_{jk}, \tag{4.51}$$

in which $\mathbf{v}_{pq}$ and $\mathbf{a}_{pq}$ denote, respectively, the velocity and the acceleration of the point p relative to the frame origin point q. Clearly, for $j = k$ in (4.51), we have $\mathbf{v}_{jj} = \mathbf{0}$ and $\mathbf{a}_{jj} = \mathbf{0}$; that is, *no particle may have a nontrivial velocity or acceleration relative to itself*. Thus, by (4.51), *the mutual velocity and acceleration vectors for each pair of origin points are equal and oppositely directed*:

$$\mathbf{v}_{ij} = -\mathbf{v}_{ji}, \qquad \mathbf{a}_{ij} = -\mathbf{a}_{ji}. \tag{4.52}$$

Finally, with the aid of (4.51), we derive the following *kinematic chain rules for simple relative motion*:

$$\begin{aligned} \mathbf{v}_{n0} &= \mathbf{v}_{n,n-1} + \mathbf{v}_{n-1,n-2} + \cdots + \mathbf{v}_{21} + \mathbf{v}_{10}, \\ \mathbf{a}_{n0} &= \mathbf{a}_{n,n-1} + \mathbf{a}_{n-1,n-2} + \cdots + \mathbf{a}_{21} + \mathbf{a}_{10}. \end{aligned} \tag{4.53}$$

The easy application of the foregoing chain rules will be demonstrated next.

Examples 4.8. The ferris wheel shown in Fig. 4.15 is turning with a constant, counterclockwise angular speed $\omega = (1/2)$ rad/sec (about 5 rpm). A ball B is thrown from the ground to a rider at A so that the ball arrives at A with the following velocity and acceleration relative to the ground G:

$$\mathbf{v}_{BG} = -15\mathbf{i} - 4\mathbf{j} \text{ ft/sec}, \qquad \mathbf{a}_{BG} = -32\mathbf{j} \text{ ft/sec}^2. \tag{4.54a}$$

What are the velocity and the acceleration of the ball apparent to the rider at the position shown? Assume that the seat does not swing to and fro about its axle.

Solution. Since the seat does not swing about its axle, as the wheel turns, the moving frame φ at A always remains parallel to the ground frame Φ at G. Therefore, bearing in mind the assigned data in (4.54a), the velocity and the acceleration of the ball B relative to the rider A are given by the chain rules (4.51):

$$\mathbf{v}_{BA} = \mathbf{v}_{BG} + \mathbf{v}_{GA}, \qquad \mathbf{a}_{BA} = \mathbf{a}_{BG} + \mathbf{a}_{GA}. \tag{4.54b}$$

It remains to determine the velocity and the acceleration of G relative to A. Since point O is fixed in Φ, the absolute velocity and acceleration of A in Φ are given by (2.27) and (2.30). Taking into account the rule (4.52), we find

$$\begin{aligned} \mathbf{v}_{AG} &= -\mathbf{v}_{GA} = \mathbf{v}_{AO} = \boldsymbol{\omega} \times \mathbf{x} = 10\mathbf{j} \text{ ft/sec}, \\ \mathbf{a}_{AG} &= -\mathbf{a}_{GA} = \mathbf{a}_{AO} = \boldsymbol{\omega} \times (\boldsymbol{\omega} \times \mathbf{x}) = -5\mathbf{i} \text{ ft/sec}^2, \end{aligned} \tag{4.54c}$$

wherein $\mathbf{x} = 20\mathbf{i}$ ft and $\boldsymbol{\omega} = (1/2)\,\mathbf{k}$ rad/sec in accordance with Fig. 4.15. Of course, $\dot{\boldsymbol{\omega}} = \mathbf{0}$ in (2.30). Thus, substitution of (4.54a) and (4.54c) into (4.54b) determines the velocity and acceleration of the ball apparent to the rider:

$$\begin{aligned} \mathbf{v}_{BA} &= -15\mathbf{i} - 14\mathbf{j} \text{ ft/sec}. \\ \mathbf{a}_{BA} &= -32\mathbf{j} + 5\mathbf{i} \text{ ft/sec}^2. \end{aligned} \tag{4.54d}$$

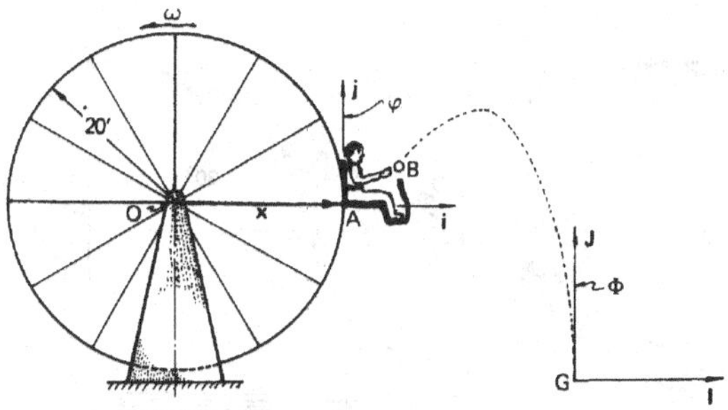

Figure 4.15. Simple relative motion of a ball apparent to a ferris wheel rider.

Since the frames are parallel, the results may be referred to either basis set with $\mathbf{i}_k = \mathbf{I}_k$. □

Example 4.9. A pin P shown in Fig. 4.16 is constrained to move in a circular groove milled to a radius of 3 ft in a large rectangular plate. The pin also slides in the straight slot of a slanted link mechanism which is moving toward the right with a constant speed of 5 ft/sec. The slot makes an angle of 30° with the horizontal drive shaft, as illustrated. Find for the instant shown the velocity and the acceleration of P relative to the plate and to the link.

Solution. Let the frame $\Phi = \{F; \mathbf{I}, \mathbf{J}, \mathbf{K}\}$ be fixed in the plate, and let $\varphi = \{O; \mathbf{i}, \mathbf{j}, \mathbf{k}\} = \{O; \mathbf{I}, \mathbf{J}, \mathbf{K}\}$ be a parallel frame fixed in the slotted link. The link frame is moving on a straight line in Φ with the assigned constant velocity

$$\mathbf{v}_{OF} = 5(\cos 30°\,\mathbf{i} - \sin 30°\,\mathbf{j}) = \frac{5}{2}(\sqrt{3}\mathbf{i} - \mathbf{j})\text{ ft/sec,} \tag{4.55a}$$

as indicated in Fig. 4.16. Of course, $\mathbf{a}_{OF} = \mathbf{0}$. Thus, with these data in mind, the absolute velocity and acceleration of the pin P relative to the plate frame at F are obtained by application of the kinematic chain rules (4.51). With the present labels, we have

$$\mathbf{v}_{PF} = \mathbf{v}_{PO} + \mathbf{v}_{OF}, \qquad \mathbf{a}_{PF} = \mathbf{a}_{PO}. \tag{4.55b}$$

Each of these vector equations involves two unknown vector quantities; hence, additional information about their components must be furnished in order to solve (4.55b) for the unknown vectors. This is done by considering the nature of the motion of the pin in the separate frames. The observer in Φ sees P move on a circle of radius $R = 3$ ft whereas the observer in φ sees P move along the straight slot; hence, at the moment of interest shown in Fig. 4.16, the unknown velocity and acceleration vectors may be written as

$$\mathbf{v}_{PF} = -v_{PF}\mathbf{j}, \qquad \mathbf{v}_{PO} = -v_{PO}\mathbf{i}, \tag{4.55c}$$

$$\mathbf{a}_{PF} = -\ddot{s}\mathbf{j} - \kappa\dot{s}^2\mathbf{i}, \qquad \mathbf{a}_{PO} \doteq -a_{PO}\mathbf{i} \tag{4.55d}$$

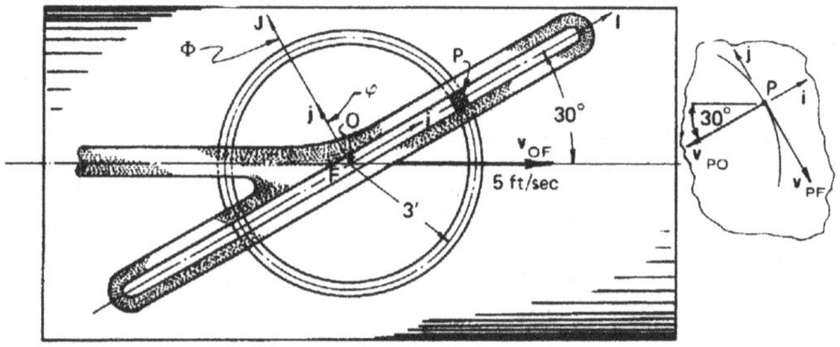

Figure 4.16. Simple relative motion of a sliding pin of a mechanism.

in terms of the unknown intrinsic variables at the position shown, and wherein $\dot{s} = |\mathbf{v}_{PF}|$ and $\kappa = 1/R = (1/3)\ \text{ft}^{-1}$. [See (1.70) and (1.71).] Thus, after collecting (4.55a), (4.55c), and (4.55d) into (4.55b) and equating the corresponding scalar components, we obtain the values $\dot{s} = v_{PF} = 5/2$, $\ddot{s} = 0$, $v_{PO} = 5\sqrt{3}/2$, and $a_{PO} = 25/12$. Therefore, with (4.55c) and (4.55d), we find at the instant of concern the velocities and the accelerations of the pin P relative to the plate F and to the link O:

$$\mathbf{v}_{PF} = -\frac{5}{2}\mathbf{j}\ \text{ft/sec}, \qquad \mathbf{v}_{PO} = -\frac{5\sqrt{3}}{2}\mathbf{i}\ \text{ft/sec}. \\ \mathbf{a}_{PF} = \mathbf{a}_{PO} = -\frac{25}{12}\mathbf{i}\ \text{ft/sec}^2. \tag{4.55e}$$

4.8. Velocity and Acceleration in Special Curvilinear Coordinates

In most of our previous studies the motion of a material point has been expressed in terms of familiar rectangular Cartesian coordinates, but this is not always desirable. In fact, it happens that many problems may be best described in other geometrically natural coordinate systems. We have already seen, for example, the simplicity achieved in problems where the intrinsic component representations for the velocity and acceleration of a particle were more appropriate. In this section, equations (4.46) and (4.48) will be used to derive expressions for the velocity and acceleration of a particle in terms of special curvilinear coordinate systems commonly called cylindrical and spherical coordinate systems. It will be evident, however, that the same procedure may be used for any orthogonal coordinate system; and it is emphasized that there is no need for the reader to memorize the special formulas derived below. On the contrary, the derivations of representations for the velocity and acceleration in cylindrical and spherical coordinate systems should be viewed only as important introductory applications of the basic equations (4.46) and (4.48), and the student's attention ought to be directed toward thorough understanding of these fundamental rules.

4.8.1. The Cylindrical Reference Frame

A coordinate system is a prescription for locating a point P in space by the specification of certain directional measure numbers along three vector lines in a reference frame consisting of an assigned origin point and a vector basis. The measure numbers are called the coordinates of P and the vector lines are called coordinate lines. The rectangular coordinate system is a familiar example of an orthogonal coordinate system in which a point P is located by specification of three directional numbers obtained by measurement of three mutually perpendicular distances along straight coordinate lines. And this rudimentary coordinate system is used to construct other orthogonal coordinate systems in which some, or possibly all, of the coordinate lines may be curved; and hence these are called curvilinear coordinate systems.

In particular, the *cylindrical coordinates* (r, ϕ, z) of a point P in a rectangular Cartesian frame $\Phi = \{F; \mathbf{I}_k\}$ are illustrated in Fig. 4.17. To locate P by its cylindrical coordinates in Φ, measure from F a radial distance r along the X axis, then trace that end point along a circular arc of radius r through an angle ϕ about the Z axis, and, finally, from its place in the XY plane, translate the point a distance z along a straight line parallel to the Z axis. This program brings us to the unique location of P in Φ. It is seen in Fig. 4.17 that for a fixed value of r the locus swept out by all such points is a cylindrical surface of radius r in Φ; therefore, as mentioned above, the three measure numbers $r \in [0, \infty)$, $\phi \in [0, 2\pi]$, and $z \in (-\infty, \infty)$ are named cylindrical coordinates. Because one of the coordinate lines traversed by the tracing point is a curved line, the cylindrical coordinate is a curvilinear coordinate system. When $z = \text{const}$, the cylindrical coordinates sometimes are called *plane polar coordinates.*

Another reference frame $\psi = \{O; \mathbf{e}_r, \mathbf{e}_\phi, \mathbf{e}_z\}$, called the *cylindrical reference frame*, may be introduced to describe this system. The origin for this frame may be chosen anywhere; however, for simplicity, we shall take O at F in Φ. The three orthogonal lines labeled r, ϕ, and z in Fig. 4.17 are parallel to the directions of the orthonormal basis for ψ, which is shown at P for convenience. The unit vector $\mathbf{e}_r$ is always in the direction of increasing values of the radius vector $\mathbf{r}$ in the XY plane; and $\mathbf{e}_z \equiv \mathbf{K}$ is the usual unit vector parallel to the z axis in the direction of increasing values of $z = Z$. The unit vector $\mathbf{e}_\phi \equiv \mathbf{e}_z \times \mathbf{e}_r$ is in the direction of increasing values of the central angle ϕ, and it

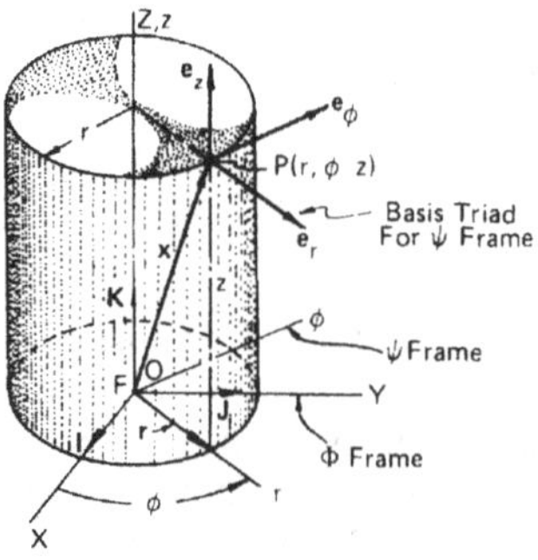

Figure 4.17. Cylindrical coordinates and the cylindrical reference frame.

is tangent at P to a circle of radius r parallel to the XY plane. Thus, referred to ψ, the position vector of a particle P is given by

$$\mathbf{x} = r\mathbf{e}_r + z\mathbf{e}_z. \tag{4.56}$$

Notice that $\mathbf{e}_z$ is a constant vector, whereas the directions of both $\mathbf{e}_r$ and $\mathbf{e}_\phi$ vary with the polar angle ϕ. The cylindrical reference frame ψ is a rotating frame whose rz plane follows the motion of the point P with the angular velocity

$$\boldsymbol{\omega}_f = \dot{\phi}\boldsymbol{K} = \dot{\phi}\mathbf{e}_z, \tag{4.57}$$

relative to the preferred frame $\Phi = \{F; \mathbf{I}, \mathbf{J}, \mathbf{K}\}$.

The absolute velocity of P referred to ψ may be obtained easily from (4.46). Since O is fixed in Φ, $\mathbf{v}_O = \mathbf{0}$; and with the aid of (4.56) and (4.57), we compute

$$\frac{\delta \mathbf{x}}{\delta t} = \dot{r}\mathbf{e}_r + \dot{z}\mathbf{e}_z, \qquad \boldsymbol{\omega}_f \times \mathbf{x} = r\dot{\phi}\mathbf{e}_\phi. \tag{4.58}$$

Therefore, (4.46) shows that the absolute velocity of a particle referred to the cylindrical reference system ψ is given by

$$\mathbf{v}_P = \dot{r}\mathbf{e}_r + r\dot{\phi}\mathbf{e}_\phi + \dot{z}\mathbf{e}_z. \tag{4.59}$$

The absolute acceleration of P referred to ψ may be derived from (4.48). Of course, $\mathbf{a}_O = \mathbf{0}$. Also, application of (4.14) to (4.57) yields $\dot{\boldsymbol{\omega}}_f = \ddot{\phi}\mathbf{e}_z$; and by aid of (4.56), (4.57), and (4.58), we obtain

$$\frac{\delta^2 \mathbf{x}}{\delta t^2} = \ddot{r}\mathbf{e}_r + \ddot{z}\mathbf{e}_z, \qquad \dot{\boldsymbol{\omega}}_f \times \mathbf{x} = r\ddot{\phi}\mathbf{e}_\phi,$$

$$\boldsymbol{\omega}_f \times (\boldsymbol{\omega}_f \times \mathbf{x}) = -r\dot{\phi}^2\mathbf{e}_r, \qquad 2\boldsymbol{\omega}_f \times \frac{\delta \mathbf{x}}{\delta t} = 2\dot{\phi}\dot{r}\mathbf{e}_\phi.$$

Thus, substitution of these terms into (4.48) yields the absolute acceleration of P referred to the cylindrical reference system ψ:

$$\mathbf{a}_P = (\ddot{r} - r\dot{\phi}^2)\,\mathbf{e}_r + (r\ddot{\phi} + 2\dot{r}\dot{\phi})\,\mathbf{e}_\phi + \ddot{z}\mathbf{e}_z. \tag{4.60}$$

Equations (4.59) and (4.60) each consist of a radial component, a tangential component, and an axial component. Notice that the tangential component of acceleration also may be abbreviated as $r^{-1}d(r^2\dot{\phi})/dt$.

Finally, for an alternative derivation of (4.59) and (4.60), the reader is invited to consider the following

Exercise 4.1. Apply (4.12) to show that the rate of rotation of each basis vector $\mathbf{e}_\alpha$ of ψ is determined by

$$\dot{\mathbf{e}}_\alpha = \boldsymbol{\omega}_f \times \mathbf{e}_\alpha. \tag{4.61}$$

Then, start with (4.56) and derive (4.59) and (4.60) by differentiation with respect to time in Φ. □

It is useful to note from (4.59) that the elemental displacement vector is given by

$$d\mathbf{x} = dr\mathbf{e}_r + rd\phi\mathbf{e}_\phi + dz\mathbf{e}_z. \tag{4.62}$$

Thus, the squared element of arc length of the particle trajectory is given in cylindrical coordinates by

$$ds^2 = d\mathbf{x} \cdot d\mathbf{x} = dr^2 + r^2 d\phi^2 + dz^2. \tag{4.63}$$

Of course, the vector $\mathbf{t} = d\mathbf{x}/ds$ is the usual intrinsic, unit vector tangent to the path of the particle.

Use of the formulas (4.59) and (4.60) will be illustrated in two introductory examples. The first problem will show the easy direct application of these equations to the helical motion of an electron. The second example will demonstrate three methods that use cylindrical coordinates to obtain the same problem solution in slightly different ways, one method being the easy direct application of (4.59) and (4.60). Afterwards, two further applications that employ cylindrical coordinates to create meaningful results will be presented.

Example 4.10. An electron E moves in a preferred frame Φ with a constant speed v along a cylindrical helix described in cylindrical coordinates by the equations $r = \alpha$, a constant, and $2\pi z = p\phi$, where p is the constant pitch. [See (1.95) and Fig. 1.16] Find the absolute acceleration of the electron.

Solution. To take advantage of the assigned constant speed condition, we first compute the velocity of E. With $r = \alpha$, $\dot{r} = 0$, and $\dot{z} = p\dot{\phi}/2\pi$, direct use of (4.59) yields the absolute velocity of E:

$$\mathbf{v}_E = \alpha\dot{\phi}\mathbf{e}_\phi + (p\dot{\phi}/2\pi)\,\mathbf{e}_z = \alpha\dot{\phi}\left(\mathbf{e}_\phi + \frac{p}{c}\,\mathbf{e}_z\right), \tag{4.64a}$$

wherein $c = 2\pi\alpha$. Since the speed $|\mathbf{v}_E| = v$ is constant, (4.64a) shows that the angular speed $\dot{\phi}$ has the constant value

$$\dot{\phi} = \frac{v}{\alpha}\left[1 + \left(\frac{p}{c}\right)^2\right]^{-1/2}. \tag{4.64b}$$

The acceleration may now be obtained easily from (4.60). Because

$\ddot{r}=\ddot{\phi}=\ddot{z}=0$, (4.60) reduces immediately to $\mathbf{a}_P=-\alpha\dot{\phi}^2\mathbf{e}_r$; and use of (4.64b) gives the absolute acceleration of the electron:

$$\mathbf{a}_E=-\frac{v^2}{\alpha}\left[1+\left(\frac{p}{c}\right)^2\right]^{-1}\mathbf{e}_r. \tag{4.64c}$$

It is seen that the acceleration has a constant magnitude and is always directed toward the axis of the helical motion as the electron spirals around it. □

Example 4.11. An insect is escaping at a constant rate of 1 in./sec along a radial line on a record turntable which is turning at a constant rate of 45 rpm, as indicated in Fig. 4.18. Find the absolute velocity and acceleration of the insect at the point 2 in. from the center of the turntable.

Solution (i). In terms of plane polar coordinates, we have $\dot{r}=1$ in./sec, $\ddot{r}=0$, $\dot{\phi}=45$ rpm $=(3\pi/2)$ rad/sec, and $\ddot{\phi}=0$. Thus, at the point of interest at $r=2$ in., (4.59) and (4.60) yield the solutions

$$\mathbf{v}_P=\mathbf{e}_r+3\pi\mathbf{e}_\phi \text{ in./sec}, \tag{4.65a}$$

$$\mathbf{a}_P=-\frac{9\pi^2}{2}\mathbf{e}_r+3\pi\mathbf{e}_\phi \text{ in./sec}^2. \tag{4.65b}$$

This solution certainly shows that (4.59) and (4.60) are very handy tools in certain circumstances. However, we shall show next that the same conclusions may be reached almost as easily but without direct substitution into the equations (4.59) and (4.60), which one really should not bother to memorize.

Solution (ii). We shall use the frame $\psi=\{O;\mathbf{e}_r,\mathbf{e}_\phi,\mathbf{e}_z\}$, as before. Then, with (4.46) and (4.47) in mind, at the moment of interest, we have $\mathbf{x}=2\mathbf{e}_r$ in,

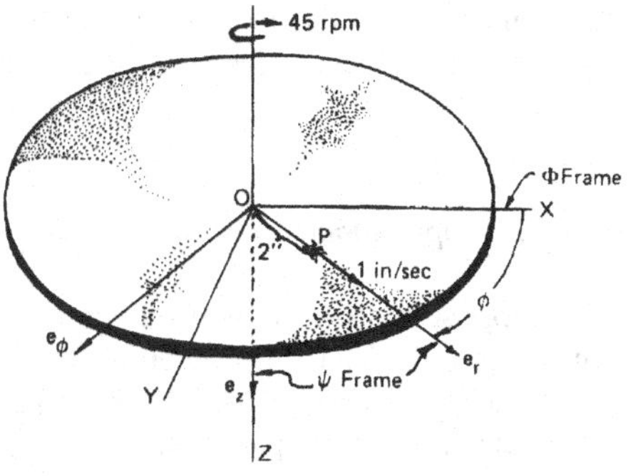

Figure 4.18. Motion referred to a cylindrical reference frame.

$\delta \mathbf{x}/\delta t = 1\mathbf{e}_r$ in./sec, $\delta^2\mathbf{x}/\delta t^2 = \mathbf{0}$, $\boldsymbol{\omega}_f = 3\pi/2\mathbf{e}_z$ rad/sec, and $\dot{\boldsymbol{\omega}}_f = \mathbf{0}$. Of course, $\mathbf{v}_O = \mathbf{0}$, $\mathbf{a}_O = \mathbf{0}$, and we determine at P

$$\boldsymbol{\omega}_f \times \mathbf{x} = 3\pi\mathbf{e}_\phi \text{ in./sec}, \qquad \boldsymbol{\omega}_f \times (\boldsymbol{\omega}_f \times \mathbf{x}) = -\frac{9\pi^2}{2}\mathbf{e}_r \text{ in./sec}^2,$$

$$2\boldsymbol{\omega}_f \times \frac{\delta \mathbf{x}}{\delta t} = 3\pi\mathbf{e}_\phi \text{ in./sec}^2.$$

Thus, upon collecting these values into (4.46) and (4.48), we shall find the same results (4.65a) and (4.65b) above. Indeed, it ought to be clear that this really is the same method that was used to derive (4.59) and (4.60).

Solution (iii). The next approach attempts to use (4.61) to deduce the results. With $\boldsymbol{\omega}_f = 3\pi/2\mathbf{e}_z$ rad/sec, (4.61) yields

$$\begin{aligned} \dot{\mathbf{e}}_r &= \boldsymbol{\omega}_f \times \mathbf{e}_r = \tfrac{3}{2}\pi\mathbf{e}_z \times \mathbf{e}_r = \tfrac{3}{2}\pi\mathbf{e}_\phi, \\ \dot{\mathbf{e}}_\phi &= \boldsymbol{\omega}_f \times \mathbf{e}_\phi = \tfrac{3}{2}\pi\mathbf{e}_z \times \mathbf{e}_\phi = -\tfrac{3}{2}\pi\mathbf{e}_r. \end{aligned} \tag{4.65c}$$

However, because *equation* (4.65a) *is not valid for all positions of* P, these relations cannot be used to compute (4.65b) by differentiation of (4.65a). Rather, it is necessary to know the formula for $\mathbf{v}_P$ valid for all times in the motion. Thus, presently, we must consider the general expression

$$\mathbf{v}_P = \mathbf{e}_r + \tfrac{3}{2}\pi r\mathbf{e}_\phi, \tag{4.65d}$$

which holds for all positions of the particle. Differentiation of (4.65d) with respect to time in Φ and use of (4.65c) for the special case at hand leads again to the result (4.65b). □

This example underscores the importance in applications of the general equations (4.46) and (4.48), or the special formulas (4.59) and (4.60). Since these equations are valid for all motions, they may be used in any problem, including those in which only the values of the variables at a specific time or in a particular position are given. The last case showed that caution is necessary when (4.61) is used in problems where only the values of vectors at a particular moment are known. Let us look at an application in which (4.61) proves useful.

4.8.1.1. Application to an Involute Mechanism

Suppose a string is wound around a circle. When the circle is fixed and the string, pulled taut, is unwound from the circumference of the circle, its end point traces a curve in the plane of the circle which is called the involute of the circle. A certain packaging machine uses this geometrical property to direct and bunch objects in preparation for their automatic packaging. The device, illustrated in Fig. 4.19, consists of a collection plate CP fastened to a slider block S that moves in a narrow slot having the shape of the involute generated by a thin, inextensible cable tightly wound on the circumference of a stationary circular pulley of radius b and attached to a swivel pin P on S. The block S is driven by a quick-return mechanism that restores it to its starting position when the collection operation is completed, and the cycle is then repeated. The cable is unwound at a constant angular speed $\omega = \dot{\phi}(t)$ during the collection cycle. We want to study the kinematics and some other geometrical properties of this mechanism.

Our intuition suggests that the instantaneous circle of radius DP is the circle of curvature of the involute at P; hence, DP is its radius of curvature at P. Moreover, the tangent to this circle at P is the tangent vector to the involute at P, and is therefore parallel to the radius OD. These seemingly evident facts will be proved below. First, the velocity and acceleration of P referred to the cylindrical reference frame $\psi = \{O; \mathbf{e}_r, \mathbf{e}_\phi, \mathbf{e}_z\}$ shown in Fig. 4.19, and the distance traveled by P as a function of the angle ϕ will be computed. Then the curvature of the involute at P, and the location of the instantaneous center of curvature will be determined. However, in this application of cylindrical coordinates, we shall make no use of the general equations (4.46) and (4.48) nor of the specialized equations (4.59) and (4.60); rather, our focus here will be on the use of (4.61).

Since the involute originates from A on the circumference of the pulley, as indicated in Fig. 4.19, it is clear that the unwound portion of the cable, namely DP, has the length $b\phi$ equal to the arc ABD. Therefore, the position vector of P in ψ is

$$\mathbf{x}(P, t) = b\mathbf{e}_r - b\phi\mathbf{e}_\phi. \tag{4.66a}$$

Time differentiation of (4.66a) in the machine frame, which is not shown, yields an equation for the velocity of P:

$$\mathbf{v}(P, t) = b\dot{\mathbf{e}}_r - b\dot{\phi}\mathbf{e}_\phi - b\phi\dot{\mathbf{e}}_\phi. \tag{4.66b}$$

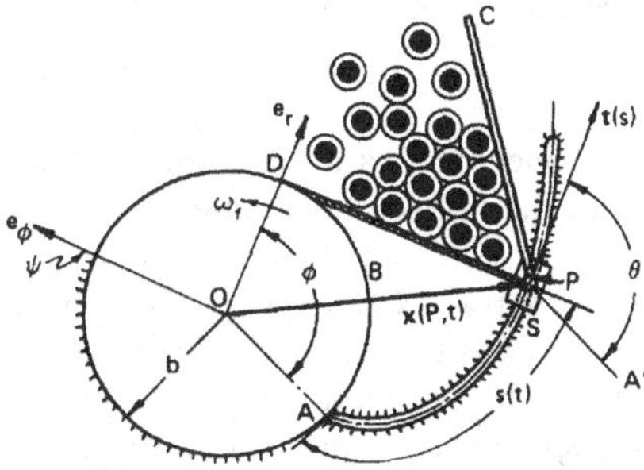

Figure 4.19. An involute mechanism of a packaging machine.

Noting that frame ψ has angular velocity $\boldsymbol{\omega}_f = \omega \mathbf{e}_z$, where $\omega = \dot{\phi}$ is an assigned constant, and recalling (4.61), we find

$$\dot{\mathbf{e}}_r = \boldsymbol{\omega}_f \times \mathbf{e}_r = \omega \mathbf{e}_\phi, \qquad \dot{\mathbf{e}}_\phi = \boldsymbol{\omega}_f \times \mathbf{e}_\phi = -\omega \mathbf{e}_r. \tag{4.66c}$$

Use of (4.66c) in (4.66b) now yields the velocity of P in the machine frame, but referred to ψ:

$$\mathbf{v}(P, t) = \omega b \phi \mathbf{e}_r. \tag{4.66d}$$

We know that the velocity vector must be parallel to the involute path of P. Therefore, the result (4.66d) and Fig. 4.19 reveal the interesting geometrical property that *the tangent vector* $\mathbf{t}(s)$ *to the involute at P is parallel to the radius vector* $\mathbf{b} = b\mathbf{e}_r$ *of the circumferential point D from O, hence also perpendicular to the cable line DP.* Indeed, since $\mathbf{v}(P, t) = \dot{s}\mathbf{t}$, it is evident from (4.66d) that $\mathbf{t} = \mathbf{e}_r$, $\mathbf{n} = \mathbf{e}_\phi$, and the speed of P on the involute is

$$\dot{s} = b\phi\omega. \tag{4.66e}$$

Thus, the distance traveled by P as a function of ϕ may now be found by integration of the differential equation (4.66e). With $\omega = \dot{\phi}$, we thereby obtain the distance $s(\phi)$ traveled by P along the involute from A:

$$s(\phi) = \tfrac{1}{2} b \phi^2. \tag{4.66f}$$

We have learned that the tangent angle θ to the involute is the same as the central angle ϕ; therefore, application of (1.68), and use of (4.66e) delivers the curvature of the involute at P:

$$\kappa = \frac{d\theta}{ds} = \frac{d\phi}{ds} = \frac{1}{b\phi}. \tag{4.66g}$$

This shows that *the radius of curvature $R = b\phi$ of the involute is simply the length DP of the unwound portion of the cable; therefore, D is the instantaneous center of curvature of the involute.*

Finally, substitution of (4.66e) and (4.66g) into (1.71) gives the acceleration of P referred to ψ:

$$\mathbf{a}(P, t) = b\omega^2(\mathbf{e}_r + \phi \mathbf{e}_\phi). \tag{4.66h}$$

To conclude this application of cylindrical coordinates, let the student consider the following exercise.

Exercise 4.2. Verify (4.66h) by direct calculation from (4.66d); and then use (4.66h) to determine the radius of curvature. Further, derive (4.66d) and (4.66h) by application of (4.46) and (4.48). What can be done in this case with the specialized equations (4.59) and (4.60)?

4.8.1.2. Application to a Spinning Pendulum Device

The motions of complex machine parts often are difficult to analyze exactly, so simpler mechanical models generally are introduced. The study of the motion of a moving thread in a textile machine obviously is complex, and thread tangling and breakage are undesirable difficulties that a textile manufacturer certainly would prefer to avoid. On the other hand, the manufacturer also wants his machines to run as rapidly as possible. Consequently, analysis of the motion of a moving thread line and of apparatus that may induce or affect its motion could be potentially useful and important to textile production interest. Needless to say, precise analysis of this kind of problem is impossible, but a simplified analysis that models special phases of the overall machine operation may be able to provide valuable information. We may imagine, for example, that a certain spinner of a textile machine may be modeled schematically as a pendulum in a rotating reference frame $1 = \{C; \mathbf{i}_k\}$ identified as the spinner housing S which turns with a constant angular velocity $\boldsymbol{\omega}$ relative to the machine frame $0 \equiv \Phi = \{C; \mathbf{I}_k\}$, as illustrated in Fig. 4.20. The moving thread T is fed from a bobbin, which is not shown here, and passes through a small hole in a ball B attached to one end of a rigid rod of length l. The other end of the rod is hinged at A to a horizontal shaft that forms part of the spinner housing S. We may expect that an investigation of the thread line motion would require knowledge of the motion of B. Therefore, this motivates our interest in finding the absolute velocity and acceleration of the pendulum ball B referred to the cylindrical reference frame $\psi = \{A; \mathbf{e}_k\}$ fixed in the pendulum as shown in Fig. 4.20.

The absolute velocity and acceleration of B may be readily determined from (4.46) and (4.48) in which the rotating frame is the cylindrical reference frame ψ fixed in the rod AB. The total angular velocity of ψ is given by use of the chain rule (4.24):

$$\boldsymbol{\omega}_f = \boldsymbol{\omega}_{20} = \boldsymbol{\omega}_{21} + \boldsymbol{\omega}_{10}. \tag{4.67a}$$

Herein $\boldsymbol{\omega}_{21} = \dot{\phi}\mathbf{e}_z$ is the angular velocity of ψ, the pendulum frame 2, relative to the spinner frame 1 fixed in S whose angular velocity relative to the

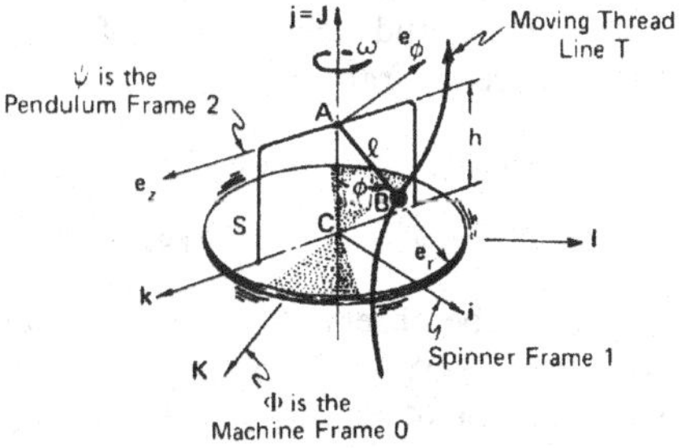

Figure 4.20. Rotating pendulum model of a textile spinner.

machine frame 0 is given as $\boldsymbol{\omega}_{10} = \boldsymbol{\omega} \equiv \omega \mathbf{j}$ in which $\mathbf{j} = -\cos\phi\, \mathbf{e}_r + \sin\phi\, \mathbf{e}_\phi$. Thus, with (4.67a), the total angular velocity of the pendulum frame 2 relative to the machine frame 0, but referred to ψ, is given by

$$\boldsymbol{\omega}_f = \omega(-\cos\phi\, \mathbf{e}_r + \sin\phi\, \mathbf{e}_\phi) + \dot{\phi}\mathbf{e}_z. \tag{4.67b}$$

It follows from use of (4.14) in (4.67b) that the angular acceleration of the pendulum relative to the machine is

$$\dot{\boldsymbol{\omega}}_f = \omega\dot{\phi}(-\sin\phi\, \mathbf{e}_r + \cos\phi\, \mathbf{e}_\phi) + \ddot{\phi}\mathbf{e}_z. \tag{4.67c}$$

Because $\mathbf{x}(B, t) = l\mathbf{e}_r$ is constant in ψ,

$$\frac{\delta \mathbf{x}}{\delta t} = \mathbf{0}, \qquad \frac{\delta^2 \mathbf{x}}{\delta t^2} = \mathbf{0}. \tag{4.67d}$$

Further, with (4.46) and (4.48) in mind, we next use (4.67b) and (4.67c) to compute

$$\boldsymbol{\omega}_f \times \mathbf{x} = \dot{\phi} l \mathbf{e}_\phi - \omega l \sin\phi\, \mathbf{e}_z, \qquad \dot{\boldsymbol{\omega}}_f \times \mathbf{x} = \ddot{\phi} l \mathbf{e}_\phi - \omega\dot{\phi} l \cos\phi\, \mathbf{e}_z; \tag{4.67e}$$

and by forming the triple product in the usual way while keeping in mind that the right-hand order of $\mathbf{e}_r$, $\mathbf{e}_\phi$, and $\mathbf{e}_z$ must be preserved so that

$$\boldsymbol{\omega}_f \times (\boldsymbol{\omega}_f \times \mathbf{x}) = \begin{vmatrix} \mathbf{e}_r & \mathbf{e}_\phi & \mathbf{e}_z \\ -\omega\cos\phi & \omega\sin\phi & \dot{\phi} \\ 0 & \dot{\phi} l & -\omega l \sin\phi \end{vmatrix},$$

we find

$$\boldsymbol{\omega}_f \times (\boldsymbol{\omega}_f \times \mathbf{x}) = -l(\dot{\phi}^2 + \omega^2 \sin^2\phi)\, \mathbf{e}_r - l\omega^2 \sin\phi\cos\phi\, \mathbf{e}_\phi - l\omega\dot{\phi}\cos\phi\, \mathbf{e}_z. \tag{4.67f}$$

Finally, noting that $\mathbf{v}_A = \mathbf{0}$ and $\mathbf{a}_A = \mathbf{0}$ for A fixed in Φ, and collecting the relations (4.67d), (4.67e), and (4.67f) into (4.46) and (4.48), we obtain the total velocity and acceleration of the pendulum ball B referred to the cylindrical reference frame ψ:

$$\begin{aligned} \mathbf{v}_B &= l\dot{\phi}\mathbf{e}_\phi - l\omega\sin\phi\, \mathbf{e}_z, \\ \mathbf{a}_B &= -l(\dot{\phi}^2 + \omega^2\sin^2\phi)\, \mathbf{e}_r + l(\ddot{\phi} - \omega^2 \sin\phi\cos\phi)\, \mathbf{e}_\phi - 2l\omega\dot{\phi}\cos\phi\, \mathbf{e}_z. \end{aligned} \tag{4.67g}$$

It will be observed that the special equations (4.59) and (4.60) cannot be used here because $\mathbf{e}_z$ is not a constant vector in Φ, as required in the derivation of (4.60). In this problem, the cylindrical reference frame has an additional rotation due to the spinner frame; nevertheless, the cylindrical coordinate representation is still appropriate. The reader may consider the following exercise.

Exercise 4.3. Apply the composition rule (4.39a) to verify (4.67c). Then start with $\mathbf{x} = l\mathbf{e}_r$ and use (4.61) to confirm the solution (4.67g).

4.8.2. The Spherical Reference Frame

The *spherical coordinates* (r, θ, ϕ) of a point P in a rectangular Cartesian frame $\Phi = \{F; \mathbf{I}_k\}$ are shown in Fig. 4.21. To locate the point P by its spherical coordinates in Φ, measure from F a radial distance r along the Z axis, then trace that end point on a circle of radius r through an angle θ about the Y axis, and, finally, from its place in the XZ plane, turn the point along a circular arc of radius $r \sin \theta$ through an angle ϕ about the Z axis. This scheme brings us to the unique location of P in Φ. It is evident in Fig. 4.21 that for a fixed value of r the locus swept out by all such points is a spherical surface of radius r in Φ; consequently, as noted above, the three measure numbers $r \in [0, \infty)$, $\theta \in [0, \pi]$, and $\phi \in [0, 2\pi]$ are called spherical coordinates. When either $\phi \equiv 0$ or $\theta \equiv \pi/2$, the spherical coordinates are the same as the plane polar coordinates described earlier.

A *spherical reference frame* $\psi = \{O; \mathbf{e}_r, \mathbf{e}_\theta, \mathbf{e}_\phi\}$ may be introduced to describe this system. For simplicity, the origin O may be chosen at F in Φ. The three perpendicular axes labeled as r, θ, ϕ in Fig. 4.21 are parallel to the directions of the orthonormal basis for ψ, which is shown at P for convenience. The unit vector $\mathbf{e}_r$ is in the direction of increasing values of the radial line r; hence, the position vector of the particle P is the simple radius vector

$$\mathbf{x} = r\mathbf{e}_r. \tag{4.68}$$

The unit vector $\mathbf{e}_\theta$ is in the direction of increasing values of the colatitude angle θ, it lies in the plane of $\mathbf{e}_r$ and $\mathbf{K}$ and is tangent at P to a meridian circle

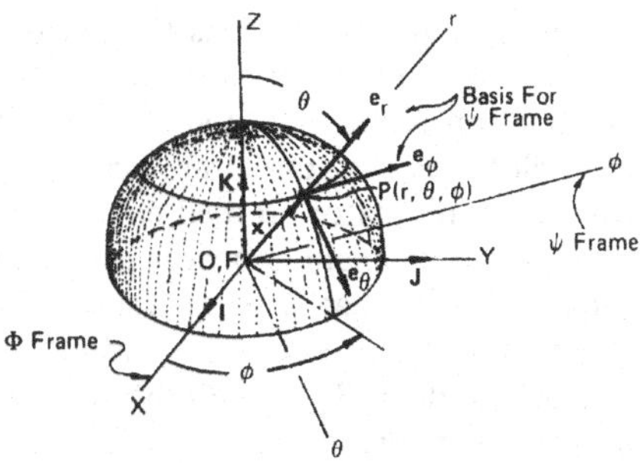

Figure 4.21. Spherical coordinates and the spherical reference frame.

whose diameter is along the Z axis. The vector $\mathbf{e}_\phi \equiv \mathbf{e}_r \times \mathbf{e}_\theta$ is in the direction of increasing values of the longitude angle ϕ, and it is tangent at P to a latitude circle parallel to the XY plane.

All of the basis vectors $\mathbf{e}_k$ vary with the angles θ and ϕ as the vector $\mathbf{e}_r$ follows the motion of the point P in Φ. Therefore, ψ is a rotating frame whose total angular velocity relative to frame $\Phi = \{F; \mathbf{I}_k\}$ is

$$\boldsymbol{\omega}_f = \dot{\phi}\mathbf{K} + \dot{\theta}\mathbf{e}_\phi.$$

Thus, with $\mathbf{K} = \cos\theta\,\mathbf{e}_r - \sin\theta\,\mathbf{e}_\theta$, the total angular velocity of ψ in Φ, but referred to ψ, is given by

$$\boldsymbol{\omega}_f = \dot{\phi}(\cos\theta\,\mathbf{e}_r - \sin\theta\,\mathbf{e}_\theta) + \dot{\theta}\mathbf{e}_\phi. \tag{4.69}$$

The velocity and acceleration of P in Φ may be obtained from (4.46) and (4.48) by the same procedure used earlier for cylindrical coordinates. It turns out from a lengthy but straightforward calculation that the absolute velocity and acceleration of a particle referred to the spherical reference frame ψ are given by

$$\mathbf{v}_P = \dot{r}\mathbf{e}_r + r\dot{\phi}\sin\theta\,\mathbf{e}_\phi + r\dot{\theta}\mathbf{e}_\theta, \tag{4.70}$$

$$\begin{aligned}\mathbf{a}_P = {}&(\ddot{r} - r\dot{\theta}^2 - r\dot{\phi}^2\sin^2\theta)\,\mathbf{e}_r + (2\dot{r}\dot{\phi}\sin\theta + 2r\dot{\phi}\dot{\theta}\cos\theta + r\ddot{\phi}\sin\theta)\,\mathbf{e}_\phi \\ &+ (r\ddot{\theta} + 2\dot{r}\dot{\theta} - r\dot{\phi}^2\sin\theta\cos\theta)\,\mathbf{e}_\theta.\end{aligned} \tag{4.71}$$

It is sometimes useful to note from (4.70) that the elemental displacement vector for spherical coordinates is

$$d\mathbf{x} = dr\mathbf{e}_r + r\sin\theta\,d\phi\mathbf{e}_\phi + rd\theta\mathbf{e}_\theta. \tag{4.72}$$

Therefore, the squared elemental arc length along the particle path is given in spherical coordinates by

$$ds^2 = d\mathbf{x}\cdot d\mathbf{x} = dr^2 + r^2\sin^2\theta\,d\phi^2 + r^2 d\theta^2. \tag{4.73}$$

The unit vector $\mathbf{t} = d\mathbf{x}/ds$ is the usual intrinsic tangent vector to the path.

The use of spherical coordinates and the description of velocity and acceleration referred to a spherical reference frame will be illustrated in three examples. The easy direct application of (4.70) and (4.71) to the motion of the robotic manipulator studied earlier will be demonstrated in the first example. The second problem concerns a speed control governor and will demonstrate three methods that use spherical coordinates in slightly different ways, one being the direct use of (4.70) and (4.71). The final illustration is an application of (4.46) and (4.48) to the motion of a point on a helicopter blade referred to a spherical reference frame.

Example 4.12. The relative angular velocity vectors for the general rotation of the manipulator claw of the robot shown in Fig. 4.7 are given in (4.40a). Recall also that the length $l(t)$ of the telescopic arm, as shown in Fig. 4.4, is a computer-controlled function of time. Find the absolute velocity and acceleration of point A on the claw referred to frame $2 = \{O; \boldsymbol{\gamma}, \mathbf{i}', \mathbf{j}\}$ shown in Fig. 4.7.

Solution. A few moments' reflection will reveal in Fig. 4.7 that with $\boldsymbol{\gamma} = \mathbf{e}_r$, $\mathbf{i}' = \mathbf{e}_\theta$, and $\mathbf{j} = \mathbf{e}_\phi$ frame 2 may be identified as the spherical reference frame $\psi = \{O; \mathbf{e}_r, \mathbf{e}_\theta, \mathbf{e}_\phi\}$. It is seen that $r = l(t)$, $\theta = \beta(t)$, and $\phi = \alpha(t)$. Therefore, the absolute velocity and acceleration of point A on the manipulator claw may be read directly from (4.70) and (4.71). We find

$$\mathbf{v}_A = \dot{l}\mathbf{e}_r + l\dot{\alpha}\sin\beta\,\mathbf{e}_\phi + l\dot{\beta}\mathbf{e}_\theta, \tag{4.74a}$$

$$\begin{aligned}\mathbf{a}_A = (\ddot{l} - l\dot{\beta}^2 - l\dot{\alpha}^2\sin^2\beta)\,\mathbf{e}_r + (2\dot{l}\dot{\alpha}\sin\beta + 2l\dot{\alpha}\dot{\beta}\cos\beta \\ + l\ddot{\alpha}\sin\beta)\,\mathbf{e}_\phi + (l\ddot{\beta} + 2\dot{l}\dot{\beta} - l\dot{\alpha}^2\sin\beta\cos\beta)\,\mathbf{e}_\theta.\end{aligned} \tag{4.74b}$$

Let the student examine the following exercise.

Exercise 4.4. Notice that $\mathbf{x}(A, t) = l(t)\,\mathbf{e}_r$ in ψ and recall (4.40a). Determine $\boldsymbol{\omega}_f$ and $\dot{\boldsymbol{\omega}}_f$ referred to ψ, and apply (4.46) and (4.48) to derive (4.74a) and (4.74b). When $\dot{\alpha} = 0.5$ rad/sec, how does the result (4.74a) compare with (4.17d) found earlier for these conditions? □

Example 4.13. The ball governor of a certain speed control device consists of an arm OA hinged at O to a vertical shaft OZ as shown in Fig. 4.22. The shaft rotates with constant angular speed $\omega_1 = 3\pi$ rad/sec in the machine frame 0, and simultaneously the arm OA is elevated at the constant rate $\omega_2 = \pi$ rad/sec in the shaft frame 1. The control ball B is attached to a light spring, but otherwise it may slide freely on the arm OA in frame 2. The designed shutoff position for the ball is set at 6 in. from O when $\theta = 45°$ and the ball is stationary on OA. To design a proper spring it is essential to know the maximum, absolute acceleration of B for the above conditions. Find the

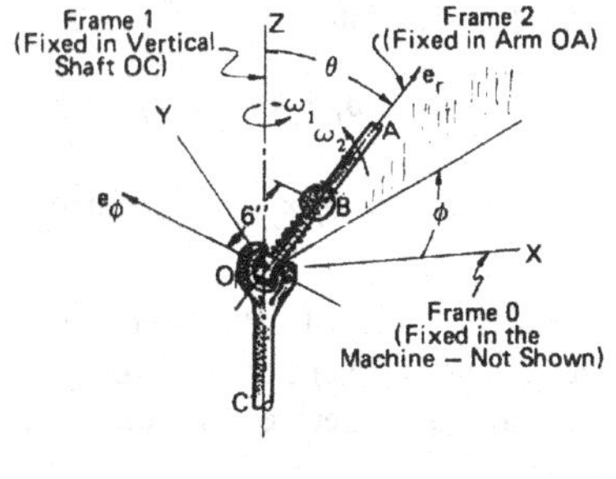

Figure 4.22. Motion of a speed control governor referred to a spherical reference frame.

absolute acceleration of B at the shutoff position. What is the absolute velocity there?

Solution (i). Since $\omega_1 = \dot{\phi} = 3\pi$ rad/sec and $\omega_2 = -\dot{\theta} = \pi$ rad/sec are assigned constants, $\ddot{\phi} = \ddot{\theta} = 0$. At the shutoff position, $r = 6$ in. and $\dot{r} = \ddot{r} = 0$. Thus, at $\theta = 45°$, (4.71) gives the absolute acceleration at the shutoff position:

$$\mathbf{a}_B = -33\pi^2 \mathbf{e}_r - 18\pi^2 \sqrt{2}\, \mathbf{e}_\phi - 27\pi^2 \mathbf{e}_\theta \text{ in./sec}^2. \tag{4.75a}$$

We see easily from (4.70) that the absolute velocity of B at shutoff is

$$\mathbf{v}_B = 9\pi \sqrt{2}\, \mathbf{e}_\phi - 6\pi \mathbf{e}_\theta \text{ in./sec.} \tag{4.75b}$$

This easy solution shows clearly that (4.70) and (4.71) may be very handy in certain situations; but it is equally plain that no one would trouble to memorize them. On the other hand, the same results also may be gotten less quickly, but just about as easily, by thoughtful use of (4.46) and (4.48). This will be shown next.

Solution (ii). Consider the motion of B in the imbedded frame $2 \equiv \psi = \{O; \mathbf{e}_r, \mathbf{e}_\theta, \mathbf{e}_\phi\}$. The total angular velocity of ψ is given by $\boldsymbol{\omega}_{20} = \boldsymbol{\omega}_{21} + \boldsymbol{\omega}_{10}$, in which

$$\boldsymbol{\omega}_{21} = -\omega_2 \mathbf{e}_\phi = -\pi \mathbf{e}_\phi, \qquad \boldsymbol{\omega}_{10} = \omega_1 \mathbf{K} = 3\pi \mathbf{K} = 3\pi(\cos\theta\, \mathbf{e}_r - \sin\theta\, \mathbf{e}_\theta)$$

referred to ψ. Thus, at $\theta = 45°$, we find

$$\boldsymbol{\omega}_f = \boldsymbol{\omega}_{20} = -\pi \mathbf{e}_\phi + \frac{3\pi\sqrt{2}}{2}(\mathbf{e}_r - \mathbf{e}_\theta) \text{ rad/sec.} \tag{4.75c}$$

Furthermore, application of the composition rule (4.39a) in which $\delta\boldsymbol{\omega}_{21}/\delta t = \mathbf{0}$ and $\delta\boldsymbol{\omega}_{10}/\delta t = \mathbf{0}$ gives

$$\dot{\boldsymbol{\omega}}_f = \dot{\boldsymbol{\omega}}_{20} = \boldsymbol{\omega}_{10} \times \boldsymbol{\omega}_{21} = \frac{3\pi^2\sqrt{2}}{2}(\mathbf{e}_\theta + \mathbf{e}_r) \text{ rad/sec}^2. \tag{4.75d}$$

Of course, $\mathbf{v}_O = \mathbf{0}$, $\mathbf{a}_O = \mathbf{0}$, $\mathbf{x} = 6\mathbf{e}_r$ in. and $\delta\mathbf{x}/\delta t = \delta^2\mathbf{x}/\delta t^2 = \mathbf{0}$; hence, bearing in mind (4.46) and (4.48) and recalling (4.75c) and (4.75d), we determine

$$\begin{gathered}\boldsymbol{\omega}_f \times \mathbf{x} = 9\pi\sqrt{2}\, \mathbf{e}_\phi - 6\pi \mathbf{e}_\theta, \qquad \dot{\boldsymbol{\omega}}_f \times \mathbf{x} = -9\pi^2\sqrt{2}\, \mathbf{e}_\phi, \\ \boldsymbol{\omega}_f \times (\boldsymbol{\omega}_f \times \mathbf{x}) = -33\pi^2 \mathbf{e}_r - 9\pi^2\sqrt{2}\, \mathbf{e}_\phi - 27\pi^2 \mathbf{e}_\theta, \quad 2\boldsymbol{\omega}_f \times \frac{\delta\mathbf{x}}{\delta t} = \mathbf{0}.\end{gathered} \tag{4.75e}$$

Finally, use of these results in (4.46) and (4.48) yields again the relations (4.75a) and (4.75b) given above. This method follows the procedure employed to derive the general equations (4.70) and (4.71).

The student will find it instructive to repeat the development of (4.70) and (4.71) starting with (4.68) and (4.69) and recalling (4.61). However, in the present case, because (4.75b) is valid only at the particular position $\theta = 45°$, (4.61) cannot be used to derive (4.75a) directly from (4.75b); rather, as in our earlier solution (iii) in Example 4.11, we must consider this more carefully.

Solution (iii). It may be seen from (4.70) that the ball velocity at the shutoff position is determined *for all time* by

$$\mathbf{v}_B = 18\pi \sin\theta\, \mathbf{e}_\phi - 6\pi \mathbf{e}_\theta \text{ in./sec;} \tag{4.75f}$$

hence, its time derivative $\mathbf{a}_B$ in the machine frame 0 can now be determined with the help of (4.61). Using (4.75c) above, we find at $\theta = 45°$

$$\begin{aligned} \dot{\mathbf{e}}_\phi &= \left[\frac{3\pi\sqrt{2}}{2}(\mathbf{e}_r - \mathbf{e}_\theta) - \pi \mathbf{e}_\phi\right] \times \mathbf{e}_\phi = -\frac{3\pi\sqrt{2}}{2}(\mathbf{e}_\theta + \mathbf{e}_r), \\ \dot{\mathbf{e}}_\theta &= \left[\frac{3\pi\sqrt{2}}{2}(\mathbf{e}_r - \mathbf{e}_\theta) - \pi \mathbf{e}_\phi\right] \times \mathbf{e}_\theta = \frac{3\pi\sqrt{2}}{2}\mathbf{e}_\phi + \pi \mathbf{e}_r. \end{aligned} \tag{4.75g}$$

Upon completion of the few details noted above, the reader may see that (4.75a) follows. The concluding remarks following Example 4.11 also are appropriate here. □

Example 4.14. A helicopter shown in Fig. 4.23 is ascending vertically with a speed v which is increasing at the rate $\sigma = \dot{v}$. The rotor blade is turning at a constant rate ω about its axis which makes an angle α with the vertical direction. At the moment of interest, the pilot is lifting the nose at the constant rate $\Omega = -\dot{\alpha}$, as shown. Relative to the ground at the instant of interest, determine the velocity and the acceleration of point B at the tip of the rotor blade, but referred to the spherical reference frame $\psi = \{O; \mathbf{e}_k\}$ imbedded in the blade.

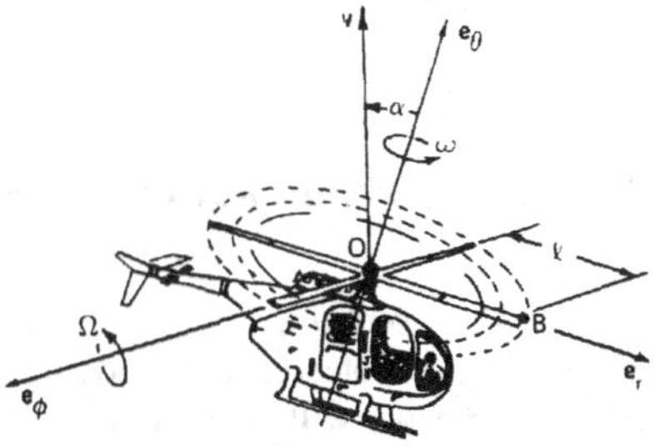

Figure 4.23. Motion of a helicopter blade point referred to a spherical reference frame.

Solution. The absolute velocity and acceleration of point O in the helicopter are given as

$$\mathbf{v}_O = v(\cos\alpha\,\mathbf{e}_\theta - \sin\alpha\,\mathbf{e}_r), \qquad \mathbf{a}_O = \sigma(\cos\alpha\,\mathbf{e}_\theta - \sin\alpha\,\mathbf{e}_r), \tag{4.76a}$$

referred to the spherical frame $\psi = \{O; \mathbf{e}_k\}$ fixed in the rotor blade as shown in Fig. 4.23.

It is evident that the total angular velocity and angular acceleration of the blade frame relative to the ground frame may be written as

$$\boldsymbol{\omega}_f = \omega\mathbf{e}_\theta + \Omega\mathbf{e}_\phi, \qquad \dot{\boldsymbol{\omega}}_f = \mathbf{0}, \tag{4.76b}$$

where the last equation is obtained by use of (4.14). Also, since $\mathbf{x}(B, t) = l\mathbf{e}_r$ is fixed in ψ,

$$\frac{\delta\mathbf{x}}{\delta t} = \mathbf{0}, \qquad \frac{\delta^2\mathbf{x}}{\delta t^2} = \mathbf{0}. \tag{4.76c}$$

With the application of (4.46) and (4.48) in mind, we next determine with the aid of (4.76b)

$$\boldsymbol{\omega}_f \times \mathbf{x} = -l\omega\mathbf{e}_\phi + l\Omega\mathbf{e}_\theta, \qquad \boldsymbol{\omega}_f \times (\boldsymbol{\omega}_f \times \mathbf{x}) = -l(\Omega^2 + \omega^2)\,\mathbf{e}_r. \tag{4.76d}$$

Thus, using (4.76a), the second equation in (4.76b), (4.76c), and (4.76d) in (4.46) and (4.48), we obtain for the moment shown in Fig. 4.23 the absolute velocity and acceleration of B referred to the spherical reference frame ψ:

$$\begin{aligned} \mathbf{v}_B &= -v\sin\alpha\,\mathbf{e}_r + (v\cos\alpha + l\Omega)\,\mathbf{e}_\theta - l\omega\mathbf{e}_\phi, \\ \mathbf{a}_B &= -[l(\Omega^2 + \omega^2) + \sigma\sin\alpha]\,\mathbf{e}_r + \sigma\cos\alpha\,\mathbf{e}_\theta. \end{aligned} \tag{4.76e}$$

Notice that in view of (4.76c), (4.46) and (4.48) reduce to (2.27) and (2.30) for description of the rigid body motion; hence these equations also may have been applied here. The following modification is left for the reader.

Exercise 4.5. Suppose $\psi = \{O; \mathbf{e}_k\}$ is fixed in the helicopter so that the blade frame turns relative to ψ with the angular velocity $\boldsymbol{\omega} = \omega\mathbf{e}_\theta$. Find for the moment shown in Fig. 4.23 the absolute velocity and acceleration of B referred to the helicopter frame ψ. Compare the results with (4.76e).

4.9. More Examples of Motion Referred to a Moving Frame

In this section, several additional examples involving motion of a material point referred to a moving reference frame will be studied, and some related concepts will be reviewed. Some problems that involve rigid bodies in motion relative to one another will be solved. To emphasize that (4.46) and (4.48) are not restricted to an isolated material point nor to a point of a rigid body, but also are applicable to a deformable solid or fluid continuum, the first problem concerns the motion of a fluid particle referred to the intrinsic frame.

Example 4.15. *Motion of a Fluid Particle in a Centrifugal Pump.* A fluid particle P shown in Fig. 4.24 moves outward along the impeller blade of a centrifugal pump with speed v and tangential acceleration a relative to the blade. The impeller rotates with a constant, clockwise angular speed ω relative to the pump casing. (a) Determine the absolute velocity and acceleration of P referred to the intrinsic frame $\mu = \{P; \mathbf{t}_k\}$. (b) In particular, let $r = R = 2$ ft and $\omega = 600$ rpm; and suppose that $v = 200$ ft/sec and $a = 100$ ft/sec^2 at the instant when the fluid particle reaches the tip T of the blade. Evaluate the absolute velocity and acceleration of P at T. The pump geometry is shown in Fig. 4.24a.

Solution. *Part (a).* Since the data are expressed in terms of intrinsic variables, it is natural to employ an intrinsic reference system at the fluid particle P, which is in motion relative to a rotating reference frame $\varphi = \{O; \mathbf{i}_k\}$ fixed in the impeller. Therefore, we shall apply (4.46) and (4.48) in which the preferred frame Φ is fixed in the pump casing.

The intrinsic velocity and acceleration of P relative to the impeller frame φ but referred to the intrinsic frame μ are given by use of (1.70) and (1.71). Thus, for the assigned parameters,

$$\frac{\delta \mathbf{x}}{\delta t} = v\mathbf{t}, \qquad \frac{\delta^2 \mathbf{x}}{\delta t^2} = a\mathbf{t} + \frac{v^2}{R}\mathbf{n}. \tag{4.77a}$$

With reference to Fig. 4.24a, we have

$$\boldsymbol{\omega}_f = \omega \mathbf{k} = -\omega \mathbf{b}, \qquad \dot{\boldsymbol{\omega}}_f = \mathbf{0}, \qquad \mathbf{x} = R \sin \psi \, \mathbf{t} - (1 - \cos \psi)\,\mathbf{n}, \tag{4.77b}$$

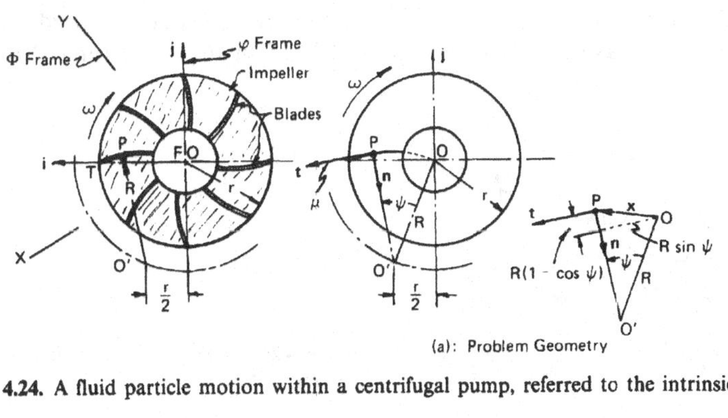

Figure 4.24. A fluid particle motion within a centrifugal pump, referred to the intrinsic frame.

in which ψ is the angular placement of P on the blade. The point O is fixed in the pump frame Φ, so $\mathbf{v}_O = \mathbf{a}_O = \mathbf{0}$. Hence, with the aid of (4.77b), the rigid body velocity and acceleration of P in Φ are given by

$$\mathbf{v}_O + \boldsymbol{\omega}_f \times \mathbf{x} = -R\omega \sin \psi \, \mathbf{n} - R\omega(1 - \cos \psi)\, \mathbf{t}, \tag{4.77c}$$

$$\mathbf{a}_O + \boldsymbol{\omega}_f \times (\boldsymbol{\omega}_f \times \mathbf{x}) + \dot{\boldsymbol{\omega}}_f \times \mathbf{x} = -R\omega^2 \sin \psi \, \mathbf{t} + R\omega^2(1 - \cos \psi)\, \mathbf{n}. \tag{4.77d}$$

The relations (4.77c) and (4.77d) are the velocity and acceleration that P would have if it were fixed to the blade.

The Coriolis acceleration is obtained with the help of the first equations in (4.77a) and (4.77b). We get

$$2\boldsymbol{\omega}_f \times \frac{\delta \mathbf{x}}{\delta t} = -2\omega v \mathbf{n}. \tag{4.77e}$$

Finally, upon substituting (4.77c) and the first equation in (4.77a) into (4.46) and using (4.77d), (4.77e) and the second equation in (4.77a) in (4.48), we derive the absolute velocity and acceleration of P referred to μ:

$$\begin{aligned} \mathbf{v}_P &= [v - R\omega(1 - \cos \psi)]\, \mathbf{t} - R\omega \sin \psi \, \mathbf{n}, \\ \mathbf{a}_P &= [a - r\omega^2 \sin \psi]\, \mathbf{t} + [v^2/R + R\omega^2(1 - \cos \psi) - 2\omega v]\, \mathbf{n}. \end{aligned} \tag{4.77f}$$

Part (b). Thus, in particular, for $\omega = 600$ rpm $= 20\pi$ rad/sec and $r = R = 2$ ft, and observing that $\psi = 60^\circ$ at the tip of the blade where $v = 200$ ft/sec and $a = 100$ ft/sec^2, we find the example values

$$\mathbf{v}_P = (137\mathbf{t} - 109\mathbf{n}) \text{ ft/sec}, \qquad \mathbf{a}_P = -(6738\mathbf{t} + 1190\mathbf{n}) \text{ ft/sec}^2.$$

Notice that in view of the blade geometry, use of a cylindrical reference frame would be equally convenient. This application is left as a student exercise. □

Example 4.16. *Motion of a Spinning Disk Referred to Various Reference Frames.* A thin disk of radius a rotates with constant angular speed ω_2 about an axle fixed in a platform that rotates with constant angular speed ω_1 about its normal axis, as illustrated in Fig. 4.25. Find in the fixed frame 0 the total

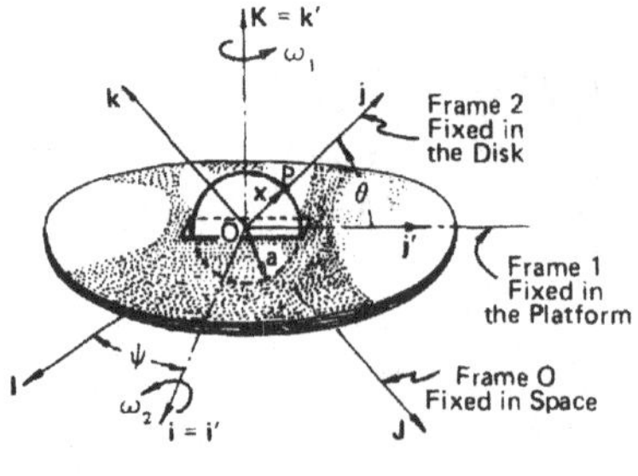

Figure 4.25. Motion of a spinning disk on a rotating platform.

velocity and acceleration of a particle P on the rim of the disk three ways: (i) referred to axes imbedded in the disk, (ii) referred to axes fixed in the platform, and (iii) referred to axes fixed in space.

Solution of (i). This case requires that we refer the motion to the disk frame $2=\{O;\mathbf{i}_k\}$ whose $\mathbf{i}$ direction is the axis of rotation of $\boldsymbol{\omega}_2$ and whose $\mathbf{j}$ direction passes through P, as shown in Fig. 4.25. Also, frame $1=\{O;\mathbf{i}'_k\}$ is fixed in the platform, and frame $0=\{O;\mathbf{I}_k\}$ denotes the preferred frame fixed in space. There is no motion of P relative to the moving frame 2, so (4.46) and (4.48) reduce to the rigid body formulas (2.27) and (2.30). Since O is fixed in frame 0, $\mathbf{v}_O=\mathbf{0}$, $\mathbf{a}_O=\mathbf{0}$, and these equations show that the total velocity and acceleration in frame 0 may be found from

$$\mathbf{v}_P=\boldsymbol{\omega}_f\times\mathbf{x},\qquad \mathbf{a}_P=\boldsymbol{\omega}_f\times(\boldsymbol{\omega}_f\times\mathbf{x})+\dot{\boldsymbol{\omega}}_f\times\mathbf{x}. \tag{4.78a}$$

The angular velocity of the disk frame 2 relative to the platform is given as $\boldsymbol{\omega}_{21}=\boldsymbol{\omega}_2=\omega_2\mathbf{i}$; and the angular velocity of frame 1 relative to the space frame 0 is $\boldsymbol{\omega}_{10}=\boldsymbol{\omega}_1=\omega_1\mathbf{K}$. Hence, with $\mathbf{K}=\sin\theta\,\mathbf{j}+\cos\theta\,\mathbf{k}$ from Fig. 4.25, it is seen that the total angular velocity of the disk frame relative to 0 may be written as

$$\boldsymbol{\omega}_f\equiv\boldsymbol{\omega}_{20}=\omega_2\mathbf{i}+\omega_1(\sin\theta\,\mathbf{j}+\cos\theta\,\mathbf{k}), \tag{4.78b}$$

referred to the disk frame. Use of (4.14) in (4.78b), which is expressed totally in terms of the disk frame, yields

$$\dot{\boldsymbol{\omega}}_f=\frac{\delta\boldsymbol{\omega}_f}{\delta t}=\omega_1\omega_2(\cos\theta\,\mathbf{j}-\sin\theta\,\mathbf{k}). \tag{4.78c}$$

The conditions $\dot{\omega}_1=\dot{\omega}_2=0$ and $\dot{\theta}=\omega_2$ were used here. Notice that this result also may be derived by application of (4.39a). In this case, $\delta\boldsymbol{\omega}_{21}/\delta t=\mathbf{0}$ in frame 1, $\delta\boldsymbol{\omega}_{10}/\delta t=\mathbf{0}$ in frame 0 and the convective term $\dot{\boldsymbol{\omega}}_{20}=\boldsymbol{\omega}_{10}\times\boldsymbol{\omega}_{21}=\omega_1\omega_2\mathbf{K}\times\mathbf{i}$ yields (4.78c).

With $\mathbf{x}=a\mathbf{j}$, the terms in (4.78a) may now be evaluated. We find

$$\begin{aligned}&\boldsymbol{\omega}_f\times\mathbf{x}=a\omega_2\mathbf{k}-a\omega_1\cos\theta\,\mathbf{i},\qquad \dot{\boldsymbol{\omega}}_f\times\mathbf{x}=a\omega_1\omega_2\sin\theta\,\mathbf{i},\\&\boldsymbol{\omega}_f\times(\boldsymbol{\omega}_f\times\mathbf{x})=a\omega_1\omega_2\sin\theta\,\mathbf{i}-a(\omega_2^2+\omega_1^2\cos^2\theta)\,\mathbf{j}+a\omega_1^2\sin\theta\cos\theta\,\mathbf{k}.\end{aligned} \tag{4.78d}$$

Putting (4.78d) into (4.78a), we reach the total velocity and acceleration of P referred to the disk frame:

$$\begin{aligned}&\mathbf{v}_P=-a\omega_1\cos\theta\,\mathbf{i}+a\omega_2\mathbf{k},\\&\mathbf{a}_P=a\omega_1\sin\theta(2\omega_2\mathbf{i}+\omega_1\cos\theta\,\mathbf{k})-a(\omega_2^2+\omega_1^2\cos^2\theta)\,\mathbf{j}.\end{aligned} \tag{4.78e}$$

Solution of (ii). This problem concerns the motion of P referred to the platform frame 1 shown in Fig. 4.25. Since P is moving relative to frame 1, the general equations (4.46) and (4.48) are used to determine the total velocity and acceleration of P in frame 0.

To begin, we note that the angular velocity and angular acceleration of the moving frame 1 in the space frame 0, but referred to the platform frame, are given by

$$\boldsymbol{\omega}_f \equiv \boldsymbol{\omega}_{10} = \boldsymbol{\omega}_1 = \omega_1 \mathbf{k}' \quad \text{and} \quad \dot{\boldsymbol{\omega}}_f = \mathbf{0}. \tag{4.79a}$$

Thus, by use of (4.79a) and with $\mathbf{v}_O = \mathbf{0}$ and $\mathbf{a}_O = \mathbf{0}$ as before, it is seen that the absolute velocity and acceleration of P for the present case are determined by

$$\mathbf{v}_P = \frac{\delta \mathbf{x}}{\delta t} + \boldsymbol{\omega}_f \times \mathbf{x}, \qquad \mathbf{a}_P = \frac{\delta^2 \mathbf{x}}{\delta t^2} + \boldsymbol{\omega}_f \times (\boldsymbol{\omega}_f \times \mathbf{x}) + 2\boldsymbol{\omega}_f \times \frac{\delta \mathbf{x}}{\delta t}. \tag{4.79b}$$

The position vector of P in frame 1, with reference to Fig. 4.25, is

$$\mathbf{x} = a(\cos\theta\, \mathbf{j}' + \sin\theta\, \mathbf{k}'). \tag{4.79c}$$

Therefore, noting that $\dot{\theta} = \omega_2$, we find from (4.79c)

$$\frac{\delta \mathbf{x}}{\delta t} = a\omega_2(-\sin\theta\, \mathbf{j}' + \cos\theta\, \mathbf{k}'), \tag{4.79d}$$

$$\frac{\delta^2 \mathbf{x}}{\delta t^2} = -a\omega_2^2(\cos\theta\, \mathbf{j}' + \sin\theta\, \mathbf{k}'). \tag{4.79e}$$

These represent the velocity and acceleration that P would have if frame 1 were fixed in frame 0; they are the velocity and acceleration of P relative to the platform.

Also, with the aid of the first equation in (4.79a), (4.79c), and (4.79d), the remaining terms in (4.79b) may be computed next. We get

$$\begin{gathered} \boldsymbol{\omega}_f \times \mathbf{x} = -a\omega_1 \cos\theta\, \mathbf{i}', \qquad \boldsymbol{\omega}_f \times (\boldsymbol{\omega}_f \times \mathbf{x}) = -a\omega_1^2 \cos\theta\, \mathbf{j}', \\ 2\boldsymbol{\omega}_f \times \frac{\delta \mathbf{x}}{\delta t} = 2a\omega_1\omega_2 \sin\theta\, \mathbf{i}'. \end{gathered} \tag{4.79f}$$

Finally, the substitution of (4.79d), (4.79e), and (4.79f) into (4.79b) yields the absolute velocity and acceleration of P referred to the platform frame:

$$\begin{aligned} \mathbf{v}_P &= -a\omega_1 \cos\theta\, \mathbf{i}' - a\omega_2(\sin\theta\, \mathbf{j}' - \cos\theta\, \mathbf{k}'), \\ \mathbf{a}_P &= 2a\omega_1\omega_2 \sin\theta\, \mathbf{i}' - a\cos\theta(\omega_1^2 + \omega_2^2)\, \mathbf{j}' - a\omega_2^2 \sin\theta\, \mathbf{k}'. \end{aligned} \tag{4.79g}$$

This method differs from the approach adopted in the solution of (i) because the particle has a motion relative to a rotating frame, and hence the angular velocity of the body and of the reference frame has to be distinguished. In part (i), there was no motion of P relative to frame 2 because frame 2 was imbedded in the body containing P; and in all such problems the basic equations (4.46) and (4.48) reduce to the rigid body formulas (2.27) and (2.30). Care must be exercised in the evaluation of $\boldsymbol{\omega}_f$ and $\dot{\boldsymbol{\omega}}_f$; otherwise, in either case, the calculations are straightforward. Let us now look at the solution to (iii).

Solution of (iii). The motion of P referred to the space frame 0 will be determined next. Since the frame 0 is fixed in space, we have

$$\mathbf{v}_O=\mathbf{0}, \qquad \mathbf{a}_O=\mathbf{0}, \qquad \boldsymbol{\omega}_f=\mathbf{0}, \qquad \dot{\boldsymbol{\omega}}_f=\mathbf{0}. \tag{4.80a}$$

Therefore, (4.46) and (4.48) show that the absolute velocity and acceleration of P are determined by the familiar equations

$$\mathbf{v}_P=\frac{\delta \mathbf{x}}{\delta t}, \qquad \mathbf{a}_P=\frac{\delta^2 \mathbf{x}}{\delta t^2}. \tag{4.80b}$$

The reader will recall that these are the basic equations (1.8) and (1.10) for the velocity and acceleration of a particle in a "fixed" frame.

The position vector of P referred to the platform frame is given be (4.79c). Therefore, use of the basis transformation

$$\mathbf{j}'=-\sin\psi\,\mathbf{I}+\cos\psi\,\mathbf{J}, \qquad \mathbf{k}'=\mathbf{K}, \tag{4.80c}$$

obtained from the geometry of Fig. 4.25, yields the position vector of P referred to frame 0:

$$\mathbf{x}=a\cos\theta(-\sin\psi\,\mathbf{I}+\cos\psi\,\mathbf{J})+a\sin\theta\,\mathbf{K}. \tag{4.80d}$$

Consequently, recalling that $\omega_1=\dot{\psi}$ and $\omega_2=\dot{\theta}$, we find by (4.80b) the absolute velocity and acceleration of P referred to the space frame 0:

$$\begin{aligned}\mathbf{v}_P&=a(\omega_2\sin\theta\sin\psi-\omega_1\cos\theta\cos\psi)\,\mathbf{I}+a\omega_2\cos\theta\,\mathbf{K}\\&\quad-a(\omega_2\sin\theta\cos\psi+\omega_1\cos\theta\sin\psi)\,\mathbf{J},\end{aligned} \tag{4.80e}$$

$$\begin{aligned}\mathbf{a}_P&=a(\omega_1^2-\omega_2^2)\cos\theta\sin\psi\,\mathbf{I}-a\omega_2^2\sin\theta\,\mathbf{K}\\&\quad-a[(\omega_1^2+\omega_2^2)\cos\theta\cos\psi-2\omega_1\omega_2\sin\theta\sin\psi]\,\mathbf{J}.\end{aligned} \tag{4.80f}$$

The three cases studied here demonstrate that the relations (4.46) and (4.48) for the velocity and acceleration of a material point referred to a moving frame contain all of the basic equations (1.8), (1.10), (2.27), and (2.30) studied earlier in Chapters 1 and 2. However, in this example, the origin of the moving frame was fixed in the preferred frame. We turn now to a similar example in which this is not the case. □

Example 4.17. *Application to the Motion of an Oscillating Fan Blade.* The blade of an oscillating fan shown in Fig. 4.26 spins about a horizontal shaft of length l with a constant, counterclockwise angular speed ω_2. The fan assembly oscillates about a vertical axis with a variable speed $\omega_1 = \omega \cos pt$, where p is the constant frequency of the fan oscillations and ω is the maximum angular speed. Find in the ground frame $0 = \{F; \mathbf{I}_k\}$ the total velocity of a point P at the tip of a blade two ways: (i) referred to the frame $1 = \{O; \mathbf{i}_k\}$ imbedded in the motor housing, and (ii) referred to frame $2 = \{O; \mathbf{i}'_k\}$ fixed in the blade. What is the total angular acceleraton of the blade for these cases?

Solution of (i). In this problem, the particle P is moving relative to the oscillating motor housing frame 1 shown in Fig. 4.26; therefore, we shall employ (4.46) to find its absolute velocity. With this objective in mind, we now determine the required terms.

The position vector of P in the motor frame 1 is given by

$$\mathbf{x} = a(\cos\phi\,\mathbf{j} + \sin\phi\,\mathbf{k}). \tag{4.81a}$$

Therefore, the relative velocity of P in frame 1 is

$$\frac{\delta\mathbf{x}}{\delta t} = a\omega_2(-\sin\phi\,\mathbf{j} + \cos\phi\,\mathbf{k}), \tag{4.81b}$$

in which $\omega_2 = \dot{\phi}$ is the constant angular speed of the blade in the motor frame.

The total angular velocity of the motor frame relative to the ground frame $0 = \{F; \mathbf{I}_k\}$ is given as

$$\boldsymbol{\omega}_f \equiv \boldsymbol{\omega}_{10} = \boldsymbol{\omega}_1 = \omega_1\mathbf{k}. \tag{4.81c}$$

Hence, with the aid of (4.81a), we obtain

$$\boldsymbol{\omega}_f \times \mathbf{x} = -a\omega_1 \cos\phi\,\mathbf{i}. \tag{4.81d}$$

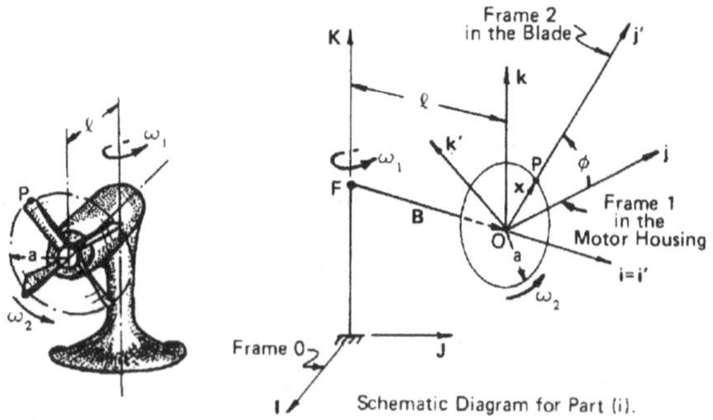

Figure 4.26. Motion of a particle at the tip of an oscillating fan blade.

It remains to find the velocity of point O referred to frame 1. The vector of O from F is $\mathbf{B}=l\mathbf{i}$, referred to frame 1. Since O is a point of a rigid body having the angular velocity $\boldsymbol{\omega}_1$ about a fixed base point at F, its velocity may be found by use of (2.27). Thus,

$$\mathbf{v}_O=\mathbf{v}_F+\boldsymbol{\omega}_1\times\mathbf{B}=\omega_1 l\mathbf{j}. \tag{4.81e}$$

Notice that the same result may be gotten by differentiation of $\mathbf{B}$ in frame 1; namely, $\mathbf{v}_O=\dot{\mathbf{B}}=l\boldsymbol{\omega}_f\times\mathbf{i}=\omega_1 l\mathbf{j}$.

Substitution of (4.81b), (4.81d), and (4.81e) into (4.46) yields the solution. Thus, the absolute velocity of P referred to the motor frame 1 is

$$\mathbf{v}_P=-a\omega_1\cos\phi\,\mathbf{i}+(\omega_1 l-a\omega_2\sin\phi)\,\mathbf{j}+a\omega_2\cos\phi\,\mathbf{k}, \tag{4.81f}$$

wherein we recall that $\omega_1=\omega\cos pt$.

Solution of (ii). This problem requires that the motion of P be referred to the blade frame $2=\{O;\mathbf{i}'_k\}$ shown in Fig. 4.26. Since there is no motion of P relative to the blade frame, (4.46) shows that the velocity is determined by the rigid body formula (2.27):

$$\mathbf{v}_P=\mathbf{v}_O+\boldsymbol{\omega}_f\times\mathbf{x}. \tag{4.82a}$$

The total angular velocity of the blade is given by

$$\begin{aligned}\boldsymbol{\omega}_f\equiv\boldsymbol{\omega}_{20}&=\boldsymbol{\omega}_{21}+\boldsymbol{\omega}_{10}=\omega_2\mathbf{i}'+\omega_1\mathbf{K}\\&=\omega_2\mathbf{i}'+\omega_1(\sin\phi\,\mathbf{j}'+\cos\phi\,\mathbf{k}').\end{aligned} \tag{4.82b}$$

Hence, using the position vector $\mathbf{x}=a\mathbf{j}'$ in frame 2, we get

$$\boldsymbol{\omega}_f\times\mathbf{x}=a\omega_2\mathbf{k}'-a\omega_1\cos\phi\,\mathbf{i}'. \tag{4.82c}$$

The velocity of the base point O is found from the first equation in (4.81e) in which $\mathbf{B}=l\mathbf{i}'$. Thus, referred to the blade frame 2, we have

$$\mathbf{v}_O=l\boldsymbol{\omega}_1\times\mathbf{i}'=l\omega_1(\cos\phi\,\mathbf{j}'-\sin\phi\,\mathbf{k}'). \tag{4.82d}$$

Substitution of (4.82c) and (4.82d) into (4.82a) yields the absolute velocity of P referred to the blade frame 2:

$$\mathbf{v}_P=-a\omega_1\cos\phi\,\mathbf{i}'+l\omega_1\cos\phi\,\mathbf{j}'+(a\omega_2-l\omega_1\sin\phi)\,\mathbf{k}'. \tag{4.82e}$$

Note that herein $\omega_1=\omega\cos pt$.

This completes the primary problem solutions. We now consider the question of the total angular acceleration of the blade for the previous cases.

The total angular acceleration of the blade referred to the motor frame 1 may be found by application of (4.39a). The angular velocity of frame 2 relative to frame 1, namely, $\boldsymbol{\omega}_{21} = \omega_2 \mathbf{i}$, is constant in the motor frame 1, hence $\delta\boldsymbol{\omega}_{21}/\delta t = \mathbf{0}$. Of course, $\boldsymbol{\omega}_{10} = \omega_1 \mathbf{K} = \omega \cos pt\, \mathbf{K}$ in frame 0; therefore, $\delta\boldsymbol{\omega}_{10}/\delta t = -\omega p \sin pt\, \mathbf{K}$. Finally, the convective term is $\boldsymbol{\omega}_{10} \times \boldsymbol{\omega}_{21} = \omega_1 \omega_2 \mathbf{j}$. Thus. use of (4.39a) yields the total angular acceleration of the fan blade referred to frame 1:

$$\dot{\boldsymbol{\omega}}_{20} = \omega\omega_2 \cos pt\, \mathbf{j} - \omega p \sin pt\, \mathbf{k}. \tag{4.83a}$$

The same result, but now referred to the blade frame 2, may be obtained from (4.83a) by a change of basis; however, since the last expression in (4.82b) gives $\boldsymbol{\omega}_{20}$ already expressed in terms of the basis of the blade frame, the total angular acceleration of the fan blade referred to that frame also may be derived directly from the last equation in (4.82b) by application of (4.14). In either case, we shall find

$$\begin{aligned}\dot{\boldsymbol{\omega}}_{20} &= \omega(-p \sin pt \sin \phi + \omega_2 \cos pt)\, \mathbf{j}' \\ &\quad - \omega(p \sin pt \cos \phi + \omega_2 \cos pt \sin \phi)\, \mathbf{k}'. \quad \square\end{aligned} \tag{4.83b}$$

Example 4.18. *Application to the Motion of an Epicyclic Gear Train.* An epicyclic gear train is a system of gears and shafting in which at least one gear moves around the circumference of other fixed or moving gears. The planet gear P_2 shown in Fig. 4.12 rolls around the circumference of gear P_1, and so the planetary train studied earlier is an epicyclic gear train. An epicyclic arrangement permits an unusual assembly of gears and shafting and it sometimes provides an uncommon angular velocity ratio while maintaining relative simplicity of the design.

The motion of the epicyclic gear train shown in Fig. 4.27 will be studied here. The train consists of three bevel gears A, C, and D and two spur gears B and E arranged so that the shaft of B is concentric with the output shaft of gear D. The gear A is fixed in the machine frame $0 = \{F; \mathbf{I}_k\}$. The spur gear E is fixed to the input shaft and drives the train through the power gear B which is cut from a special casting that serves as the bearing housing for the epicyclic gear C and as a bearing for the power gear assembly consisting of B and C. The power gear assembly revolves around the concentric supporting output shaft with a specified total angular velocity $\boldsymbol{\omega}_{10} = \omega \mathbf{I}$ in the machine frame, as indicated in Fig. 4.27. We wish to determine by vector methods (a) the angular velocity of the epicyclic gear C relative to the power gear B, (b) the absolute angular velocity and angular acceleration of C referred to frame $1 = \{B; \mathbf{i}_k\}$ fixed in B, and (c) the angular speed of the output gear D. The design, as shown in Fig. 4.27, requires that the pitch angles ϕ and θ satisfy

$$\phi + 2\theta = \pi/2. \tag{4.84}$$

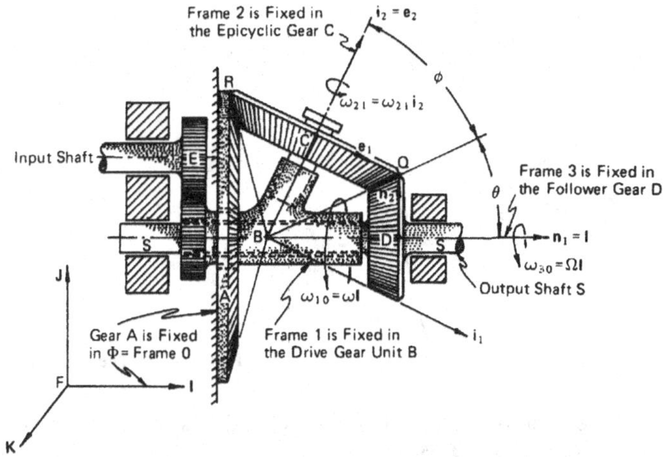

Figure 4.27. An epicyclic gear train assembly.

Solution: *Part* (*a*). Let frame $3 = \{D; \mathbf{n}_k\}$ be fixed in the output gear D whose absolute angular velocity is denoted by $\boldsymbol{\omega}_{30} = \Omega \mathbf{n}_1 = \Omega \mathbf{I}$; and let frame $2 = \{C; \mathbf{e}_k\}$ be fixed in the epicyclic gear C, as shown in Fig. 4.27. Then the angular velocity of the epicyclic frame 2 relative to the power gear frame 1 may be written as $\boldsymbol{\omega}_{21} = \omega_{21}\mathbf{e}_2 = \omega_{21}\mathbf{i}_2$, and the total angular velocity of C in the machine will be given by

$$\boldsymbol{\omega}_{20} = \boldsymbol{\omega}_{21} + \boldsymbol{\omega}_{10}. \tag{4.85a}$$

In addition, the three absolute angular speeds of gears B, C, and D are related through the rolling constraint imposed by the gear teeth. Of course, the point Q on gear C is turning relative to the rotating frame 1 in gear B. However, since the reference frames are fixed in the gears, their material points have no motion relative to these frames. Hence, the basic equations (4.46) and (4.48) reduce for each gear to the rigid body formulas (2.27) and (2.30) referred to frames whose angular velocities in the machine frame are $\boldsymbol{\omega}_{10}$, $\boldsymbol{\omega}_{20}$, and $\boldsymbol{\omega}_{30}$, as noted before. Thus, for the point R on the fixed gear A, the contact point at R on C must satisfy

$$\mathbf{v}_R = \mathbf{v}_C + \boldsymbol{\omega}_{20} \times \mathbf{r} = \mathbf{0}, \tag{4.85b}$$

in which $\mathbf{r}$ is the position vector of R from C, as shown in Fig. 4.28. Of course, point C belongs also to the extension of the power gear B for which $\mathbf{v}_B = \mathbf{0}$; hence, with $\mathbf{r} = -\mathbf{q}$, as shown, we find with (4.85b) the absolute velocity of point C:

$$\mathbf{v}_C = \boldsymbol{\omega}_{20} \times \mathbf{q} = \boldsymbol{\omega}_{10} \times \mathbf{b}, \tag{4.85c}$$

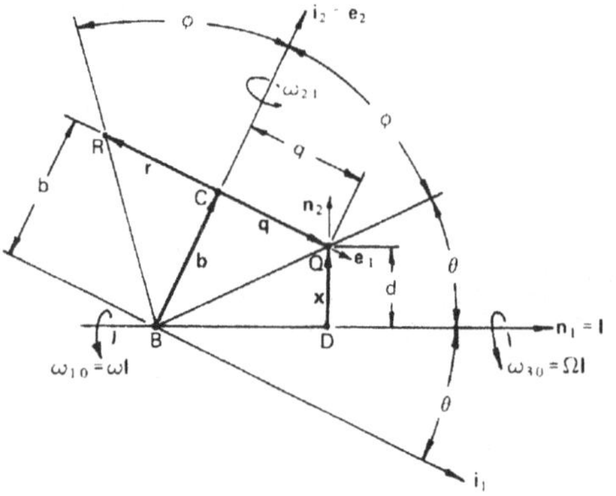

Figure 4.28. The gear train geometry.

in which $\mathbf{q}$ is the position vector of Q from C and $\mathbf{b}$ is the position vector of C from B. Use of (4.85a) in (4.85c) yields

$$\boldsymbol{\omega}_{21} \times \mathbf{q} = \boldsymbol{\omega}_{10} \times (\mathbf{b} - \mathbf{q}). \tag{4.85d}$$

Using the geometry of Fig. 4.28 and the rule (4.84), we find from (4.85d) the relative angular velocity component

$$\omega_{21} = -\omega[\sin\theta + (b/q)\sin(\phi + \theta)] = -\omega(\sin\theta + \cot\phi\cos\theta).$$

But the last term may be written as the ratio $\cos(\phi - \theta)/\sin\phi$; hence, with the use of (4.84), the angular velocity of the epicyclic gear C relative to the power gear B may be expressed as

$$\boldsymbol{\omega}_{21} = \omega_{21}\mathbf{i}_2 = -\omega\frac{\sin 3\theta}{\cos 2\theta}\mathbf{i}_2. \tag{4.85e}$$

Part (b). The total angular velocity of the epicyclic gear in the machine frame may be derived by substitution of (4.85e) into (4.85a). We use an elementary trigonometric identity and thus find

$$\boldsymbol{\omega}_{20} = \omega\cos\theta(\mathbf{i}_1 - \tan 2\theta\,\mathbf{i}_2), \tag{4.86a}$$

referred to the power gear frame 1.

The absolute angular acceleration of C in the machine frame may be derived by use of (4.39a). Equation (4.85e) shows that $\boldsymbol{\omega}_{21}$ is constant in frame 1; and $\boldsymbol{\omega}_{10}$ is constant in frame 0. Hence, (4.39a) reduces to $\dot{\boldsymbol{\omega}}_{20} = \boldsymbol{\omega}_{10} \times \boldsymbol{\omega}_{21} = \omega\omega_{21}\mathbf{n}_1 \times \mathbf{i}_2$; and with the aid of (4.85e), we obtain the angular acceleration of the epicyclic gear in the machine frame, but referred to the power gear frame 1:

$$\dot{\boldsymbol{\omega}}_{20} = -\omega^2\frac{\sin 3\theta\cos\theta}{\cos 2\theta}\mathbf{i}_3. \tag{4.86b}$$

The reader also may confirm this result by differentiation of (4.86a).

Part (c). It remains to determine the angular velocity of the output gear D. Since the rolling contact point Q belongs to both C and D, and $\mathbf{v}_D = \mathbf{0}$, we have

$$\mathbf{v}_Q = \boldsymbol{\omega}_{30} \times \mathbf{x} = \mathbf{v}_C + \boldsymbol{\omega}_{20} \times \mathbf{q}, \tag{4.87a}$$

wherein $\mathbf{x}$ is the position vector of Q from D, as shown in Fig. 4.28. Therefore, use of (4.85c) in (4.87a) yields

$$\boldsymbol{\omega}_{30} \times \mathbf{x} = 2\boldsymbol{\omega}_{20} \times \mathbf{q} = 2\boldsymbol{\omega}_{10} \times \mathbf{b}. \tag{4.87b}$$

It seen from Fig. 4.28 that (4.87b) yields the component relation

$$\Omega = 2\omega(b/d)\sin(\phi + \theta). \tag{4.87c}$$

Further, with the aid of the rule (4.84) and the triangle relations $b/c = \cot\phi$ and $c/\sin\phi = d/\sin\theta$, it follows from (4.87c) that the angular speed of the output gear D is

$$\Omega = 4\omega\cos^2\theta. \tag{4.87d}$$

This completes the solution of the problem. It may be seen from (4.84) that $0 < \theta < \pi/4$. It thus follows from (4.87d) that the ratio of the output angular speed to the input angular speed decreases with increasing θ so that $\Omega/\omega \in (4, 2)$; and according to (4.85e), the ratio of the relative angular speed of the epicyclic gear to the input angular speed grows with θ so that $\omega_{12}/\omega \in (0, \infty)$. Specifically, for $\theta = \phi = 30°$, it may be seen that $\Omega/\omega = 3$ and $\omega_{12}/\omega = 2$. It is an exercise for the student to plot polar graphs of these speed ratios as functions of the pitch angle θ. □

Example 4.19. *Motion of a Rolling Disk.* A thin, circular disk D of radius r, shown in Fig. 4.29, rolls without slipping on a simple, smooth path P in the

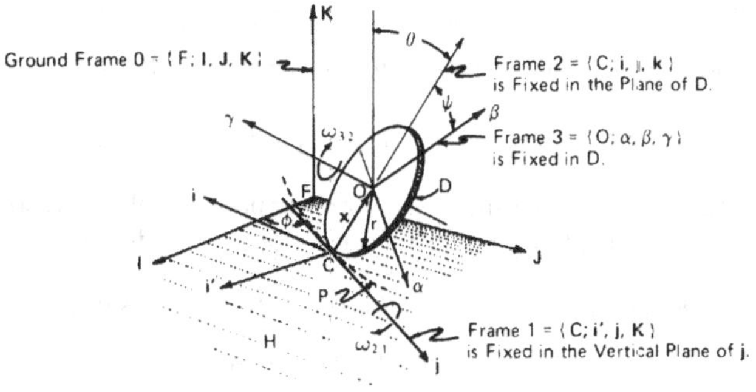

Figure 4.29. A thin disk rolling on a horizontal surface.

horizontal plane H of the ground frame $0 = \{F; \mathbf{I}_k\}$. The angle ψ describes the rotation of the disk about its axle $\boldsymbol{\gamma}$ at O; the tilt of the plane of the disk from the vertical axis $\mathbf{K}$ is denoted by θ; and the rotation of the tangent vector $\mathbf{j}$ to the path P is measured by the angle ϕ. Of course, the motion of the disk depends upon the forces and torques that act on it; and it will be learned further on that these are related by basic dynamical laws to the acceleration of the center of the disk and to its total angular velocity and angular acceleration. Find the total angular velocity and angular acceleration of the disk referred to frame $2 = \{C; \mathbf{i}, \mathbf{j}, \mathbf{k}\}$ defined by the plane of the disk through the tangent line of the point C of rolling contact, but not fixed in the disk. Determine the absolute acceleration of the center of D referred to frame 2.

Solution. Let frame $1 = \{C; \mathbf{i}', \mathbf{j}, \mathbf{K}\}$ be fixed in the vertical plane through the tangent line to P at C, and let frame $3 = \{O; \boldsymbol{\alpha}, \boldsymbol{\beta}, \boldsymbol{\gamma}\}$ be fixed in the disk. Then for the assigned frames, $\boldsymbol{\omega}_{32} = -\dot{\psi}\boldsymbol{\gamma}$ is the angular velocity of the disk frame 3 relative to the tilted frame 2; $\boldsymbol{\omega}_{21} = -\dot{\theta}\mathbf{j}$ denotes the angular velocity of the tilted frame in the plane of the disk relative to frame 1; and $\boldsymbol{\omega}_{10} = \dot{\phi}\mathbf{K}$ is the angular velocity of the vertical tangent plane relative to the ground frame 0. Thus, in accordance with (4.25b), the total angular velocity of the disk is given by

$$\boldsymbol{\omega}_{30} = \boldsymbol{\omega}_{32} + \boldsymbol{\omega}_{21} + \boldsymbol{\omega}_{10} = -\dot{\psi}\boldsymbol{\gamma} - \dot{\theta}\mathbf{j} + \dot{\phi}\mathbf{K}. \tag{4.88a}$$

Since $\boldsymbol{\gamma} = \mathbf{i}$ and $\mathbf{K} = \sin\theta\,\mathbf{i} + \cos\theta\,\mathbf{k}$, the total angular velocity (4.88a) referred to the tilted frame 2 is

$$\boldsymbol{\omega}_{30} = (\dot{\phi}\sin\theta - \dot{\psi})\,\mathbf{i} - \dot{\theta}\mathbf{j} + \dot{\phi}\cos\theta\,\mathbf{k}. \tag{4.88b}$$

The total angular acceleration of the disk may be determined by use of (4.36). We leave this for the reader, and here derive the result from the basic equation (4.11). We note that $\mathbf{U} = \boldsymbol{\omega}_{30}$ is a vector referred to the moving frame 2 whose total angular velocity is $\boldsymbol{\omega}_f = \boldsymbol{\omega}_{20} = -\dot{\theta}\mathbf{j} + \dot{\phi}\mathbf{K}$. This method yields from (4.88b)

$$\begin{gathered}\frac{\delta\boldsymbol{\omega}_{30}}{\delta t} = (\ddot{\phi}\sin\theta + \dot{\phi}\dot{\theta}\cos\theta - \ddot{\psi})\,\mathbf{i} - \ddot{\theta}\mathbf{j} + (\ddot{\phi}\cos\theta - \dot{\phi}\dot{\theta}\sin\theta)\,\mathbf{k},\\ \boldsymbol{\omega}_{20}\times\boldsymbol{\omega}_{30} = \boldsymbol{\omega}_{20}\times\boldsymbol{\omega}_{32} = [(\dot{\phi}\sin\theta - \dot{\psi})\,\mathbf{i} - \dot{\theta}\mathbf{j} + \dot{\phi}\cos\theta\,\mathbf{k}]\times(-\dot{\psi}\mathbf{i}).\end{gathered} \tag{4.88c}$$

Thus, substitution of (4.88c) into (4.11) and expansion of the vector product yields the required absolute angular acceleration of the disk:

$$\begin{aligned}\dot{\boldsymbol{\omega}}_{30} = {}&(\ddot{\phi}\sin\theta + \dot{\phi}\dot{\theta}\cos\theta - \ddot{\psi})\,\mathbf{i} - (\ddot{\theta} + \dot{\phi}\dot{\psi}\cos\theta)\,\mathbf{j}\\ &+ [\ddot{\phi}\cos\theta - \dot{\theta}(\dot{\psi} + \dot{\phi}\sin\theta)]\,\mathbf{k}.\end{aligned} \tag{4.88d}$$

To determine the absolute acceleration of the center of the disk, it is convenient to first find the absolute velocity of O referred to frame 2. Since the contact point C has zero velocity in frame 0, the velocity of O according to (2.27) may be found from $\mathbf{v}_O = \boldsymbol{\omega}_{30} \times \mathbf{x}$ in which $\mathbf{x} = r\mathbf{k}$. Hence, use of (4.88b) yields the absolute velocity of O referred to frame 2:

$$\mathbf{v}_O = -r[\dot{\theta}\mathbf{i} + (\dot{\phi} \sin\theta - \dot{\psi})\,\mathbf{j}]. \tag{4.88e}$$

We recall the basic rule (4.11) in which $\mathbf{U} = \mathbf{v}_O$ and $\boldsymbol{\omega}_f = \boldsymbol{\omega}_{20}$ is the angular velocity of the frame to which $\mathbf{v}_O$ is referred in (4.88e). The reader may show by this method that the absolute acceleration of the center of the disk referred to frame 2 is given by

$$\begin{aligned}\mathbf{a}_O = &-r[\ddot{\theta} + \dot{\phi}\cos\theta(\dot{\psi} - \dot{\phi}\sin\theta)]\,\mathbf{i} - r(\ddot{\phi}\sin\theta + 2\dot{\theta}\dot{\phi}\cos\theta - \ddot{\psi})\,\mathbf{j}\\ &-r[\dot{\theta}^2 + \dot{\phi}\sin\theta(\dot{\phi}\sin\theta - \dot{\psi})]\,\mathbf{k}.\end{aligned} \tag{4.88f}$$

4.10. Motion Referred to an Earth Frame

An especially important example of motion relative to a moving frame is the motion of a particle referred to a reference frame fixed in the earth. Our daily experience suggests that so far as the observable motions of cams, gears, tops, cars, and trains are concerned, the earth's motion is of little apparent consequence. But unnoticeably small effects may accumulate. One may find after a time, for example, that contacting surfaces, like train wheels and tracks, suffer unusual wear induced by the earth's rotation. Other effects may be more evident. The long flight times at large speeds over great distances traveled by supersonic aircraft, ballistic missiles, earth satellites, planetary vehicles, and heavenly bodies exemplify increasingly the importance of our accounting for the earth's motion in tracking the motions of such bodies.

Between these extremes, other subtle and powerful effects caused by the motion of the earth are observable. The circulation of the winds and weather variation are vivid and sometimes awesome examples. The motion of a Foucault pendulum, a device contrived in 1851 by the French physicist Jean L. Foucault (1819–1868), exhibits a subtle phenomenon induced by the rotation of the earth. To an observer in the northern hemisphere, the plane of oscillation of the pendulum appears to rotate clockwise once each day. However, the pendulum actually swings continuously in the same plane while the earth rotates counterclockwise beneath it. In the southern hemisphere, the relative pendulum rotation is counterclockwise. This phenomenon is due to the Coriolis acceleration produced by the earth's rotation as the pendulum swings back and forth. It has been reported also that during the famous battle of the Falkland Islands in World War I, the British naval gunners were surprised to see their shells fall consistently 100 or so yards forward or aft of the German ships. Although the gun sight design accounted for the Coriolis deflection of a shell, apparently the design was based on the assumption that combat most likely would occur near 50°N latitude, never near 50°S latitude. Consequently, the initial British barrage missed its mark by a distance equal to nearly twice the Coriolis deflection!

These few examples show that although it is quite natural to choose a reference frame fixed in the earth to study the motion of material objects on or near its surface, one must bear in mind that the earth itself has a complex motion compounded of many rotations. And it is important to know the extent to which this motion affects velocity and acceleration terms, and to know what error may be committed when the earth's motion is neglected. Therefore, we shall examine in this section expressions for the apparent velocity and acceleration when the earth's motion is taken into account. But what shall we use for the earth's motion?

4.10.1. Motion of the Earth

The principal motions of the earth are its axial rotation and its revolution around the sun. However, these are not the only motions experienced by the earth. The earth's rotational axis, called the celestial pole, differs slightly from the geometric pole of symmetry. Astronomical measurements based on dynamical theory have determined that the geometric pole traces around the celestial pole a cone whose base diameter at the north pole is about 25 ft. This nutational disturbance, known as the Eulerian motion, has an observed period of 428 days, but it has a negligible effect on the earth's principal motions. Therefore, as usual, we shall ignore it.

The earth's axis also rotates about a line through its center and perpendicular to the plane of the ecliptic, the earth's orbital plane of motion. This precessional motion, a phenomenon similar to that observed in an ordinary spinning top, is extremely slow, with a period of roughly 26,000 years. In addition, the inclination of the earth's axis to the plane of the ecliptic changes very slightly; the period of this nutational disturbance is about 40,000 years. Further, the plane of the ecliptic rotates slowly around the sun, and the earth's orbital path also is subject to periodic changes. The period of orbital precession around the sun varies from 60,000 to 120,000 years, while the eccentricity of the earth's orbit, according to celestial mechanics, is now two-and-a-half times smaller than it was 180,000 years ago. It is clear that such small perturbations do not contribute significantly to the effects of the principal motions of the earth even over a 100-year period, which by comparison with the aforementioned figures is an extremely short time. Therefore, we shall omit these.

But there is more. The milky way is rotating about its central axis and moving through space relative to other galaxies in the universe; and the earth is being dragged along with it. Although the revolutionary period of the milky way is approximately 200 million years, the sun, being 30,000 light years from its center, has an orbital speed of about 140 miles/sec relative to the galaxy axis; and the orbital speed of the earth around the sun is 18.5 miles/sec, that is, 66,000 mph. To trace the motion of a body over interstellar distances, it is not inconceivable that the would have to account for some of these additional motions of the earth. However, even the interplanetary distances that cover the solar system are infinitesimal by comparison with interstellar distances, so it is reasonable to ignore any variations in the relative motions of the planets caused by galactic motion. Therefore, for all practical purposes, it is sensible to restrict attention to the effects of the earth's principal motions only, as described next.

Let $\varphi' = \{C; \mathbf{i}'_k\}$ be a frame imbedded in the earth E at any point C, as shown in Fig. 4.30. Let $\Phi = \{S; \mathbf{I}_k\}$ be another frame with origin at the center of the sun S but having no rotation with respect to the distant, so-called "fixed stars." Then the total rotation of the earth frame φ' relative to the preferred, astronomical frame Φ is equal to the sum of the earth's principal rotations $\omega_1 = 2\pi$ rad/day about the earth's polar axis and $\omega_2 = \omega_1 \div 365.25$ rad/day about an axis at S normal to the plane of the ecliptic. We thus write $\boldsymbol{\omega}_f \equiv \boldsymbol{\Omega} = \boldsymbol{\omega}_1 + \boldsymbol{\omega}_2$ for the total angular velocity of the earth frame relative to the distant stars. The angle between $\boldsymbol{\omega}_1$ and $\boldsymbol{\omega}_2$ is the tilt angle of the earth's axis, which is very nearly 23.5° from a line perpendicular to the ecliptic. Consequently, a comparison of the magnitudes ω_1 and ω_2 in a vector diagram shows that the total angular velocity vector $\boldsymbol{\Omega}$ is directed very nearly along the earth's axis of rotation and has a magnitude equal very closely to the sum of the magnitudes $\omega_1 + \omega_2 = 7.29 \times 10^{-5}$ rad/sec. (See

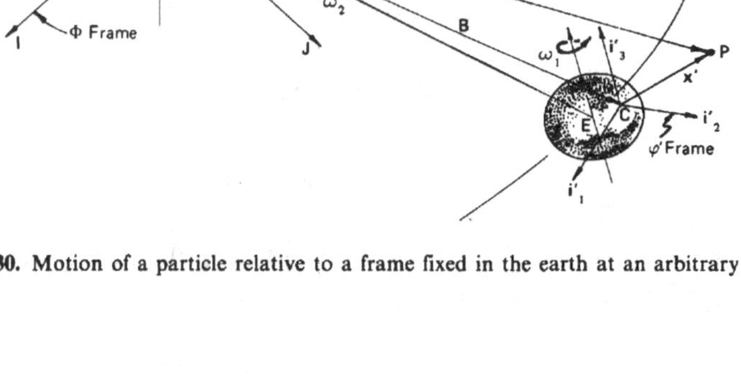

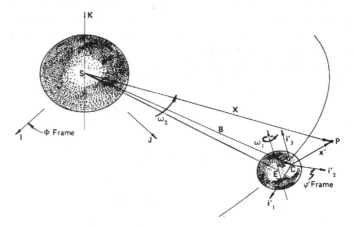

Figure 4.30. Motion of a particle relative to a frame fixed in the earth at an arbitrary point C.

Problem 4.114.) Moreover, since both $\boldsymbol{\omega}_1$ and $\boldsymbol{\omega}_2$ are virtually constant, the total angular acceleration of the earth frame relative to Φ is due mainly to the change in the direction of the earth's axis of rotation as the earth revolves about the sun. Hence, the angular acceleration has the closely approximated magnitude $|\dot{\boldsymbol{\Omega}}| = 5.77 \times 10^{-12}$ rad/sec^2. Therefore, we may conclude that to a high order of accuracy the total angular velocity and angular acceleration of the earth frame may be closely estimated by

$$\boldsymbol{\Omega} = \Omega \mathbf{n} = 7.29 \times 10^{-5} \mathbf{n} \text{ rad/sec}, \qquad \dot{\boldsymbol{\Omega}} = \mathbf{0}, \tag{4.89}$$

wherein $\mathbf{n}$ is a unit vector in the north pole direction.

4.10.2. Velocity and Acceleration Relative to the Earth

For the conditions set in (4.89), the velocity and acceleration of a particle relative to the earth frame may be easily derived. We consider a particle P moving in space with position vector $\mathbf{x}'$ from C in φ' and $\mathbf{X}$ from S in Φ, as shown in Fig. 4.30. Then by use of (4.89) in (4.46) and (4.48), we obtain the relative velocity and acceleration of P in φ':

$$\begin{aligned} \mathbf{v} &\equiv \frac{\delta \mathbf{x}'}{\delta t} = \mathbf{v}_P - (\mathbf{v}_C + \boldsymbol{\Omega} \times \mathbf{x}'), \\ \mathbf{a} &\equiv \frac{\delta^2 \mathbf{x}'}{\delta t^2} = \mathbf{a}_P - [\mathbf{a}_C + \boldsymbol{\Omega} \times (\boldsymbol{\Omega} \times \mathbf{x}') + 2\boldsymbol{\Omega} \times \mathbf{v}], \end{aligned} \tag{4.90}$$

wherein $\mathbf{v}_C$ and $\mathbf{a}_C$, and $\mathbf{v}_P$ and $\mathbf{a}_P$ denote the total velocity and acceleration in frame Φ of the origin point C in E and of the particle P, respectively.

It is convenient in (4.90) to take C at the geometrical center of the earth, but still be able to refer the motion to a reference frame φ fixed at a convenient site O in the earth's surface (or any other point in E), as shown in Fig. 4.31. Since the vector $\mathbf{r}$ of O from C is fixed in the earth, hence also in both φ and φ', we have $\delta \mathbf{r}/\delta t = \mathbf{0}$ and $\delta^2 \mathbf{r}/\delta t^2 = \mathbf{0}$. As a consequence, a shift to a reference frame φ at O in no way affects $\mathbf{v}$ and $\mathbf{a}$ in (4.90). Indeed, with $\mathbf{x}' = \mathbf{r} + \mathbf{x}$, where $\mathbf{x}$ is the vector of P from O in φ, as shown in Fig. 4.31, we see that $\delta \mathbf{x}'/\delta t = \delta \mathbf{x}/\delta t$ and $\delta^2 \mathbf{x}'/\delta t^2 = \delta^2 \mathbf{x}/\delta t^2$. Therefore, (4.90) also may be written as

$$\begin{aligned} \mathbf{v} &= \frac{\delta \mathbf{x}}{\delta t} = \mathbf{v}_P - [\mathbf{v}_C + \boldsymbol{\Omega} \times (\mathbf{r} + \mathbf{x})], \\ \mathbf{a} &= \frac{\delta^2 \mathbf{x}}{\delta t^2} = \mathbf{a}_P - \{\mathbf{a}_C + \boldsymbol{\Omega} \times [\boldsymbol{\Omega} \times (\mathbf{r} + \mathbf{x})] + 2\boldsymbol{\Omega} \times \mathbf{v}\}. \end{aligned} \tag{4.91}$$

When P moves on or near the earth's surface in the vicinity of O so that

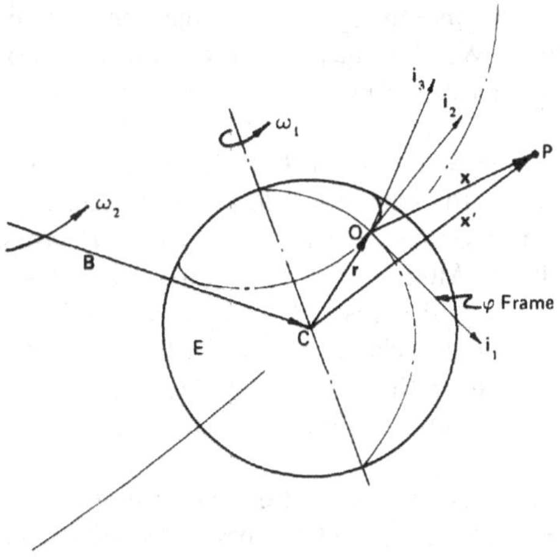

Figure 4.31. A particle motion relative to a frame fixed in the earth's surface.

$|\mathbf{x}| \ll |\mathbf{r}| = 4000$ miles, approximately, then $\mathbf{x}' = \mathbf{r}$, very nearly, and equations (4.90) or (4.91) may be replaced by

$$\mathbf{v} = \frac{\delta \mathbf{x}}{\delta t} = \mathbf{v}_P - (\mathbf{v}_C + \boldsymbol{\Omega} \times \mathbf{r}),$$
$$\mathbf{a} = \frac{\delta^2 \mathbf{x}}{\delta t^2} = \mathbf{a}_P - [\mathbf{a}_C + \boldsymbol{\Omega} \times (\boldsymbol{\Omega} \times \mathbf{r}) + 2\boldsymbol{\Omega} \times \mathbf{v}]. \tag{4.92}$$

In equations (4.90)–(4.92), the terms preceded by the negative sign represent in each case the principal source of errors that would be committed by adopting the assumption that the earth is fixed in space relative to the distant stars, i.e., by assuming that the relative and the absolute velocities and accelerations are the same. In particular, we shall prove in Chapter 6 that the Coriolis acceleration $2\boldsymbol{\Omega} \times \mathbf{v}$ accounts for the slow apparent rotation of the plane of oscillation of the Foucault pendulum and for the deflection of the British artillery shells mentioned earlier. Hence, if this term is neglected, potentially serious errors may result and opportunity for the discovery of other subtle effects may be lost. For most ordinary problems, however, the deviation in the motion of the particle, which is not predicted when the less realistic idealization of the earth as a nonrotating body is used, is of little consequence because of the smallness of $|\boldsymbol{\Omega}|$. Indeed, the maximum intensity of the Coriolis acceleration is estimated as $2\Omega v = 1.5 \times 10^{-4} v \text{ sec}^{-1}$. Although the Coriolis acceleration usually is minute, its effects are difficult to predict without a careful analysis, and for certain effects, it may not be negligible. In particular, it is clear that for bodies such as ballistic missiles and earth satellites, for which the apparent speed v is large, the Coriolis correction takes on greater importance. We shall have more to say later about effects induced by the principal motions of the earth.

Example 4.20. A river flows south with uniform speed v relative to the earth. We shall describe the absolute velocity and acceleration of a river particle P relative to the center of the earth when P crosses northern and southern latitude lines. Afterwards, the results will be evaluated at 30°N latitude, and compared with those of a satellite traveling similarly with an orbital speed of 17,000 mph relative to the earth at an altitude of 100 miles. Finally, the contribution to the total acceleration due to the earth's motion around the sun will be estimated. See Fig. 4.32.

Solution. The earth being nearly spherical, it is natural to use a spherical reference frame $\psi = \{C; \mathbf{e}_r, \mathbf{e}_\theta, \mathbf{e}_\phi\}$ at the center of the earth. Then, in general terms, r is constant, $\dot{r}=0$, $\ddot{r}=0$; $\dot{\phi}=\Omega$ is the total angular speed of the earth frame; and $v=r\dot{\theta}$ is the speed of the river particle relative to ψ. Therefore, $\ddot{\phi}=\ddot{\theta}=0$ also. Direct use of (4.70) and (4.71) yields the absolute velocity and acceleration of P relative to the center C, that is.

$$\begin{aligned} \mathbf{v}_P-\mathbf{v}_C &= r\Omega \sin\theta\, \mathbf{e}_\phi + v\mathbf{e}_\theta, \\ \mathbf{a}_P-\mathbf{a}_C &= -(r\Omega^2 \sin^2\theta + v^2/r)\, \mathbf{e}_r + 2\Omega v \cos\theta\, \mathbf{e}_\phi \\ &\quad - r\Omega^2 \sin\theta \cos\theta\, \mathbf{e}_\theta. \end{aligned} \tag{4.93a}$$

The component $2\Omega v \cos\theta$ is the Coriolis acceleration, and v^2/r is the acceleration of P relative to the earth and arises from the earth's curvature. The remaining centripetal acceleration terms have the magnitude $r\Omega^2 \sin\theta$. It will be shown in the next chapter that the centripetal acceleration affects only the apparent weight of the particle.

The foregoing description of the terms in (4.93a) is made more apparent

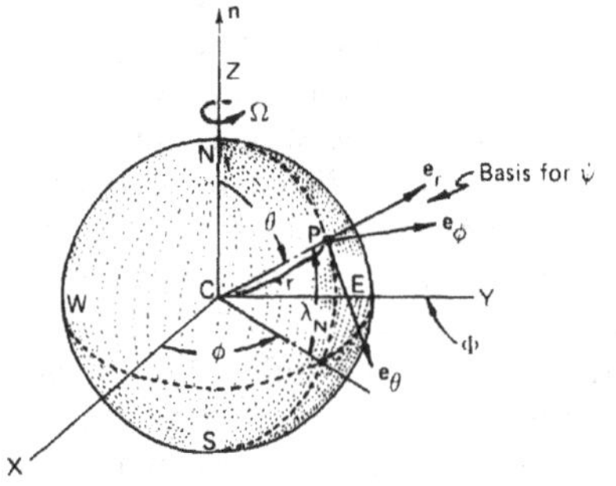

Figure 4.32. A fluid particle flowing southward on the earth.

by its derivation from (4.90). The relative velocity and acceleration of P are given by

$$\mathbf{v} = \frac{\delta \mathbf{x}'}{\delta t} = v\mathbf{e}_\theta, \qquad \mathbf{a} = \frac{\delta^2 \mathbf{x}'}{\delta t^2} = -\kappa \dot{s}^2 \mathbf{e}_r = -\frac{v^2}{r}\mathbf{e}_r, \tag{4.93b}$$

in which the effect of the earth's curvature is explicit.

Writing $\mathbf{x}' = r\mathbf{e}_r$ and noting that $\mathbf{\Omega} = \Omega \mathbf{n}$, we obtain $\mathbf{\Omega} \times \mathbf{x}' = r\Omega \sin\theta\, \mathbf{e}_\phi$. We thus find that the centripetal acceleration is directed from P toward the axis of rotation and is given by

$$\mathbf{\Omega} \times (\mathbf{\Omega} \times \mathbf{x}') = -r\Omega^2 \sin\theta(\sin\theta\, \mathbf{e}_r + \cos\theta\, \mathbf{e}_\theta). \tag{4.93c}$$

This shows that the tangential component of (4.93c) is negative in the northern hemisphere where $\sin\theta \cos\theta > 0$ and positive in the southern hemisphere where the inequality is reversed. Hence, the earth's rotation tends to retard the southerly river flow above the equator and enhance it below. This component vanishes at the equator.

And finally, the Coriolis acceleration is

$$2\mathbf{\Omega} \times \mathbf{v} = 2\Omega v \cos\theta\, \mathbf{e}_\phi. \tag{4.93d}$$

Notice that the Coriolis acceleration is directed toward the east in Fig. 4.32. Therefore, in its flow southward, the river particle is deflected eastward in the northern hemisphere where $\cos\theta > 0$ and westward in the southern hemisphere where $\cos\theta < 0$. The Coriolis acceleration vanishes at the equator.

Collecting (4.93b), (4.93c), and (4.93d) into (4.90) yields the same formula (4.93a) found in the foregoing review of spherical coordinates. This closes the present description of the flow.

Let us consider the magnitudes of the terms in (4.93a) when P is at the latitude $\lambda_N = 30°$N so that $\theta = 60°$. Recalling (4.89) and noting that $r = 4000$ miles, $v = 5$ mph, and $\Omega = 0.263$ rad/hr, we find for the indicated units

$$\begin{aligned} \mathbf{v}_P - \mathbf{v}_C &= 910\mathbf{e}_\phi + 5\mathbf{e}_\theta \text{ mph} = 1335\mathbf{e}_\phi + 7.29\mathbf{e}_\theta \text{ ft/sec}, \\ \mathbf{a}_P - \mathbf{a}_C &= -(207 + 0.006)\,\mathbf{e}_r + 1.32\mathbf{e}_\phi - 119.8\mathbf{e}_\theta \text{ m/hr}^2 \\ &= -(0.085 + 2.4 \times 10^{-6})\,\mathbf{e}_r + 5.4 \times 10^{-4}\mathbf{e}_\phi - 0.049\mathbf{e}_\theta \text{ ft/sec}^2. \end{aligned} \tag{4.93e}$$

We thus see in a typical case that the acceleration effects of the earth's rotation are quite small. However, they might not be negligible in certain situations. Of course, as noted earlier, the Coriolis effect is small; however, it is still 100 times larger than the relative acceleration of P resulting from the earth's curvature. Of course, it is evident that such small quantities generally are not noticeable in our daily lives. Since the acceleration is virtually zero, the particle velocity is nearly constant. Notice, however, the huge transport term in the first equation in (4.93e). This term has no effect on the velocity apparent to an earth observer.

On the other hand, for increasingly larger values of the relative speed, it is clear that the relative acceleration and the Coriolis acceleration will be the dominant terms in (4.93a). Let us consider, for example, an earth satellite moving southward with a relative speed $v = 17{,}000$ mph at an altitude of 100 miles above the surface of the earth ($r = 4100$ miles) at the latitude $\lambda_N = 30°$N. Then (4.93a) yields the values

$$\mathbf{v}_P - \mathbf{v}_C = 933.1\mathbf{e}_\phi + 17{,}000\mathbf{e}_\theta \text{ mph} = 1369\mathbf{e}_\phi + 24{,}933\mathbf{e}_\theta \text{ ft/sec},$$

$$\mathbf{a}_P - \mathbf{a}_C = -(212.7 + 7.05 \times 10^4)\,\mathbf{e}_r + 4471\mathbf{e}_\phi - 122.8\mathbf{e}_\theta \text{ m/hr}^2 \qquad (4.93\text{f})$$

$$= -(0.087 + 28.72)\,\mathbf{e}_r + 1.82\mathbf{e}_\phi - 0.05\mathbf{e}_\theta \text{ ft/sec}^2.$$

A significant increase in the second and third terms of (4.93f) compared with those in the second equation in (4.93e) is evident. Notice the small change in the centripetal acceleration, i.e., the first and last terms in (4.93f), compared with the second equation in (4.93e). This is due only to the change in the path radius; and it shows that the approximation in (4.92) for motion on or near the earth's surface will be quite good, the error being roughly 2.3% in 100 miles.

Finally, to estimate the contribution to the total acceleration that is due to the earth's motion about the sun, let us assume that the earth's elliptical orbit is nearly circular with radius $R = 96 \times 10^6$ miles from the center of the sun. Then the orbital speed of C is $\dot{s} = R\omega_2$, very nearly; and because ω_2 is constant, we have $\ddot{s} = 0$. Therefore, the earth's rotation around the sun contributes a centripetal acceleration of magnitude $|\mathbf{a}_C| = (\dot{s}^2/R) = R\omega_2^2$ directed toward the sun. Thus, with the value $\omega_2 = 2 \times 10^{-7}$ rad/sec, this yields the estimate

$$|\mathbf{a}_C| = 49.25 \text{ m/hr}^2 = 0.02 \text{ ft/sec}^2, \qquad (4.93\text{g})$$

which is small though possibly not negligible. This term will be studied further in Chapter 5.

This completes our study of applications of the equations for motion referred to a moving reference frame. The reader should be thoroughly familiar with the principal equations developed and applied throughout this chapter. These important equations are summarized below.

4.11. Summary of the Principal Equations

The main equations developed in this chapter were derived by the application of the fundamental rule (4.11):

$$\dot{\mathbf{U}}(t)=\frac{\delta \mathbf{U}(t)}{\delta t}+\boldsymbol{\omega}_f(t)\times \mathbf{U}(t) \qquad \text{[cf. (4.11)]}.$$

This gives the absolute time rate of change of a vector $\mathbf{U}(t)$ in a preferred frame Φ when $\mathbf{U}(t)$ is referred to a reference frame φ having an angular velocity $\boldsymbol{\omega}_f$ relative to Φ.

Equation (4.11) was rewritten in the form (4.21) and applied to derive the composition rule (4.22) for several angular velocity vectors:

$$\boldsymbol{\omega}_{ik}=\boldsymbol{\omega}_{ij}+\boldsymbol{\omega}_{jk}, \qquad i, j, k=0, 1, 2,\ldots, n \qquad \text{[cf. (4.22)]}.$$

And the simple relative angular velocity rule relating the mutual relative angular velocities of any two frames also was deduced:

$$\boldsymbol{\omega}_{ij}=-\boldsymbol{\omega}_{ji} \qquad \text{[cf. (4.23)]}.$$

The subsequent application of (4.11) in (4.22) and (4.23) led to the corresponding mutual angular acceleration rule for any two frames, namely,

$$\dot{\boldsymbol{\omega}}_{ij}=-\dot{\boldsymbol{\omega}}_{ji} \qquad \text{[cf. (4.32)]},$$

and the composition rule for several angular acceleration vectors:

$$\dot{\boldsymbol{\omega}}_{ik}=\left[\frac{\delta \boldsymbol{\omega}_{ij}}{\delta t}+\boldsymbol{\omega}_{jk}\times\boldsymbol{\omega}_{ij}\right]+\dot{\boldsymbol{\omega}}_{jk} \qquad \text{[cf. (4.37)]}.$$

It was shown that the rules (4.22), (4.23), (4.32), and (4.37) are useful in the solutions of problems that involve the simultaneous rotations of several rigid bodies. The applications commonly involved two or three relatively rotating bodies, and for these special cases (4.22) and (4.37) yield the following kinematic chain and composition rules:

$$\boldsymbol{\omega}_{20}=\boldsymbol{\omega}_{21}+\boldsymbol{\omega}_{10} \qquad \text{[cf. (4.24)]},$$

$$\dot{\boldsymbol{\omega}}_{20}=\frac{\delta \boldsymbol{\omega}_{21}}{\delta t}+\boldsymbol{\omega}_{10}\times\boldsymbol{\omega}_{21}+\frac{\delta \boldsymbol{\omega}_{10}}{\delta t} \qquad \text{[cf. (4.39a)]};$$

$$\boldsymbol{\omega}_{30}=\boldsymbol{\omega}_{32}+\boldsymbol{\omega}_{21}+\boldsymbol{\omega}_{10} \qquad \text{[cf. (4.24) and (4.25a)]},$$

$$\dot{\boldsymbol{\omega}}_{30}=\left[\frac{\delta \boldsymbol{\omega}_{32}}{\delta t}+\boldsymbol{\omega}_{20}\times\boldsymbol{\omega}_{32}\right]+\left[\frac{\delta \boldsymbol{\omega}_{21}}{\delta t}+\boldsymbol{\omega}_{10}\times\boldsymbol{\omega}_{21}\right]+\frac{\delta \boldsymbol{\omega}_{10}}{\delta t}$$

$$\text{[cf. (4.35) and (4.36)]}.$$

Notice that the first pair of these equations is contained in the second pair and may be obtained from the second set upon replacement of 3 by 2 and use of the rule $\boldsymbol{\omega}_{jj} = \mathbf{0}$. Several applications of these equations have been studied.

The fundamental equations (4.46) and (4.48) relating the velocities and accelerations of a particle in two frames in relative motion were derived next. We recall

$$\mathbf{v}_P = \frac{\delta \mathbf{x}}{\delta t} + \mathbf{v}_O + \boldsymbol{\omega}_f \times \mathbf{x} \qquad [\text{cf. } (4.46)],$$

$$\mathbf{a}_P = \frac{\delta^2 \mathbf{x}}{\delta t^2} + \mathbf{a}_O + \boldsymbol{\omega}_f \times (\boldsymbol{\omega}_f \times \mathbf{x}) + \dot{\boldsymbol{\omega}}_f \times \mathbf{x} + 2\boldsymbol{\omega}_f \times \frac{\delta \mathbf{x}}{\delta t} \qquad [\text{cf. } (4.48)].$$

It was learned that these equations contain the basic formulas (2.27) and (2.30) derived in Chapter 2 for the velocity and acceleration of a point of a rigid body, and the rudimentary defining equations (1.8) and (1.10) established in Chapter 1 as the foundation for much of the work that followed. The Coriolis acceleration, the last term in (4.48), arises from the motion of a particle in a rotating reference frame.

The reduction of (4.46) and (4.48) to the important special case of a simple relative motion for which $\boldsymbol{\omega}_f = \mathbf{0}$ was studied. Equations (4.46) and (4.48) were then used to derive the special formulas (4.59), (4.60) and (4.70), (4.71) for use in problems where cylindrical or spherical coordinates arise naturally. All of these special formulas were applied to a variety of problems of relative motion and motion referred to a moving reference frame. The application of (4.46) and (4.48) to the description of motion relative to a reference frame fixed in the earth also was introduced.

The foregoing equations constitute the principal equations studied in this chapter. These will be encountered many times in our future studies of dynamics. Some special topics that require that the reader be familiar with the content of Chapter 3 will be presented next. Therefore, the reader who may have skipped over this material in a first reading may now wish to move ahead to the next chapter. No loss of continuity will be suffered.

4.12. Some Special Advanced Topics

Three special topics will be discussed briefly. The first is an application of the vector component transformation law (3.107) to an earlier example. The derivative of a tensor is introduced next, and the rule for the derivative of the generalized product of two or more tensors is described. The product rule is then applied to determine the time derivative of an orthogonal tensor. This result, which is related to the time rate of change of the rotation tensor studied in the last chapter, will be applied to derive the basic formulas (2.27) and (2.30) for the velocity and acceleration of a rigid body particle. The third topic will present the time derivative of a tensor referred to a moving reference frame. Frame-indifferent transformation laws for vectors and tensors are investigated; and the linear transformation for a change of frame is used to derive in a different manner other forms of the basic velocity and acceleration equations (4.46) and (4.48).

4.12.1. Application of the Vector Component Transformation Law

The reader will recall from Chapter 3 that a change of basis induces a change in the scalar components of a vector. It will be shown in this section that the vector component transformation law (3.107) has application to some of the problems studied earlier in this chapter. As an illustration, we recall the equations (4.78e) for the velocity and acceleration of a point P on the rim of the spinning disk described in Example 4.16 and illustrated in Fig. 4.25. These vectors are referred to a basis $\mathbf{i}_k$ fixed in the disk. We wish to determine the velocity and acceleration of P referred to a basis $\mathbf{i}'_k$ which is fixed in the rotating platform.

Since the scalar components of $\mathbf{v}_P$ and $\mathbf{a}_P$ are known with respect to the disk basis $\mathbf{i}_k$, their scalar components with respect to the platform basis $\mathbf{i}'_k$ may be computed by the vector component transformation law (3.107a). Accordingly, we write

$$v'_P = Av_P, \qquad a'_P = Aa_P; \tag{4.94}$$

and we note from Fig. 4.25 that the transformation matrix defined by (3.96) is given by

$$A = [\cos\langle \mathbf{i}'_k, \mathbf{i}_l\rangle] = \begin{bmatrix} 1 & 0 & 0 \\ 0 & \cos\theta & -\sin\theta \\ 0 & \sin\theta & \cos\theta \end{bmatrix}. \tag{4.95}$$

The matrix components v_P of $\mathbf{v}_p$ and a_P of $\mathbf{a}_P$ in the disk frame $2 = \{O; \mathbf{i}_k\}$ are obtained from (4.78e); and the transformation laws (4.94) yield their corresponding scalar components referred to the platform frame $1 = \{O; \mathbf{i}'_k\}$:

$$v'_P = \begin{bmatrix} 1 & 0 & 0 \\ 0 & \cos\theta & -\sin\theta \\ 0 & \sin\theta & \cos\theta \end{bmatrix} \begin{bmatrix} -a\omega_1 \cos\theta \\ 0 \\ a\omega_2 \end{bmatrix} = \begin{bmatrix} -a\omega_1 \cos\theta \\ -a\omega_2 \sin\theta \\ a\omega_2 \cos\theta \end{bmatrix},$$

$$a'_P = \begin{bmatrix} 1 & 0 & 0 \\ 0 & \cos\theta & -\sin\theta \\ 0 & \sin\theta & \cos\theta \end{bmatrix} \begin{bmatrix} 2a\omega_1\omega_2 \sin\theta \\ -a(\omega_2^2 + \omega_1^2 \cos^2\theta) \\ a\omega_1^2 \sin\theta \cos\theta \end{bmatrix}$$

$$= \{2a\omega_1\omega_2 \sin\theta,\ -a\cos\theta(\omega_1^2+\omega_2^2),\ -a\omega_2^2 \sin\theta\}.$$

Hence, the velocity and acceleration vectors referred to the platform frame $1 = \{O; \mathbf{i}'_k\}$ are found to be

$$\mathbf{v}_P = -a\omega_1 \cos\theta\, \mathbf{i}' - a\omega_2(\sin\theta\, \mathbf{j}' - \cos\theta\, \mathbf{k}'),$$

$$\mathbf{a}_P = 2a\omega_1\omega_2 \sin\theta\, \mathbf{i}' - a\cos\theta(\omega_1^2+\omega_2^2)\, \mathbf{j}' - a\omega_2^2 \sin\theta\, \mathbf{k}' \qquad \text{[cf. (4.79g)]}.$$

It is seen that these are the same as (4.79g) derived by the method described for part (ii) in Example 4.16. Other applications of the transformation laws will be encountered in later chapters.

Exercise 4.6. Use the transformation law (3.107) to derive (4.80e) and (4.80f) from the equations (4.79g).

4.12.2. Derivative of a Tensor Function of a Scalar Variable

A tensor quantity $\mathbf{T}$ may be a function of one or more scalar variables. For example, the Euler rotation tensor $\mathbf{R}$ studied in Chapter 3 may vary with the time t so that $\mathbf{R} = \mathbf{R}(t)$ is a tensor-valued function of the scalar time variable t. Consequently, it is natural to inquire about its time rate of change and to learn its connection with our previous studies. Similarly, the rates of change with respect to time, temperature, or mass density of other tensor quantities such as stress, strain, and electrical or thermal conductivity would be useful in the study of the mechanics of solids and fluids. These applications also would involve the use of the derivative of a tensor that varies continuously with respect to some scalar variable σ, say. The definition of this derivative and some rules for its application are presented below.

The change $\Delta\mathbf{T}(\sigma)$ in the tensor $\mathbf{T}(\sigma)$ due to a change $\Delta\sigma$ in the scalar variable σ is defined by $\Delta\mathbf{T}(\sigma) = \mathbf{T}(\sigma + \Delta\sigma) - \mathbf{T}(\sigma)$. The tensor function $\mathbf{T}(\sigma)$ is said to be *continuous* at the value σ if $\Delta\mathbf{T}(\sigma) \to 0$ as $\Delta\sigma \to 0$, i.e., if $\mathbf{T}(\sigma + \Delta\sigma) \to \mathbf{T}(\sigma)$. Let I be an assigned interval of values of σ that contains the value Σ. Then, in accordance with the usual definition of the foregoing limit, $\mathbf{T}(\sigma)$ is continuous at $\sigma = \Sigma$ if for every $\varepsilon > 0$ there exists a $\delta > 0$ such that $|\mathbf{T}(\sigma) - \mathbf{T}(\Sigma)| < \varepsilon$ for every $\sigma \in I$ that satisfies $|\sigma - \Sigma| < \delta$. It is apparent, therefore, that the definition of the magnitude of a tensor is needed here. This definition was given in Problem 3.14. Thus, the derivative of $\mathbf{T}(\sigma)$ with respect to σ, denoted by $d\mathbf{T}(\sigma)/d\sigma$, is defined as the limit of the difference quotient $\Delta\mathbf{T}(\sigma)/\Delta\sigma$ as $\Delta\sigma \to 0$:

$$\frac{d\mathbf{T}(\sigma)}{d\sigma} \equiv \lim_{\Delta\sigma \to 0} \frac{\Delta\mathbf{T}(\sigma)}{\Delta\sigma} = \lim_{\Delta\sigma \to 0} \frac{\mathbf{T}(\sigma+\Delta\sigma)-\mathbf{T}(\sigma)}{\Delta\sigma}. \tag{4.96}$$

If the limit exists for the value σ, then $\mathbf{T}(\sigma)$ is said to be *differentiable* at σ. If the function $\mathbf{T}(\sigma)$ is differentiable at σ, it also is continuous there. Obviously, the derivative of a tensor is another tensor, and its higher derivatives, provided they exist, are generated similarly.

This definition generalizes the definition (A.16) for the derivative of a vector function of σ; and it is easy to show that the same kinds of rules studied in Appendix A for vector functions hold for tensor functions. A tensor whose scalar components T_{ij} with respect to at least one constant basis $\mathbf{e}_k$ are constants is called a constant tensor. Clearly, if $\mathbf{T}(\sigma) = \boldsymbol{\tau}$ is a constant tensor, then $\Delta\mathbf{T}(\sigma) = \mathbf{0}$ is the zero tensor and (4.96) yields $d\mathbf{T}(\sigma)/d\sigma = d\boldsymbol{\tau}/d\sigma = \mathbf{0}$. It follows that a constant tensor has a constant magnitude, and its derivative is the zero tensor. Moreover, it follows from the tensor component transformation law (3.108) that if the derivative of a tensor vanishes in one basis, it does so in every basis. Similarly, the rules for the derivative of the sum of two tensors $\mathbf{A}(\sigma)$ and $\mathbf{B}(\sigma)$ and of the product of a scalar $a(\sigma)$ and a tensor $\mathbf{B}(\sigma)$ are given by the familiar rules (A.17), (A.18), and (A.19) established for vectors. Of course, it follows from the representation (3.29) of a tensor $\mathbf{T}(\sigma)$ referred to a constant basis $\mathbf{e}_{ij} = \mathbf{e}_i \otimes \mathbf{e}_j$ that

$$\frac{d\mathbf{T}(\sigma)}{d\sigma} = \frac{dT_{ij}(\sigma)}{d\sigma}\mathbf{e}_{ij}. \tag{4.97}$$

Hence, in the same manner as for vectors, the scalar components of the derived tensor relative to a constant basis $\mathbf{e}_k$ are the usual derivatives of its nine scalar components $T_{ij}(\sigma)$ with respect to σ. The rule for the derivative of a tensor function referred to a moving reference frame will be derived further on.

Exercise 4.7. Prove that the derivative of the transpose of a tensor $\mathbf{T}(\sigma)$ referred to a fixed basis is equal to the transpose of the derivative of the tensor:

$$\frac{d\mathbf{T}^T(\sigma)}{d\sigma} = \left[\frac{d\mathbf{T}(\sigma)}{d\sigma}\right]^T. \qquad \square \tag{4.98}$$

The rules for the derivative of the products (3.11), (3.36) and the tensor product of two vector functions of σ are summarized by the generalized product rule (A.26) for the quantities $\alpha(\sigma)$ and $\beta(\sigma)$. In particular, the derivative of the tensor product of two vectors $\mathbf{a}(\sigma)$ and $\mathbf{b}(\sigma)$ is given by the following rule:

$$\frac{d}{d\sigma}[\mathbf{a}(\sigma)\otimes\mathbf{b}(\sigma)] = \mathbf{a}(\sigma)\otimes\frac{d\mathbf{b}(\sigma)}{d\sigma} + \frac{d\mathbf{a}(\sigma)}{d\sigma}\otimes\mathbf{b}(\sigma). \tag{4.99}$$

Exercise 4.8. (a) Let $\mathbf{u}(\sigma)$, $\mathbf{v}(\sigma)$, and $\mathbf{w}(\sigma)$ be three differentiable vector-valued functions of σ. Determine the derivative of the product $[\mathbf{u}(\sigma)\otimes\mathbf{v}(\sigma)]\,\mathbf{w}(\sigma)$. (b) Derive (4.98) from the rule (3.42) for arbitrary constant vectors. □

The foregoing rules for the derivatives of tensors will be applied below. The derivative of an orthogonal tensor will be studied next; and afterwards its connection with the angular velocity and acceleration of a rigid body particle will be explored. Finally, tensorial forms of the equations for the velocity and acceleration of a rigid body particle will be derived.

4.12.2.1. *Derivative of an Orthogonal Tensor Function*

An orthogonal tensor $\mathbf{Q}$ may vary with the time t so that $\mathbf{Q} = \mathbf{Q}(t)$ is an orthogonal tensor function of t which has the properties (3.68) for all t, namely,

$$\mathbf{Q}(t)\,\mathbf{Q}^T(t) = \mathbf{Q}^T(t)\,\mathbf{Q}(t) = \mathbf{1}. \tag{4.100}$$

Notice that the product (4.100) is the constant identity tensor; hence, the derivative of this product must vanish. The generalized product rule (A.26) and use of (4.98) yield the relations

$$\mathbf{Q}\dot{\mathbf{Q}}^T + \dot{\mathbf{Q}}\mathbf{Q}^T = \mathbf{0} \quad \text{and} \quad \mathbf{Q}^T\dot{\mathbf{Q}} + \dot{\mathbf{Q}}^T\mathbf{Q} = \mathbf{0}, \tag{4.101}$$

where the superimposed dot denotes the usual time derivative d/dt and the explicit indication of the time dependence has been suppressed for brevity. It follows from (4.101) that

$$\dot{\mathbf{Q}}\mathbf{Q}^T = -\mathbf{Q}\dot{\mathbf{Q}}^T = -(\dot{\mathbf{Q}}\mathbf{Q}^T)^T \quad \text{and} \quad \mathbf{Q}^T\dot{\mathbf{Q}} = -\dot{\mathbf{Q}}^T\mathbf{Q} = -(\mathbf{Q}^T\dot{\mathbf{Q}})^T.$$

Therefore, we learn that the tensors $\mathbf{S}$ and $\mathbf{S}^*$ defined by

$$\mathbf{S} \equiv \dot{\mathbf{Q}}\mathbf{Q}^T \quad \text{and} \quad \mathbf{S}^* \equiv \mathbf{Q}^T\dot{\mathbf{Q}} \tag{4.102}$$

are skew tensors. Hence, it follows from (4.102) that the derivative of an orthogonal tensor $\mathbf{Q}(t)$ may be written as

$$\dot{\mathbf{Q}} = \mathbf{S}\mathbf{Q} = \mathbf{Q}\mathbf{S}^*, \tag{4.103}$$

where $\mathbf{S}$ and $\mathbf{S}^*$ are similar, antisymmetric tensors with the properties

$$\mathbf{S} = -\mathbf{S}^T, \qquad \mathbf{S}^* = -\mathbf{S}^{*T}, \quad \text{and} \quad \mathbf{S}^* = \mathbf{Q}^T\mathbf{S}\mathbf{Q}. \tag{4.104}$$

Let $\mathbf{u}$ (or $\mathbf{v}$) be an arbitrarily assigned vector and recall that associated with each skew tensor $\mathbf{S}$ there exists a vector $\boldsymbol{\Omega}$ for which (3.60) holds. Thus, for the skew tensors (4.104), there exist vectors $\boldsymbol{\Omega}$ and $\boldsymbol{\Omega}^*$ defined by (3.59) such that

$$\mathbf{S}\mathbf{u} = \boldsymbol{\Omega} \times \mathbf{u}, \qquad \mathbf{S}^*\mathbf{v} = \boldsymbol{\Omega}^* \times \mathbf{v}. \tag{4.105}$$

It can be shown that $\boldsymbol{\Omega} = \mathbf{Q}\boldsymbol{\Omega}^*$ and $\mathbf{u} = \mathbf{Q}\mathbf{v}$. The meaning of $\mathbf{S}^*$ will be examined later.

4.12.2.2. *The Angular Velocity and Angular Acceleration Tensors*

It was shown in Chapter 3 that the rotation tensor $\mathbf{R}$ is an orthogonal tensor that describes the rotation of a rigid body about a line; and Euler's theorem showed that $\mathbf{R}$ is the same as the orthogonal basis transformation tensor $\mathbf{Q}$ defined in (3.103) and having the components (3.106). The results are summarized in (3.123). Therefore, in accordance with (4.103) and (4.104), there exists a skew tensor $\mathbf{S} = \mathbf{W} = -\mathbf{W}^T$, say, for which

$$\dot{\mathbf{R}}(t) = \mathbf{W}\mathbf{R} = \mathbf{R}\mathbf{W}^*, \tag{4.106a}$$

$$\mathbf{W}^* = \mathbf{R}^T\mathbf{W}\mathbf{R}. \tag{4.106b}$$

Hence, the tensor $\mathbf{W}$, or $\mathbf{W}^*$, may be called the *angular velocity tensor.*

The second derivative of (4.106) also may be computed similarly. We find the formulas

$$\ddot{\mathbf{R}} = (\mathbf{W}^2 + \dot{\mathbf{W}})\,\mathbf{R} \tag{4.107a}$$

$$= \mathbf{R}(\mathbf{W}^{*2} + \dot{\mathbf{W}}^*); \tag{4.107b}$$

hence, $\dot{\mathbf{W}}$, or $\dot{\mathbf{W}}^*$, may be named the *angular acceleration tensor.*

Exercise 4.9. Prove that the angular acceleration tensors $\dot{\mathbf{W}}$ and $\dot{\mathbf{W}}^*$ are antisymmetric, similar tensors so that

$$\dot{\mathbf{W}} = -\dot{\mathbf{W}}^T, \qquad \dot{\mathbf{W}}^* = -\dot{\mathbf{W}}^{*T}, \quad \text{and} \quad \dot{\mathbf{W}}^* = \mathbf{R}^T\dot{\mathbf{W}}\mathbf{R}. \tag{4.108}$$

Show also that

$$\mathbf{W}^{*2} = \mathbf{R}^T\mathbf{W}^2\mathbf{R}. \tag{4.109}$$

4.12.2.3. *Velocity and Acceleration of a Rigid Body Particle*

The foregoing results will be used below to obtain tensor equations for the velocity and acceleration of a rigid body particle. Let us recall the formula (3.125) for the general rigid body displacement of a particle P: and let us write $\hat{\mathbf{X}}(P) \equiv \mathbf{X}(P, t)$ and $\hat{\mathbf{B}}(O) \equiv \mathbf{B}(O, t)$ for the position vectors of the rigid body particle P and the base point O in the preferred frame $\Phi = \{F; \mathbf{I}_k\}$ at time t. Further, let $\hat{\mathbf{x}}(P) \equiv \mathbf{x}(P, t)$ denote the position vector of P from O in frame Φ at time t, so that in (3.125) $\mathbf{x}(P) = \mathbf{x}(P, t_0) \equiv \mathbf{x}_0(P)$ is the position vector of P from O in Φ at the initial or reference time t_0. And, finally, let $\mathbf{R} \equiv \mathbf{R}(t)$. Then the general motion (3.125) of any particle P of a rigid body as a function of time in the preferred frame Φ may be written as

$$\mathbf{X}(P, t) = \mathbf{B}(O, t) + \mathbf{R}(t)\, \mathbf{x}_0(P). \tag{4.110}$$

Moreover, in the present notation, (3.86) shows that

$$\mathbf{x}(P, t) = \mathbf{R}(t)\, \mathbf{x}_0(P), \tag{4.111}$$

wherein $\mathbf{R}(t_0) = \mathbf{1}$.

The velocity and acceleration in Φ of the particle P is obtained by differentiation of (4.110). In familiar notation, we find

$$\mathbf{v}_P(t) = \mathbf{v}_O(t) + \dot{\mathbf{R}}(t)\, \mathbf{x}_0, \tag{4.112a}$$

$$\mathbf{a}_P(t) = \mathbf{a}_O(t) + \ddot{\mathbf{R}}(t)\, \mathbf{x}_0. \tag{4.112b}$$

With the aid of (4.106a), (4.107a), and (4.111), the last terms in (4.112a) and (4.112b) may be written as

$$\begin{aligned} \dot{\mathbf{R}}\mathbf{x}_0 &= \mathbf{W}\mathbf{R}\mathbf{x}_0 = \mathbf{W}\mathbf{x} = \boldsymbol{\omega} \times \mathbf{x}, \\ \ddot{\mathbf{R}}\mathbf{x}_0 &= (\mathbf{W}^2 + \dot{\mathbf{W}})\, \mathbf{R}\mathbf{x}_0 = (\mathbf{W}^2 + \dot{\mathbf{W}})\, \mathbf{x} = \boldsymbol{\omega} \times (\boldsymbol{\omega} \times \mathbf{x}) + \dot{\boldsymbol{\omega}} \times \mathbf{x}, \end{aligned} \tag{4.113}$$

in which $\boldsymbol{\omega}$ and $\dot{\boldsymbol{\omega}}$ are the vectors of the skew tensors $\mathbf{W}$ and $\dot{\mathbf{W}}$, respectively. Finally, upon substitution of (4.113) into (4.112), we derive the familiar equations (2.27) and (2.30):

$$\mathbf{v}_P = \mathbf{v}_O + \boldsymbol{\omega} \times \mathbf{x} \qquad \text{[cf. (2.27)]},$$

$$\mathbf{a}_P = \mathbf{a}_O + \boldsymbol{\omega} \times (\boldsymbol{\omega} \times \mathbf{x}) + \dot{\boldsymbol{\omega}} \times \mathbf{x} \qquad \text{[cf. (2.30)]}.$$

The tensor equations corresponding to these rules may be read from (4.112) and the third of each relation in (4.113):

$$\mathbf{v}_P = \mathbf{v}_O + \mathbf{W}\mathbf{x}, \qquad \mathbf{a}_P = \mathbf{a}_O + \mathbf{W}^2\mathbf{x} + \dot{\mathbf{W}}\mathbf{x}. \tag{4.114}$$

Thus, we see that the familiar angular velocity and angular acceleration vectors $\boldsymbol{\omega}$ and $\dot{\boldsymbol{\omega}}$ are the respective vectors of the antisymmetric, angular velocity, and angular acceleration tensors $\mathbf{W}$ and $\dot{\mathbf{W}}$ introduced above. Analogous results may be obtained from the second forms of the derivatives (4.106b) and (4.107b), but the procedure is somewhat awkward and is omitted here. The meaning of $\mathbf{W}^*$ and $\dot{\mathbf{W}}^*$ will be described further on. The one missing step is left for the reader in the following exercise.

Exercise 4.10. Prove that the vector of the skew tensor $\dot{\mathbf{W}}$ is the time derivative of the vector of the skew tensor $\mathbf{W}$, as indicated in (4.113).

4.12.3. Derivative of a Tensor Referred to a Moving Frame

The total time derivative of a tensor referred to a moving frame will be studied in this section. It will be shown that the apparent and absolute time derivatives of the angular velocity tensor are the same. Tensorial forms of the basic equations for the velocity and acceleration of a particle in motion relative to a moving reference frame will be derived by two methods, one of which involves introduction to frame-indifferent transformations. We shall start with the derivation of the tensor form of the familiar equation for the derivative of a vector referred to a moving frame.

Let $\varphi = \{O; \mathbf{i}_k\}$ be a reference frame whose motion relative to a preferred frame $\Phi = \{F; \mathbf{I}_k\}$ is described by the time-dependent basis transformation (3.105) so that

$$\mathbf{i}_k(t) = \mathbf{Q}(t)\,\mathbf{I}_k \tag{4.115a}$$

$$\equiv \mathbf{R}_f(t)\,\mathbf{I}_k, \tag{4.115b}$$

where we introduce $\mathbf{R}_f(t) \equiv \mathbf{Q}(t)$ for the rotation of frame φ relative to frame Φ. Hence, $\mathbf{R}_f$ may be identified as the rotation tensor described in (3.123). Thus, differentiation of (4.115) and use of (4.106a) yields

$$\frac{d\mathbf{i}_k(t)}{dt} = \dot{\mathbf{R}}_f(t)\,\mathbf{I}_k = \mathbf{W}_f \mathbf{R}_f \mathbf{I}_k, \tag{4.116}$$

where $\mathbf{W}_f$ is the angular velocity of frame φ relative to frame Φ. Therefore, with the aid of (4.115) and the first result in (4.113), we derive the familiar rule

$$\frac{d\mathbf{i}_k(t)}{dt} = \mathbf{W}_f \mathbf{i}_k \tag{4.117a}$$

$$= \boldsymbol{\omega}_f \times \mathbf{i}_k, \tag{4.117b}$$

in which $\boldsymbol{\omega}_f$, the vector of the skew tensor $\mathbf{W}_f$, is the angular velocity of frame φ relative to Φ. (See Problem 4.117.)

Let $\mathbf{U}(t) = u_k(t)\,\mathbf{i}_k(t)$ be a vector referred to φ. Then differentiation of this equation and use of (4.117a) yields the total time rate of change of $\mathbf{U}$:

$$\dot{\mathbf{U}}(t) = \frac{\delta \mathbf{U}(t)}{\delta t} + \mathbf{W}_f \mathbf{U}(t). \tag{4.118}$$

This is, of course, the tensor form of the basic equation (4.11), which follows immediately from (4.117b).

The same idea may be used to deduce the time derivative of a tensor referred to a moving frame. To discover the rule for this case, let us consider the derivative (4.99) applied to the tensor basis $\mathbf{i}_{pq} = \mathbf{i}_p \otimes \mathbf{i}_q$. Then with the aid of (4.117a), we find easily the rule

$$\frac{d\mathbf{i}_{pq}(t)}{dt} = \mathbf{W}_f \mathbf{i}_{pq} - \mathbf{i}_{pq} \mathbf{W}_f. \tag{4.119}$$

The formula $\mathbf{T}\mathbf{v} = \mathbf{v}\mathbf{T}^T$ for an arbitrary vector $\mathbf{v}$ also was used in (4.119), so that $\mathbf{W}_f \mathbf{v} = -\mathbf{v}\mathbf{W}_f$ holds for the skew tensor $\mathbf{W}_f$. Proof of this rule follows from extension of the definition of the tensor product of vectors $\mathbf{a}$ and $\mathbf{b}$ so that

$$\mathbf{c}(\mathbf{a} \otimes \mathbf{b}) = (\mathbf{c} \cdot \mathbf{a})\,\mathbf{b} \tag{4.119a}$$

holds for all vectors $\mathbf{c}$. The verification of this step is left for the reader.

Let us now consider the time derivative in Φ of the tensor $\mathbf{T}(t) = T_{pq}(t)\,\mathbf{i}_{pq}(t)$ referred to φ; we have

$$\dot{\mathbf{T}}(t) = \frac{dT_{pq}}{dt}\mathbf{i}_{pq} + T_{pq}\frac{d\mathbf{i}_{pq}}{dt}. \tag{4.120}$$

Thus, using (4.119) and introducing the *apparent time derivative* of $\mathbf{T}(t)$ defined by

$$\frac{\delta \mathbf{T}(t)}{\delta t} \equiv \frac{dT_{pq}(t)}{dt}\mathbf{i}_{pq}(t), \tag{4.121}$$

we obtain the following formula for *the total time derivative of a tensor* $\mathbf{T}(t)$ *in the preferred frame* Φ *when* $\mathbf{T}(t)$ *is referred to a frame* φ *which is rotating with the angular velocity* $\mathbf{W}_f$ *relative to* Φ:

$$\dot{\mathbf{T}}(t) = \frac{\delta \mathbf{T}(t)}{\delta t} + \mathbf{W}_f(t)\,\mathbf{T}(t) - \mathbf{T}(t)\,\mathbf{W}_f(t). \tag{4.122}$$

The last two terms in (4.122) form the *convective time rate of change* of $\mathbf{T}(t)$. Higher-order derivatives of $\mathbf{T}$ may be obtained similarly.

For an illustration, let $\mathbf{T} = \mathbf{W}_f$. Then (4.122) shows that

$$\dot{\mathbf{W}}_f(t) \equiv \frac{d\mathbf{W}_f(t)}{dt} = \frac{\delta \mathbf{W}_f(t)}{\delta t}. \tag{4.123}$$

This is the tensor analog of the vector rule (4.14) in which $\boldsymbol{\omega}_f$ is the vector of the skew tensor $\mathbf{W}_f$. Let the reader show that, in fact, (4.14) may be derived from (4.123).

4.12.3.1. *Velocity and Acceleration Referred to a Moving Frame, Revisited*

The formula for the derivative of a vector referred to a moving frame is given in (4.118). We shall apply this result to derive the tensor analog of the equations for the absolute velocity and acceleration of a particle referred to a moving frame. Specifically, we shall consider the relation (4.44) in which the position vector $\mathbf{x}(t)$ of the particle P from O is referred to the moving frame $\varphi = \{O; \mathbf{i}_k\}$. Then (4.118) shows that the absolute time rates of change of $\mathbf{x}(t)$ in Φ are related to the apparent time rates of change in φ by the following tensorial equations:

$$\dot{\mathbf{x}}(t) = \frac{\delta \mathbf{x}}{\delta t} + \mathbf{W}_f \mathbf{x}, \tag{4.124}$$

$$\ddot{\mathbf{x}}(t) = \frac{\delta^2 \mathbf{x}}{\delta t^2} + \mathbf{W}_f^2 \mathbf{x} + \dot{\mathbf{W}}_f \mathbf{x} + 2\mathbf{W}_f \frac{\delta \mathbf{x}}{\delta t}. \tag{4.125}$$

Thus, by use of (4.43), (4.124), and (4.125), the tensor formulas for the absolute velocity and acceleration of P referred to the moving frame φ are provided by

$$\mathbf{v}_P = \frac{\delta \mathbf{x}}{\delta t} + \mathbf{v}_O + \mathbf{W}_f \mathbf{x}, \tag{4.126}$$

$$\mathbf{a}_P = \frac{\delta^2 \mathbf{x}}{\delta t^2} + \mathbf{a}_O + \mathbf{W}_f^2 \mathbf{x} + \dot{\mathbf{W}}_f \mathbf{x} + 2\mathbf{W}_f \frac{\delta \mathbf{x}}{\delta t}. \tag{4.127}$$

The last term in (4.127) is identified as the Coriolis acceleration of P. Thus, bearing in mind the equations (4.114) for the rigid body velocity and acceleration of P in Φ, we have the following familiar conclusions: *(i) The absolute velocity of a particle P is equal to the relative velocity of P in φ plus the rigid body velocity of P in Φ. (ii) The absolute acceleration of P is the sum of the relative acceleration of P in φ, the rigid body acceleration of P in Φ and the Coriolis acceleration of P in Φ.* The reader may show easily that (4.126) and (4.127) are equivalent to the fundamental relations (4.46) and (4.48), respectively.

The same results may be derived by other methods. An especially useful and elegant procedure of particular importance in areas of continuum mechanics is presented below. This will require a brief introduction to frame-indifferent transformations, so these will be studied first.

4.12.3.2. *Frame-Indifferent Transformations*

To begin, let us remember the basis transformation tensor (3.103) expressed as

$$\mathbf{Q}(t) = \mathbf{i}_k(t) \otimes \mathbf{I}_k. \tag{4.128}$$

We recall that this may be viewed as the Euler rotation from frame Φ into frame φ. For clarity in the subsequent construction, let us write $\mathbf{x}_\varphi(t)$ for the position vector in Φ of a point P from the origin of φ, but referred to φ, so that with the basis transformation (4.115a), we have

$$\mathbf{x}_\varphi(t) = x_k(t)\,\mathbf{i}_k(t) \tag{4.129a}$$

$$= \mathbf{Q}(x_k(t)\,\mathbf{I}_k). \tag{4.129b}$$

Let us define the position vector

$$\mathbf{x}_\Phi(t) \equiv x_k(t)\,\mathbf{I}_k. \tag{4.130}$$

Then use of (4.130) in (4.129) yields the following transformation law for the relative position vector:

$$\mathbf{x}_\varphi(t) = \mathbf{Q}(t)\,\mathbf{x}_\Phi(t). \tag{4.131}$$

It is seen in (4.130) that $\mathbf{x}_\Phi$ has the same scalar components in frame Φ as the vector $\mathbf{x}_\varphi$ in (4.129a). Thus, $\mathbf{x}_\Phi$ is the same vector seen in Φ as the vector $\mathbf{x}_\varphi$ seen in φ. Of course, the point P cannot occupy two places at the same moment, so we think of $\mathbf{x}_\Phi$ as the position vector of the image of P in Φ; it is the position vector of another point in Φ from which we may consider that P was rotated by the instantaneous transformation $\mathbf{Q}(t)$. Then, because each observer perceives his own basis as relatively fixed, the two observers in φ and Φ agree, independently of the specific point P, that they perceive the same vector in their individual frame. Therefore, a transformation of the type (4.131) is said to be *frame indifferent*, or *objective*.

Suppose, for example, that $\mathbf{x}_\Phi(t) = a(t)\,\mathbf{I}_1$; and consider an arbitrary transformation (4.128). Then according to (4.131), we find

$$\mathbf{x}_\varphi(t) = (\mathbf{i}_k(t) \otimes \mathbf{I}_k)\,a(t)\,\mathbf{I}_1 = a(t)\,\mathbf{i}_k(t)\,\delta_{k1} = a(t)\,\mathbf{i}_1(t),$$

wherein we recall that $\mathbf{I}_k \cdot \mathbf{I}_q = \delta_{kq}$. Thus, at the time t, the observer in φ sees exactly the same vector along his $\mathbf{i}_1$ direction as seen by the observer Φ along his $\mathbf{I}_1$ direction. The orthogonal tensor $\mathbf{Q}$ preserves the properties of the vector by a pure rotation about the origin point O of φ.

The time derivatives in φ of the vector $\mathbf{x}_\varphi(t)$ behave in the same way. Differentiation of (4.130) in Φ gives $\dot{\mathbf{x}}_\Phi = \dot{x}_k(t)\,\mathbf{I}_k$; and recalling the definition of the δ-derivative in (4.10), we find by (4.129a) and (4.115a)

$$\frac{\delta \mathbf{x}_\varphi}{\delta t} = \dot{x}_k(t)\,\mathbf{i}_k(t) = \mathbf{Q}(\dot{x}_k(t)\,\mathbf{I}_k).$$

Therefore, we have the following transformation law for the apparent time derivative:

$$\frac{\delta \mathbf{x}_{\varphi}(t)}{\delta t} = \mathbf{Q}(t)\, \dot{\mathbf{x}}_{\Phi}(t). \tag{4.132}$$

Higher-order apparent derivatives follow the same rule; for example,

$$\frac{\delta^2 \mathbf{x}_{\varphi}(t)}{\delta t^2} = \mathbf{Q}(t)\, \ddot{\mathbf{x}}_{\Phi}(t). \tag{4.133}$$

Thus, the observers in φ and Φ confirm the same time derivatives for the vectors of (4.131); hence, the apparent time rates (4.132) and (4.133) are objective quantities. Observe that the transformation laws (4.132) and (4.133) show that the apparent time derivatives of (4.131) are obtained by holding $\mathbf{Q}(t)$ fixed. It is seen from (4.128) that this is equivalent to holding the basis vectors $\mathbf{i}_k$ fixed in the original definition (4.10).

Any vector $\mathbf{v}_{\varphi}(t) = v_k(t)\, \mathbf{i}_k(t)$ referred to the moving frame φ may be written in the same manner as (4.131). Therefore, we have the following *frame-indifferent transformation law for a vector referred to a moving frame*:

$$\mathbf{v}_{\varphi}(t) = \mathbf{Q}(t)\, \mathbf{v}_{\Phi}(t), \tag{4.134}$$

wherein $\mathbf{v}_{\Phi}(t) = v_k(t)\, \mathbf{I}_k$. Notice that $\mathbf{v}_{\varphi}$ share the same scalar components, only their basis directions are changed by the orthogonal transformation $\mathbf{Q}$. These vectors are frame indifferent. Moreover, their nth-order apparent time derivatives also are objective:

$$\frac{\delta^n \mathbf{v}_{\varphi}(t)}{\delta t^n} = \mathbf{Q}(t)\, \overset{(n)}{\mathbf{v}}_{\Phi}(t). \tag{4.135}$$

The transformation (4.128) also induces a transformation on tensors referred to a moving frame. To derive the transformation law for a tensor $\mathbf{T}_{\varphi} = T_{pq}(t)\, \mathbf{i}_{pq}(t)$ referred to the moving frame φ, we use (4.115a) to obtain

$$\mathbf{T}_{\varphi}(t) = T_{pq}(t)\, \mathbf{i}_p(t) \otimes \mathbf{i}_q(t) = \mathbf{Q}(t)[T_{pq}(t)\, \mathbf{I}_p \otimes \mathbf{I}_q]\, \mathbf{Q}^T(t).$$

It follows that $\mathbf{T}_{\varphi}$ must obey the following *frame-indifferent transformation law for a tensor referred to a moving frame*:

$$\mathbf{T}_{\varphi}(t) = \mathbf{Q}(t)\, \mathbf{T}_{\Phi}(t)\, \mathbf{Q}^T(t), \tag{4.136}$$

wherein, by definition, $\mathbf{T}_{\Phi}(t) = T_{\mathrm{pq}}(t)\, \mathbf{I}_{pq}$ in frame Φ. It seen again that $\mathbf{T}_{\varphi}$ and $\mathbf{T}_{\Phi}$ share the same scalar components though they refer to different directions related by the transformation $\mathbf{Q}$. Therefore, the observer Φ sees exactly the same entity as seen by the observer φ. As a consequence, any tensor quantity that respects the rule (4.136) is said to be frame indifferent, or objective. (See Problem 4.119.)

For an illustration, we may recall the tensor $\mathbf{S}^*$ defined in (4.102) and having the properties (4.104), the last of which shows that $\mathbf{S} = \mathbf{Q}\mathbf{S}^*\mathbf{Q}^T$. It is seen easily from (4.102) that $\mathbf{S}$ is referred to the moving frame φ. Hence, $\mathbf{S}^*$ is the tensor whose components seen in Φ with respect to $\mathbf{I}_{pq}$ are the same as those of $\mathbf{S}$ seen in φ with respect to $\mathbf{i}_{pq}$. The same thing holds *mutatis mutandis* for the relative angular velocity tensor $\mathbf{W}^*$ in (4.106) and its derivative in (4.108). Therefore, the tensors $\mathbf{S}$, $\mathbf{W}$, and $\dot{\mathbf{W}}$ encountered earlier are objective. It is not suprising that $\mathbf{W}^2$ in (4.109) also is frame-indifferent. It follows readily from (4.136) that the Nth power $\mathbf{T}_\varphi^N$ of an objective tensor is objective.

It is evident that the rule (4.136) holds for all apparent time derivatives of (4.136); for, holding the basis $\mathbf{i}_k(t)$ fixed in the δ-derivative implies in (4.128) that $\mathbf{Q}(t)$ is to be held fixed in (4.136). Thus, the transformation law for the nth-order apparent time derivative of a tensor referred to a moving frame is given by

$$\frac{\delta^n \mathbf{T}_\varphi(t)}{\delta t^n} = \mathbf{Q}(t) \overset{(n)}{\mathbf{T}}_\Phi(t) \, \mathbf{Q}^T(t). \tag{4.137}$$

For $n = 1$, this rule coincides with (4.121), but herein expressed directly in terms of $\mathbf{Q}$. (See Problem 4.120.)

4.12.3.3. *A Change of Frame, Velocity, and Acceleration. The End.*

A *change of frame* is characterized by an orthogonal, linear transformation that preserves distances and time intervals. This change is exhibited in terms of the position vectors $\mathbf{X}_\Phi$ and $\mathbf{x}_\varphi$ of the same point P from the origins F and O of the respective reference frames $\Phi = \{F; \mathbf{I}_k\}$ and $\varphi = \{O; \mathbf{i}_k(t)\}$ in accordance with the equation

$$\mathbf{X}_\Phi(t) = \mathbf{B}_\Phi(t) + \mathbf{x}_\varphi(t) \tag{4.138a}$$

$$= \mathbf{B}_\Phi(t) + \mathbf{Q}(t)\, \mathbf{x}_\Phi(t). \tag{4.138b}$$

Herein $\mathbf{B}_\Phi(t)$ is the position vector of O from F, $\mathbf{Q}(t)$ is the rigid rotation of frame φ relative to Φ, and (4.131) applies. A change of frame usually includes a time shift $\mathbf{t}_\Phi = t_\varphi + \tau$, where τ is a constant. This expresses the condition that the clock in frame φ may be ahead or behind the clock in the preferred frame. However, we shall lose nothing by supposing that all observers use the same universal clock so that $\tau = 0$; and we shall continue to write t for the universal time. It is evident, of course, that (4.138a) is the transformation introduced differently in Section 4.6.1 and diagrammed in Fig. 4.14, but expressed in different, though parallel, notation. The reader will recall that equations (4.46) and (4.48) were derived by starting from (4.138a) and by using the definition of the δ-derivative in (4.10). In this section, it will be shown that the same relations may be obtained by starting with (4.138b).

The absolute velocity and acceleration of P in Φ may be gotten by differentiation of (4.138b) as usual. This yields

$$\mathbf{v}_P \equiv \dot{\mathbf{X}}_\Phi = \mathbf{Q}\dot{\mathbf{x}}_\Phi + \dot{\mathbf{B}}_\Phi + \dot{\mathbf{Q}}\mathbf{x}_\Phi, \tag{4.139}$$

$$\mathbf{a}_P \equiv \ddot{\mathbf{X}}_\Phi = \mathbf{Q}\ddot{\mathbf{x}}_\Phi + \ddot{\mathbf{B}}_\Phi + \ddot{\mathbf{Q}}\mathbf{x}_\Phi + 2\dot{\mathbf{Q}}\dot{\mathbf{x}}_\Phi. \tag{4.140}$$

Clearly, $\mathbf{v}_O \equiv \dot{\mathbf{B}}_\Phi$ and $\mathbf{a}_O \equiv \ddot{\mathbf{B}}_\Phi$ are the absolute velocity and acceleration of the origin O in Φ. In the right-hand side of (4.139) and (4.140), the first terms are the relative velocity and acceleration vectors (4.132) and (4.133), respectively; the second pair of right-hand terms may be identified as the rigid body velocity and acceleration contributions resulting from the rotation $\mathbf{Q}(t)$; and the final term in (4.140) is the Coriolis acceleration. Indeed, recalling that $\mathbf{Q}(t) \equiv \mathbf{R}_f(t)$ is the Euler rotation tensor, and introducing (4.106a), (4.107a), (4.132), and (4.133), the reader may show easily that (4.139) and (4.140) coincide with (4.126) and (4.127) derived earlier; and, of course, these are equivalent to the fundamental relations (4.46) and (4.48) which are the central equations of this chapter. It is seen that (4.139) and (4.140) do not obey the general vector rule (4.134). Plainly, the total velocity and acceleration of a particle are not objective quantities; different observers perceive different totals.

In general, the total time derivative of a vector $\mathbf{v}_\varphi$ or a tensor $\mathbf{T}_\varphi$ referred to a moving frame is not objective. The rule (4.134) provides

$$\dot{\mathbf{v}}_\varphi = \mathbf{Q}\dot{\mathbf{v}}_\Phi + \dot{\mathbf{Q}}\mathbf{v}_\Phi = \frac{\delta \mathbf{v}_\varphi}{\delta t} + \mathbf{W}_f \mathbf{v}_\varphi, \tag{4.141}$$

in which we recall $\mathbf{W}_f \equiv \dot{\mathbf{Q}}\mathbf{Q}^T$ and the rule (4.135) for $n = 1$. The last of (4.141) is to be compared with (4.118). Similarly, the total time derivative of the tensor $\mathbf{T}_\varphi$ in (4.136) yields the formula (4.122) by application of the rule (4.137). Thus, total time derivatives of frame-indifferent vectors and tensors generally are not objective.

This concludes our study of some special advanced topics in kinematics. The tensor equations derived throughout are useful in advanced theoretical applications of kinematics to studies of deformable solids and fluids and in the theory of the mechanical response of materials under the action of forces and torques in both statics and dynamics. Although it would be natural to continue our studies into these interesting areas, these topics extend too far beyond the scope of our future needs in the present work. We shall begin study of the analysis of motion under the action of forces and torques in the next volume.

References

1. BONNER, W., *The Mystery of the Expanding Universe*, Macmillan, New York, 1964. This informative, entertaining and authoritative account of the steady-state and relativistic theories of the motion of the universe, written by a mathematician, but without mathematics, is an excellent introductory book for the general reader. The scale of the universe and of the earth's motion within it are described; and, from this beginning, the stage is set for the author's lucid presentation and objective evaluation of two rival theories of cosmology.
2. BRAND, L., *Vectorial Mechanics*, Wiley, New York, 1930. See Chapter 10. See also *Vector and Tensor Analysis*, Wiley, New York, 1947. Some kinematics of rigid body motion and the Darboux vector are discussed in Chapter 3.
3. CROUCH, T., *Matrix Methods Applied to Engineering Rigid Body Mechanics*, Pergamon, New York, 1981. The notation used in this text is awkward, but the problems present good applications of the equations for rigid body motion, the composition rules for angular velocities and accelerations, and of the equations for relative motion.
4. KANE, T. R. *Dynamics*, Holt, Reinhart and Winston, New York, 1968. This text concerns more advanced topics and applications in Langrangian mechanics. Remarks analogous to those for Ref. 3 are appropriate here. The composition rules are developed in Chapter 2.
5. MARION, J., *Classical Dynamics of Particles and Systems*, Academic, New York, 1970. Chapter 11 contains a nice discussion of some problems involving motion relative to the earth.
6. MERIAM, J. L., *Dynamics*, 2nd Edition, Wiley, New York, 1975. Chapters 2, 7, and 8 contain many additional examples and excellent parallel problems for collateral study.
7. SHAMES, I., *Engineering Mechanics*, Vol. 2, *Dynamics*, 2nd Edition, Prentice-Hall, Englewood Cliffs, New Jersey, 1966. Chapters 11 and 15 contain additional examples and similar problems for auxiliary study.
8. SYNGE, J. L., and GRIFFITH, B. A., *Principles of Mechanics*, 2nd Edition, McGraw-Hill, New York, 1949. Chapters 4 and 12 cover some topics of relative motion in a plane and in space.

Problems

4.1. At the instant shown, a rigid body has an angular velocity $\boldsymbol{\omega} = 3\mathbf{j} - 6\mathbf{k}$ rad/sec relative to the preferred frame $\Phi = \{F; \mathbf{I}_k\}$, but referred to the imbedded frame $\varphi = \{O; \mathbf{i}_k\}$. (a) If the base point O has the velocity $\mathbf{v}_O = 6\mathbf{i} + 3\mathbf{j} - 2\mathbf{k}$ m/sec in Φ, what is its velocity referred to Φ at the instant of interest? (b) Find at this instant the absolute velocity of the particle P at $\mathbf{X} = 2\mathbf{I} + \mathbf{J}$ m when O is at $\mathbf{B} = 4\mathbf{I} - 3\mathbf{J} + \mathbf{K}$ m in Φ.

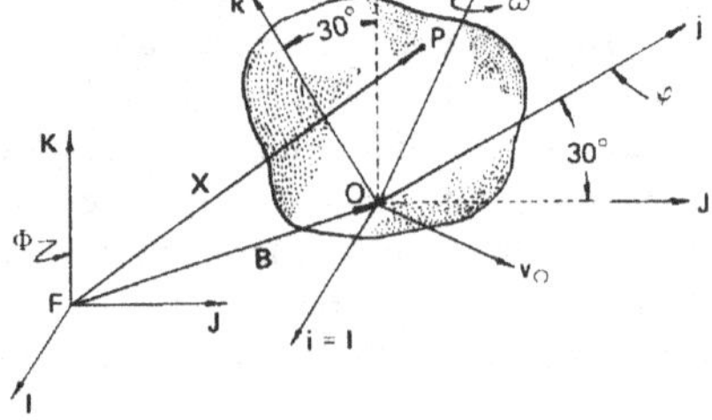

Problem 4.1.

4.2. A vector $\mathbf{U}(t) = 6t\mathbf{i} - 3t^2\mathbf{j} + 6\mathbf{k}$ units is referred to a reference frame $\varphi = \{O; \mathbf{i}_k\}$ which has a constant angular velocity $\boldsymbol{\omega}_f = 3\mathbf{i} + 2\mathbf{j} + 6\mathbf{k}$ units relative to a preferred frame $\Phi = \{F; \mathbf{I}_k\}$. Find the time rate of change of $\mathbf{U}(t)$ apparent to observers in the two frames, and referred to φ.

4.3. The rigid body shown in the figure is moving relative to a preferred frame $\Phi = \{F; \mathbf{I}_k\}$. Referred to an imbedded frame $\varphi = \{O; \mathbf{i}_k\}$, the body has a constant angular velocity $\boldsymbol{\omega} = 5\mathbf{i} + 6\mathbf{j} + 4\mathbf{k}$ rad/sec, and the base point O has a velocity $\mathbf{v}_O = 12\mathbf{i} + 5\mathbf{j} - 6\mathbf{k}$ m/sec, both relative to Φ. (a) Find the time rate of change in Φ of the position vector $\mathbf{x}$ of the particle whose place is $(2, -3, 4)$ m in φ. (b) What is its rate of change in φ? (c) What is the absolute acceleration of point O?

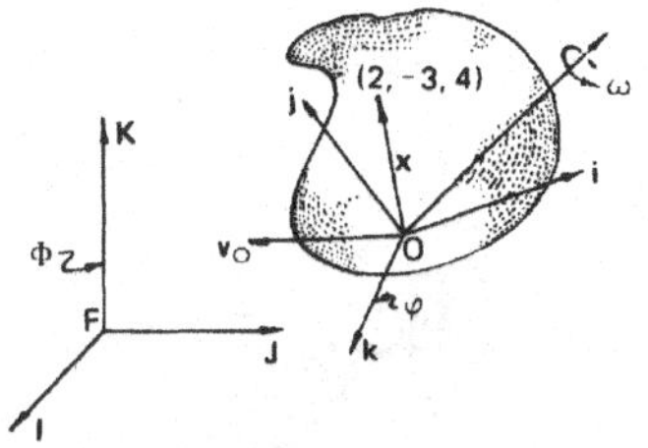

Problem 4.3.

4.4. A body imbedded reference frame $\varphi = \{O; \mathbf{i}_p\}$ has a translational velocity $\mathbf{v}_O = 2\mathbf{i} - 6\mathbf{j} - 10\mathbf{k}$ ft/sec, an angular velocity $\boldsymbol{\omega}_f = 4\mathbf{i} + 6\mathbf{j} - 10\mathbf{k}$ rad/sec and an angular acceleration $\dot{\boldsymbol{\omega}}_f = 2\mathbf{i} - 4\mathbf{j} + 11\mathbf{k}$ rad/sec^2 relative to a frame $\Phi = \{F; \mathbf{I}_j\}$ at the instant t_0 when $\mathbf{i}_k = \mathbf{I}_k$. Compute $d\mathbf{i}_k/dt$ and $d^2\mathbf{i}_k/dt^2$ in Φ at t_0.

4.5. Let $\mathbf{x}$ denote the position vector of the point $P = (5, 3, 7)$ ft from the point $Q = (4, 5, -6)$ ft in the body frame φ whose motion at the instant t_0 is described in the previous problem. Determine the absolute, instantaneous rates $\dot{\mathbf{x}}$ and $\ddot{\mathbf{x}}$ referred to φ. Does $\mathbf{v}_O$ affects the results?

4.6. A 40-ft-high antenna tower is located in the preferred frame $\Phi = \{F; \mathbf{I}_k\}$, as shown. The position vector of the top of the antenna from the station F is denoted by $\mathbf{X}$. Another frame $\varphi = \{O; \mathbf{e}_k\}$ has the absolute velocity

$$\mathbf{v}_O = 6\mathbf{e}_1 - 4\mathbf{e}_2 + 7\mathbf{e}_3 \text{ ft/sec}$$

at an instant t_0 when $\mathbf{e}_k = \mathbf{I}_k$. The angular velocity and angular acceleration of φ relative to Φ at time t_0 are given by

$$\boldsymbol{\omega} = 4\mathbf{e}_1 + 2\mathbf{e}_2 - 3\mathbf{e}_3 \text{ rad/sec}, \qquad \dot{\boldsymbol{\omega}} = 3\mathbf{e}_1 - 5\mathbf{e}_2 \text{ rad/sec}^2.$$

Compute the first and second time rates of change of $\mathbf{X}$ as seen from φ at the instant of interest. Does $\mathbf{v}_O$ affect the results?

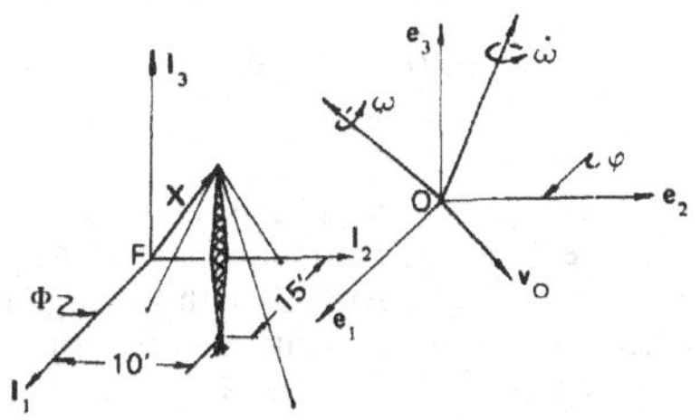

Problem 4.6.

4.7. Consider the hinged joint and slider block shown in Fig. 2.20. Let frame $1 = \{B; \boldsymbol{\beta}, \boldsymbol{\gamma}, \mathbf{k}\}$ be fixed in the slider block B, as shown, and introduce another frame $2 = \{B; \boldsymbol{\alpha}, \boldsymbol{\beta}, \boldsymbol{\lambda}\}$ fixed in the hinged yoke of the connecting rod AB so that $\boldsymbol{\lambda}$ is parallel to l. (a) Determine as functions of y the angular velocity of the rod relative to B and the absolute angular velocity of B in the frame $\Phi = \{F; \mathbf{i}_k\}$ shown in Fig. 2.19. (b) What is the angular speed of the rod relative to the slider block? (c) What is the absolute angular velocity of the rod referred to frame 1?

4.8. Two gears A and B are held in rolling contact by a link of length l between their centers. The gear A rotates with an angular velocity $\boldsymbol{\omega}_A$ relative to the fixed frame $\Phi = \{O; \mathbf{i}, \mathbf{j}\}$, while the gear B moves around the periphery of A with an angular velocity $\boldsymbol{\omega}_B$ relative to Φ. Find the angular velocity of each gear relative to the link.

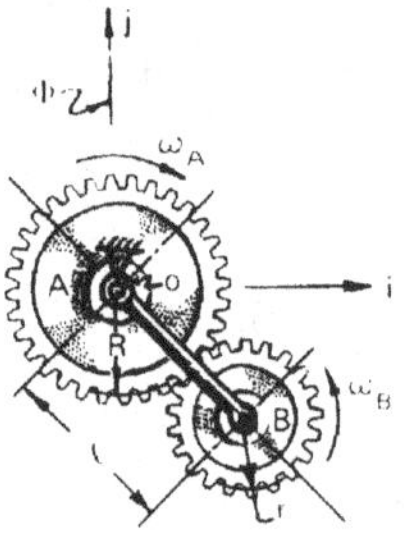

Problem 4.8.

4.9. The control rod $0A$ of a gear mechanism shown in the figure is rotating with an angular speed $\omega_1 = 15$ rad/sec about the center of a fixed ring gear G. Find the angular velocity of the pinion gear P relative to the control rod and relative to the ring gear. What is the velocity of point B relative to the ring gear at the instant shown?

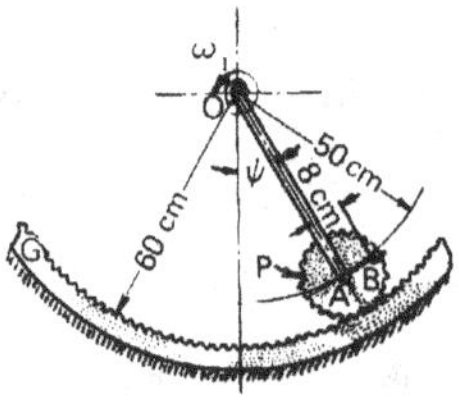

Problem 4.9.

4.10. An aerial ride in the amusement park consists of a passenger cockpit C that revolves with an angular velocity $\boldsymbol{\omega}_2$ relative to its supporting arm OB, while OB rotates with an angular velocity $\boldsymbol{\omega}_1$ relative to the control tower, as shown. A reference frame $\varphi = \{O; \mathbf{i}_p\}$ is fixed in OB. Find the absolute angular velocity and angular acceleration of the cockpit referred to φ at the instant for which

$$\omega_1 = 0.8 \text{ rad/sec}, \qquad \omega_2 = 0.4 \text{ rad/sec},$$
$$\dot{\omega}_1 = 0.1 \text{ rad/sec}^2, \qquad \dot{\omega}_2 = 0.2 \text{ rad/sec}^2.$$

4.11. A disk A of radius 2 ft is turning as shown with a constant angular speed $\omega_A = 5$ rad/sec relative to a large circular plate P. At the instant τ shown in the figure, the plate is rotating in a ring bearing R with an angular speed $\omega_P = 3$ rad/sec, which is increasing at the rate of 2 rad/sec^2 relative to the ring. The ring bearing also is spinning with an angular speed $\omega_R = 10$ rad/sec, which is decreasing at the rate of 4 rad/sec^2

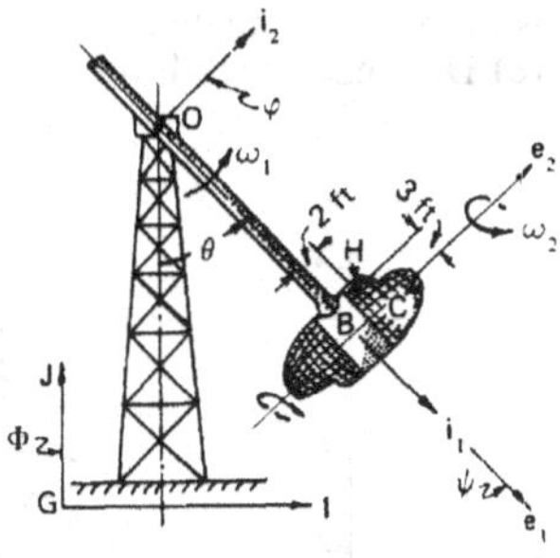

Problem 4.10.

relative to the ground frame. Find at time τ the total angular velocity and angular acceleration of A in Φ.

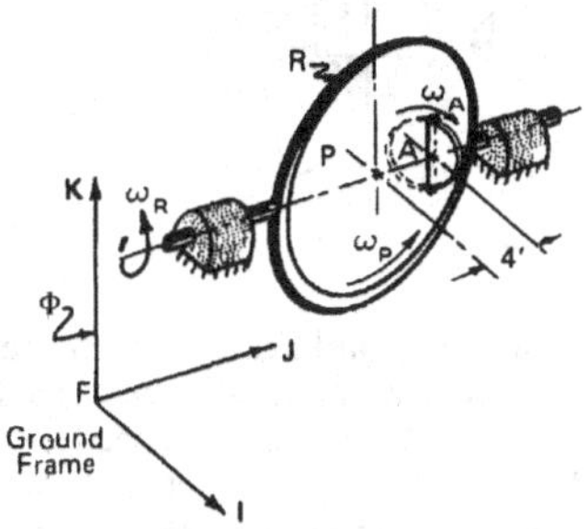

Problem 4.11.

4.12. At the instant shown in the figure, a crate C is being lifted by a crane at a cable speed $v = 1.8$ m/sec, which is increasing at the rate of 1.2 m/sec^2 relative to the crane. The crane is turning simultaneously with an angular speed $\omega = 2$ rad/sec, which is decreasing at the rate $\dot{\omega} = -1$ rad/sec^2 relative to the ground frame $\Phi = \{F; \mathbf{I}_k\}$. If there is no slippage between the cable A and the pulley P, what are the absolute angular velocity and angular acceleration of P in Φ at the instant of interest?

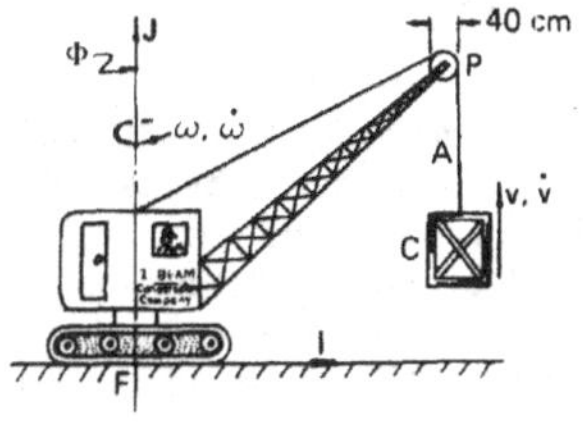

Problem 4.12.

4.13. A cross-sectional view of a motor-driven bevel gear mounted on a rotating table is shown in the figure. This assembly corresponds to the device shown pictorially in Fig. 4.1. The bevel gear G has a pitch angle α and it engages another bevel gear B whose axis of rotation coincides with the axis of the table. The motor drives G at a constant angular velocity $\boldsymbol{\omega}$, as indicated. The gear B may be driven independently at a constant angular velocity $\boldsymbol{\Omega} = \Omega\mathbf{K}$ by the engagement of its drive clutch C in the machine. (a) What is the total angular velocity of the table T when the clutch is engaged? (b) What is the absolute angular velocity of T when B is fixed in the machine frame $\Phi = \{F; \mathbf{I}_k\}$? (c) Determine the absolute angular acceleration of G for each case above.

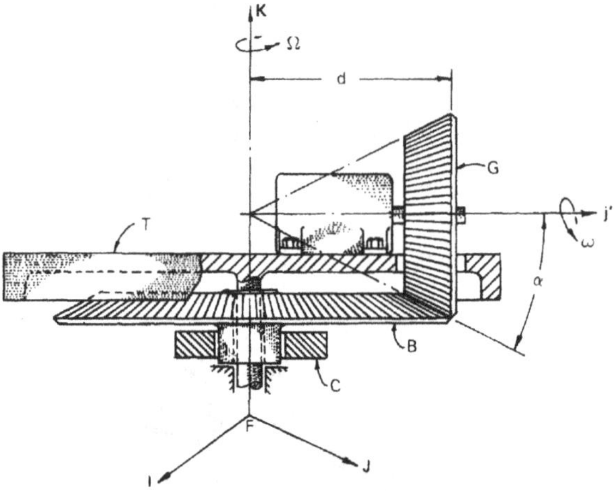

Problem 4.13.

4.14. The four-bar space linkage shown in the figure consists of two half yokes of a universal joint connected by a circular link AB set in roller bearings at the joints A and B. (a) Find the ratio of the angular speed ω_2 of the follower shaft to the angular speed ω_1 of the drive shaft. What are the maximum and minimum values of their ratio? (b) Determine the angle of rotation ψ of the follower shaft as a function of the angle of rotation θ of the drive shaft. Write a computer program to plot $\psi = \psi(\theta; \phi)$ for assigned shaft angles ϕ, and execute it for $\phi = 0°$, $10°$, $50°$.

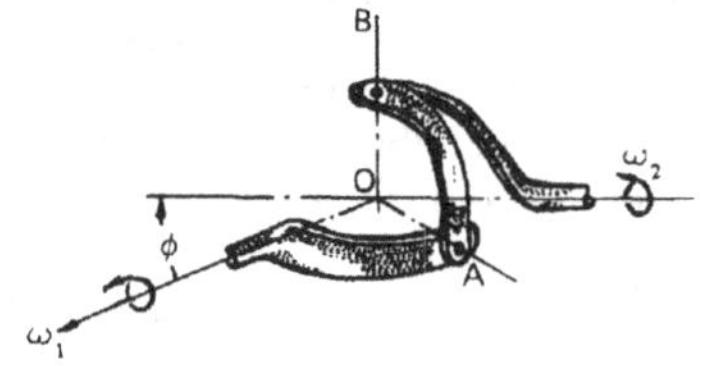

Problem 4.14.

4.15. Variation in the angular speed between the driver and follower shafts of a universal joint may be eliminated by use of a double universal joint shown in the figure. The second shaft makes the same angle ϕ with the connecting shaft. If the cross-link arms CD and EF of the connecting shaft are parallel, the angular speed ω_3 of the follower shaft will be the same as the angular speed $\omega_1$of the drive shaft. This fact may be readily verified from the analysis in the text. (a) Determine ω_3 for a double joint

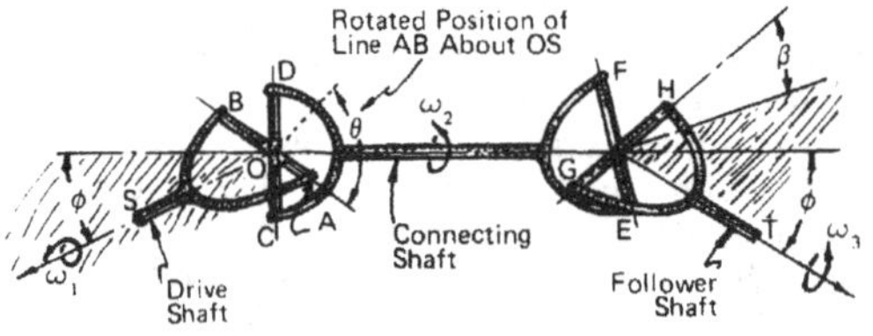

Problem 4.15.

whose cross-link arm GH makes an angle β with the plane ABS of the drive yoke. (b) How is the angle of rotation of the follower shaft related to the angle of rotation of the drive shaft?

4.16. The differential of an automobile is a gear train that allows the rear wheels to rotate at different angular speeds when the vehicle is moving on a curved road. The drive shaft from the transmission turns a bevel gear D with an angular velocity $\boldsymbol{\omega}$ relative to the vehicle; and D drives a beveled ring gear R at an angular speed Ω, as illustrated. The pitch angle of D is θ, and the axles of D and R intersect in a right angle. In addition, two (more commonly four) identical epicyclic bevel gears $B_1$and B_2, with a pitch angle ϕ, rotate in opposite directions about axles $A_1$and A_2 fixed in bearings in R, but they turn with the same angular speed ω_3 relative to R. These gears mesh with beveled gears that are fixed to the rear wheel drive shafts S_1 and S_2. The angular velocities of the wheel shafts $S_1$and S_2 relative to the car are denoted by $\boldsymbol{\omega}_1$ and $\boldsymbol{\omega}_2$, as shown. Show that relative to the car, the angular speeds of the drive shaft, the gears B_1 and B_2, and the ring gear R, are related to the angular speeds of the wheels by the relations

$$2\omega = (\omega_1 + \omega_2)\cot\theta, \qquad 2\omega_3 = (\omega_2 - \omega_1)\cot\phi, \qquad 2\Omega = \omega_1 + \omega_2.$$

If the automobile has a speed v as it turns on a circular curve of radius R, what will be the angular speeds of the wheels relative to the ground?

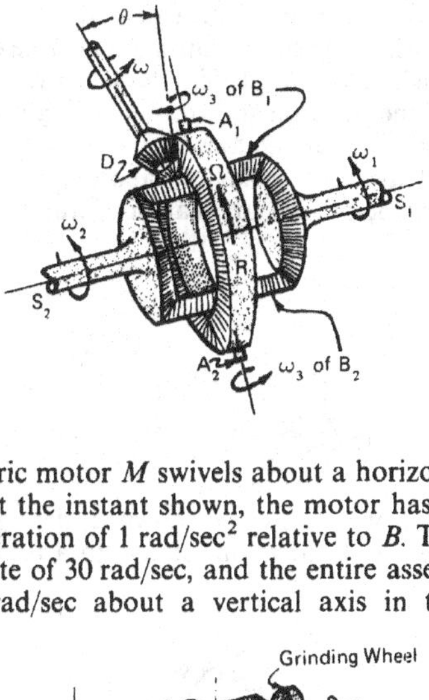

Problem 4.16.

4.17. The housing of an electric motor M swivels about a horizontal shaft supported by bearings in a bracket B. At the instant shown, the motor has an angular speed of 3 rad/sec and an angular acceleration of 1 rad/sec^2 relative to B. The motor drives a grinding wheel G at a constant rate of 30 rad/sec, and the entire assembly rotates with a constant angular speed of 5 rad/sec about a vertical axis in the ground frame

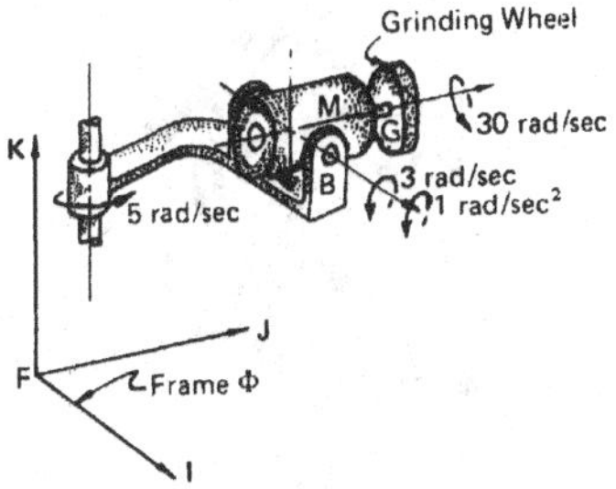

Problem 4.17.

$\Phi = \{F; \mathbf{I}_k\}$, as indicated. Find the absolute angular velocity and angular acceleration of G at the moment shown. Define carefully all reference frames used and refer the solution to Φ.

4.18. A centrifuge is used to simulate flight conditions in the training of pilots and astronauts. The structural beam AB of the centrifuge shown in the figure has an angular speed $\omega_1 = 2$ rad/sec which is increasing at the rate of $\dot{\omega}_1 = 1$ rad/sec^2 about a fixed vertical axis at the moment t_0. The capsule is rotating, as shown, at a constant rate $\omega_2 = 2$ rad/sec relative to the beam, and the seat is turning within the capsule at a constant rate $\omega_3 = 1$ rad/sec relative to the capsule, as indicated. Determine in the laboratory frame $\Phi = \{F; \mathbf{I}_k\}$ the total angular velocity and angular acceleration of a pilot strapped to the seat in the configuration shown at the instant t_0.

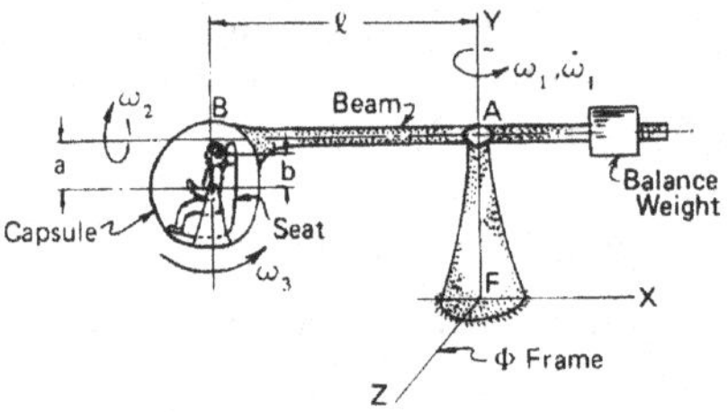

Problem 4.18.

4.19. At the instant of interest, a disk D is spinning about its axis AB with an angular speed $\omega_1 = 20$ rad/sec and an angular acceleration $\dot{\omega}_1 = 2$ rad/sec^2 relative to the aircraft in which AB is fixed. Simultaneously, the aircraft is executing a roll and turn maneuver. In the position shown, the rolling rates of the plane are $\omega_2 = 2$ rad/sec and $\dot{\omega}_2 = 0.2$ rad/sec^2 about the axis of the aircraft, and the plane's turning rates are $\omega_3 = 2$ rad/sec and $\dot{\omega}_3 = -0.6$ rad/sec^2 about a vertical axis in space. Compute the absolute angular velocity and angular acceleration of the disk in the configuration shown.

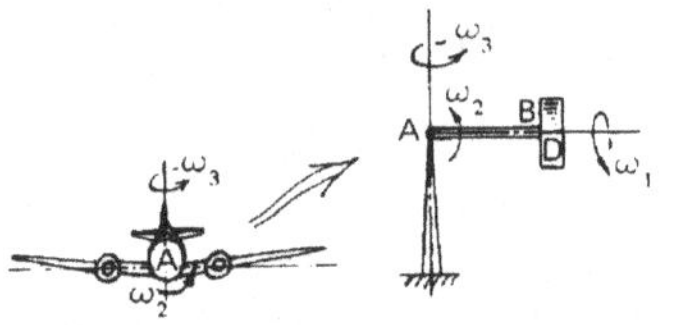

Problem 4.19.

4.20. The rotor of a gyroscope shown in the figure is spinning at a constant rate of 15 rad/sec about the **k** direction relative to a frame $\varphi = \{O; \mathbf{i}_q\}$ fixed in the gimbal

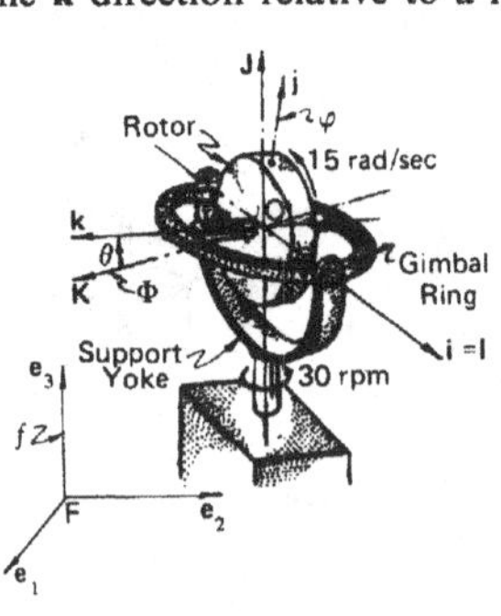

Problem 4.20.

ring. The angle θ between the plane of the ring and the horizontal plane of the frame $\Phi = \{O; \mathbf{I}_p\}$ fixed in the support yoke is growing at a constant rate of 5 rad/sec. The entire assembly is precessing about the **J** direction with a constant angular speed of 30 rpm relative to a fixed frame f. Find in rpm measure units the magnitudes of the absolute angular velocity and angular acceleration of the rotor when $\theta = 30°$.

4.21. An armored missile carrier is preparing to launch its weapon. At the instant illustrated, the missile M is being elevated at an angular speed $\dot{\alpha} = 0.5$ rad/sec relative to its launching pad L, and this rate is decreasing at 0.2 rad/sec each second when $\alpha = 30°$. The pad is rotating concurrently at an angular speed $\dot{\beta} = 0.3$ rad/sec relative to the carrier C, and this rate is increasing at 0.1 rad/sec^2 when the pad is positioned at $\beta = 10°$. At the same instant, C is turning as shown at the constant rate $\dot{\gamma} = 0.2$ rad/sec relative to the ground frame Φ. Determine the absolute angular velocity and angular acceleration of the missile referred to Φ.

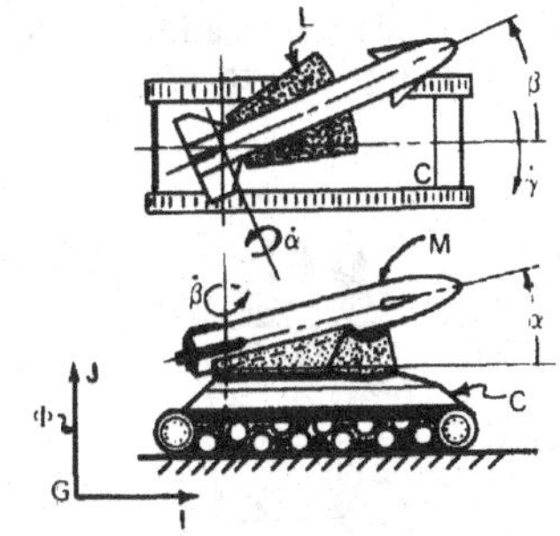

Problem 4.21.

4.22. A submarine diving simulator shown in the figure consists of a command bridge B which can rotate within a drum D to simulate rolling, while the drum is supported in swivel bearings in an A-frame to simulate diving pitch about a horizontal axis. To model yaw and translation, the A-frame is mounted on an air table with directional air jets so that it can turn and move freely over a smooth horizontal surface in the test facility. The controllor's reactions are monitored by television and by observation through a one-way glass screen. In a typical exercise, the A-frame rotates (yaw) over the horizontal surface with a constant angular speed $\omega_1 = 0.25$ rad/sec in the laboratory frame $\Phi = \{F; \mathbf{e}_k\}$, as shown; and, relative to the A-frame, the drum pitches at a constant rate $\omega_2 \equiv \dot{\alpha} = 0.18$ rad/sec about the **i** direction of the frame $\varphi = \{D; \mathbf{i}_k\}$

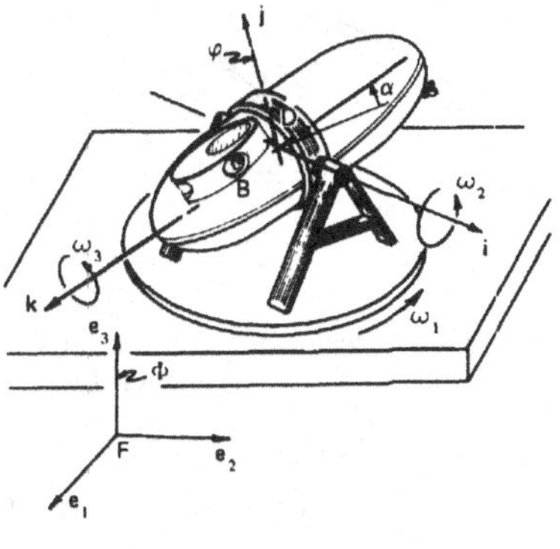

Problem 4.22.

fixed in the drum. At the same time, the bridge rolls within the drum with the relative angular speed $\omega_3 = 0.65 \cos(0.14t)$ rad/sec. Find as functions of time t in Φ the absolute angular velocity and angular acceleration of the command bridge referred to φ. Determine their values at the initial instant when $\alpha = 0$.

4.23. A cutting wheel W is driven at a constant angular speed $\omega = 300$ rpm by an electric motor which is mounted in swivel bearings attached to a rotating table. At the instant shown, the motor is turning about the swivel bearing axis with an angular speed of 5 rpm which is decreasing at the constant rate of 2 rpm^2 relative to the table. The table has a constant angular speed of 30 rpm relative to the machine frame $\Phi = \{G; \mathbf{I}_k\}$. (a) Determine the angular velocity and angular acceleration of W as seen by a machine operator M and referred to Φ at the moment of interest. (b) Identify in the calculations the rates of change of the angular velocity of the wheel relative to the motor, the motor relative to the table, and the table relative to the machine, as seen by M. (c) What is the angular acceleration of W at the moment of concern, as seen by an observer fixed in the table? (d) Derive a formula for the time rate of change in Φ of the angular acceleration of W apparent to M.

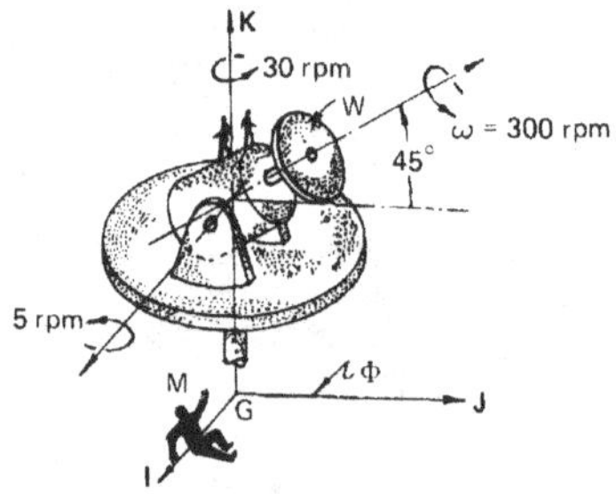

Problem 4.23.

4.24. A disk D of radius 1 ft rotates as shown with a constant angular speed $\omega_2 = 10$ rad/sec relative to a platform which is turning at a constant rate $\omega_1 = 20$ rad/sec relative to the ground frame $\Phi = \{G; \mathbf{m}_k\}$. Let $\mathbf{r}$ be the radius vector of a point P on the rim of D referred to a suitable frame φ in D. (a) Find in Φ the first and second time derivatives of $\mathbf{r}$ at the instant t_0 when the platform is in the horizontal position and $\mathbf{r} = \mathbf{m}_3$. (b) What is the total angular acceleration of D relative to Φ at the time t_0? (c) Verify the results for (a) by an alternative method.

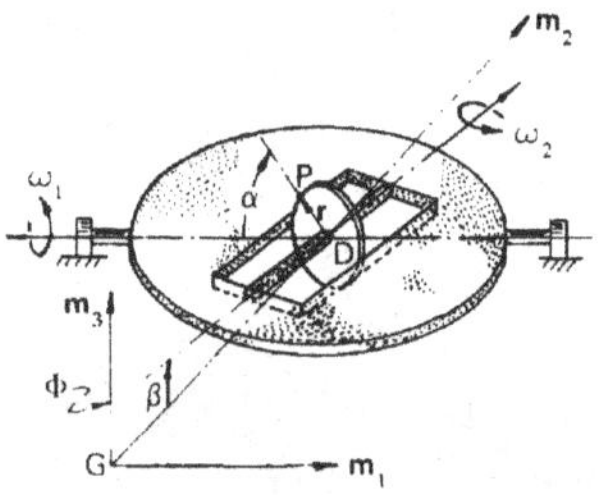

Problem 4.24.

4.25. (a) Show that the scalar α and the relative angular speeds in the analysis of the universal joint of Section 4.2.2.1. are given by

$$\alpha = (1 - \sin^2\phi \cos^2\theta)^{-1/2},$$

$$\omega_{31} = -\omega_2 \sin\phi \cos\theta, \qquad \omega_{32} = -\omega_1 \alpha \sin\phi \sin\theta.$$

Thus, find the absolute angular velocity $\boldsymbol{\omega}_{30}$ of the cross link referred to the cross-link frame φ_3 in Fig. 4.9, and expressed in terms of the angular speed ω_1 of the drive shaft. (b) Find the form of the result referred to the fixed frame Φ. (c) What are the absolute velocities of the end points A and D on the cross link? (d) Describe an alternative scheme for the determination of the results found here.

4.26. (a) Use the results of the previous problem to determine the absolute angular acceleration of the cross link of the universal joint referred to the cross-link frame φ_3. Assume that the angular velocity $\boldsymbol{\omega}_1$ of the drive shaft is constant in Φ. (b) What are absolute accelerations of the end points A and D on the cross link? (c) Describe an alternative plan for the derivation of these results.

4.27. The position vector of a particle P is given by $\mathbf{X} = 3t^2\mathbf{I}$ in a preferred frame $\Phi = \{F; \mathbf{I}, \mathbf{J}, \mathbf{K}\}$. Find the velocity and acceleration of P in Φ. Then consider another frame $\varphi = \{F; \mathbf{i}, \mathbf{j}, \mathbf{K}\}$ which is rotating with a constant angular velocity $\boldsymbol{\omega}_f = \omega\mathbf{K}$ relative to Φ. Use equations (4.46) and (4.48) to find the absolute velocity and acceleration of P referred to φ. Show that the velocity and acceleration determined in this manner are the same as the velocity and acceleration determined first in the preferred frame Φ and afterwards projected upon the axes of the moving frame φ.

4.28. (a) The Darboux vector is defined by

$$\boldsymbol{\delta} = \kappa\mathbf{b} - \tau\mathbf{t}$$

in terms of the intrinsic basis $\{\mathbf{e}_k\} = \{\mathbf{t}, \mathbf{n}, \mathbf{b}\}$, the path curvature κ, and the torsion τ. Show that the Serret–Frenet equations (1.108) may be abbreviated by

$$\frac{d\mathbf{e}_k}{ds} = \boldsymbol{\delta} \times \mathbf{e}_k. \tag{A}$$

(b) Note that $\boldsymbol{\delta}$ has the physical dimensions of reciprocal length: $[\boldsymbol{\delta}] = [L^{-1}]$. Hence, $\dot{s}$ being the particle speed, the vector $\boldsymbol{\omega} = \dot{s}\boldsymbol{\delta}$ has the dimensions of reciprocal time. Therefore, it is seen from (A) that

$$\frac{d\mathbf{e}_k}{dt} = \boldsymbol{\omega} \times \mathbf{e}_k. \tag{B}$$

What is the physical interpretation of the Darboux vector?

(c) Prove that the absolute acceleration of a particle referred to the intrinsic frame $\psi = \{P; \mathbf{e}_k\}$ may be derived from the equation

$$\mathbf{a} = \frac{\delta \mathbf{v}}{\delta t} + \boldsymbol{\omega} \times \mathbf{v}. \tag{C}$$

Discuss the physical interpretation of the scalar components of $\boldsymbol{\omega}$, and describe its orientation to the osculating plane.

4.29. A plane flying northwest experiences an apparent wind (relative to the plane) from the west. When it flies northeast with the same speed relative to the ground, the apparent wind is from the east, but twice as brisk as before. Set up a reference frame $\Phi = \{F; \mathbf{I}_k\}$ fixed in the ground, describe all velocity vectors in terms of Φ, and find the true wind direction in Φ.

4.30. While making a landing approach from the north, an aircraft has a speed of v mph relative to the air which is blowing steadily from the northwest at $2kv$ mph. At the moment shown, the plane is decreasing its speed relative to the ground at the rate of α mph each second. The propellor, which is r ft long, is rotating at ω rad/min. Find the absolute velocity (in mph) and acceleration (in mph/sec) of the tip T of the propellor at the instant of interest.

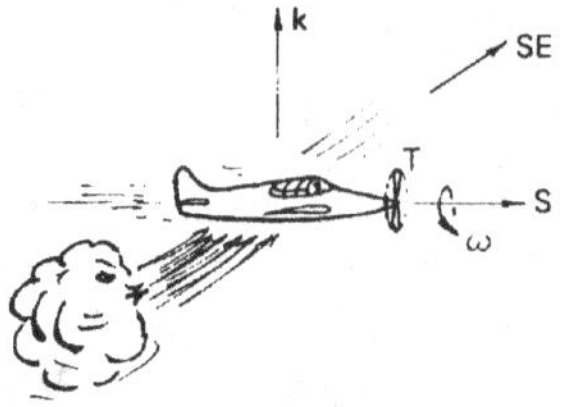

Problem 4.30.

4.31. Let ψ denote the slant angle of the Scotch crank whose operation is described in Problem 1.15. Apply equations (4.51) to find the velocity and acceleration of the piston point Q as functions of ψ and the crank angle $\theta(t)$.

4.32. A circular cylinder of radius a rolls without slipping on a horizontal table which is suspended, as shown, by two hinged links of equal length L. The cylinder has the relative angular velocity $\boldsymbol{\omega}_2$, and the links have angular velocity $\boldsymbol{\omega}_1$ as the table swings back and forth. Find the absolute velocity and acceleration of the center C of the cylinder. What is the angular velocity of the table? What are the velocity and acceleration of C relative to the table?

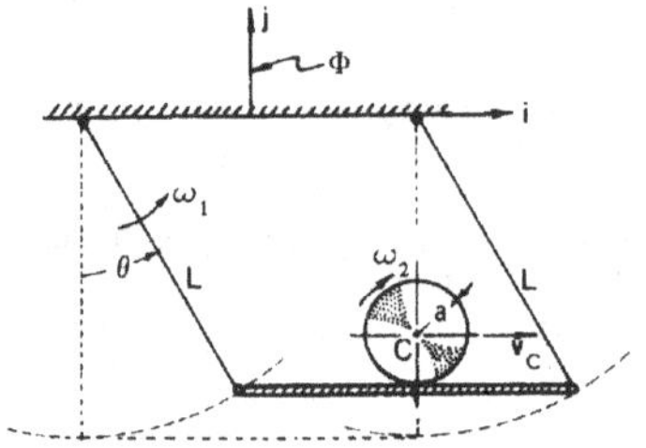

Problem 4.32.

4.33. The motion of the pins A and B of a Scotch mechanism shown in the figure are controlled by the link C which moves to the right with a constant speed of 6 ft/sec during a period of its motion. The pins are constrained to move in a parabolic track

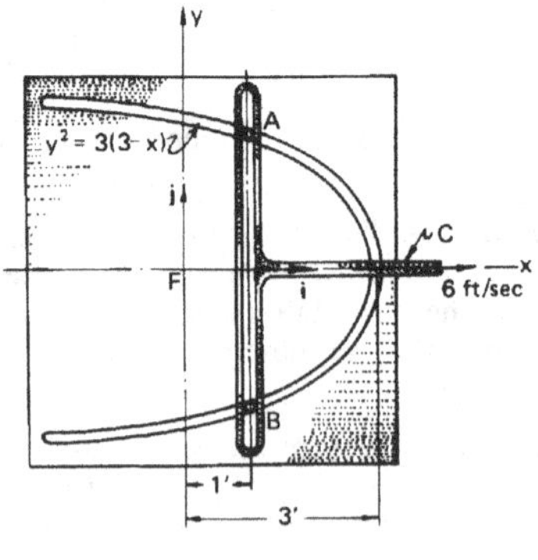

Problem 4.33.

described by $y^2 = 3(3-x)$ in the frame $\Phi = \{F; \mathbf{i}_k\}$. Find the velocity and acceleration of pin A in frame Φ when the link is in the configuration shown. What are the velocity and acceleration of A relative to B at this instant?

4.34. Suppose that the mechanism described in Example 4.9 has the initial position shown in Fig. 4.16. Find the velocity and acceleration of the pin P relative to the frames Φ and φ at the instant $t = 0.4$ sec.

4.35. A pin P controls the motion of two slotted links so that they move on guide rods at right angles to one another. At the instant illustrated, link A has a speed to the right of 15 cm/sec and is decelerating at the rate of 50 cm/sec^2. Concurrently, the link B is moving upward with a speed of 20 cm/sec and is slowing down at the rate of 75 cm/sec each second. What is the radius of curvature of the trajectory of P at this instant?

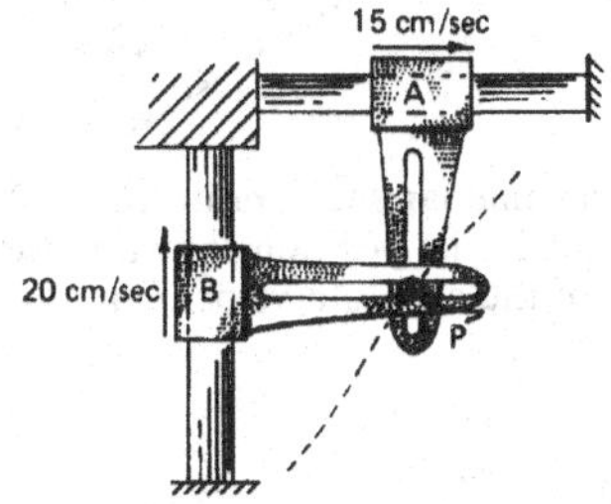

Problem 4.35.

4.36. The vertical motion of a piston is controlled by a cam ABC attached to a conveyor belt that moves horizontally with constant velocity $\mathbf{v}$, as illustrated. The face of the cam has the shape of a full cosine wave of length L and amplitude β. Find the velocity and acceleration of the piston in the fixed frame $\Phi = \{F; \mathbf{n}_k\}$. What is the time interval for which these results will hold?

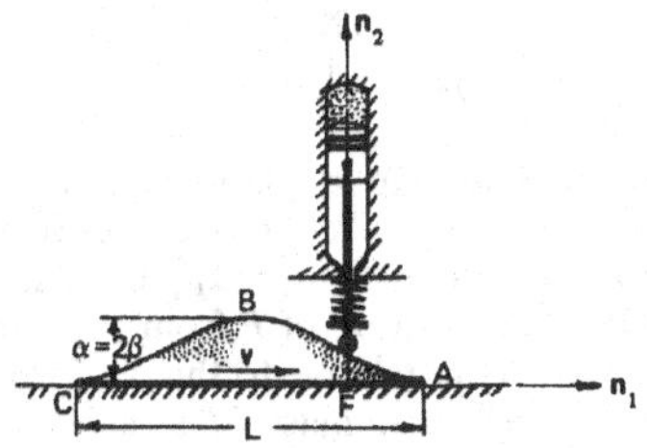

Problem 4.36.

4.37. A motor boat leaves from the point P on a river bank to reach a point Q on the opposite side but distance W downstream, as shown. The boat moves at $2v_0$ ft/sec relative to the water which has a uniform downstream flow rate of v_0 ft/sec. An observer at P sees the boat travel on a straight line in the ground frame. Find the time required for the boat to reach Q. Evaluate the result for the case when $W = 750$ ft and $v_0 = 5$.

4.38. A pin P is constrained to move in the slots of two sliding links A and B. The mechanism is designed so that the circular link B may turn freely about F and at the same time constrain the path of the pin P without affecting its motion, which is controlled by link A. During an interval of its motion, the link A moves toward the right with a constant speed $v_A = 4\omega$, while the circular link B of radius 5 units turns as

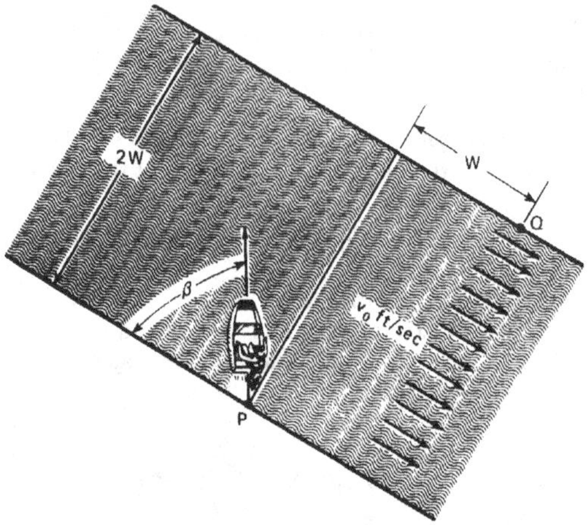

Problem 4.37.

shown with a constant angular speed ω in frame $\Phi = \{F; \mathbf{i}_k\}$. At an instant t_0 during this period, the device is in the position shown in the figure. Find as functions of ω the instantaneous values of the absolute velocity and acceleration of the pin P referred to frame Φ at the instant t_0.

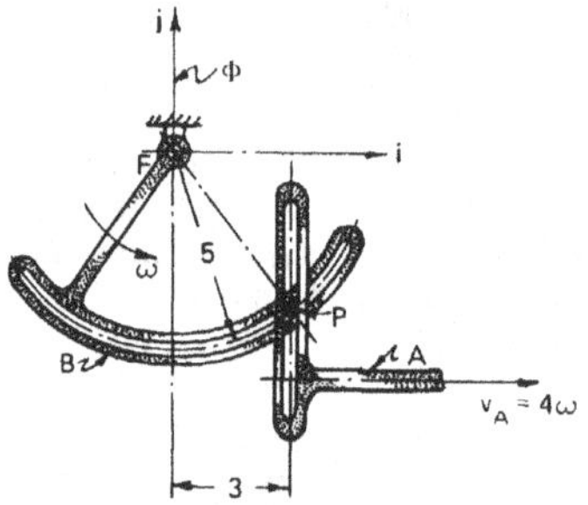

Problem 4.38.

4.39. A motor is bolted to a platform that is supported by springs and constrained by roller bearings to move vertically with a sinusoidal motion $B(t) = a + b \sin pt$ in the frame $\Phi = \{F; \mathbf{I}_k\}$ fixed in the foundation. Here a, b, and p are constants and B is the position of P from F, as shown in the figure. The motor has a constant angular velocity $\boldsymbol{\omega}$ relative to the platform; but its rotor shaft is unbalanced. This is modeled by an eccentric mass M attached to the rotor at a radial distance r

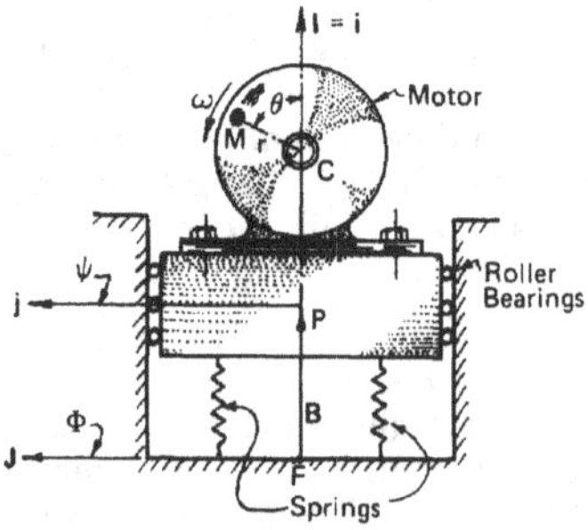

Problem 4.39.

from the center C of the shaft, as shown. Find the absolute velocity and acceleration of M referred to the frame $\psi = \{P; \mathbf{i}_k\}$ fixed in the platform. What is the velocity of M relative to P in ψ?

4.40. A river flow has an eastward directed, parabolic velocity distribution that varies from zero at both banks, which are 600 m apart, to a maximum value of 1 m/sec at midstream, as shown. A motor boat moves across the river at the constant rate of 10 m/sec relative to the water and always is directed toward the north bank. Determine the absolute velocity and acceleration of the boat expressed as functions of the parameter z defined by $z = y/300$; and find the path $x = x(z)$ traveled by the boat. How far downstream does the boat travel in crossing the river?

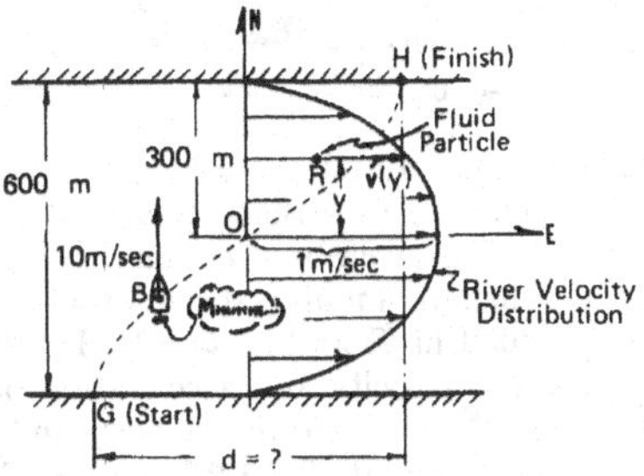

Problem 4.40.

4.41. Two pins A and B shown in the figure are constrained to move in circular slots of radius 15 cm. One slot is cut in a fixed flat plate and the other is in the curved link of a mechanism. During the interval of interest, the link moves as shown with a constant speed $v = 9$ cm/sec. Initially, the point D on the link is at F, the origin of the fixed circular slot. (a) Determine the initial velocity and acceleration of the pin A in the fixed frame $\Phi = \{F; \mathbf{I}_k\}$. (b) What are the velocity and acceleration of the pin A relative to the pin B at the initial instant?

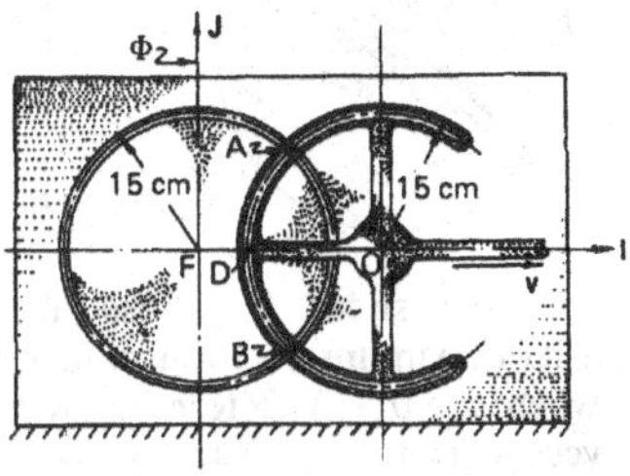

Problem 4.41.

4.42. Suppose that the circular slots of the mechanism described in the previous problem have radius r and that the curved link has a constant velocity $v\mathbf{I}$ during an interval of its motion in which the point D is initially at F in Φ $\{F; \mathbf{I}_k\}$, as before. Let ϕ be defined by $\cos\phi = \frac{1}{2}[1 + (v/r)\,t]$. (a) Find as functions of ϕ the velocity and acceleration of the pin A in Φ. (b) Determine as functions of ϕ the velocity and acceleration of pin A relative to pin B. (c) Compute the foregoing solution quantities at the times $t = 0$ and $t = 1$ sec for the special case when $r = 15$ cm and $v = 9$ cm/sec.

4.43. A cam mechanism shown in the figure is to be designed so that the upward motion of the spring-loaded push rod is uniform with speed v over the entire cam rise r. The downward return stroke of the rod also is to be uniform but twice as rapid. The cam is attached to a conveyor belt that moves horizontally with a constant velocity $\mathbf{V}$, as indicated. Find the dimensions a and b, and determine the cam profile $y = y(x)$ in terms of the given design parameters. The reader who may be familiar with singularity functions should solve this design problem by the singularity methods described at the end of Chapter 1.

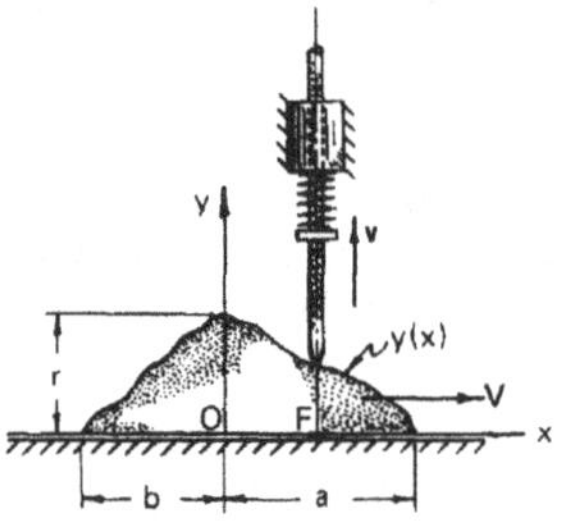

Problem 4.43.

4.44. Two slider blocks A and B are hinged at the ends of a rigid connecting rod of length $2a$. The blocks slide in perpendicular slots in a rotating table T so that the position of B relative to T is given by $x(t) = a \sin pt$, where a and p are constants. The table turns with a constant angular speed $\omega = 10$ rad/sec, as shown. Introduce a suitable frame φ imbedded in T, and let $\Phi = \{0; \mathbf{I}_k\}$ denote the preferred frame in the ground. (a) What are the velocity and acceleration of the slider B relative to T? (b) Referred to both φ and Φ, what are the velocity and acceleration in Φ that B would have if it were locked in its slot at x? (c) What is the Coriolis acceleration of B referred to φ? (d) What is the velocity of the moving block B apparent to an observer in Φ, and referred to both φ and Φ? (e) How are the foregoing solutions related to the absolute velocity and acceleration of B?

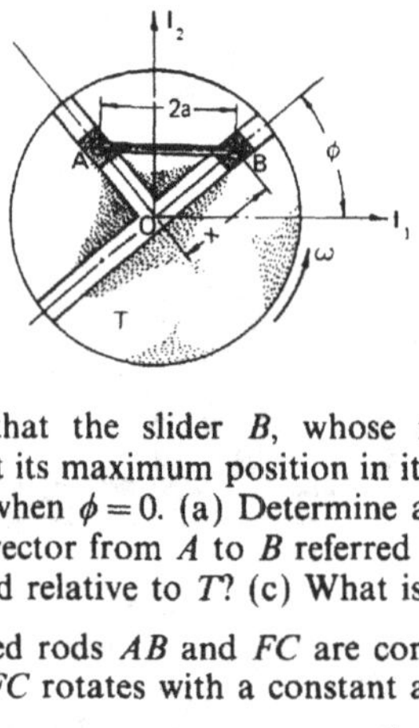

Problem 4.44.

4.45. Suppose that the slider B, whose motion is described in the previous problem, is locked at its maximum position in its slot. Use the previous data, and consider the instant t_0 when $\phi = 0$. (a) Determine at t_0 the first and second time rates of change in Φ of the vector from A to B referred to φ. (b) What is the angular velocity of the connecting rod relative to T? (c) What is the total angular velocity of the rod?

4.46. Two hinged rods AB and FC are connected by a slider block B that slides along FC. The rod FC rotates with a constant angular velocity $\boldsymbol{\omega} = 12\mathbf{k}$ rad/sec. What

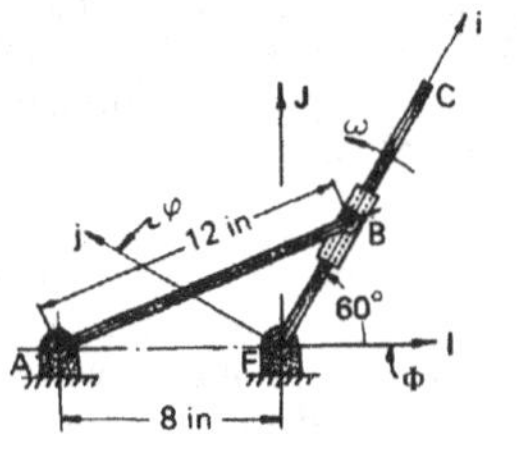

Problem 4.46.

is the absolute velocity of the slider B at the position shown in the figure and referred to the frame $\varphi = \{F; \mathbf{i}_k\}$ imbedded in the rod FC?

4.47. The motion of a telescopic link AB is controlled by the simultaneous rotation of the rigid links OA and BC. When the device is in the position shown, OA has a counterclockwise angular speed $\omega_1 = 10$ rad/sec and the telescopic rod is rotating clockwise at $\omega_2 = 4$ rad/sec in the machine frame $\Phi = \{O; \mathbf{i}_k\}$. What is the angular velocity of the link BC in Φ at this instant?

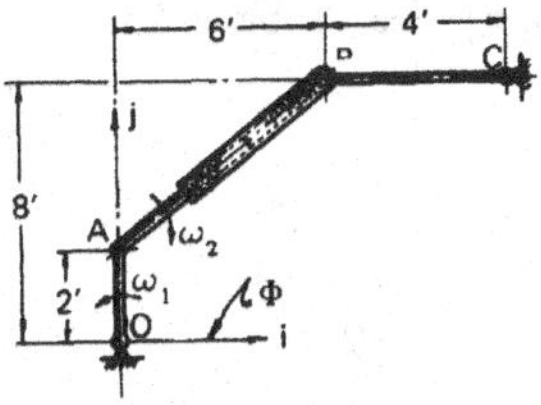

Problem 4.47.

4.48. A pendulum bob B is attached by a hinged rod of length l to a slider block S that moves in the vertical direction so that $x = a + b \sin pt$, where a, b, and p are constants. The bob oscillates in the vertical plane shown in the figure. Determine the acceleration of B in the ground frame $\Phi = \{O; \mathbf{I}_k\}$, but referred to the intrinsic frame that follows B.

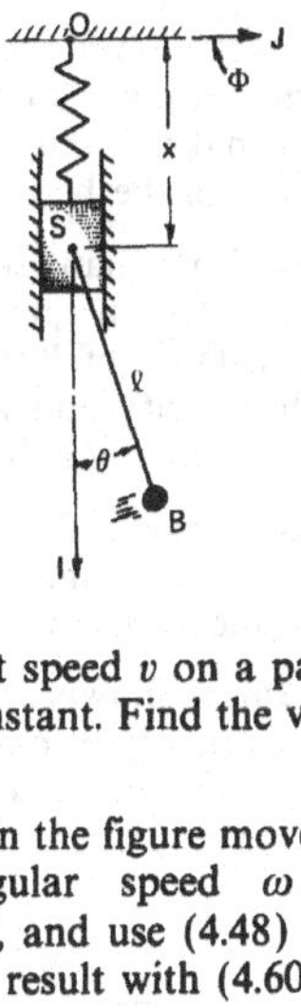

Problem 4.48.

4.49. A particle P moves with constant speed v on a path described by the plane polar equation $r = m \sin \phi$, where m is a constant. Find the velocity and acceleration of P as functions of ϕ.

4.50. The slider P of a linkage shown in the figure moves along a rod that rotates about its vertical shaft with an angular speed $\omega = \dot{\phi}$. Consider a frame $f = \{O; \mathbf{e}_r, \mathbf{e}_\phi, \mathbf{e}_z\}$ fixed in the rotating rod, and use (4.48) to derive the equation for the acceleration of the slider. Compare the result with (4.60).

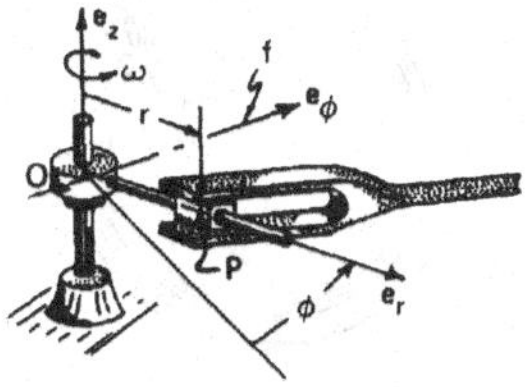

Problem 4.50.

4.51. Show that the curvature κ of a plane curve defined in polar coordinates by the equation $r=r(\phi)$ is given by

$$\kappa=\left|\frac{r^2+2r'^2-rr''}{(r^2+r'^2)^{3/2}}\right| \qquad \text{with } ' \equiv d/d\phi.$$

4.52. An electron E moves with constant speed c along an Archimedean spiral $r=a\phi$ shown in the figure. The scalar a is a constant. Find the velocity and acceleration of E referred to the cylindrical frame $\varphi=\{F; \mathbf{e}_r, \mathbf{e}_\phi, \mathbf{e}_z\}$ and to the intrinsic frame $\psi=\{E; \mathbf{t}, \mathbf{n}, \mathbf{b}\}$. See Problem 4.51.

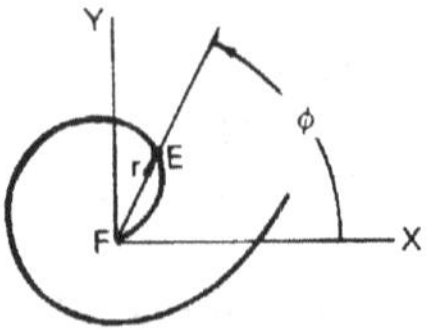

Problem 4.52.

4.53. The motion of the center R of the roller of a limacon cam is described in Problem 1.17. Determine the velocity and acceleration of R referred to the cylindrical reference frame $\psi=\{O; \mathbf{e}_r, \mathbf{e}_\phi\}$ shown in the problem figure.

4.54. A small bead moves with a constant relative speed v along a thin circular ring of radius a. The ring moves simultaneously with a constant speed μ parallel to the positive $\mathbf{K}$ direction, which makes an angle α with a line normal to the plane of the ring. What is the acceleration of the bead?

4.55. A particle P moves in a plane with an acceleration which is directed always toward a fixed point F in Φ. The magnitude of the acceleration is inversely proportional to the square of the distance of P from F. Prove that the hodograph is a circle. Hint: Find the hodograph velocity and acceleration in cylindrical coordinates, and determine the curvature of the hodograph.

4.56. The trajectory of a particle P is a limacon described by the plane polar equation $r=a-b\cos\phi(t)$, where a and b are constants and $a>b$. Sketch the path, and determine the velocity and acceleration of P as functions of ϕ and its derivatives.

4.57. A shifting mechanism shown in the figure moves a slider block S in a circular slot of radius b. The shifting lever passes through the slider and rotates about the hinge 0 with a constant angular velocity $\boldsymbol{\omega}$. Find the velocity and acceleration of S in terms of cylindrical coordinates.

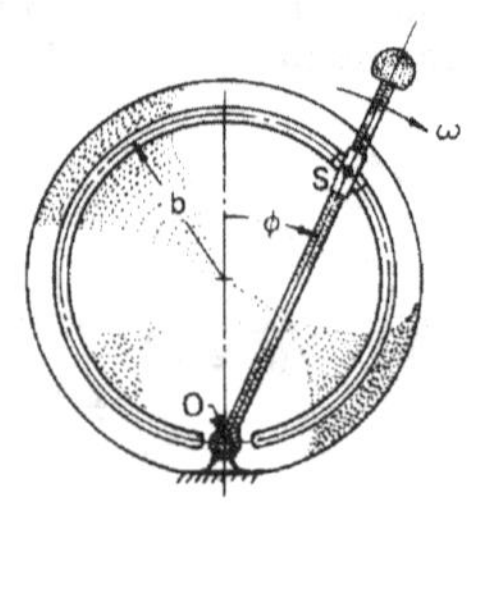

Problem 4.57.

4.58. The slotted drive link FA of a mechanism moves a pin P in a spiral groove $r = a\phi$ cut in a fixed plate. The link starts from rest at $\phi = \pi/2$ rad and moves with a constant angular acceleration α. What is the acceleration of the pin P and the angular velocity of the link when $\phi = 5\pi/4$ rad?

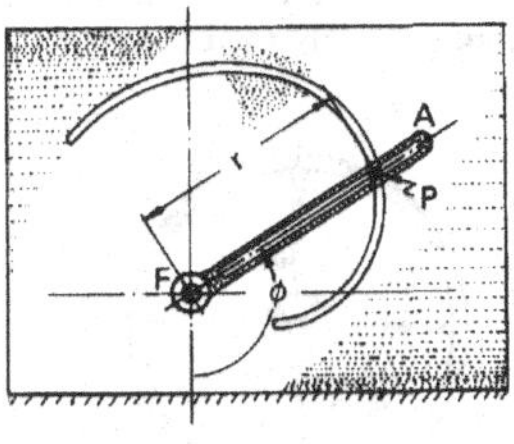

Problem 4.58.

4.59. The agitator in a washing machine has a periodic vertical motion $z = a \sin pt$ and a simultaneous periodic rotary motion $\omega = b \sin qt$ about its vertical axis, where a, b, p, and q are constants that characterize the motion. Find the acceleration of a point P on the agitator at a fixed perpendicular distance r from the vertical axis.

4.60. A search light at F in $\Phi = \{F; \mathbf{n}_k\}$ has its beam fixed on an aircraft A which is flying horizontally with a constant velocity $\mathbf{v}$ at an altitude h in Φ. Determine the angular velocity $\boldsymbol{\omega}$ of the search light in tracking the aircraft. Express the solution in terms of the geometrical parameters shown in the figure.

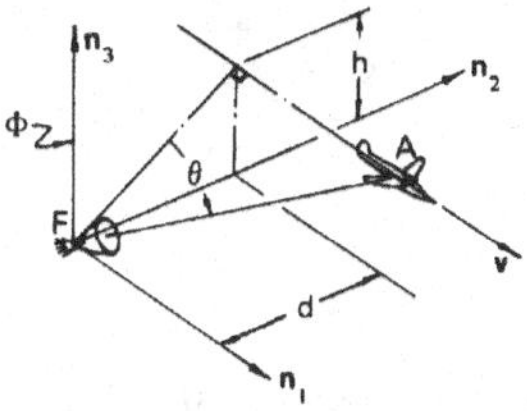

Problem 4.60.

4.61. During an interval of its motion, a crane arm turns about an axis at O with a constant angular speed $\omega = 0.5$ rad/sec. The crane carries a cable system that consists of a movable pulley P, a rewind engine R, and a hoisting cable PAB attached to P and to an object at B which is to be moved. When the rewind engine clutch is disengaged, the taut cable is unreeled by raising the crane arm, as shown. If $r = 0$ and $\phi = 0$ initially, calculate the acceleration of P when $\phi = 30°$ and $D = 40$ ft. What is the greatest angle through which this machine may operate?

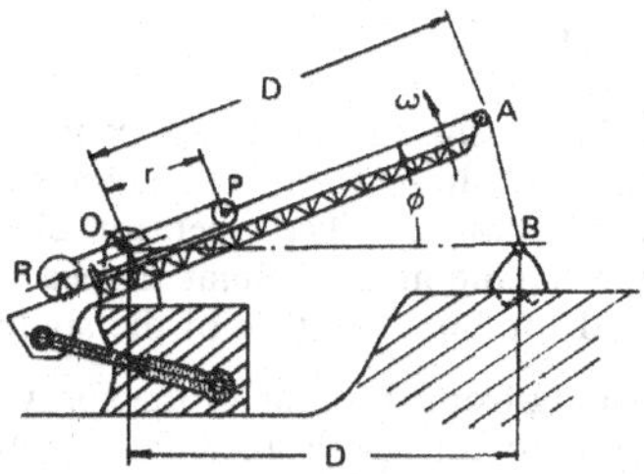

Problem 4.61.

4.62. A flywheel with a threaded hub rotates, as shown, on a right-hand screw having a pitch of 2 cm. At the instant of interest, the wheel is turning at an angular speed ω which is increasing at the rate $\dot{\omega}$. At the same instant, a sleeve S is moving along a wheel spoke with a constant, central directed speed of 5 cm/sec and is 25 cm from the screw axis. Find the total velocity and acceleration of the sleeve in the ground frame. The angular measure is in radians.

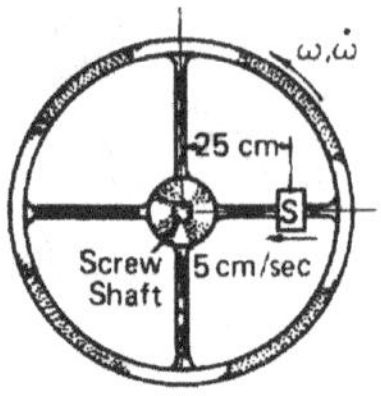

Problem 4.62.

4.63. A circular platform rotates as shown with an angular velocity $\boldsymbol{\omega}$ and an angular acceleration $\dot{\boldsymbol{\omega}}$ about F in the fixed frame $\Phi = \{F; \mathbf{I}_k\}$. A pin P of a mechanism (not shown) moves in a radial groove so that its distance $a(t)$ from the center F is a function of time t. Find the absolute velocity and acceleration of the pin referred to a frame fixed in the platform.

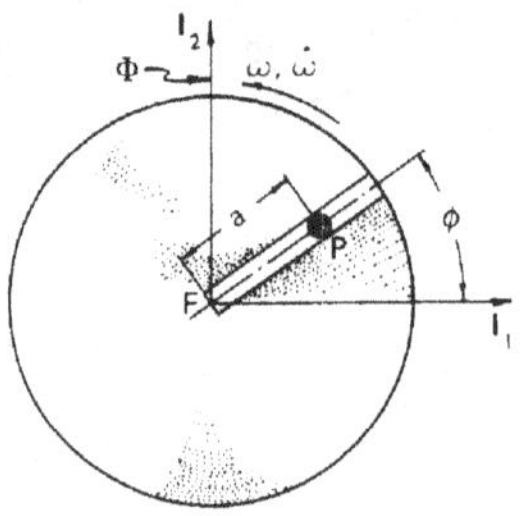

Problem 4.63.

4.64. A valve handle of diameter 30 cm advances at the constant rate $v = 3$ cm/sec along a right-handed screw shaft of pitch 2 cm, as shown. Compute the acceleration of a point on the rim of the handle, referred to a cylindrical reference frame.

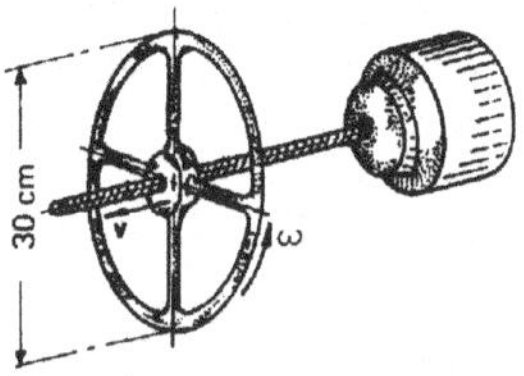

Problem 4.64.

4.65. Use the data assigned in the last problem, and suppose further that at the instant of interest t_0, the handle is turning at the rate $\omega = 3\pi$ rad/sec, which is increasing at the rate $\dot{\omega} = 2\pi$ rad/sec^2. The operator observes a spider escaping clockwise along the rim of the handle at an estimated speed of 1 cm/sec relative to the handle. What is the observed total acceleration of the spider at the instant t_0?

4.66. Water issuing from the ends of the spray wheel of a garden sprinkler shown in the figure causes it to turn with an angular velocity $\boldsymbol{\omega}(t)$. Assume that the water travels through the sprinkler with a constant relative speed v. Determine the absolute velocity and acceleration of the fluid particle P at d from O in the straight portion of the arm. What is the absolute acceleration of P at the exit E?

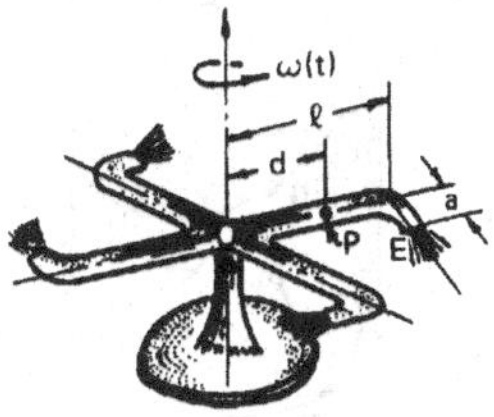

Problem 4.66.

4.67. A fluid is discharged from a nozzle and runs in a thin sheet down a cylindrical surface, as illustrated. The velocity of the fluid particle at any point P on the surface of the cylinder is given by $\mathbf{v} = ch^{1/2}\mathbf{e}_\phi$, where c is a constant and h is the vertical distance of P below the nozzle, as shown. Determine the acceleration of the fluid particle as a function of the colatitude angle ϕ.

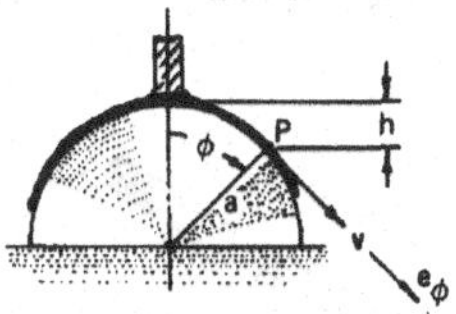

Problem 4.67.

4.68. A disk having a circular slot rotates about its axis at O with a constant angular velocity $\boldsymbol{\omega} = 10\mathbf{k}$ rad/sec, as shown. During an interval of its motion, a pin P moves in the slot with a constant speed $v = 75$ cm/sec relative to the disk. Compute the absolute acceleration of P by two methods: (i) by use of (4.48), and (ii) by use of (4.60).

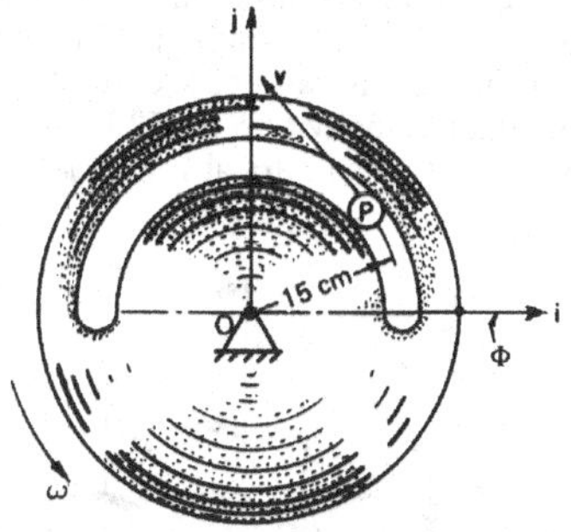

Problem 4.68.

4.69. An electron, initially at rest, is emitted from a cathode at a point E in an electromagnetic field. The field drives the electron along a cylindrical helix of radius r and pitch p, as illustrated, so that its tangential acceleration component is proportional to $\cos\psi$, where ψ is the constant helix angle. (See Example 1.14.) What are the velocity and acceleration of the electron when it passes the field point P. Refer the vectors to a cylindrical reference frame.

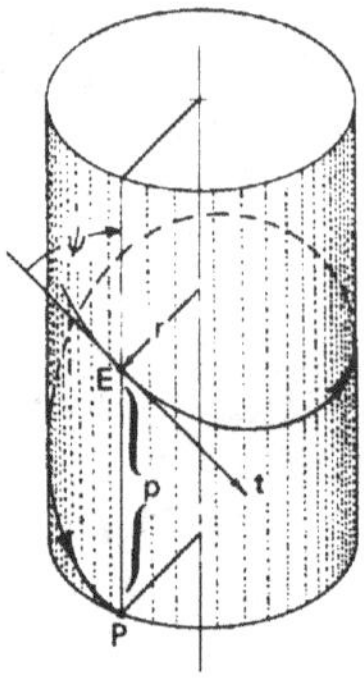

Problem 4.69.

4.70. The slotted link of a mechanism is rotating with an angular speed $\omega = 2$ rad/sec and an angular acceleration $\dot{\omega} = 4$ rad/sec^2 at the instant when the link is at the 30° position shown in the figure. At that moment, the slider B is 20 cm from the hinge pin center at O, and this distance is increasing at the constant rate of 15 cm/sec. Find the absolute velocity and acceleration of B referred to the fixed frame $\Phi = \{O; \mathbf{I}_k\}$ and to the frame $\varphi = \{O; \mathbf{i}_k\}$ imbedded in the link.

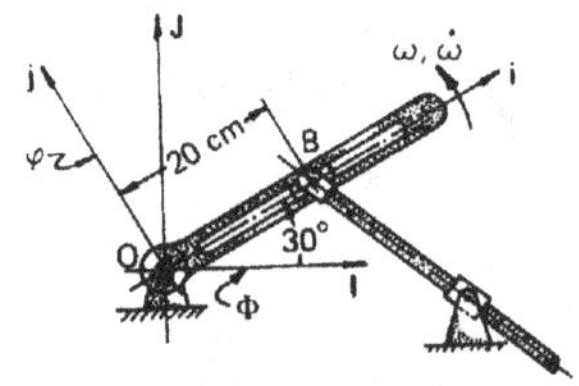

Problem 4.70.

4.71. Two wheels of a vehicle are modelled as thin disks A and B of radius r mounted in bearings perpendicular to the ends of a shaft S of length $2L$, as shown in the figure. Both disks roll without slipping on a horizontal plane H in the ground frame $\Phi = \{G; \mathbf{I}_k\}$. Let α and β denote the disk rotation angles, as shown; and let ϕ be the angle between $\mathbf{I}_1$ and the line parallel to S through the points of rolling contact at C and D in H. Find the velocity and acceleration of the center O of S at $\mathbf{X} = (X, Y, r)$ in Φ. What is the angular velocity of S in Φ? Express the results as functions of r, L, ϕ, $\dot{\alpha}$, and $\dot{\beta}$, and referred to the cylindrical frame $\psi = \{O; \mathbf{e}_k\}$ fixed in S with $\mathbf{e}_z = \mathbf{I}_3$.

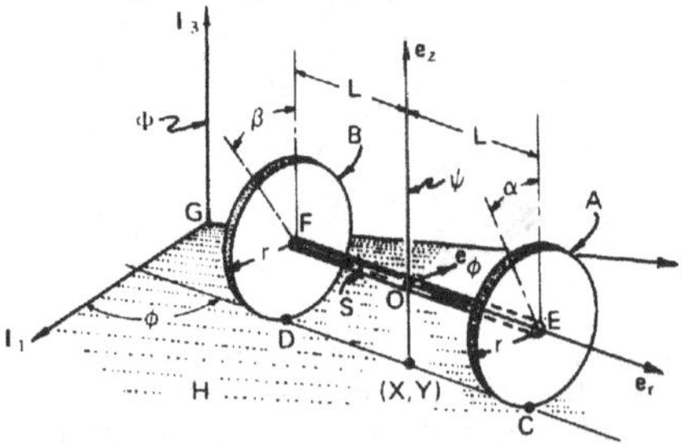

Problem 4.71.

4.72. A sorting hopper of a certain automatic assembly machine is designed to move small electronic components along a spiral path on the surface of a right circular cone of base radius b and altitude h. The motion is induced by the hopper vibrations. Let the frame $\Phi = \{O; \mathbf{i}_k\}$ be fixed in the base of the hopper. Suppose that a typical tiny component P, initially at rest at $\mathbf{x}_0 = b\mathbf{i}$, is turned counterclockwise about the cone axis with a constant angular speed $\dot{\phi}(t) = \omega$ as it advances in the direction of the axis at a constant rate $\dot{z}(t) = A$ in Φ. Find the motion, the velocity, and the acceleration of P as functions of time t and referred to the cylindrical reference frame $\psi = \{P; \mathbf{e}_r, \mathbf{e}_\phi, \mathbf{e}_z\}$. Express the results in terms of b, h, ω, and the pitch p of the spiral trajectory.

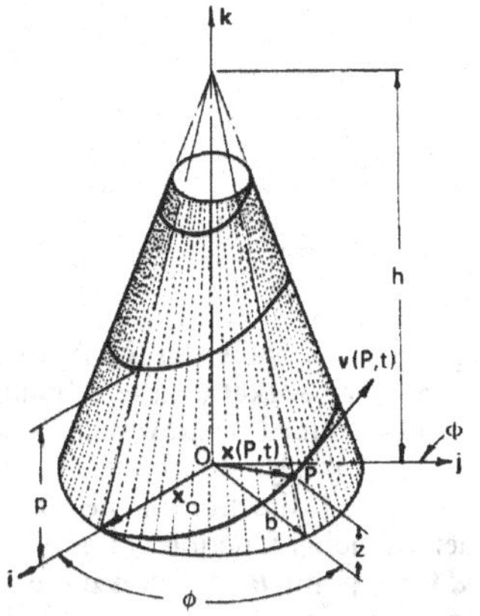

Problem 4.72.

4.73. Begin with the equations (4.68) and (4.69), and apply the basic relation (4.61) to derive the formulas (4.70) and (4.71) for the velocity and acceleration of a particle referred to a spherical reference frame. Show that for plane polar coordinates these yield the same formulas as do (4.59) and (4.60).

4.74. A particle P, initially at the place with the spherical coordinates $(r, \pi/2, \pi)$, moves on a sphere of radius r in such a way that $\theta = \theta_0 \cos 2\omega t$ and $\phi = \phi_0(1 - \sin \omega t)$, where ω is a constant. Determine the velocity and acceleration of P at $t = 0$.

4.75. The fluid particle P shown in the figure is moving in a test tube with a constant relative velocity $\mathbf{v}$. The tube is held at a fixed angle θ in a centrifuge which is spinning with a constant angular velocity $\boldsymbol{\omega}$, as shown. Find the total velocity and acceleration of P at a distance a along the tube, referred to the frame $\psi = \{F; \mathbf{i}_k\}$ fixed in the tube.

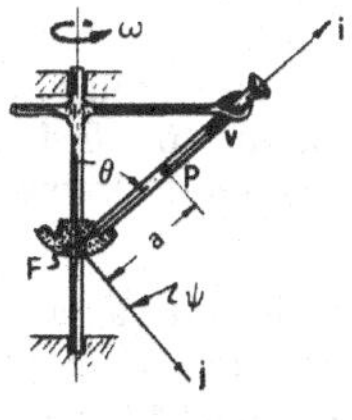

Problem 4.75.

4.76. Recall the data given for the shutoff position of the ball B of the speed control device described in Example 4.13. Determine the absolute velocity and acceleration of B referred to frame $1 = \{O; \mathbf{i}, \mathbf{j}, \mathbf{K}\}$ fixed in the vertical shaft, as noted in Fig. 4.22.

4.77. A bead B slides as shown on a thin, circular hoop of radius 20 cm. The bead moves with a constant speed of 40 cm/sec relative to the hoop which spins about its vertical diameter with a constant angular speed $\omega = 10$ rad/sec. Use two methods, one based upon the application of spherical coordinates, to determine the absolute velocity and acceleration of B at the position shown.

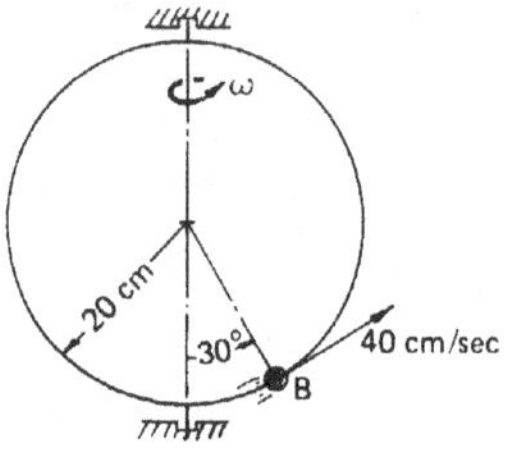

Problem 4.77.

4.78. A telescopic rod OA shown in the figure has a ball and socket fitting, called a ball joint, at each end. The socket on the tubular member is fixed in the machine at O, while the end A is constrained to move with a constant speed v in a horizontal circular slot of radius a. Let $\psi \in [0, 2\pi]$ be the counterclockwise rotation angle of the slider A measured from the $\mathbf{i}$ direction in the plane of the circular slot. Write a computer program to determine as functions of ψ the spherical components of the scaled velocity ratio $a\mathbf{v}_B/vd$ for a point B at distance d along the tubular member, referred to the spherical reference frame $\varphi = \{O; \mathbf{e}_r, \mathbf{e}_\theta, \mathbf{e}_\phi\}$. Use the values $a = b = c/2 = h/6$ for the calculation, and plot the programmed functions. What can be said about the largest and smallest values of the scaled velocity components of B for this case?

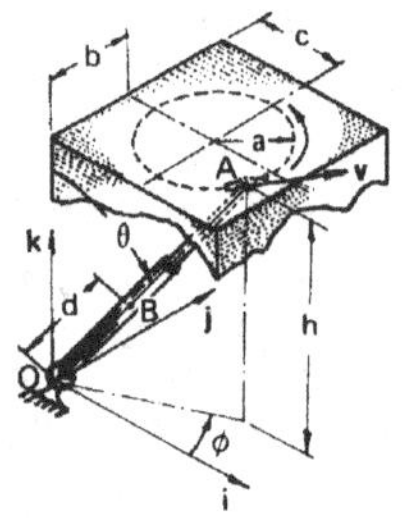

Problem 4.78.

4.79. The motion of a textile spinner was modeled in Section 4.8.1.2 and described in Fig. 4.20. For the conditions specified in the example, find the total velocity and acceleration of the ball B referred to frame $1 = \{C; \mathbf{i}_k\}$ fixed in the spinner housing, as shown in Fig. 4.20.

4.80. The driver A of a Geneva intermittent motion mechanism shown in the figure rotates with a constant angular speed ω. As the drive pin P enters and leaves the slot of the follower wheel B, called a Geneva wheel, it generates an intermittent motion of B. (a) Find the velocity of P relative to B; determine the angular velocity of the Geneva wheel; and find the period of the intermittent motion of B. (b) Express the center distance b in terms of the radius a, and compute the quantities in part (a) at the instant when P crosses the center line AB.

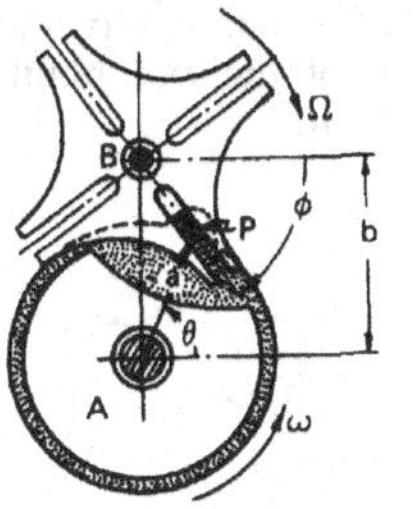

Problem 4.80.

4.81. A thin gear S of radius 2 cm drives a thin gear G of radius 5 cm so that G has a constant angular speed $\omega_2 = 20$ rad/sec, as shown, relative to a thin, square plate P to which both gears are fixed in bearings. At the instant τ shown in the figure, the plate is turning about a vertical axle with an angular speed $\omega_1 = 10$ rad/sec, which is increasing at the rate of 5 rad/sec each second in the preferred frame $\Phi = \{F; \mathbf{I}_k\}$. At the instant τ, a point M on the rim of G has the placement $\theta = \tan^{-1}(3/4)$. (a) Find the velocity at τ of the point M as seen by an observer at O in the frame $\psi = \{O; \mathbf{i}_k\}$ fixed in the square plate, as shown. (b) What is the absolute velocity of M in Φ but referred to ψ at time τ? (c) Find the total acceleration of the gear point M referred to ψ at the moment of interest. (d) Determine the total angular velocity and angular acceleration of G referred to frame ψ.

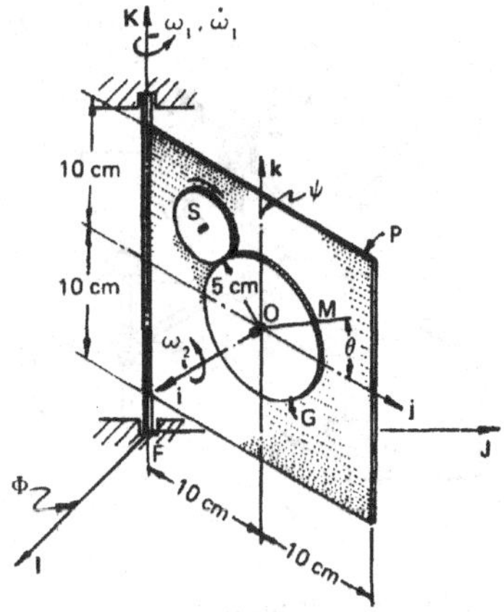

Problem 4.81.

4.82. The instant of concern is described in the previous problem. What are the total velocity and acceleration of the gear point M referred to a reference frame $f = \{O; \mathbf{e}_k\}$ fixed in the gear G?

4.83. Consider the instant described in Problem 4.81. What are the absolute angular velocity and angular acceleration of the gear S referred to the plate frame $\psi = \{O; \mathbf{i}_k\}$?

4.84. The landing gear of an aircraft is retracted and stored in the fuselage by rotating the gear with a constant angular speed $\omega_1 = 1$ rad/sec about its hinge through O, as shown. At the beginning of the motion when the gear is still in the vertical position, the wheel is spinning at the rate $\omega_2 = 80$ rad/sec, which is decreasing at the rate $\dot{\omega}_2 = 2\ \text{rad/sec}^2$. Determine, relative to the aircraft at the initial instant, the velocity and acceleration of the rim point P on the wheel, referred to the frame $\psi = \{O; \mathbf{i}_k\}$ fixed in the strut.

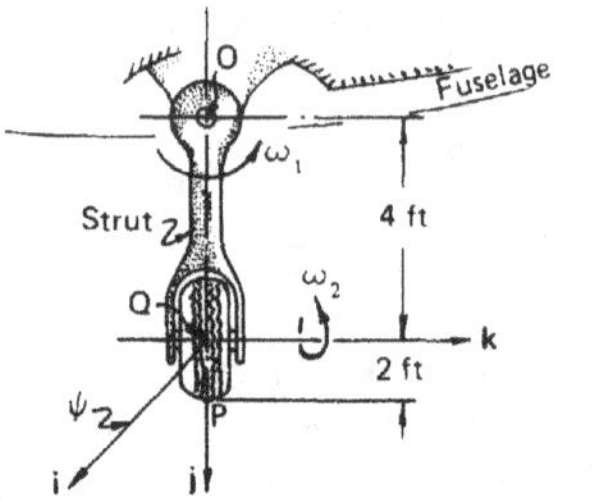

Problem 4.84.

4.85. A ride at an amusement park consists of an assembly S of four cabins attached to each of three rigid carriage arms, as illustrated. The carriage C turns clockwise at a constrant rate $\dot{\beta} = 15$ rpm relative to the ground frame Φ. Each assembly S rotates about an axle A with a constant, counterclockwise angular speed $\dot{\alpha} = 25$ rpm relative to Φ. Find the absolute velocity and acceleration of a passenger P referred to a frame $\varphi = \{A; \mathbf{i}_k\}$ in S, at the instant shown.

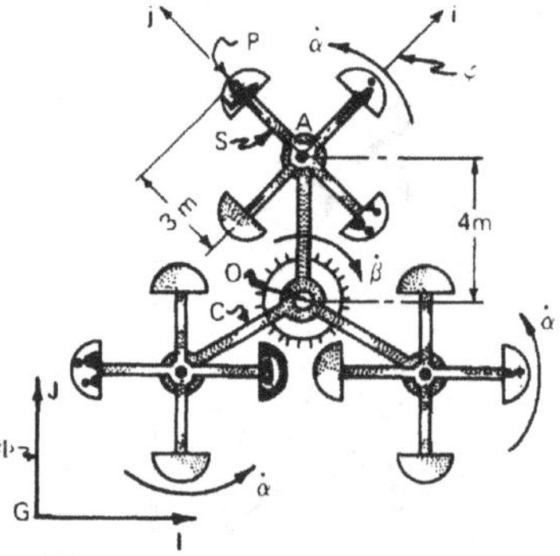

Problem 4.85.

4.86. A gear G of radius 2 cm rotates with a constant angular speed $\alpha = 3$ rad/sec relative to a large circular plate P, as shown. The plate has a constant angular speed $\beta = 2$ rad/sec relative to a vertical shaft S. At the instant illustrated, S is turning with an angular speed $\gamma = 1$ rad/sec but is slowing down at the rate of $1\ \text{rad/sec}^2$ relative to the ground frame $\Phi = \{F; \mathbf{I}_k\}$. Let R be a particle at the rim of G. (a) Find for the instant of interest the total tangential and centripetal accelerations of R in Φ, but referred to the frame $\varphi = \{G; \mathbf{i}_k\}$ fixed in G. (b) Determine for the instant of concern the total acceleration of R referred to φ.

4.87. Use the data for the instant described in the previous problem to determine for the moment of interest the total tangential and centripetal accelerations of R referred to a frame $\psi = \{O; \mathbf{i}'_k\}$ fixed in the plate P. What is the absolute acceleration of R referred to ψ?

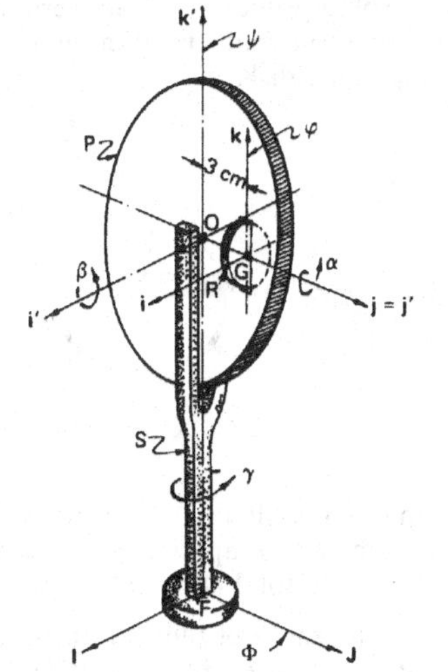

Problem 4.86.

4.88. Suppose that at an instant of interest the passenger cockpit C of the aerial ride illustrated for Problem 4.12 is revolving with an angular speed $\omega_2 = 0.5$ rad/sec which is increasing at the rate of 0.2 rad/sec^2 relative to the supporting arm OB. At the same time, OB has an absolute angular speed $\omega_1 = 0.8$ rad/sec which is decreasing at the rate of 0.1 rad/sec^2 relative to the control tower. The figure shows an occupant's head situated at H. Determine for the moment of interest (a) the absolute Coriolis and tangential accelerations of H referred to the frame $\psi = \{B; \mathbf{e}_k\}$ fixed in the cockpit C, and (b) the velocity and acceleration of H relative to frame $\varphi = \{B; \mathbf{i}_k\}$ fixed in OB.

4.89. The circular wheel shown in the figure drives a control link AR through a roller R whose motion in a groove in the wheel is regulated by a screw so that the distance $a(t)$ of R from O is a controlled function of the time t. When the roller is at its extreme position b in the groove, the link AR is perpendicular to the screw axis. During the operation of the system, the wheel rotates slowly about O with a constant angular speed ω. Find the absolute velocity and acceleration of the roller center, and the angular velocity and angular acceleration of the control link, at the moment when R reaches its extreme position.

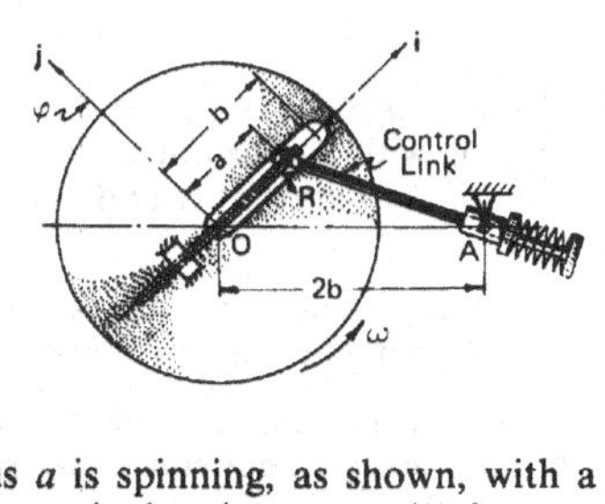

Problem 4.89.

4.90. A circular disk of radius a is spinning, as shown, with a constant angular velocity $\boldsymbol{\omega}_2$ relative to its supporting yoke bearings at A. At the same time, the yoke is rotating about the axis of its supporting rod with a constant angular velocity $\boldsymbol{\omega}_1$, as indicated. Find the absolute velocity and acceleration of a point P on the rim of the disk two ways: (i) referred to a frame fixed in the rotating yoke rod, and (ii) referred to a frame fixed in the spinning disk.

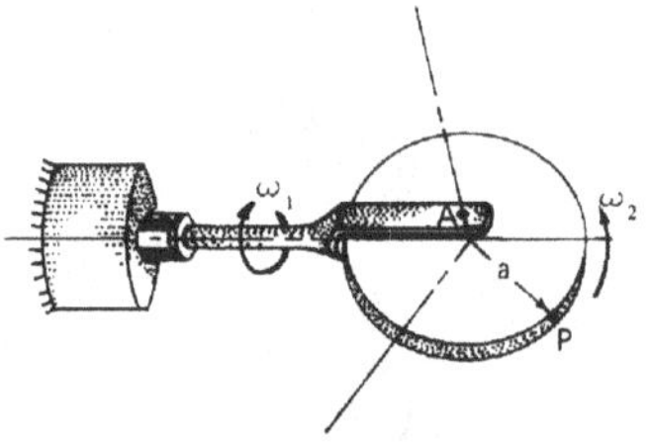

Problem 4.90.

4.91. The motion of an oscillating fan blade is described in Example 4.17. Use the data assigned in the example, and apply the two methods described there to find in the ground frame $0 = \{F; \mathbf{I}_k\}$ the total acceleration of the blade point P. See Fig. 4.26.

4.92. A motor-driven gear G of radius a turns, as shown, with a constant angular speed ω_1 about a horizontal axle fixed in a table T that concurrently rotates with a constant angular speed ω_2 about a vertical axis fixed in the machine structure S. The center of G is at a distance b from the center O of the table. Find the total velocity and acceleration of the rim point P on G referred to (i) a reference frame fixed in T at the center of G and (ii) a frame fixed in G.

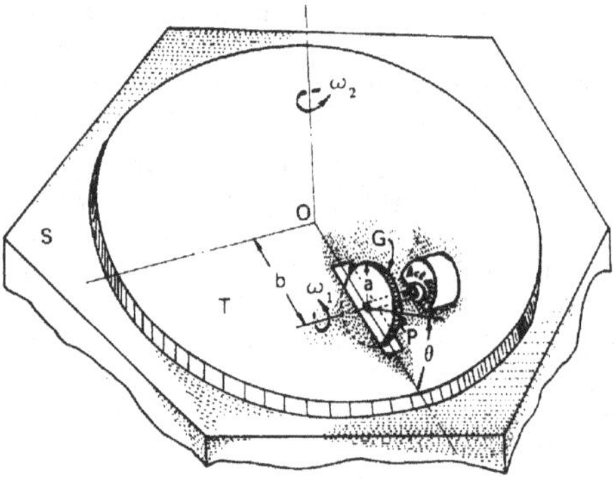

Problem 4.92.

4.93. A gear G of radius a rolls, as shown, on a rack gear R cut in a table T that turns about its normal axis with a constant angular speed ω. Determine the absolute velocity and acceleration of the points at A and B on the rim of G, referred to a frame fixed in T.

4.94. Solve the previous problem for the velocity and acceleration of the gear points at A and B, but referred to a frame fixed in G.

4.95. During a period of its motion, a slider block moves as shown with a constant speed $v = 2$ cm/sec along a straight slot in a disk A that rotates in a ring bearing with a constant, counterclockwise angular speed $\omega_2 = 30$ rad/sec relative to the ring. At the same time, the ring bearing, which is supported by a horizontal shaft B, rotates as

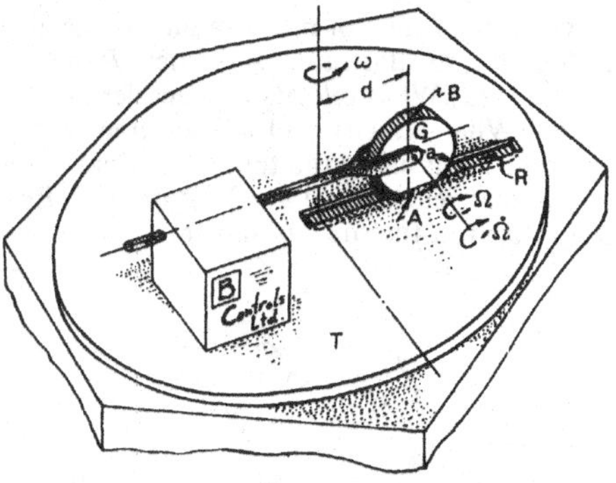

Problem 4.93.

indicated with a constant angular speed $\omega_1 = 50$ rad/sec relative to the machine frame Φ. Assume that $\alpha = 30°$ and $r = 1$ cm for the position shown. Find the absolute velocity and acceleration of the slider block at the position of interest and referred to a frame fixed in A.

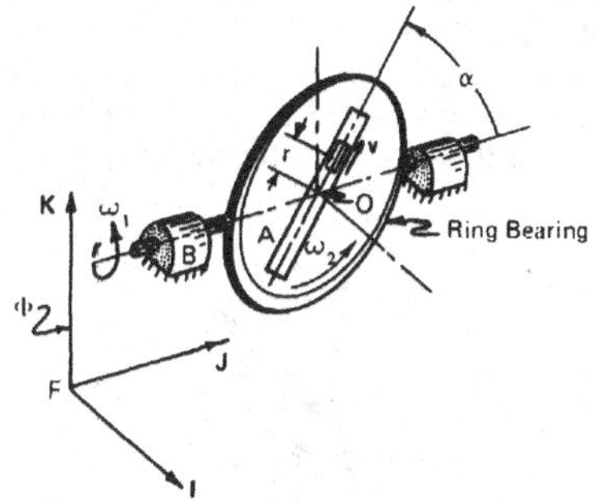

Problem 4.95.

4.96. The slider shown in the previous problem oscillates relative to the disk so that its position in the slot from the center O is given by $r = a \sin \Omega t$, where Ω is the constant frequency of the oscillation and a is the maximum radial displacement of the slider. Assume that ω_1 and ω_2 are constant. Determine the absolute velocity and acceleration of the slider. What are its values at the positions $r = 0$ and $r = a$?

4.97. A slider has a pin P that is constrained by a curved link to move in a circular slot of radius 25 cm as the slider moves vertically along its guide rod, as shown. During an interval of its motion, the slider moves upward at the constant rate of 150 cm/sec. What are the angular velocity and angular acceleration of the curved link at the instant shown in the figure?

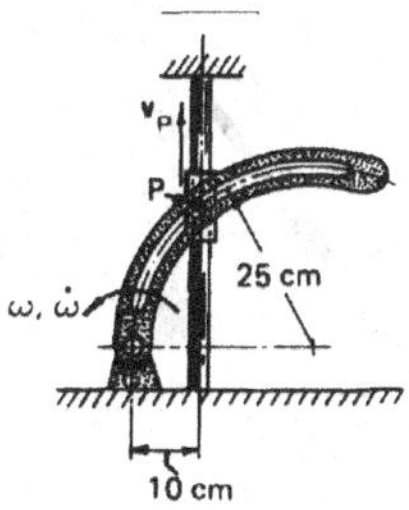

Problem 4.97.

4.98. A circular glass tube of mean radius a rotates about a fixed axle with constant angular speed ω_1, as illustrated. A particle P moves within the tube with a constant relative speed v. (a) What velocity and acceleration would P have if the tube were fixed in space? (b) What velocity and acceleration would P have if it were frozen in place at an arbitrary point along the tube? (c) What is the Coriolis acceleration of P? (d) Determine the total velocity and acceleration of P. Refer all solutions to a reference frame $\varphi = \{O; \mathbf{i}_k\}$ fixed in the tube as shown.

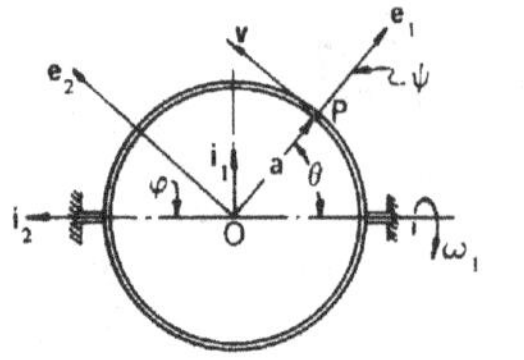

Problem 4.98.

4.99. Rework the last problem for the case when the motion is referred to a frame $\psi = \{O; \mathbf{e}_k\}$ that follows the particle.

4.100. The main beam AB of a derrick structure shown in the figure is 12 m long. At the moment of interest, the structure is rotating with an angular speed $\omega_1 = 2$ rad/sec and an angular acceleration $\dot{\omega}_1 = 1$ rad/sec^2, as illustrated. At the same instant, the beam is being lowered at the rate $\dot{\theta} = 1$ rad/sec which is increasing at the rate of 2 rad/sec^2, and $\theta = 45°$. Find the velocity and acceleration of the end point B of the beam.

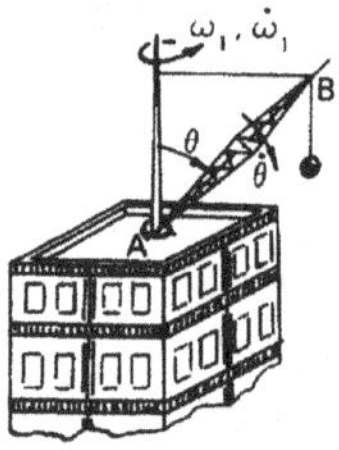

Problem 4.100.

4.101. The crank OA of a space mechanism shown in the figure rotates with a constant angular velocity $\boldsymbol{\omega} = -20\mathbf{K}$ rad/sec about an axis perpendicular to the XY plane of the machine frame Φ. The link FB, which is constrained from rotating about its axis $\mathbf{j}$, is mounted in a ball joint at F and passes through a slider attached to the crank by a ball joint at A. (a) Determine the absolute angular velocity and angular acceleration of the link at the position $\theta = 0$ and referred to its imbedded frame $\varphi = \{F; \mathbf{i}_k\}$. (b) Find the absolute velocity of point B; and compute for the moment of interest the relative velocity and acceleration of the slider A referred to φ.

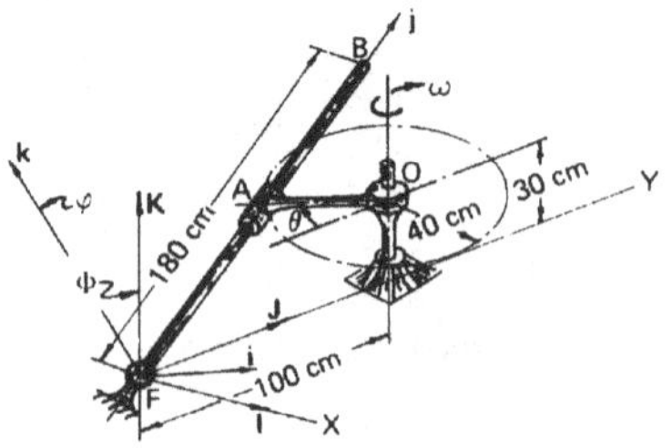

Problem 4.101.

4.102. A connecting rod bearing surface of radius a is attached to a crankshaft of length $4a$. The connecting rod, during an interval of its motion, rotates as shown with a constant angular speed ω_2 relative to the crankshaft, which turns as shown with a constant angular speed ω_1 in the machine frame $\Phi = \{F; \mathbf{I}_k\}$. A droplet P of lubricating oil moves in the bearing gap with constant speed v relative to the connecting rod, as shown. Determine the instantaneous, absolute velocity and acceleration of P with respect to frame Φ.

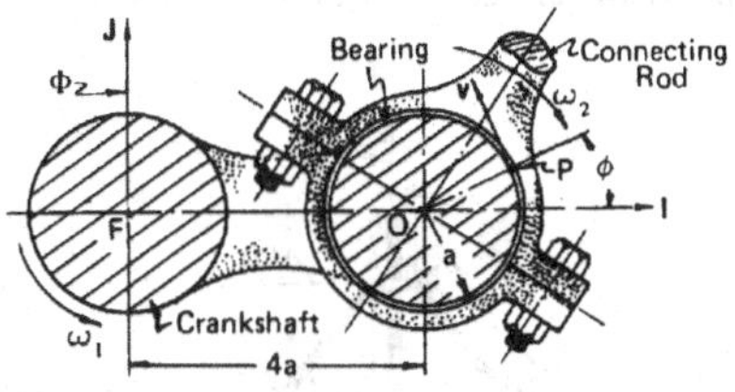

Problem 4.102.

4.103. In a viscous fluid flow situation two concentric circular cylinders of radii r and R rotate about their common axis, as shown. The outer cylinder A has a constant angular velocity $\boldsymbol{\omega}_1$, and the inner cylinder B, which is dragged along by the adherence of the viscous fluid, has an angular velocity $\boldsymbol{\omega}_2(t)$ relative to A. Measurements of the rotation rates and torques that act on the cylinders may be used to determine certain fluid viscosity properties. Find the absolute velocity and acceleration of a fluid particle P on the inner cylinder wall by two methods: (i) referred to a frame fixed in A, and (ii) referred to a frame fixed in B.

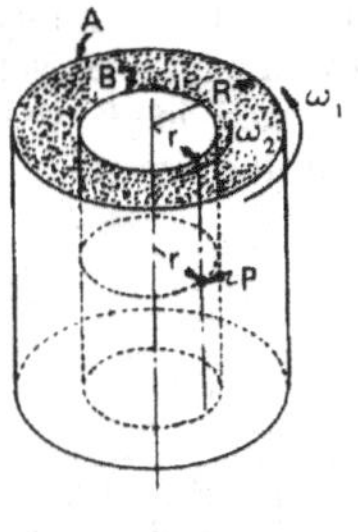

Problem 4.103.

4.104. A helicopter blade oscillates about a hinged joint A with a constant "flapping" frequency p. The flapping is described in the figure by the angle $\phi(t) = \phi_0 + \phi_1 \sin pt$, where ϕ_0 is the natural blade position and ϕ_1 is the flapping amplitude. The rotor turns the blade about the vertical axis with a constant angular velocity $\boldsymbol{\omega}$. What are the absolute velocity and acceleration of the blade point B apparent to the pilot and referred to the frame $\mu = \{A; \mathbf{i}_k\}$ fixed in the rotor?

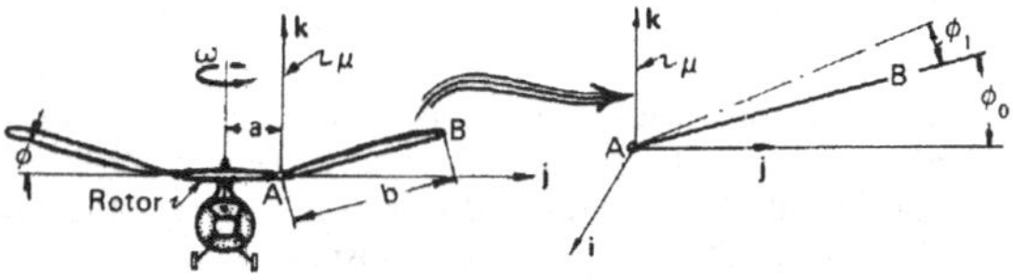

Problem 4.104.

4.105. A flyball governor is shown in the figure. The sleeve A is fixed to the vertical shaft, but the sleeve C may slide up and down. The balls B are held by links AB and BC of equal length l hinged at B and to the sleeves at a distance a from the center of the shaft. During the motion of the machine, the device rotates with a constant angular velocity $\boldsymbol{\omega}$ about a vertical axis, as indicated; and each ball oscillates so that the angle shown varies according to $\psi(t)=\psi_0+\psi_1 \sin pt$, where ψ_0, ψ_1, and p are constants. Find the absolute velocity and acceleration of the center of a ball B by two procedures: (i) referred to a frame $f=\{A; \mathbf{i}_k\}$ fixed in the sleeve A, as shown, and (ii) referred to a spherical reference frame $\psi=\{A; \mathbf{e}_r, \mathbf{e}_\theta, \mathbf{e}_\phi\}$.

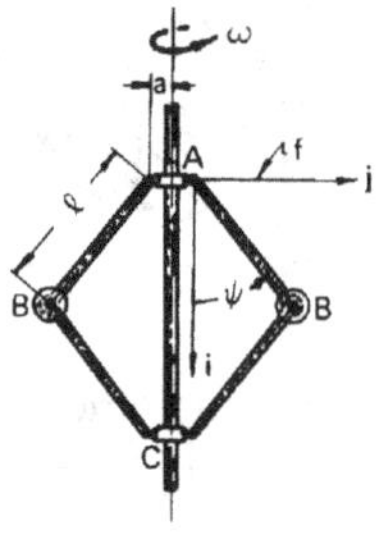

Problem 4.105.

4.106. A disk D of radius 2 ft spins with a constant angular speed of 3 rad/sec relative to a plate P, as illustrated. The plate turns about a horizontal axis with a constant angular speed of 2 rad/sec relative to a table T that rotates about a fixed vertical axis with a constant angular speed of 1 rad/sec relative to the ground frame $\Phi=\{F; \mathbf{I}_k\}$. At the instant t_0, the assembly is in the configuration shown in the figure. (a) Find at time t_0 the absolute angular velocity and angular acceleration of D referred to the frame $\varphi=\{O; \mathbf{i}_k\}$ fixed in P. (b) Find for the instant shown the absolute velocity and acceleration of the rim point M at 45° on D, referred to the frame φ.

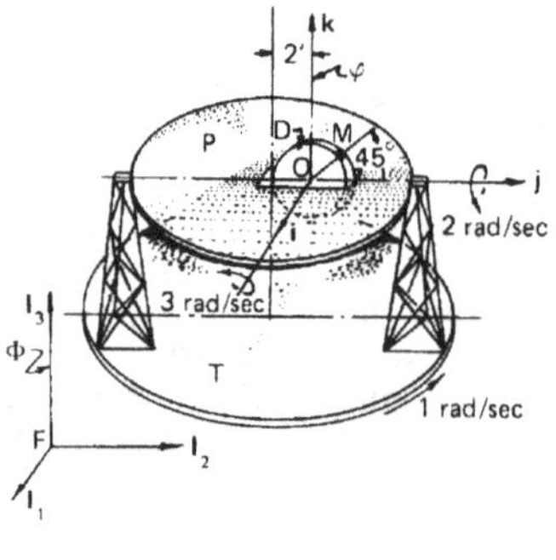

Problem 4.106.

4.107. A disk of radius r turns as shown with an angular speed $\omega(t)$ in a fixed frame $\Phi=\{O; \mathbf{i}_k\}$. At the same time, the lever arm OA rotates oppositely with the same angular speed $\Omega=\omega$ in Φ, and it moves a slider block B in a slot cut in the disk. Determine the acceleration of B at the position shown.

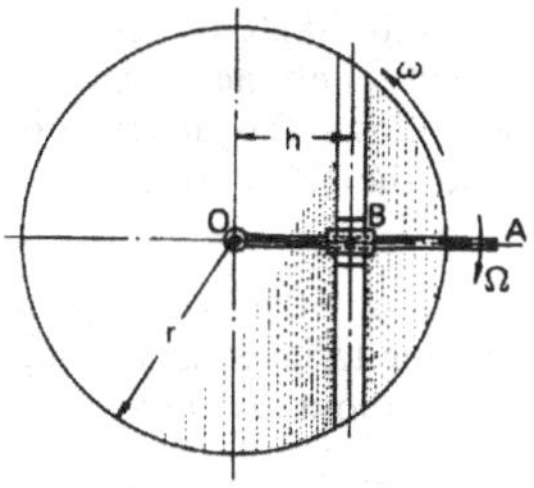

Problem 4.107.

4.108. Suppose that the cam described in Problem 1.17 has a constant angular velocity $\boldsymbol{\omega} = 60\mathbf{i}_3$ rpm relative to the ground, while the arm OR rotates with a constant angular velocity $\boldsymbol{\Omega} = -30\mathbf{i}_3$ rpm relative to the cam. Note that the frame $\psi = \{O; \mathbf{e}_k\}$ is fixed in the rod and the frame $\Phi = \{O; \mathbf{I}_k\}$ is fixed in the cam, as shown in the figure for Problem 1.17. (a) What is the velocity of the center of the roller R relative to the arm OR and referred to ψ? (b) What is its acceleration relative to the cam and referred to ψ? (c) Find the absolute acceleration of R referred to ψ when the cam and roller have the relative positions shown in the figure.

4.109. Helical gears A and B are arranged at right angles on a platform that rotates with a constant angular speed $\dot{\alpha} = 20$ rad/sec about the axle CC, as indicated. The drive gear A turns relative to the platform with a constant angular speed $\dot{\beta} = 15$ rad/sec, as shown. Determine the absolute velocity and acceleration of a point P on the gear tooth circle of A when it contacts B and the platform is in the horizontal position shown.

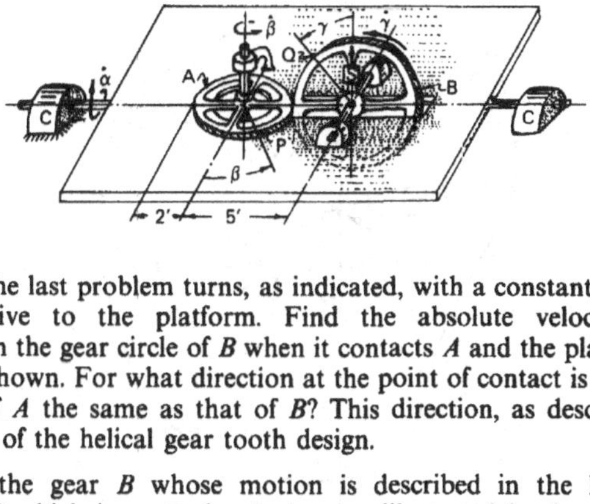

Problem 4.109.

4.110. The gear B in the last problem turns, as indicated, with a constant angular speed $\dot{\gamma} = 10$ rad/sec relative to the platform. Find the absolute velocity and acceleration of a point Q on the gear circle of B when it contacts A and the platform is in the horizontal position shown. For what direction at the point of contact is the normal velocity component of A the same as that of B? This direction, as described in Example 2.8, is a property of the helical gear tooth design.

4.111. The spoke of the gear B whose motion is described in the last two problems carries a slider S which is part of a device not illustrated in the previous figure. (a) In the vertical position at a place 2 ft from the axle of gear B, the slider has a speed of 2 ft/sec and an acceleration of 1 ft/sec^2 relative to B. Determine the velocity and acceleration of the slider in the ground frame when the platform is in the horizontal position shown before. (b) At the instant of interest, the gear B has an angular acceleration $\ddot{\gamma} = 5$ rad/sec^2 and the platform is accelerating at the rate $\ddot{\alpha} = 10$ rad/sec^2, but the angular speeds are the same as before. What will be the effect of these rates on the total velocity and acceleration of the slider S found in (a)?

4.112. Consider the motion of the oil droplet P described in Problem 4.102. Rework the problem for a general configuration of the engine apparatus in which the crankshaft line FO makes an angle θ, say, with the vector $\mathbf{I}$ in the fixed frame Φ. (a) Find the absolute velocity and acceleration of P in Φ, but referred to a frame $\alpha = \{O; \mathbf{e}_r, \mathbf{e}_\phi, \mathbf{e}_z\}$ that follows the droplet. (b) Determine the form of the solution referred to frame Φ; and thereby derive the solution to Problem 4.102 as the special case for which $\theta = 0$.

4.113. Compute in ft/sec^2 the acceleration components of a point of the earth's surface at 45°N latitude due to its constant angular velocity about the polar axis alone. The radius of the earth is about 4000 miles.

4.114. What is the angle between $\boldsymbol{\omega}_1$, the angular velocity of the earth about its polar axis, and the total principal angular velocity vector $\boldsymbol{\Omega} = \boldsymbol{\omega}_1 + \boldsymbol{\omega}_2$ described in the text? And what is $|\boldsymbol{\Omega}|$?

4.115. A small ball P moves in a smooth slot cut in a flat plate as shown in the figure. The plate rotates with a constant angular speed $\dot{\theta} = \omega$ about an axle at O in the frame $\Phi = \{O; \mathbf{I}, \mathbf{J}\}$ fixed in the plane space. The ball is attached to a spring that exerts a force on it; and relative to the frame $\psi = \{F; \mathbf{e}, \mathbf{f}\}$ fixed in the plate, as shown, $x(t)$ denotes the total stretch of the spring from its undeformed state at $x = 0$. The total acceleration of P referred to ψ must satisfy the rule

$$\mathbf{a}_P = N\mathbf{e} - p^2 x\mathbf{f},$$

in which p is a constant and $N(t)$ is an unknown force. Find the absolute acceleration of P referred to ψ, and thereby derive two equations that may be used to determine $x(t)$ and $N(t)$ as functions of t. The solution of such equations is investigated in Chapter 6.

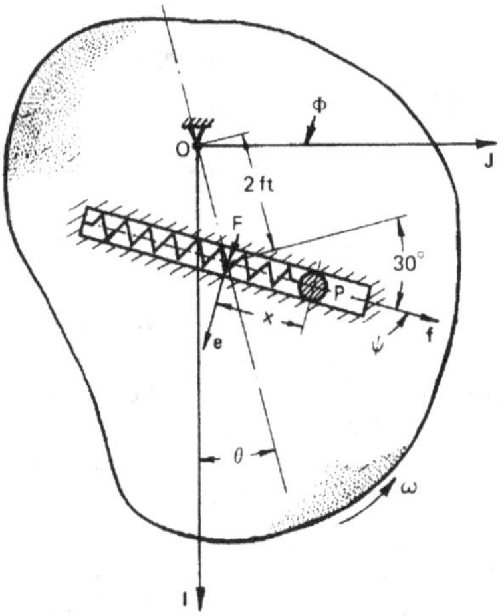

Problem 4.115.

4.116. (a) Apply the vector component transformation law (3.107) to derive (4.81f) from (4.82e) in Example 4.17. (b) Use the assigned data to derive from (4.81f) the absolute acceleration of the blade point referred to the motor frame 1. Apply the transformation law to determine from this result the absolute acceleration of P referred to the blade frame 2. See Fig. 4.26.

4.117. Begin with the vector equation (4.8) and prove, conversely, that the time rate of change of the Euler rotation tensor is given by (4.106a).

4.118. (a) Let $\mathbf{T}(t)$ be a tensor in a preferred frame Φ but referred to a moving frame $\varphi = \{O; \mathbf{i}_k\}$ whose angular velocity tensor relative to Φ is $\mathbf{W}_f$. Determine the second total time derivative $\ddot{\mathbf{T}}(t)$. (b) Find the formula for $\ddot{\mathbf{W}}_f$, and prove that $\ddot{\mathbf{W}}_f$ is a skew tensor. (c) Show that the tensor equation for the derivative $\ddot{\mathbf{W}}_f$ implies the equivalent vector formula

$$\ddot{\boldsymbol{\omega}}_f = \frac{\delta^2 \boldsymbol{\omega}_f}{\delta t^2} + \boldsymbol{\omega}_f \times \frac{\delta \boldsymbol{\omega}_f}{\delta t}$$

for the angular velocity vector $\boldsymbol{\omega}_f$, which is easily confirmed by (4.11).

4.119. Let $\mathbf{u}_\varphi$ and $\mathbf{v}_\varphi$ be arbitrary vectors referred to a moving frame φ, and suppose that they are related by the linear transformation $\mathbf{v}_\varphi = \mathbf{T}_\varphi \mathbf{u}_\varphi$, where $\mathbf{T}_\varphi$ is a tensor referred to φ. Let $\mathbf{u}_\Phi$ and $\mathbf{v}_\Phi$ be the corresponding vectors for which (4.134) holds and write $\mathbf{v}_\Phi = \mathbf{T}_\Phi \mathbf{u}_\Phi$. Determine the relation connecting $\mathbf{T}_\Phi$ and $\mathbf{T}_\varphi$.

4.120. The absolute rotational rates $\dot{\mathbf{R}}_f = \dot{\mathbf{R}}(t)$ and $\ddot{\mathbf{R}}_f = \ddot{\mathbf{R}}(t)$ of the frame φ relative to frame Φ are given by (4.106a) and (4.107a). Use the result of Problem 4.118(a) to show that the apparent time derivatives of $\mathbf{R}_f = \mathbf{R}(t)$ are determined by

$$\frac{\delta \mathbf{R}_f}{\delta t} = \mathbf{R}_f \frac{d\mathbf{R}_f}{dt} \mathbf{R}_f^T, \qquad \frac{\delta^2 \mathbf{R}_f}{\delta t^2} = \mathbf{R}_f \frac{d^2 \mathbf{R}_f}{dt^2} \mathbf{R}_f^T.$$

Appendix A

The Elements of Vector Calculus

Use of this book requires that the reader be familiar with the elements of vector calculus. These powerful tools are used because they provide greater physical insight into the mechanics, and because they lend clarity and elegance to the development of the subject matter. The following summary of some definitions, rules, and vector operations essential to the study of the principles of mechanics is intended to serve as a review of these topics.

A.1. Vector Algebra

A *vector* $\mathbf{v}$ is a geometrical object identified as a directed line segment of length v, called its *magnitude*. If $v = 0$, the vector is called the *zero vector*, and we write $\mathbf{v} = \mathbf{0}$. A vector of unit magnitude $v = 1$ is called a *unit vector*. It is important to know that *any vector* $\mathbf{v}$ *may be expressed as the product of its magnitude with a unit vector*. Indeed, if $\mathbf{e}$ is a unit vector, then $\mathbf{v} = v\mathbf{e}$ is a vector of magnitude v in the direction $\mathbf{e}$; therefore, we say that $\mathbf{v}$ is *parallel* to $\mathbf{e}$ in the same direction, or sense. The vector $\mathbf{u} = -u\mathbf{e}$ has magnitude u and is directed opposite to $\mathbf{e}$; hence, $\mathbf{u}$ is also parallel to $\mathbf{e}$, but in the opposite direction, or sense. Therefore, $\mathbf{u} = -\mathbf{v}$ means that $\mathbf{u}$ and $\mathbf{v}$ are oppositely directed vectors of equal magnitude.

These few definitions and simple results demonstrate the necessity to use a special notation to distinguish vectors from scalars. For this reason, **boldface type** is reserved for all vector quantities. In handwritten work, a wavy underline $\sim$ or other distinguishing symbol must be used to identify a vector quantity. Let us continue.

The elementary rules for vector addition and subtraction, and par-

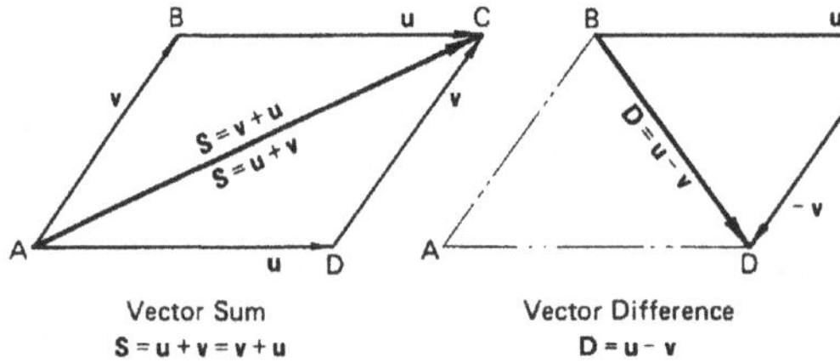

Figure A.1. The sum and difference of two vectors and their parallelogram interpretation showing the commutative property of addition.

ticularly their geometrical interpretation in a parallelogram construction, are useful. These are shown in Fig. A.1 for the reader's review.

Any vector $\mathbf{v}$ can be identified uniquely by its three perpendicular, directed projections $\{v_k\} = (v_1, v_2, v_3)$ upon the three mutually perpendicular axes x_1, x_2, x_3 (or x, y, z) of a rectangular Cartesian coordinate system, as shown in Fig. A.2. The positive directions of these coordinate lines, which are assumed to be arranged in a right-hand sense as usual, are themselves vector lines identified by three unit vectors, say, $\mathbf{e}_1, \mathbf{e}_2, \mathbf{e}_3$. The set $\{\mathbf{e}_k\}$ is called a *basis*, and the three orthogonal projections v_k are known as the *scalar components* of the vector $\mathbf{v}$ referred to $\{\mathbf{e}_k\}$. Thus, when $\mathbf{v}$ is referred to this basis, it has the unique representation

$$\mathbf{v} = \sum_{k=1}^{3} v_k \mathbf{e}_k = v_1 \mathbf{e}_1 + v_2 \mathbf{e}_2 + v_3 \mathbf{e}_3, \tag{A.1}$$

which is illustrated in Fig. A.2. This means that only vectors equal to $\mathbf{v}$ can have the same scalar components when referred to $\{\mathbf{e}_k\}$. That is, $\mathbf{u} = \mathbf{v}$ *if and*

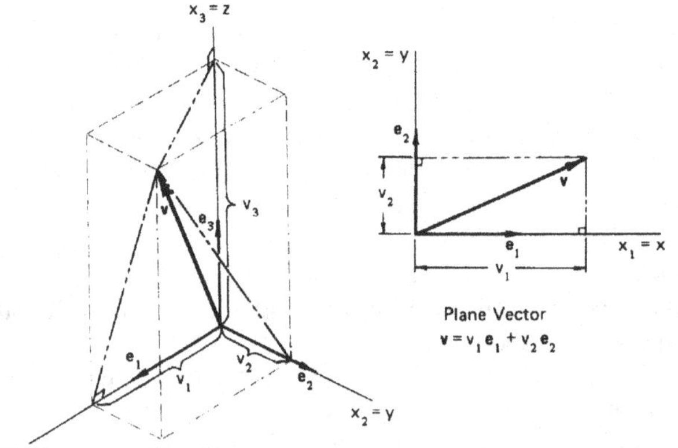

Figure A.2. Scalar component representations of a vector $\mathbf{v}$ in a plane and in space.

only if the corresponding scalar components u_k *and* v_k *are equal when* **u** *and* **v** *are referred to* $\{\mathbf{e}_k\}$: $u_k = v_k$ for $k = 1, 2, 3$. This rule for equality of two vectors **u** and **v** referred to the same basis vectors $\mathbf{e}_k$ is used often in applications to determine unknown components in vector equations.

The same vector **v** may be referred to any other basis $\{\mathbf{e}'_k\}$, say. Of course, its scalar components $\{v'_k\}$ in this basis will be different from those in (A.1), but its representation has the same form as (A.1), namely, $\mathbf{v} = \sum_{k=1}^{3} v'_k \mathbf{e}'_k$.

The foregoing geometrical description of the scalar components shown in Fig. A.2 implies that *the magnitude of a vector* **v** *is equal to the square root of the sum of the squares of its scalar components*:

$$v \equiv |\mathbf{v}| = (v_1^2 + v_2^2 + v_3^2)^{1/2}. \tag{A.2}$$

Of course, the length of a vector must be the same regardless of the basis used to represent it.

A.1.1. The Dot Product

The *dot product* of two vectors **u**, **v** is a scalar defined by

$$\mathbf{u} \cdot \mathbf{v} \equiv uv \cos\langle \mathbf{u}, \mathbf{v} \rangle \tag{A.3}$$

where $u = |\mathbf{u}|$, $v = |\mathbf{v}|$, and $\cos\langle \mathbf{u}, \mathbf{v}\rangle$ is the cosine of the angle between the vectors **u** and **v**. Hence, *if neither* **u** *nor* **v** *is the zero vector, then* $\mathbf{u} \cdot \mathbf{v} = 0$ *implies that* **u** *is perpendicular to* **v**. Since $\cos\langle \mathbf{v}, \mathbf{u}\rangle = \cos\langle \mathbf{u}, \mathbf{v}\rangle$, it is clear that *the dot product is commutative*: $\mathbf{u} \cdot \mathbf{v} = \mathbf{v} \cdot \mathbf{u}$.

An *orthonormal basis* $\{\mathbf{e}_k\}$ is a set of mutually perpendicular unit vectors; hence, by (A.3) these satisfy

$$\begin{gathered} \mathbf{e}_1 \cdot \mathbf{e}_1 = 1, \qquad \mathbf{e}_2 \cdot \mathbf{e}_2 = 1, \qquad \mathbf{e}_3 \cdot \mathbf{e}_3 = 1, \\ \mathbf{e}_1 \cdot \mathbf{e}_2 = \mathbf{e}_2 \cdot \mathbf{e}_1 = 0, \qquad \mathbf{e}_3 \cdot \mathbf{e}_1 = \mathbf{e}_1 \cdot \mathbf{e}_3 = 0, \qquad \mathbf{e}_2 \cdot \mathbf{e}_3 = \mathbf{e}_3 \cdot \mathbf{e}_2 = 0. \end{gathered} \tag{A.4}$$

With the aid of these relations and the representation (A.1), it may be seen that

$$\mathbf{v} \cdot \mathbf{e}_k = v_k \qquad \text{for} \quad k = 1, 2, 3. \tag{A.5}$$

That is, $\mathbf{v} \cdot \mathbf{e}_1 = v_1$, for example. Thus, using (A.3) in (A.5), we have

$$v_k = \mathbf{v} \cdot \mathbf{e}_k = v \cos\langle \mathbf{v}, \mathbf{e}_k \rangle, \qquad k = 1, 2, 3. \tag{A.6}$$

The three angles $\theta_k \equiv \langle \mathbf{v}, \mathbf{e}_k \rangle$, the angles between **v** and the three positive coordinate directions $\mathbf{e}_k$, are called the *direction angles* of **v**; and their cosines in (A.6) are known as the *direction cosines* of **v**. If **v** is a unit vector, $v = 1$ and

(A.6) shows that *the scalar components of a unit vector are its direction cosines. Hence, the sum of the squares of the direction cosines of a vector is equal to one.*

Finally, let $\mathbf{u} = u_1\mathbf{e}_1 + u_2\mathbf{e}_2 + u_3\mathbf{e}_3$. Then, with the aid of (A.4), its scalar product with the vector $\mathbf{v}$ in (A.1) also may be written as

$$\mathbf{u} \cdot \mathbf{v} = u_1 v_1 + u_2 v_2 + u_3 v_3. \tag{A.7}$$

In particular, when $\mathbf{u} = \mathbf{v}$, (A.7) becomes the squared magnitude of $\mathbf{v}$ defined in (A.2). Hence, $v = |\mathbf{v}| = (\mathbf{v} \cdot \mathbf{v})^{1/2}$.

A.1.2. The Cross Product

The *cross product* of two vectors $\mathbf{u}$, $\mathbf{v}$ is a vector defined by

$$\mathbf{u} \times \mathbf{v} = uv \sin\langle \mathbf{u}, \mathbf{v} \rangle \mathbf{n}, \tag{A.8}$$

in which $\mathbf{n}$ is a unit vector perpendicular to both $\mathbf{u}$ and $\mathbf{v}$ in the sense of a right-hand screw turning from $\mathbf{u}$ toward $\mathbf{v}$ and advancing along the direction $\mathbf{n}$. Also, $\sin\langle \mathbf{u}, \mathbf{v} \rangle$ is the sine of the smaller angle between $\mathbf{u}$ and $\mathbf{v}$. Reversing the product in (A.8) reverses the sense of $\mathbf{n}$; therefore, we see that *the cross product is not commutative, rather*

$$\mathbf{v} \times \mathbf{u} = -\mathbf{u} \times \mathbf{v}. \tag{A.9}$$

Hence, (A.9) shows that $\mathbf{u} \times \mathbf{u} = \mathbf{0}$. More generally, according to (A.8), *if neither* $\mathbf{u}$ *nor* $\mathbf{v}$ *is the zero vector, then* $\mathbf{u} \times \mathbf{v} = \mathbf{0}$ *means that* $\mathbf{u}$ *and* $\mathbf{v}$ *are parallel vectors; hence, there exists in this case a constant* $k \neq 0$ *such that* $\mathbf{u} = k\mathbf{v}$.

In particular, the orthonormal basis $\{\mathbf{e}_k\}$ satisfies the right-hand rule mentioned above; and (A.8) and (A.9) yield

$$\begin{aligned} &\mathbf{e}_1 \times \mathbf{e}_2 = \mathbf{e}_3, && \mathbf{e}_2 \times \mathbf{e}_3 = \mathbf{e}_1, && \mathbf{e}_3 \times \mathbf{e}_1 = \mathbf{e}_2, \\ &\mathbf{e}_2 \times \mathbf{e}_1 = -\mathbf{e}_3, && \mathbf{e}_3 \times \mathbf{e}_2 = -\mathbf{e}_1, && \mathbf{e}_1 \times \mathbf{e}_3 = -\mathbf{e}_2, \\ &\mathbf{e}_1 \times \mathbf{e}_1 = \mathbf{0}, && \mathbf{e}_2 \times \mathbf{e}_2 = \mathbf{0}, && \mathbf{e}_3 \times \mathbf{e}_3 = \mathbf{0}. \end{aligned} \tag{A.10}$$

Forming the cross product of the vectors $\mathbf{u} = u_1\mathbf{e}_1 + u_2\mathbf{e}_2 + u_3\mathbf{e}_3$ and $\mathbf{v} = v_1\mathbf{e}_1 + v_2\mathbf{e}_2 + v_3\mathbf{e}_3$ and using (A.10), the reader may show that

$$\mathbf{u} \times \mathbf{v} = (u_2 v_3 - u_3 v_2)\,\mathbf{e}_1 - (u_1 v_3 - u_3 v_1)\,\mathbf{e}_2 + (u_1 v_2 - u_2 v_1)\,\mathbf{e}_3. \tag{A.11}$$

However, this result is more easily remembered by its determinant format:

$$\mathbf{u} \times \mathbf{v} = \begin{vmatrix} \mathbf{e}_1 & \mathbf{e}_2 & \mathbf{e}_3 \\ u_1 & u_2 & u_3 \\ v_1 & v_2 & v_3 \end{vmatrix}. \tag{A.12}$$

Of course, the cross product is a vector, not a determinant; but this format is a useful aid in the easy recovery of (A.11) when (A.12) is expanded by the first row in the same sense as a regular determinant.

A.1.3. Triple Products

Two important vector identities involving products of three vectors $\mathbf{u}$, $\mathbf{v}$, $\mathbf{w}$ are useful. The *scalar triple product* $\mathbf{u} \cdot (\mathbf{v} \times \mathbf{w})$ satisfies the rule

$$\mathbf{u} \cdot (\mathbf{v} \times \mathbf{w}) = (\mathbf{u} \times \mathbf{v}) \cdot \mathbf{w}. \tag{A.13}$$

And the *vector triple product* $\mathbf{u} \times (\mathbf{v} \times \mathbf{w})$ is equivalent to

$$\mathbf{u} \times (\mathbf{v} \times \mathbf{w}) = (\mathbf{u} \cdot \mathbf{w})\mathbf{v} - (\mathbf{u} \cdot \mathbf{v})\mathbf{w}. \tag{A.14}$$

Notice on the left-hand side in (A.14) that $\mathbf{u}$ is outside of the parentheses, $\mathbf{v}$ is adjacent to $\mathbf{u}$ and $\mathbf{w}$ is remote from $\mathbf{u}$. Then, with these identifications, the rule (A.14) may be easily recalled by the mnemonic relation outer $\times$ (adjacent $\times$ remote) = (outer $\cdot$ remote) adjacent − (outer $\cdot$ adjacent) remote. Recalling (A.9), we see that the left-hand side of (A.14) may be rewritten as

$$\mathbf{u} \times (\mathbf{v} \times \mathbf{w}) = -(\mathbf{v} \times \mathbf{w}) \times \mathbf{u} = (\mathbf{w} \times \mathbf{v}) \times \mathbf{u}. \tag{A.15}$$

Therefore, the same mnemonic tool may be used also when the vector triple product has the reversed form indicated by the last term in (A.15).

It is customary to use the unit basis vectors $\mathbf{i}_1 = \mathbf{i}$, $\mathbf{i}_2 = \mathbf{j}$, and $\mathbf{i}_3 = \mathbf{k}$ for the direction vectors of the familiar rectangular Cartesian axes $x_1 = x$, $x_2 = y$, $x_3 = z$, respectively. The same convention will be used here; but flexibility in our choice of basis sets also will be essential. Therefore, the foregoing results have been described in terms of the general orthonormal basis $\{\mathbf{e}_k\}$. Of course, the same relations must hold *mutatis mutandis* for the orthonormal set $\{\mathbf{i}_k\} = \{\mathbf{i}, \mathbf{j}, \mathbf{k}\}$, or for any other. The usual Cartesian basis will be used later on in conjunction with the foregoing definitions, rules, and operations applied in demonstration of the rules for differentiation and integration of a vector function of a scalar variable.

A.2. Derivative of a Vector Function of a Scalar Variable

Applications that involve rates of change of position and direction with respect to time, arc length, or an angular variable, for example, require the use of the derivative of a vector $\mathbf{u}(\sigma)$ that varies continuously with respect to a scalar variable σ. The definition of this derivative and some rules for its application will be reviewed below, and the results will be demonstrated in a few examples.

The change $\Delta\mathbf{u}(\sigma)$ in the vector $\mathbf{u}(\sigma)$ due to a change $\Delta\sigma$ in the scalar variable σ is defined by $\Delta\mathbf{u}(\sigma) \equiv \mathbf{u}(\sigma+\Delta\sigma) - \mathbf{u}(\sigma)$. The vector function $\mathbf{u}(\sigma)$ is said to be *continuous* at the value σ if $\Delta\mathbf{u}(\sigma) \to \mathbf{0}$ as $\Delta\sigma \to 0$, i.e., if $\mathbf{u}(\sigma+\Delta\sigma) \to \mathbf{u}(\sigma)$. The *derivative* of $\mathbf{u}(\sigma)$ with respect to σ, denoted by $d\mathbf{u}(\sigma)/d\sigma$, is defined as the limit of the difference quotient $\Delta\mathbf{u}(\sigma)/\Delta\sigma$ as $\Delta\sigma \to 0$; thus,

$$\frac{d\mathbf{u}(\sigma)}{d\sigma} \equiv \lim_{\Delta\sigma \to 0} \frac{\Delta\mathbf{u}(\sigma)}{\Delta\sigma} = \lim_{\Delta\sigma \to 0} \frac{\mathbf{u}(\sigma+\Delta\sigma) - \mathbf{u}(\sigma)}{\Delta\sigma}. \tag{A.16}$$

If the limit exists for the value σ, then $\mathbf{u}(\sigma)$ is said to be *differentiable* at σ. If the function $\mathbf{u}(\sigma)$ is differentiable at σ, it also is continuous there.

A vector that has both constant magnitude and constant direction is called a *constant vector*. If $\mathbf{u}(\sigma) = \mathbf{a}$ is a constant vector, then clearly $\Delta\mathbf{u} = \mathbf{0}$ is the zero vector and (A.16) yields $d\mathbf{a}/d\sigma = \mathbf{0}$: *the derivative of a constant vector is the zero vector.*

The following additional rules also may be easily proved by use of (A.16):

Rule 1: The derivative of the vector sum of two vectors $\mathbf{A}(\sigma)$ and $\mathbf{B}(\sigma)$ is equal to the sum of their derivatives:

$$\frac{d}{d\sigma}[\mathbf{A}(\sigma) + \mathbf{B}(\sigma)] = \frac{d\mathbf{A}(\sigma)}{d\sigma} + \frac{d\mathbf{B}(\sigma)}{d\sigma}. \tag{A.17}$$

Rule 2: The derivative of the product of a scalar $a(\sigma)$ and a vector $\mathbf{B}(\sigma)$ is equal to the product of the scalar times the derivative of the vector plus the product of the derivative of the scalar times the vector:

$$\frac{d}{d\sigma}[a(\sigma)\,\mathbf{B}(\sigma)] = a(\sigma)\frac{d\mathbf{B}(\sigma)}{d\sigma} + \frac{da(\sigma)}{d\sigma}\mathbf{B}(\sigma). \tag{A.18}$$

If $\mathbf{B}$ is a constant vector, $d\mathbf{B}/d\sigma = \mathbf{0}$ and

$$\frac{d}{d\sigma}[a(\sigma)\mathbf{B}] = \frac{da(\sigma)}{d\sigma}\mathbf{B}. \tag{A.19}$$

Let us consider the representation (A.1) of a vector $\mathbf{v}(\sigma)$ referred to a constant basis $\{\mathbf{e}_k\}$:

$$\mathbf{v}(\sigma) = \sum_{k=1}^{3} v_k(\sigma)\,\mathbf{e}_k = v_1(\sigma)\,\mathbf{e}_1 + v_2(\sigma)\,\mathbf{e}_2 + v_3(\sigma)\,\mathbf{e}_3.$$

Then, application of (A.17) and (A.19) yields

$$\frac{d\mathbf{v}(\sigma)}{d\sigma} = \sum_{k=1}^{3} \frac{dv_k(\sigma)}{d\sigma} \mathbf{e}_k; \tag{A.20}$$

thus, the scalar components of the derived vector relative to the constant basis $\{\mathbf{e}_k\}$ are the ordinary derivatives of its three scalar components $v_k(\sigma)$ with respect to σ.

Example A.1. The vector sum

$$\mathbf{x}(t) = R[\cos(\omega t)\mathbf{i} + \sin(\omega t)\mathbf{j}] + At\mathbf{k}, \tag{A.21}$$

describes the motion of a particle in space. In (A.21), R, ω, A are constant scalars and $\{\mathbf{i}_k\} = \{\mathbf{i}, \mathbf{j}, \mathbf{k}\}$ are the constant, unit basis vectors of a rectangular Cartesian coordinate system. In addition, in order to eliminate excessive use of parentheses in vector functions of the kind (A.21), we shall agree to write all trigonometric functions in a form similar to the following:

$$\cos \omega t\, \mathbf{i} \equiv \cos(\omega t)\mathbf{i} \equiv (\cos \omega t)\mathbf{i}.$$

The same thing will be done with hyperbolic and exponential functions, such as

$$\sinh \omega t\, \mathbf{i} = \sinh(\omega t)\mathbf{i} = (\sinh \omega t)\mathbf{i},$$

$$\exp \omega t\, \mathbf{i} = \exp(\omega t)\mathbf{i} = (\exp \omega t)\mathbf{i}.$$

Of course, $\cos \mathbf{v}$, $\sinh \mathbf{v}$, and $\exp \mathbf{v}$, for example, are not defined for vectors; therefore, it is not possible, for example, to confuse $\cos \omega t\, \mathbf{i}$ with $\cos(\omega t \mathbf{i})$, for the latter is not defined.

With this convention in mind, let us find the derivative of (A.21) with respect to the scalar variable t. Application of the rule (A.17) to the function (A.21), wherein t is identified as the σ variable, yields

$$\frac{d\mathbf{x}(t)}{dt} = \frac{d}{dt}[R(\cos \omega t\, \mathbf{i} + \sin \omega t\, \mathbf{j})] + \frac{d}{dt}(At\mathbf{k}).$$

Repeating the application of (A.17) to the vector sum in the first bracket, we find

$$\frac{d\mathbf{x}(t)}{dt} = \frac{d}{dt}(R \cos \omega t\, \mathbf{i}) + \frac{d}{dt}(R \sin \omega t\, \mathbf{j}) + \frac{d}{dt}(At\mathbf{k});$$

and by use of (A.19) this may be reduced further to

$$\frac{d\mathbf{x}(t)}{dt} = \frac{d}{dt}(R \cos \omega t)\mathbf{i} + \frac{d}{dt}(R \sin \omega t)\mathbf{j} + \frac{d}{dt}(At)\mathbf{k}. \tag{A.22}$$

Notice that because $\{\mathbf{i}_k\}$ is a constant basis, application of the formula (A.20) to the motion vector (A.21) yields (A.22) immediately. We thus reach

$$\frac{d\mathbf{x}(t)}{dt} = -R\omega \sin \omega t\, \mathbf{i} + R\omega \cos \omega t\, \mathbf{j} + A\mathbf{k}. \tag{A.23}$$

The second derivative may be computed similarly. This time, by direct use of (A.20) in (A.23), we have

$$\frac{d^2\mathbf{x}(t)}{dt^2} = \frac{d}{dt}(-R\omega \sin \omega t)\mathbf{i} + \frac{d}{dt}(R\omega \cos \omega t)\mathbf{j} + \frac{dA}{dt}\mathbf{k}.$$

Hence, finally,

$$\frac{d^2\mathbf{x}(t)}{dt^2} = -R\omega^2 \cos \omega t\, \mathbf{i} - R\omega^2 \sin \omega t\, \mathbf{j} = -R\omega^2(\cos \omega t\, \mathbf{i} + \sin \omega t\, \mathbf{j}). \qquad \square$$

We shall require also the following rules for the derivatives of the scalar and vector products of two vector functions $\mathbf{A}(\sigma)$ and $\mathbf{B}(\sigma)$:

Rule 4:

$$\frac{d}{d\sigma}[\mathbf{A}(\sigma) \cdot \mathbf{B}(\sigma)] = \mathbf{A}(\sigma) \cdot \frac{d\mathbf{B}(\sigma)}{d\sigma} + \frac{d\mathbf{A}(\sigma)}{d\sigma} \cdot \mathbf{B}(\sigma) \tag{A.24}$$

Rule 5:

$$\frac{d}{d\sigma}[\mathbf{A}(\sigma) \times \mathbf{B}(\sigma)] = \mathbf{A}(\sigma) \times \frac{d\mathbf{B}(\sigma)}{d\sigma} + \frac{d\mathbf{A}(\sigma)}{d\sigma} \times \mathbf{B}(\sigma). \tag{A.25}$$

It will be observed that the rules (A.18), (A.24), and (A.25) have the same form. Hence, in different notation, all of these product rules may be summarized for quantities $\alpha(\sigma)$ and $\beta(\sigma)$ by

$$\frac{d}{d\sigma}[\alpha(\sigma) * \beta(\sigma)] = \alpha(\sigma) * \frac{d\beta(\sigma)}{d\sigma} + \frac{d\alpha(\sigma)}{d\sigma} * \beta(\sigma), \tag{A.26}$$

i.e., *the derivative of a product (denoted by* $*$*) is equal to the first quantity times the derivative of the second plus the derivative of the first quantity times the second.* It is easier to remember this general rule than the several special product formulas.

Examples A2. Let $\mathbf{A}(\sigma) = 16\sigma^2\mathbf{i} + 10\sigma\mathbf{j} - 16\mathbf{k}$ and $\mathbf{B} = 6\mathbf{i} + 3\mathbf{j}$. Find the derivatives with respect to σ of $\mathbf{A} \cdot \mathbf{B}$ and $\mathbf{B} \times \mathbf{A}$.

Solution. We may solve this problem in two ways: first, by computing the products and differentiating the single scalar or vector function that results; or second, by differentiating **A** and **B** separately and using the rules (A.24) and (A.25). The results must be the same in both cases. Let us consider both methods in their turn.

Method 1. We recall (A.7) and (A.12) to calculate

$$\mathbf{A}\cdot\mathbf{B}=96\sigma^2+30\sigma,$$

$$\mathbf{B}\times\mathbf{A}=\begin{vmatrix}\mathbf{i} & \mathbf{j} & \mathbf{k}\\ 6 & 3 & 0\\ 16\sigma^2 & 10\sigma & -16\end{vmatrix}=-48\mathbf{i}+96\mathbf{j}+(60\sigma-48\sigma^2)\mathbf{k}.$$

Then we differentiate these resultant scalar and vector functions to obtain

$$\frac{d}{d\sigma}(\mathbf{A}\cdot\mathbf{B})=192\sigma+30,\qquad \frac{d}{d\sigma}(\mathbf{B}\times\mathbf{A})=(60-96\sigma)\mathbf{k}. \tag{A.27}$$

Method 2. We recall the rule (A.20) to derive

$$\frac{d\mathbf{A}(\sigma)}{d\sigma}=32\sigma\mathbf{i}+10\mathbf{j}\quad\text{and}\quad\frac{d\mathbf{B}}{d\sigma}=\mathbf{0}.$$

Now we apply the rules (A.24) and (A.25), and compute the appropriate products using (A.7) and (A.12) to obtain

$$\frac{d}{d\sigma}(\mathbf{A}\cdot\mathbf{B})=\mathbf{A}\cdot\frac{d\mathbf{B}}{d\sigma}+\frac{d\mathbf{A}}{d\sigma}\cdot\mathbf{B}=\frac{d\mathbf{A}}{d\sigma}\cdot\mathbf{B}=192\sigma+30,$$

$$\frac{d}{d\sigma}(\mathbf{B}\times\mathbf{A})=\mathbf{B}\times\frac{d\mathbf{A}}{d\sigma}+\frac{d\mathbf{B}}{d\sigma}\times\mathbf{A}=\mathbf{B}\times\frac{d\mathbf{A}}{d\sigma}$$

$$=\begin{vmatrix}\mathbf{i} & \mathbf{j} & \mathbf{k}\\ 6 & 3 & 0\\ 32\sigma & 10 & 0\end{vmatrix}=\mathbf{k}(60-96\sigma).$$

These conclusions coincide with (A.27) above. The first method often is the more economical procedure for direct calculations of known vector functions, whereas the second, which is based upon (A.24) and (A.25), is more useful in theoretical work where the vector functions are arbitrary quantities. The example shows that the end results in either case are the same.

A.3. Integration of a Vector Function of a Scalar Variable

To integrate a vector function $\mathbf{v}(\sigma)$ that varies continuously with respect to the scalar variable σ, we use (A.1) to express the function in terms of its scalar components with respect to a constant basis; then we integrate the separate scalar component functions in the usual manner. Suppose that the basis $\{\mathbf{e}_k\}$ in (A.1) is a constant basis, i.e., independent of σ. Then the integral of the vector function $\mathbf{v}(\sigma)$ for $t_0 \leqslant \sigma \leqslant t$ is determined by

$$\int_{t_0}^{t} \mathbf{v}(\sigma)\, d\sigma = \mathbf{e}_1 \int_{t_0}^{t} v_1(\sigma)\, d\sigma + \mathbf{e}_2 \int_{t_0}^{t} v_2(\sigma)\, d\sigma + \mathbf{e}_3 \int_{t_0}^{t} v_3(\sigma)\, d\sigma. \qquad \text{(A.28)}$$

Example A.3. Let us consider the integral of the vector function $\mathbf{v}(\sigma) = 2\sin 2\sigma\,\mathbf{i} + (60 - 96\sigma)\mathbf{k}$ for $0 \leqslant \sigma \leqslant t$ in the constant basis $\{\mathbf{i}, \mathbf{j}, \mathbf{k}\}$. Then applying (A.28), we obtain

$$\begin{aligned} \int_0^t \mathbf{v}(\sigma)\, d\sigma &= \int_0^t [2\sin 2\sigma\,\mathbf{i} + (60 - 96\sigma)\mathbf{k}]\, d\sigma \\ &= \mathbf{i}\int_0^t 2\sin 2\sigma\, d\sigma + \mathbf{k}\int_0^t (60 - 96\sigma)\, d\sigma \\ &= \mathbf{i}(-\cos 2\sigma\,|_0^t) + \mathbf{k}(60\sigma - 48\sigma^2)\,|_0^t + c\mathbf{j}\,|_0^t, \end{aligned}$$

where c is an unknown constant of integration which contributes nothing to the definite integral. Thus, finally,

$$\int_0^t \mathbf{v}(\sigma)\, d\sigma = (1 - \cos 2t)\mathbf{i} + 12t(5 - 4t)\mathbf{k}.$$

Other methods of integration of vector functions will be discussed when the need arises within the text.

A.4. Maximum and Minimum Vector Values

The terms "larger" and "smaller," or "greater" and "less" imply an ordered relationship between the relative "size" of two things; and only scalar quantities obey the implied rules of inequality. Therefore, the terms maximum and minimum used in regard to vector values apply strictly to their scalar magnitudes. Nevertheless, at the place where the vector attains its greatest or least magnitude, the vector also has a specific direction. Consequently, specification of the maximum and minimum values of a vector requires that both the magnitude and the direction be provided. If it may be possible in a particular problem that the value of a vector at one or more places may be zero, then the zero vector is its absolute minimum value among all others. A negatively directed vector is neither "smaller" than the zero vector nor "less than" the oppositely directed vector having the same magnitude at a different place. The following example will illustrate this point.

Example A.4. The motion of a particle P that moves along an elliptical path is described by the position vector

$$\mathbf{x}(P, t) = a \cos \omega t \, \mathbf{i} + b \sin \omega t \, \mathbf{j}, \tag{A.29}$$

wherein a, b, and ω are constants and t is the scalar time variable. We shall assume that $a > b$. The velocity vector of P is defined by the time derivative of (A.29):

$$\mathbf{v}(P, t) \equiv \frac{d\mathbf{x}(P, t)}{dt} = \omega(-a \sin \omega t \, \mathbf{i} + b \cos \omega t \, \mathbf{j}). \tag{A.30}$$

Determine the maximum and minimum values of the velocity vector, and locate the points on the ellipse where these occur on the interval $0 \leqslant \omega t \leqslant 2\pi$.

Solution. We notice at once that there is no point on the ellipse at which the velocity vector vanishes. Since $a > b$, it also is evident from (A.30) that for $\omega t = \pi/2$ and $3\pi/2$, $\mathbf{v}(P, t) = \mp a\omega \mathbf{i}$ has a magnitude that is larger than the magnitude of its value $\mathbf{v}(P, t) = \pm b\omega \mathbf{j}$ at $\omega t = 0$ and π. We shall establish that these values are indeed the extremes sought.

The squared magnitude of (A.30) is given by

$$v^2 \equiv \mathbf{v} \cdot \mathbf{v} = \omega^2(a^2 \sin^2 \omega t + b^2 \cos^2 \omega t). \tag{A.31}$$

The extreme values we seek, if any exist (and we know for this example that they do) are determined from the condition

$$\frac{dv^2}{dt} = \omega^2(a^2 - b^2) 2 \sin \omega t \cos \omega t = \omega^2(a^2 - b^2) \sin 2\omega t = 0. \tag{A.32}$$

This holds if and only if $2\omega t = n\pi$ for $n = 0, 1, 2, 3, \ldots$. However, only values of $n \leqslant 3$ need be considered on the interval of interest; for, when $n = 4$, the particle passes the same place for which $n = 0$. Substitution of these four distinct values of ωt into (A.29) and (A.30) shows that

(i) when $\omega t = 0, \pi$: $\mathbf{v} = \pm \omega b \mathbf{j}$ at $\mathbf{x} = \pm a \mathbf{i}$, and
(ii) when $\omega t = \pi/2, 3\pi/2$: $\mathbf{v} = \mp \omega a \mathbf{i}$ at $\mathbf{x} = \pm b \mathbf{j}$.

These results coincide with our initial observations.

Since $a > b$, the velocity vector has its largest values $\mathbf{v} = \mp \omega a \mathbf{i}$ at the two points at $\mathbf{x} = \pm b\mathbf{j}$ on the minor axis of the ellipse; and its minimum values $\mathbf{v} = \pm\omega b\mathbf{j}$ occur at $\mathbf{x} = \pm a\mathbf{i}$, the end points of the major axis. In particular, at $\mathbf{x} = a\mathbf{i}$, the particle velocity is directed upward, $\mathbf{v} = \omega b\mathbf{j}$, whereas at $\mathbf{x} = -a\mathbf{i}$, the velocity is directed downward, $\mathbf{v} = -\omega b\mathbf{j}$; but it has the same smallest magnitude at both places. The description of the maximum velocity behavior is similar.

Problems

A.1. (a) Determine the derivatives with respect to the scalar variable s of the sum and difference of the vectors

$$\mathbf{u}(s) = (4s^3 + 6)\mathbf{j} - 5s\mathbf{k}, \qquad \mathbf{v}(s) = -3\mathbf{i} + 6s^2\mathbf{k}.$$

(b) Find the derivatives of the dot and cross products of these vectors. Use two methods.

A.2. Let $\mathbf{A}(t) = 5t^2\mathbf{i} + 3\mathbf{j} - 2t\mathbf{k}$. What is the value of the derivative with respect to t of the vector function $\mathbf{B}(t) = 10t^2\mathbf{A}(t)$ when $t = 2$?

A.3. (a) Show that a vector function $\mathbf{u}(t)$ of the scalar t has a constant magnitude if and only if $\mathbf{u} \cdot d\mathbf{u}/dt = 0$. (b) Prove that the vector $\mathbf{u}(t)$ remains parallel to a fixed line with the unit direction vector $\mathbf{e}$ when and only when $\mathbf{u} \times d\mathbf{u}/dt = \mathbf{0}$.

A.4. Derive from the definition of the derivative in (A.16) the rule (A.26) for the derivative of a generalized product of two vectors.

A.5. A particle P has a motion $\mathbf{x}(P, t) = \mathbf{a} \cos \omega t + \mathbf{b} \sin \omega t$, where $\mathbf{a}$, $\mathbf{b}$ are constant vectors, ω is a constant scalar, and the scalar variable t is the time. Show that $\mathbf{x}$ satisfies the vector differential equations

$$\mathbf{x} \times \frac{d\mathbf{x}}{dt} = \omega\mathbf{a} \times \mathbf{b} \quad \text{and} \quad \frac{d^2\mathbf{x}}{dt^2} + \omega^2\mathbf{x} = \mathbf{0}.$$

A.6. The motion of a particle along a space path C is given by the position vector $\mathbf{x}(s) = \boldsymbol{\alpha}s + \boldsymbol{\alpha} \times \mathbf{f}(s)$, in which $\boldsymbol{\alpha}$ is a constant vector and $\mathbf{f}(s)$ is a certain continuously differentiable vector function of the scalar arc length variable s. The unit tangent vector to C is defined by $\mathbf{t} \equiv d\mathbf{x}/ds$, and the unit principal normal vector to the path is defined by $\mathbf{n} = R\, d\mathbf{t}/ds$, where $R \equiv 1/|d\mathbf{t}/ds|$. Show that the tangent vector to C makes a constant angle with $\boldsymbol{\alpha}$, and the principal normal vector is perpendicular to $\boldsymbol{\alpha}$. Let $\mathbf{b} \equiv \mathbf{t} \times \mathbf{n}$, and draw a sketch showing the geometrical relation of the four vectors $\mathbf{t}$, $\mathbf{n}$, $\mathbf{b}$, and $\boldsymbol{\alpha}$.

A.7. Let $d^2\mathbf{x}(t)/dt^2 = \mathbf{a}$, a constant vector; and suppose that $d\mathbf{x}(0)/dt = \mathbf{v}_0$ and $\mathbf{x}(0) = \mathbf{x}_0$ when the scalar $t = 0$. Find by integration of the given equation the vector functions $\mathbf{v}(t) \equiv d\mathbf{x}/dt$ and $\mathbf{x}(t)$.

A.8. Find the integral of the sum of the vectors assigned in Problem A.1, when $2 \leqslant s \leqslant 6$.

Appendix B

The Elements of Matrix Algebra

Computations and theoretical discussions involving systems of equations often are simplified by the introduction of matrices. This is particularly true in the theory of finite rigid body motions; consequently, the reader interested in the study of this topic is expected to have some experience with the tools of elementary matrix algebra. An outline of the relevant definitions and theorems with some examples for the reader's review of results needed for use in this text is provided below. Additional details may be found in standard books on matrices, a few of which are listed in the References.

B.1. Matrix Notation

A *matrix* M of type $p \times q$ is an ordered rectangular array of elements arranged in p rows and q columns:

$$M = \begin{bmatrix} M_{11} & M_{12} & \cdots & M_{1q} \\ M_{21} & M_{22} & \cdots & M_{2q} \\ \vdots & \vdots & & \vdots \\ M_{p1} & M_{p2} & & M_{pq} \end{bmatrix}. \tag{B.1}$$

The elements may be real or complex numbers, variables, or even other matrices, for example. The *index notation* described in Section 3.2 of the text will be used to facilitate our writing of matrix elements and to assist our understanding of terms resulting from various matrix operations indicated more concisely by use of *direct notation*. Thus, the element in the ith row and

jth column of the matrix M in (B.1) is denoted by M_{ij}, where the range of the indices is specified by the description of the matrix type, namely, $i = 1, 2, \ldots, p$ rows and $j = 1, 2, \ldots, q$ columns. Sometimes the arrays are abbreviated as $M = [M_{ij}]$.

The special matrix M of type $n \times n$ is called a *square matrix*. However, the matrix of type 1×1 is a scalar quantity written without brackets or index notation.

An array $[M_{1j}]$ of p elements arranged in a row is named a *row matrix* of type $1 \times p$; and an array $[M_{j1}]$ of q elements arranged in a column is known as a *column matrix* of type $q \times 1$. The elements of a row or column matrix m, say, may be conveniently written as $m_j = M_{1j}$ or M_{j1}, respectively; and we sometimes write $m = [m_j]$ for both types. For illustration, a row matrix $x = [x_k]$ and a column matrix $a = [a_j]$ are shown below:

$$x = [x_1, x_2, \ldots, x_p], \qquad a = \begin{bmatrix} a_1 \\ a_2 \\ \vdots \\ a_q \end{bmatrix}.$$

In a square matrix M, the diagonal that contains the elements M_{11}, M_{22}, $M_{33}, \ldots$ is called the *principal* or *leading diagonal*. If $M_{ij} = 0$ when $i \neq j$, so that all elements not on the leading diagonal are zero, M is called a *diagonal matrix*. An *identity matrix* I is a diagonal matrix whose principal diagonal elements are all equal to unity; therefore, its elements, as shown in (3.2), are described by the Kronecker delta symbol δ_{ij} extended to indices of range n. In particular, the identity matrix of type 3×3 is defined by

$$I = [\delta_{ij}] = \begin{bmatrix} 1 & 0 & 0 \\ 0 & 1 & 0 \\ 0 & 0 & 1 \end{bmatrix}. \tag{B.2}$$

A matrix M (or m) all of whose elements $M_{ij} = 0$ (or $m_i = 0$) is titled a *zero matrix* of the same type. It is written only in direct notation as 0.

B.2. The Elementary Operations

The primary matrix operations are addition and multiplication. These operations are defined in this section. The transpose and trace operations also are discussed. And from time to time throughout our study of matrices we shall steal an opportunity to explore other uses of the index notation.

B.2.1. Equality of Matrices

Two matrices of the same type are equal if and only if their corresponding elements are the same. Thus, if A and B are matrices of type $n \times n$, for example, then the relation $A = B$ is equivalent to a set of n^2 equalities $A_{jk} = B_{jk}$. Similarly, the matrix equation $x = m$ is equivalent to n equations $x_k = m_k$.

B.2.2. Multiplication by a Scalar

The product λM or $M\lambda$ of a matrix M by a scalar λ is a matrix S whose elements are defined by $S_{ij} = \lambda M_{ij}$. Thus, $S = \lambda M = M\lambda = [\lambda M_{ij}]$. For example,

$$5\begin{bmatrix} 0 & -1 \\ 2 & 3 \end{bmatrix} = \begin{bmatrix} 0 & -1 \\ 2 & 3 \end{bmatrix} 5 = \begin{bmatrix} 0 & -5 \\ 10 & 15 \end{bmatrix}.$$

When $\lambda = -1$, we shall write $-M \equiv -1M$. Similarly, $s = \lambda m = m\lambda = [\lambda m_i]$ and $-m \equiv -1\mathrm{m}$ for a row or column matrix m. Since row and column matrices are merely special kinds of matrices of type $p \times q$, henceforth, these will not be distinguished explicitly.

B.2.3. Addition of Matrices

The *sum* of matrices A and B of type $p \times q$ is another $p \times q$ matrix $C = A + B$ whose elements are defined by $C_{ij} = A_{ij} + B_{ij}$. As usual, subtraction of matrices of the same type is defined by $A - B \equiv A + (-B)$.

Example B.1. As an illustration, consider the following matrix with polynomial elements:

$$A = \begin{bmatrix} 3x^4 + 2x^3 - x^2 + x - 1 & x^2 + 3x + 2 \\ 3x^3 + 2x^2 - x & x^4 + x + 1 \end{bmatrix}.$$

By application of the definitions of addition and scalar multiplication for matrices, the matrix A may be written

$$A = \begin{bmatrix} 3 & 0 \\ 0 & 1 \end{bmatrix} x^4 + \begin{bmatrix} 2 & 0 \\ 3 & 0 \end{bmatrix} x^3 + \begin{bmatrix} -1 & 1 \\ 2 & 0 \end{bmatrix} x^2$$
$$+ \begin{bmatrix} 1 & 3 \\ -1 & 1 \end{bmatrix} x + \begin{bmatrix} -1 & 2 \\ 0 & 1 \end{bmatrix}.$$

Therefore, the matrix A of polynomial elements may be expressed similarly as a polynomial $A = \sum_{n=0}^{4} A_n x^n$ whose coefficients A_n are matrices. Notice that A_4 is a diagonal matrix.

B.2.4. Multiplication of Matrices

Let A be a matrix of type $p \times l$ and B a matrix of type $l \times q$. Then the *product* of A and B is a matrix $P = AB$ of type $p \times q$ whose elements are defined by

$$P_{ij} = A_{ik} B_{kj}. \tag{B.3}$$

Observe the use in (B.3) of the *summation convention* introduced in Section 3.2, namely, an index that appears exactly twice in any term is to be summed over its prescribed range. In (B.3), k is the dummy index; its range is l. It is important to notice that *the product of the matrices A and B is defined only if the number of columns of A equals the number of rows of B.* Any two matrices that have this feature are said to be *conformable.* Only conformable matrices can be multiplied together.

Example B.2. Consider the matrices

$$A = \begin{bmatrix} 1 & 2 & 3 \\ 0 & 1 & 5 \\ 0 & 0 & 1 \end{bmatrix}, \qquad B = \begin{bmatrix} 3 & 0 \\ 1 & 1 \\ 2 & 0 \end{bmatrix}, \qquad C = \begin{bmatrix} 1 & 0 \\ 2 & 0 \end{bmatrix}.$$

Since A and B are conformable, they may be multiplied to yield a product

$$P = AB = \begin{bmatrix} 1 & 2 & 3 \\ 0 & 1 & 5 \\ 0 & 0 & 1 \end{bmatrix} \begin{bmatrix} 3 & 0 \\ 1 & 1 \\ 2 & 0 \end{bmatrix} = \begin{bmatrix} 11 & 2 \\ 11 & 1 \\ 2 & 0 \end{bmatrix}.$$

To see this, we apply the rule (B.3) to write out the elements of P. In particular, for the summed index range $k = 1, 2, 3$, we have

$$\begin{aligned} P_{12} = A_{1k} B_{k2} &= A_{11} B_{12} + A_{12} B_{22} + A_{13} B_{32} \\ &= (1)[0] + (2)[1] + (3)[0] = 2. \end{aligned}$$

Notice that the numbers in parentheses correspond to the elements in the *first row of A*, those in brackets are the elements in the *second column* of B, and the result is the product element P_{12} in the *first row and second column* of P. More generally, to compute the element P_{ij}, multiply the elements in the ith row of

A consecutively with the elements in the jth column of B and add the results. Thus, further,

$$P_{32}=(0)[0]+(0)[1]+(1)[0]=0$$

is the element in the third row and second column of P; and so on.

Notice that the matrices B and C also are conformable. Their product BC is given by

$$M=BC=\begin{bmatrix}3 & 0\\ 1 & 1\\ 2 & 0\end{bmatrix}\begin{bmatrix}1 & 0\\ 2 & 0\end{bmatrix}=\begin{bmatrix}3 & 0\\ 3 & 0\\ 2 & 0\end{bmatrix}.$$

The matrices A and C, however, are not conformable; consequently, their product cannot be formed. □

As a consequence of the definition of matrix multiplication, *the product of two matrices is in general not commutative*; that is, in general

$$AB\neq BA, \qquad A_{ij}B_{jk}\neq B_{ij}A_{jk}. \tag{B.4}$$

In particular, notice in the example above that the products BA and CB cannot even be formed. It is also plain that products of matrices of the same type $p\times q$ cannot be formed unless $p=q=n$. Thus, square matrices of the same type always are conformable and their product is a square matrix of the same type. Two matrices A and B are called *commutable* if $AB=BA$. Trivially, every square matrix is commutable with itself. In particular, $II=I$. In fact, an identity matrix of type $n\times n$ commutes with any $n\times n$ matrix. If M is a square matrix,

$$IM=MI=M, \qquad \delta_{ij}M_{jk}=M_{ij}\delta_{jk}=M_{ik}. \tag{B.5}$$

If $x=[x_j]$ is a $1\times p$ matrix and $y=[y_i]$ is a $q\times 1$ matrix, the product $M=yx$ is a matrix of type $q\times p$ with elements $M_{ij}\equiv y_i x_j$. However, the product xy is defined only when $p=q$, in which case the result is a scalar $\lambda\equiv x_i y_i$.

It is interesting that the product of two matrices may be a zero matrix though neither of the multipliers is. For example,

$$\begin{bmatrix}2 & 0\\ 1 & 0\end{bmatrix}\begin{bmatrix}0 & 0\\ 1 & 3\end{bmatrix}=\begin{bmatrix}0 & 0\\ 0 & 0\end{bmatrix}.$$

However, if $AB=0$ or $BA=0$ for an arbitrary conformable matrix B, then $A=0$ is the only solution of this equation. Indeed, we need only choose $B=I$ to reveal this.

Matrix multiplication has the following *associative* and *distributive* properties: If A, B, C are (conformable) matrices and λ is a scalar, then

$$A(BC) = (AB)C, \tag{B.6a}$$

$$A(B+C) = AB + AC, \tag{B.6b}$$

$$\lambda(AB) = (\lambda A)B = A(\lambda B). \tag{B.6c}$$

B.2.5. Transpose of a Matrix

The matrix of type $q \times p$ formed from the elements of a $p \times q$ matrix M by turning all the rows into corresponding columns (or all columns into corresponding rows) is named the *transpose* of M. We denote it by M^T. Thus, the element in the ith row and jth column of M^T is equal to the element in the jth row and ith column of M:

$$(M^T)_{ij} = M_{ji}. \tag{B.7}$$

When no confusion may result, (B.7) also may be written as $M_{ij}^T = M_{ji}$. Suppose, for example, that

$$M = \begin{bmatrix} a & b & c \\ d & e & f \end{bmatrix}.$$

Then

$$M^T = \begin{bmatrix} a & d \\ b & e \\ c & f \end{bmatrix}.$$

Evidently, the transpose of a row (column) matrix is a column (row) matrix.

It is easy to verify the following rules of transposition for any matrices A and B of the same type:

$$(A^T)^T = A, \tag{B.8a}$$

$$(\lambda A + \mu B)^T = \lambda A^T + \mu B^T, \tag{B.8b}$$

where λ and μ are scalars. We also have the following useful theorem: *The transpose of the product of any two square matrices is equal to the product of their transposes in reverse order*:

$$(AB)^T = B^T A^T. \tag{B.9}$$

A square matrix M is said to be *symmetric* if $M^T = M$; it is called *antisymmetric* or *skew* if $M^T = -M$. It thus follows from (B.7) that

$$M = M^T \Leftrightarrow M_{ij} = M_{ji}, \tag{B.10a}$$

$$M = -M^T \Leftrightarrow M_{ij} = -M_{ji}. \tag{B.10b}$$

In particular, every diagonal matrix is symmetric; the identity is an example: $I = I^T$. We notice also that (B.10b) shows that all of the principal diagonal elements of a skew matrix are zero. Specifically, for symmetric and skew matrices of type 3×3, we have from (B.10)

$$M = M^T = \begin{bmatrix} M_{11} & M_{12} & M_{13} \\ M_{12} & M_{22} & M_{23} \\ M_{13} & M_{23} & M_{33} \end{bmatrix}, \qquad M = -M^T = \begin{bmatrix} 0 & M_{12} & M_{13} \\ -M_{12} & 0 & M_{23} \\ -M_{13} & -M_{23} & 0 \end{bmatrix}.$$

B.2.6. Trace of a Square Matrix

The *trace* of a square matrix S, denoted by $\operatorname{tr} S$, is the sum of the principal diagonal elements of S:

$$\operatorname{tr} S = S_{kk}. \tag{B.11}$$

Equation (B.10b) shows that the trace of every skew matrix is zero. This is evident for the skew matrix M in the example. Also, with the aid of (3.2), we have $\operatorname{tr} I = \delta_{kk} = n$ for the identity matrix of type $n \times n$, for example. Indeed, for the 3×3 identity in (B.2), $\operatorname{tr} I = 3$.

It is easily seen that the trace operation obeys the addition rule

$$\operatorname{tr}(\lambda A + \mu B) = \lambda \operatorname{tr} A + \mu \operatorname{tr} B, \tag{B.12}$$

for matrices A and B and scalars λ and μ.

Finally, it may be shown that *the trace of the product of square matrices A and B is equal to the trace of their commuted product*:

$$\operatorname{tr}(AB) = \operatorname{tr}(BA). \tag{B.13}$$

We recall that in general $AB \neq BA$; however, (B.13) holds always. Indeed, we need only note that $\operatorname{tr}(AB) = A_{ij}B_{ji} = B_{ji}A_{ij} = \operatorname{tr}(BA)$.

B.3. Determinant of a Matrix

The theory of determinants, as we shall see shortly, is inextricably connected to that of matrices. It is essential, therefore, that the reader recall the more important basic properties of determinants, especially the Laplace formula for the cofactor expansion of a determinant by its rows and columns. These topics will be reviewed in this section in preparation for their future use.

The determinant formed from the elements D_{ij} of a square matrix D is called the *determinant of the matrix*. We denote it by

$$\det D = \det[D_{ij}].$$

Although all of the subsequent results hold for matrices of type $n \times n$, we shall consider for simplicity only those of type 3×3. In this case, $\det D$ is the number Δ defined by

$$\Delta = \det \begin{bmatrix} D_{11} & D_{12} & D_{13} \\ D_{21} & D_{22} & D_{23} \\ D_{31} & D_{32} & D_{33} \end{bmatrix} \equiv D_{11}D_{22}D_{33} + D_{31}D_{12}D_{23} + D_{21}D_{32}D_{13}$$

$$- D_{31}D_{22}D_{13} - D_{21}D_{12}D_{33} - D_{11}D_{32}D_{23}. \tag{B.14}$$

This familiar expansion rule may be abbreviated with the use of the permutation symbol defined by (3.3) in the text:

$$\det D = \varepsilon_{pqr} D_{p1} D_{q2} D_{r3} \tag{B.15a}$$

$$= \varepsilon_{pqr} D_{1p} D_{2q} D_{3r}. \tag{B.15b}$$

The first of (B.15) is just the usual expansion of $\det D$ by columns, and the second is its expansion by rows. These index representations for $\det D$ may be used to derive in only a few lines the elementary properties of determinants.

B.3.1. Basic Properties of Determinants

It is clear from (B.15) that if all the elements in any row or column of D are zero, then $\det D = 0$. It is also seen that equality of the row and column expansions of $\det D$ implies that

$$\det D = \det D^T. \tag{B.16}$$

That is, the value of $\det D$ is unaltered if all the rows are interchanged with corresponding columns. Though in general $D \neq D^T$, the result (B.16) holds always.

The following additional properties may be easily confirmed by application of (B.15) and use of (3.3). Let the reader show the following:

(a) If B is the matrix obtained from a matrix D by interchanging any two rows or columns, then $\det B = -\det D$. It thus follows that when any two rows or columns of D are identical, $\det D = 0$.

(b) If B is formed from D by multiplying all the elements in any row or column of D by a number λ, then $\det B = \lambda \det D$.

(c) If any two rows or columns of D are proportional, then $\det D = 0$.

(d) If $B = \lambda D$ are 3×3 matrices, then $\det B = \lambda^3 \det D$.

Finally, it is important to recall the product rule: *The determinant of the product of two square matrices is equal to the product of their determinants*:

$$\det(AB) = (\det A)(\det B). \tag{B.17}$$

Notice that even if A and B are not commutable, $\det(AB) = \det(BA)$ holds always.

B.3.2. Cofactors of a Determinant

The column expansion of $\det D$ in (B.15a) also may be written

$$\det D = D_{p1} C_{p1}, \tag{B.18}$$

where the quantities $C_{p1} \equiv \varepsilon_{pqr} D_{q2} D_{r3}$ are the signed determinants formed in the familiar column expansion procedure by deleting the elements in the pth row and first column in $\det D$. To see this, consider the determinant (B.14) expanded as just described:

$$\det D = D_{11} \begin{vmatrix} D_{22} & D_{23} \\ D_{32} & D_{33} \end{vmatrix} - D_{21} \begin{vmatrix} D_{12} & D_{13} \\ D_{32} & D_{33} \end{vmatrix} + D_{31} \begin{vmatrix} D_{12} & D_{13} \\ D_{22} & D_{23} \end{vmatrix}.$$

Comparison with (B.18) shows that

$$C_{11} = \begin{vmatrix} D_{22} & D_{23} \\ D_{32} & D_{33} \end{vmatrix}, \qquad C_{21} = -\begin{vmatrix} D_{12} & D_{13} \\ D_{32} & D_{33} \end{vmatrix}, \qquad C_{31} = \begin{vmatrix} D_{12} & D_{13} \\ D_{22} & D_{23} \end{vmatrix},$$

which are the signed determinants mentioned above. Since the same factoring procedure may be repeated for any column elements, we may write (B.18) more generally as

$$\det D = D_{pk} C_{pk}, \qquad \text{no sum on } k = 1, 2, 3. \tag{B.19}$$

The quantity C_{pk} is called the *cofactor* of the element D_{pk} in the column expansion of $\det D$. A similar procedure may be applied to the row expansion of $\det D$ to obtain

$$\det D = D_{kp} C_{kp}, \qquad \text{no sum on } k = 1, 2, 3. \tag{B.20}$$

Notice in (B.19) and (B.20) that the repeated index k has been explicitly excluded from the summation rule and therefore is to be treated as a free index, whereas the index p is to be summed over its range as usual.

Because the product $D_{pl}C_{p1} = \varepsilon_{pqr}D_{pl}D_{q2}D_{r3} = 0$ if $l \neq 1$ and $=\det D$ when it is, we may write this as $\delta_{l1} \det D = D_{pl}C_{p1}$. But any column may be used for the expansion besides the first; hence, more generally it is seen that (B.19) and (B.20), in a similar manner, may be expressed as

$$\delta_{lm} \det D = D_{pl}C_{pm} = D_{lp}C_{mp}. \tag{B.21}$$

This is known as the *Laplace formula* for the expansion of det D by columns and rows. Let $C = [C_{ij}]$ denote the matrix of the cofactors of det D, and recall the rules (B.3) and (B.7). Then (B.21) may be neatly expressed as

$$I \det D = D^T C = DC^T. \tag{B.22}$$

Since I is symmetric, (B.8) and (B.9) show that (B.22) may be written in several ways with $D^TC = C^TD = DC^T = CD^T$.

B.4. The Inverse of a Matrix

Let M be a given matrix. If there exists a matrix B such that $BM = MB = I$, then the matrix M is said to be *invertible*, and B is titled the *inverse* of M. The inverse, when one exists, will be denoted by $B \equiv M^{-1}$, so that we may write

$$MM^{-1} = I, \qquad M^{-1}M = I. \tag{B.23}$$

Notice that both M and M^{-1} must be square matrices of the same type, hence only square matrices can have inverses. But it does not follow that an inverse exists for each square matrix. Indeed, let us consider any two *nonzero* square matrices M and B with the property that $MB = 0$; and let us assume that M^{-1} exists. Then, with the aid of (B.6a) and (B.23), we find

$$B = IB = (M^{-1}M)B = M^{-1}(MB) = M^{-1}0 = 0,$$

which is a contradiction. Therefore, in this case, M^{-1} does not exist.

On the other hand, we can show easily that an invertible matrix has exactly one inverse. Suppose that M is invertible and that S is a second inverse of M so that, in addition to (B.23), $SM = MS = I$ holds. Then multiplication of this equation by M^{-1} and use of (B.23) shows that $S = M^{-1}$. Hence, M has a unique inverse. As consequence of this fact, if we may somehow construct an inverse matrix M^{-1} that satisfies (B.23), it will be the only one we need to find. But how can we know when a given square matrix actually has an inverse before we start looking for it? The problem is somewhat like the task of finding a needle in a stack of hay. There is little comfort in our knowing that only one needle has to be found, if there is no needle in the stack in the first place. Of course, a trip to the local Rent-All Shop to get a metal detector would seem like a good idea; then we would be able to test the haystack before we begin work, and perhaps an auxiliary tool, a magnet, may enable us to find the needle quickly. So, before we begin our search for the unique inverse of a given square matrix M, we would like to subject M to a simple test to determine in advance, if, in fact, its inverse exists. The inverse matrix detector we need to fix the existence of an inverse is provided by the following easy test:

A matrix M has an inverse that satisfies (B.23) *if and only if* $\det M \neq 0$. Thus, this important theorem shows that the determinant of M is the inverse matrix detector we need: if $\det M \neq 0$, the inverse of M exists; but if $\det M = 0$, the inverse does not exist.

Indeed, imagine that M has an inverse so that (B.23) holds. Then (B.17) shows that

$$\det(MM^{-1}) = (\det M)(\det M^{-1}) = \det I = 1.$$

Hence, $\det M \neq 0$ is necessary for existence of M^{-1}. Conversely, if $\det M \neq 0$, we may define a matrix $B = C^T/\det M$ in which $C = [C_{ij}]$ is the matrix of the cofactors of $\det M$. Then with $D = M$ in (B.22), we have $MB = MC^T/\det M = I$; and bearing in mind the symmetry of (B.22), also $BM = C^T M/\det M = I$. Therefore, M is invertible with the unique inverse $B \equiv M^{-1}$ given by

$$M^{-1} = C^T/\det M. \tag{B.24}$$

The matrix C of the cofactors of $\det M$ is named the *adjoint matrix* of the matrix M. The inverse elements are written as $(M^{-1})_{ji} \equiv M_{ji}^{-1}$, and (B.24) yields the index formula

$$M_{ji}^{-1} = C_{ij}/\det M = \langle \text{cofactor of } M_{ij} \text{ in } \det M \rangle / \det M. \tag{B.25}$$

A matrix M whose determinant is nonzero is said to be *nonsingular*; otherwise, M is called *singular*. Thus, the foregoing theorem says that a matrix has an inverse if and only if the matrix is nonsingular.

Example B.3. To find the inverse of the matrix

$$M = \begin{bmatrix} \cos\theta & -\sin\theta & 0 \\ \sin\theta & \cos\theta & 0 \\ 0 & 0 & 1 \end{bmatrix},$$

we first compute $\det M$. Since $\det M = 1$, our inverse matrix detector shows that M is nonsingular and hence really has an inverse. The unique inverse is determined by (B.24), which in the present case yields $M^{-1} = C^T$. We next compute the adjoint matrix C, whose elements, we recall, are the cofactors of det M:

$$C=[C_{kl}]=\begin{bmatrix} C_{11} & C_{12} & C_{13} \\ C_{21} & C_{22} & C_{23} \\ C_{31} & C_{32} & C_{33} \end{bmatrix}=\begin{bmatrix} \cos\theta & -\sin\theta & 0 \\ \sin\theta & \cos\theta & 0 \\ 0 & 0 & 1 \end{bmatrix}.$$

Thus,

$$M^{-1}=\begin{bmatrix} \cos\theta & \sin\theta & 0 \\ -\sin\theta & \cos\theta & 0 \\ 0 & 0 & 1 \end{bmatrix}.$$

Notice in this example that $M=C$; but this will not happen in general. The important special case in which $M^{-1}=M^T$ will be studied further on.

B.4.1. Properties of Inverse Matrices

The theory of determinants has shown that $\det(\lambda M)=\lambda^n \det M$ for a matrix M of type $n\times n$ and a scalar multiplier λ. Hence, if M is nonsingular, so is λM; and $(\lambda M)(\lambda M)^{-1}=I$ holds. It thus follows that

$$(\lambda M)^{-1}=\lambda^{-1}M^{-1}. \tag{B.26}$$

We also recall that the determinant of the inverse of a matrix is equal to the reciprocal of the determinant of that matrix: $\det M^{-1}=1/\det M=(\det M)^{-1}$; hence, when M is nonsingular, so is M^{-1}. We may therefore construct the matrix $(M^{-1})^{-1}$, which satisfies the relation

$$(M^{-1})^{-1}=M. \tag{B.27}$$

Also, it can be proved that

$$(M^{-1})^T=(M^T)^{-1}. \tag{B.28}$$

Finally, we recall the following theorem: *If M and B are invertible matrices of the same type, then the inverse of their product is equal to the product of their inverses in reverse order*:

$$(MB)^{-1}=B^{-1}M^{-1}. \tag{B.29}$$

B.4.2. Orthogonal Matrices

An invertible matrix M having the property

$$M^{-1}=M^T, \qquad M_{ij}^{-1}=M_{ji}, \tag{B.30}$$

is called an *orthogonal* matrix. It follows from (B.17) and (B.23) that every orthogonal matrix satisfies the relations

$$M^T M = I, \qquad MM^T = I, \qquad \det M = \pm 1. \tag{B.31}$$

It is clear that the value of $\det M = \pm 1$ is necessary but not sufficient to characterize M as orthogonal. As an example we may consider the matrix

$$M = \begin{bmatrix} 1 & 0 \\ 3 & 1 \end{bmatrix}.$$

This has $\det M = 1$, but

$$M^{-1} = \begin{bmatrix} 1 & 0 \\ -3 & 1 \end{bmatrix} \neq M^T.$$

Hence, M is not an orthogonal matrix.

An orthogonal matrix M is called *proper orthogonal* if $\det M = +1$, and *improper orthogonal* if $\det M = -1$. The rules (B.24), (B.30), and (B.31) show that for an orthogonal matrix the cofactors are given by $C = \pm M$, where the plus sign is used for proper M, the minus sign for improper M. The identity matrix is an easy example of a proper orthogonal matrix: $I^{-1} = II^{-1} = I = I^T$. The matrix M of Example B.3 is another.

It is shown in Chapter 3 that proper orthogonal matrices represent finite rigid body rotations. We shall have no need for improper orthogonal matrices which arise in transformations of basis vectors of opposite hand; these involve reflection in addition to rotation. We shall find in later chapters that proper orthogonal matrices play a key role in the transformation equations for the moments and products of inertia of a rigid body.

References

1. CROUCH, T., *Matrix Methods Applied to Engineering Rigid Body Problems*, Pergamon, New York, 1981. An excellent variety of applications of matrices to problems in kinematics may be found in Chapter 3. Unfortunately, the reader must be prepared to first wade through a sea of awkward notation.
2. FRAZER, R. A., DUNCAN, W. J., and COLLAR, A. R., *Elementary Matrices*, Cambridge U. P., Cambridge, 1957. This is an advanced text on the theory of matrices and its applications to dynamics and to vibration problems in aeronautical engineering.
3. MURDOCH, D. C., *Linear Algebra for Undergraduates*, Wiley, New York, 1957. The focus of this book is on the mathematics of matrices, but it is much less abstract than similar works.
4. PIPES, L. A., *Matrix Methods for Engineers*, Prentice-Hall, Englewood Cliffs, New Jersey, 1963. This book is an excellent source for applications of matrices to problems of engineering interest. Matrix algebra and calculus, including the elements of determinants, are developed thoroughly in the first four chapters, and the remainder of the text covers applications of matrices to elasticity theory, the analysis of structures, classical mechanics (see Chapter 7), the theory of vibrations, and topics in electrical engineering circuits.

Problems

B.1. Two square matrices A and B are defined by

$$A = \begin{bmatrix} 2 & 1 & -3 \\ 0 & 5 & 1 \\ -1 & 1 & 1 \end{bmatrix}, \qquad B = \begin{bmatrix} 2 & -1 & 3 \\ 1 & -1 & 0 \\ 4 & 2 & 1 \end{bmatrix}.$$

Find scalars α and β so that the matrix $C = \alpha A + \beta B$ has the properties that $C_{13} = C_{31}$ and $\operatorname{tr} C = (\operatorname{tr} A)^2$; and thus determine the matrix C.

B.2. Find for the matrices given in the last problem (a) AB and $\operatorname{tr}(AB)$, (b) BA and $\operatorname{tr}(BA)$.

B.3. Compute all products that can be formed from the following matrices alone, exclusive of products with themselves:

$$A = \begin{bmatrix} 1 & 0 & 2 & 0 \\ 0 & 1 & 0 & 0 \\ -2 & 0 & -1 & 0 \\ 0 & 0 & 0 & 3 \end{bmatrix}, \qquad B = \begin{bmatrix} 3 & -4 \\ 1 & 0 \\ 5 & -1 \\ 2 & 1 \end{bmatrix}, \qquad C = \begin{bmatrix} 1 & -2 \\ 2 & 1 \end{bmatrix}.$$

B.4. Use index notation to derive the rules (B.8) and (B.9) for the transposition of matrices.

B.5. A nonzero square matrix H satisfies the equation $\alpha H + \beta H^T = 0$ in which α and β are nonzero constants. Prove that either $\alpha = -\beta$ and H is symmetric, or $\alpha = \beta$ and H is skew.

B.6. The elements of two nonzero symmetric matrices U and V of type 3×3 must be chosen to satisfy the relation

$$U_{kj} V_{in} - U_{nk} V_{ij} + U_{ji} V_{kn} - U_{in} V_{jk} = 0.$$

Show that this is possible if and only if $U = \lambda V$, and find the scalar λ. Is λ unique? Hint: Introduce a summation on one pair of indices, convert the result to a matrix equation, and derive the specified relation.

B.7. (a) Let M be a square matrix. Show that M may be written uniquely as the sum of a symmetric matrix S and an antisymmetric matrix A, and thus determine the matrices S and A. The matrices $M_S \equiv S$ and $M_A \equiv A$ are called the *symmetric and skew parts* of the matrix M. (b) Find $\operatorname{tr} M_S$ and $\operatorname{tr} M_A$. (c) Apply the results of this theorem to find the symmetric and skew parts of the matrices given in Problem B.1.

B.8. (a) Prove that $\operatorname{tr}(AB) = \operatorname{tr}(B^T A^T)$. (b) If B is a square matrix and A is a symmetric matrix of the same type, show that $\operatorname{tr}(BA) = \operatorname{tr}(B_S A)$. (c) If A is a skew matrix, show that $\operatorname{tr}(BA) = \operatorname{tr}(B_A A)$. (d) What can be said about $\operatorname{tr}(A_S B_A)$ and $\operatorname{tr}(A_A B_S)$? The subscripts are defined in the previous problem.

B.9. Find the constants α, β, and γ; and determine the trace and the determinant of the symmetric matrix

$$M = \begin{bmatrix} 1 & \alpha/2 & 2 \\ 1 & 3 - \alpha + 2\gamma & 1 \\ \gamma + 2 & 3\alpha - 5\beta & \beta - 1 \end{bmatrix}.$$

B.10. The matrix T is antisymmetric. Determine the constants α, β,..., and thereby find the matrix

$$T=\begin{bmatrix} \alpha & \alpha+\beta/2 & 2 \\ \delta & 2-\beta+2\gamma^2 & \varepsilon \\ \gamma-2 & 3\beta-5\phi & \phi-1 \end{bmatrix}.$$

B.11. Show that (B.15) is equivalent to (B.14). Apply (B.15) to prove (B.16) and the four properties that follow it in the text.

B.12. Prove the following two additional basic principles of determinants: (a) If each element of a row, or a column, of a matrix B is expressed as the sum of two (or more) terms, the det B equals the sum of two (or more) corresponding determinants. (b) If a multiple of one row (or column) of a matrix D is added to another row (or column) to form another matrix B, then $\det B = \det D$. (c) Use the theorem (b) to show that

$$\begin{vmatrix} \alpha & 1 & 1 \\ 1 & \alpha & 1 \\ 1 & 1 & \alpha \end{vmatrix} = (\alpha-1)^2(\alpha+2).$$

B.13. Apply the fundamental properties of determinants to show that (B.15) also may be written as

$$\varepsilon_{pqr} \det D = \varepsilon_{ijk} D_{ip} D_{jq} D_{kr} = \varepsilon_{ijk} D_{pi} D_{qj} D_{rk}.$$

Use this result together with (B.15) to derive (B.17).

B.14. Determine for the matrix B in Problem B.3 the possible products among BB, B^TB, and BB^T, and compute their determinants, if they exist. Do these determinants obey the rule (B.17)? Why?

B.15. For the given matrices, compute $\det(PQ)$ two ways:

$$P=\begin{bmatrix} 1 & -2 & 3 \\ 0 & 0 & 2 \\ -1 & 0 & 0 \end{bmatrix}, \qquad Q=\begin{bmatrix} 4 & -1 & 7 \\ 6 & 5 & -4 \\ 0 & 0 & 2 \end{bmatrix}.$$

B.16. Show that the adjoint matrix C of a symmetric $n\times n$ matrix D is symmetric. However, if D is skew and n is odd, prove that its adjoint C is symmetric and $\det D = 0$. What can be said in the case when n is even?

B.17. Consider a system of n homogeneous linear equations in n unknowns x_k given by $B_{jk}x_k = 0$. Prove that this system has nontrivial solutions when and only when $\det B = 0$. On the other hand, show that a similar system of nonhomogeneous equations $B_{jk}x_k = \alpha_j$ has a unique solution if and only if $\det B \neq 0$. Find the unique matrix solution for this case, and apply the result to solve the system of linear equations

$$\begin{aligned} 3x-2y+z &= -2, \\ 2x+y-z &= 5, \\ x-2y+3z &= -6. \end{aligned}$$

B.18. Derive the rules (B.27), (B.28), and (B.29).

B.19. Prove that B^{-1} is a symmetric matrix when and only when B is symmetric. What may be said about B^{-1} when B is skew?

B.20. Determine $(M^T)^{-1}$ and M^{-1} for the matrix

$$M = \begin{bmatrix} 1 & 2 & 3 & 1 \\ 2 & 0 & 0 & 0 \\ 3 & 0 & 1 & 0 \\ 1 & 0 & 0 & 1 \end{bmatrix}.$$

B.21. Is it possible to find any symmetric or antisymmetric orthogonal matrices of type 3×3? Explain.

B.22. Let Q be an orthogonal matrix. Are Q^T and Q^{-1} orthogonal matrices? Is the product of two (or more) orthogonal matrices also an orthogonal matrix? Provide proof.

B.23. (a) Write down the index representations of the rule (B.31) for orthogonal matrices, and write out one diagonal and one nondiagonal element of the product for 3×3 matrices. What simple corresponding vector interpretation can you assign to these particular products and others like them? (b) Classify the following matrices according as they may be proper or improper orthogonal, or neither. Explain briefly your conclusion for each case.

$$A = \begin{bmatrix} 0 & -1 & 1/2 \\ -1 & 0 & 0 \\ 1/2 & 0 & 1 \end{bmatrix}, \qquad B = \begin{bmatrix} 1 & 0 & 0 \\ 0 & \sin\theta & \cos\theta \\ 0 & -\cos\theta & \sin\theta \end{bmatrix},$$

$$C = \begin{bmatrix} 1/\sqrt{2} & 1/\sqrt{2} & 0 \\ \sqrt{6} & 0 & 0 \\ 1/\sqrt{3} & -1/\sqrt{3} & -1/\sqrt{3} \end{bmatrix}, \qquad D = \begin{bmatrix} 0 & 0 & -1 \\ 1 & 0 & 0 \\ 0 & 1 & 0 \end{bmatrix}.$$

B.24. Let M be a given $n \times n$ proper orthogonal matrix. Show that its adjoint matrix also is proper orthogonal. What may be said about its adjoint when M is improper orthogonal?

B.25. Suppose that C is the adjoint matrix of a nonsingular $n \times n$ matrix M. Show that $\det C = (\det M)^{n-1}$. How are the cofactors of $\det C$ related to the elements of M? What may be said about the $\det C$ when M is either proper or improper orthogonal?

Answers to Selected Problems

Chapter 1

1.1. (b) $\mathbf{x}(S, 0) = \mathbf{0}$, $\mathbf{v}(S, 2) = 8(40\mathbf{i} + 3\mathbf{j} - 6\mathbf{k})$.
1.3. $\mathbf{a}(Q, 2) = 9\mathbf{i}$ cm/sec^2.
1.5. (a) $s(t) = 5t$; (b) $\mathbf{a}(P, t) = -8(\sin 2t\, \mathbf{i} + \cos 2t\, \mathbf{j})$, $\quad \mathbf{a} \cdot \mathbf{v} = 0$.
1.7. (a) $t_1 = 2$ sec; (b) $\mathbf{a}(P, t_1) = 8\mathbf{e}$ ft/sec^2.
1.9. (b) $\ddot{s}_1 = 72t^3/(9t^4 + 1)^{1/2}$, $|\mathbf{a}_1| = 24t$.
1.11. $\mathbf{a}(P, t) = -a\omega^2 \cos \psi\, \mathbf{i}$.
1.13. $\mathbf{v}(B, t) = -(R\omega \sin \theta + r\dot{\beta} \sin \beta)\, \mathbf{I} + (R\omega \cos \theta + r\dot{\beta} \cos \beta)\, \mathbf{J}$.
1.15. $\mathbf{a}(Q, t) = -a\omega^2(\cos \omega t - 3^{-1/2} \sin \omega t)\, \mathbf{i}$, $\mathbf{v}_{\max} = \pm 2a\omega \sqrt{3}/3\mathbf{i}$.
1.17. $\mathbf{v}(R, 2) = -5\pi(\frac{3}{2}\mathbf{i}_1 + \mathbf{i}_2)$ cm/sec.
1.21. $\dot{s}(2) = 5\pi \sqrt{2}/8$ cm/sec, $\mathbf{x}(P, 4) = 6\mathbf{J}$ cm.
1.23. $\dot{s}(A) = \sqrt{21}$ in./sec at $x = 3$ in., $\mathbf{x}(A, t) = 3t\mathbf{i} + 4[1 - t^2/4]^{1/2}\, \mathbf{j}$ in.
1.25. (b) $E = 2/3$, (d) $\mathbf{v}_S = -32\mathbf{i}$ in./sec at $\theta = \pi/2$.
1.27. $\mathbf{v}(Q, 2) = 13\mathbf{i} - 12\mathbf{j} + (\pi \sqrt{3}/6)\, \mathbf{k}$ m/sec.
1.29. $a = \dfrac{2d(t_2 - t_1)}{t_1 t_2 (t_1 + t_2)}$
1.31. $\dot{s} = b\omega \sqrt{2}$.
1.33. $\mathbf{x}(S, 5) = 1250\mathbf{i} + 910\mathbf{j}$ m.
1.35. $t = 2l/\beta$.
1.37. $\tau = 3h/5v_0$.
1.39. $\mathbf{a}(P, t) = \frac{8}{3}\mathbf{i} - 8\mathbf{j}$ cm/sec^2 at $y = 2$ cm.
1.41. $t = 2$ sec.
1.45. $p = \dfrac{l}{v^*} - \dfrac{v^*(a + a^*)}{2aa^*}$, $E = 7/16$.
1.47. (b) $r(\theta) = a + \dfrac{3b}{2}\left(1 - \dfrac{\theta}{2\pi}\right)$ for $2\pi/3 \leqslant \theta \leqslant 2\pi$.
1.49. (a) $r(\theta) = a + \dfrac{b}{2}\left(3 - \dfrac{\theta}{\pi}\right)$ for $\pi \leqslant \theta \leqslant 2\pi$.
1.51. (c) $R = 25/8$ cm, $\mathbf{a} = 8\mathbf{n}$ cm/sec^2.

1.55. $\mathbf{v}(P, t_0) = 13\mathbf{t}$ ft/sec, $R(t_0) = 169\sqrt{2}/6$ ft.

1.57. $\mathbf{a}(P, t) = \dfrac{\sqrt{5}}{135}(-\mathbf{t} + 2\mathbf{n})$ in./sec^2 at $x = 3$ in.

1.59. $\mathbf{v}_A = 15\mathbf{j}$ cm/sec, $\mathbf{a}_A = -72\mathbf{j}$ cm/sec^2 at $y = 2$ cm.

1.61. $R_B = 3/\sqrt{11}$ m.

1.63. (a) $\mathbf{a}_{\max} = (135/4)\,\mathbf{n}$ cm/sec^2; (b) $\omega = 1$ rad/sec at B and D.

1.65. $\mathbf{a}(P, 3) = 9(\mathbf{i} + \mathbf{j})$, $\kappa = 0$.

1.67. $\kappa = R/(R^2 + P^2)$.

1.69. $\mathbf{a}(P, t) = (3ct^2 + b)\,\mathbf{t} + t(2bc)^{1/2}\,\mathbf{n}$.

1.71. (c) $R(1) = 9/2$ and $\mathbf{a}(P, 1) = 2(\mathbf{t} + \mathbf{n})$ at $t = 1$ sec.

1.73. $\mathbf{a}(P, t) = \dfrac{4\sqrt{2}}{5}(7\mathbf{t} + 24\mathbf{n})$ cm/sec^2 at $\theta = 45°$.

1.75. (a) $\mathbf{a} = 6(t\mathbf{t} + \mathbf{n})$ cm/sec^2; (b) $R = 54$ cm at $t = 2$ sec.

1.77. $\mathbf{v}(s) = (v_0 + \alpha s)\,\mathbf{t}$, $t = \log\left(\dfrac{v_0 + \alpha s}{v_0}\right)^{1/\alpha}$.

1.83. (a) $\mathbf{x}_H = 4(1 + x^{-4})^{-1/2}(\mathbf{i} - x^{-2}\mathbf{j})$; (c) $\mathbf{a} = 16\kappa\mathbf{n}$.

1.85. (a) $\mathbf{v}(E, t) = (k/\omega)[\sin \omega t\,\mathbf{i} + (\cos \omega t - 1)\,\mathbf{j}]$; (b) a circle.

1.89. (a) $F(x) = \frac{1}{3}(x^3 - 1)\langle x - 1\rangle^0$; (c) $F(x) = \frac{1}{3}\langle x - 1\rangle^3 - \langle x - 1\rangle^2 + \langle x - 1\rangle^1$.

1.91. $\mathbf{v}(P, t) = 15(\langle t - 0\rangle^2 - \langle t - 2\rangle^2)\,\mathbf{i}$ ft/sec,
$\mathbf{x}(P, t) = 5[3 + t^3\langle t - 0\rangle^0 - (t - 2)^3\langle t - 2\rangle^0]\,\mathbf{i} + 6\mathbf{j} + 25\mathbf{k}$ ft.

1.93. (a) $\mathbf{x}(P, \theta) = a\mathbf{i} + \dfrac{\alpha}{2\omega^2}(\langle\theta - 0\rangle^2 - 2\langle\theta - \pi\rangle^2)\,\mathbf{i}$.

1.95. $r(\theta) = a + (v/\omega)(\langle\theta - 0\rangle^1 - \frac{3}{2}\langle\theta - 2\pi/3\rangle^1)$.

1.97. $r(\theta) = a + b(\langle\theta/\pi - 0\rangle^1 - \frac{3}{2}\langle\theta/\pi - 1\rangle^1)$.

1.99. (b) $s(t) = vt - \frac{1}{2}\alpha\langle t - \tau\rangle^2$.

Chapter 2

2.1. A circle.

2.7. $3N$.

2.9. (a) 4; (c) 3.

2.11. (a) $\mathbf{d}(B) = \dfrac{\Delta\theta\sqrt{3}}{3}(\mathbf{I}_1 - \mathbf{I}_3)$; (c) no.

2.15. $\mathbf{a}_M = (-a\omega^2 \cos\psi - 2a\dot{\theta}^2 \sin\theta + 2a\ddot{\theta} \cos\theta)\,\mathbf{i} + (2a\dot{\theta}^2 \cos\theta + 2a\ddot{\theta} \sin\theta)\,\mathbf{j}$.

2.17. $\mathbf{a}_B = -30\mathbf{i} + 80\mathbf{j}$ ft/sec^2 in $\Phi = \{F; \mathbf{i}, \mathbf{j}\}$.

2.19. $\mathbf{v}_B = \omega[h(2r - h)]^{1/2}\,\mathbf{i}$, $\mathbf{v}_B|_{\max} = r\omega\mathbf{i}$.

2.21. $\mathbf{a}_A = -12\mathbf{j}$ ft/sec^2, $\boldsymbol{\omega} = 2\mathbf{k}$ rad/sec.

2.23. $\mathbf{v}_B = 30\mathbf{i}$ cm/sec, $\dot{\boldsymbol{\omega}} = 3\mathbf{k}$ rad/sec^2.

2.25. $\mathbf{a}_B = -40\sqrt{3}\,\mathbf{j}$ cm/sec^2, $\dot{\boldsymbol{\omega}} = \sqrt{3}\,\mathbf{k}$ rad/sec^2.

2.27. $\mathbf{a}_P = -48\mathbf{j} - 112\sqrt{3}\,\mathbf{i}$ cm/sec^2, path of C: $x^2 + y^2 = 16$.

2.29. $\dot{\boldsymbol{\omega}} = 21\mathbf{k}$ rad/sec^2.

2.31. $\mathbf{v}_C = -3.69(\mathbf{i} + \mathbf{j})$ m/sec, $d = 1.016$ m.

2.33. (b) $\mathbf{v}_B(t_0) = v \cot\phi_0 \sin\theta_0\mathbf{j}$, $l = b \tan\theta_0 \sin\theta_0/2 \tan\phi_0 \sin\phi_0$.

2.35. (b) $d_A = 3/4$ in.
2.37. (b) $\mathbf{v}_O = 2\mathbf{v}$, $\mathbf{v}_Q = v[(2+\sin\gamma)\,\mathbf{i} + \cos\gamma\,\mathbf{j}]$, a cycloid.
2.41. $\boldsymbol{\omega}_P = -\dfrac{\omega}{2}\mathbf{K}$, $\boldsymbol{\omega}_S = \dfrac{\omega}{4}\mathbf{K}$.
2.43. $\boldsymbol{\omega}_P = [(a\omega - r\Omega)/2b]\,\mathbf{K}$, (i) if and only if $\boldsymbol{\Omega} = -a\boldsymbol{\omega}/(a+2b)$.
2.45. $\mathbf{v}_C = -(ac\omega/b)\,\mathbf{t}$.
2.47. $\mathbf{a}_O = -ra/(R-r)\,\mathbf{i}$, $\dot{\boldsymbol{\omega}} = -a/(R-r)\,\mathbf{k}$.
2.49. $\boldsymbol{\omega} = -41\mathbf{i}_3$ rad/sec, $\dot{\boldsymbol{\omega}} = 75\mathbf{i}_3$ rad/sec^2.
2.53. (a) $\dot{\boldsymbol{\omega}}_G = -3\mathbf{k}$ rad/sec^2, (d) $\boldsymbol{\Omega} = 2\mathbf{k}$ rad/sec.
2.55. $\boldsymbol{\omega} = 48\mathbf{k}$ rad/sec, $\mathbf{a}_G = 61.65\mathbf{n}_1 + 103.92\mathbf{n}_2$ m/sec^2.
2.57. $\mathbf{v}_Q = 25\mathbf{i} - 15\sqrt{2}\,\mathbf{j} + 52\mathbf{k}$ ft/sec, $\dfrac{x-2}{14} = \dfrac{y-3}{8} = \dfrac{z}{5\sqrt{2}}$.
2.59. (b) $\boldsymbol{\omega} = \frac{8}{11}\mathbf{i} + \frac{8}{11}\mathbf{j} - \frac{35}{11}\mathbf{k}$ rad/sec.
2.61. $\mathbf{v}_B = -8.66\mathbf{i}$ m/sec, $\boldsymbol{\Omega} = 43.3\mathbf{j}$ rad/sec.
2.63. $\mathbf{a}_A = \dfrac{r}{R-r}\,[r\omega^2\mathbf{n} + (R-r)\,\dot{\omega}\mathbf{t}]$.
2.65. (a) $\boldsymbol{\omega} = 10\cos^2\theta\,\mathbf{K}$ rad/sec; (b) $\mathbf{a}_C = 50(3\mathbf{I} + \mathbf{J})$ ft/sec^2.
2.67. $\mathbf{a}_A = 40\mathbf{t} + 32\mathbf{n}$ cm/sec^2, $\dot{\boldsymbol{\omega}}_2 = -4\mathbf{k}$ rad/sec^2.
2.69. $\mathbf{v}_P = 20\sqrt{5}\,\mathbf{t}$ ft/sec, $R = 5\sqrt{10}/2$ ft.
2.71. $r = v_0/\omega$.
2.73. (a) $b = (r + l - h)/\sin\theta$ on the line perpendicular to OB; (c) $\boldsymbol{\omega} = 1.5\mathbf{k}$ rad/sec.
2.75. $v_C = a\omega\,\dfrac{\sin(\theta-\psi)}{\sin(\phi+\psi)}$, $\omega_{BC} = \dfrac{a\omega}{b}\,\dfrac{\sin(\theta+\psi)}{\sin(\phi+\psi)}$.
2.77. (a) $\dot{\boldsymbol{\omega}} = \omega^2\tan\psi\,\mathbf{K}$, (b) $x^2 + (y-l)^2 = l^2$ referred to φ, $\boldsymbol{\omega}$.
2.79. The space and body centrodes are circles centered at O, the wall center, and at A, respectively.

Chapter 3

3.1. (a) $v_2 = \omega_3 x_1 - \omega_1 x_3$
3.3. (a) $c_j = N_{jk} M_{kq} b_q$, (c) $\lambda = A_{kp} v_p v_k$
3.5. (a) ε_{pqk}, (c) 0.
3.9. (b) $\mathbf{x} = \dfrac{a_0}{D}\,[-\dot{\omega}_1^2\mathbf{i} + \dot{\omega}_1\dot{\omega}_2\mathbf{j} + (\dot{\omega}_2\omega^2 - \dot{\omega}_1\dot{\omega}_3)\,\mathbf{k}$ in φ, with $D \equiv -\omega^2(\dot{\omega}_1^2 + \dot{\omega}_2^2)$.
3.19. $2T_{32} = -2T_{23} = T_{13} = -T_{31} = (3/5)^{1/2}$, $T_{12} = T_{21} = 1/5$.
3.21. $3(x-2) = 3y/2 = 3(z+1)/2$,
$\mathbf{d}(C) = \frac{1}{8}[(18 - 6\sqrt{3})\,\mathbf{I} - 3\sqrt{3}\,\mathbf{J} + 3(2\sqrt{3} - 3)\,\mathbf{K}]$ ft.
3.23. (a) $\boldsymbol{\alpha} = \dfrac{10}{[1 + 16\sqrt{2}]^{1/2}}\,[\mathbf{i} + (1 - \sqrt{2})\,\mathbf{j} - 2\mathbf{k}]$;
(b) $\mathbf{d}(P) = -5\mathbf{i} + (\sqrt{2} - 1)\,\mathbf{j} + 8\mathbf{k}$ ft.
3.25. (a) $\hat{\mathbf{x}} = \sqrt{2}\,(-\mathbf{i} + 5\mathbf{k})$ m; (c) $\Delta\mathbf{x} = -\sqrt{2}\,\mathbf{i} + 4\mathbf{j} + (5\sqrt{2} - 6)\,\mathbf{k}$ m.
3.27. $A_{12} = A_{21} = \sqrt{3}\,A_{11}$, $A_{22} = -A_{11} = 1/2$, $A_{33} = 1$, others zero.
3.29. $\boldsymbol{\alpha} = \pm(\sqrt{2}/2)(\mathbf{i}_2 + \mathbf{i}_3)$, $\mathbf{d}(A) = 20\mathbf{i}_1 + 30\mathbf{i}_2 + 10\mathbf{i}_3$ ft.

3.31. $\psi = \sin^{-1}(\sqrt{2451}/50) = 81.95°$, not the same.
3.33. $\hat{\mathbf{X}}(P) = 6\mathbf{I} - \mathbf{J} - 2\mathbf{K}$ in Φ, $\mathbf{d}(Q)\cdot\boldsymbol{\alpha} = 4/\sqrt{3}$.
3.35. $\theta^* = \cos^{-1}[(\sqrt{2}-2)/4] = 98.42°$, $\boldsymbol{\alpha}^* = [5+2\sqrt{2}]^{-1/2}[-\mathbf{i}+\mathbf{j}+(1+\sqrt{2})\,\mathbf{k}]$.
3.39. $\mathbf{r} = \frac{1}{2}(7+2\sqrt{3})\,\mathbf{i} + \frac{1}{2}(4+3\sqrt{3})\,\mathbf{j}$ ft.
3.41. $\mathbf{B}^* = 18.227\mathbf{I} - 3.398\mathbf{J} - 5\mathbf{K}$ cm locates O^* from F in Φ, $p = -48/\pi$, not a reversed screw.
3.43. $\mathbf{r} = -\frac{2}{3}(\mathbf{j}+\mathbf{k})$ ft from B, $p = 2\sqrt{3}/\pi$, $\mathbf{b}\cdot\boldsymbol{\alpha} = 4/\sqrt{3}$.
3.45. $\theta^* = 120°$, $\mathbf{d}(T) = 2\mathbf{I} + 3\mathbf{J} + \mathbf{K}$ ft, $p = \dfrac{\sqrt{3}}{\pi}$ ft/rad, $\mathbf{b}_\alpha = \dfrac{2\boldsymbol{\alpha}}{\sqrt{3}}$ ft.
3.47. $R^*_{22} = R^*_{12} = R^*_{13} = -R^*_{23} = -\sqrt{2}/2$, $R^*_{31} = -1$, others zero.
3.49. $R^* = R_1 R_2 R_3$.
3.51. (a) $-A^*_{13} = A^*_{22} = A^*_{31} = 1$, others zero; (b) no, $\boldsymbol{\alpha}^* = \mathbf{i}_2$.
3.53. $\boldsymbol{\alpha}^* = (\sqrt{3}/3)(\mathbf{i}_2 + \sqrt{2}\,\mathbf{i}_3)$ in φ.
3.55. $\boldsymbol{\alpha}^* = -(1/\sqrt{23})[(2+\sqrt{3})\,\mathbf{i}_1 + (2\sqrt{3}-1)\,\mathbf{i}_2 + \sqrt{3}\,\mathbf{i}_3]$,
$\theta^* = \cos^{-1}(-7/16) = 115.94°$.
3.57. $|\mathbf{d}(P)| = 2.032$.

Chapter 4

4.1. (a) $\mathbf{v}_O = 6\mathbf{I} + \left(1 + \dfrac{3\sqrt{3}}{2}\right)\mathbf{J} + \left(\dfrac{3}{2} - \sqrt{3}\right)\mathbf{K}$ m/sec;

(b) $\mathbf{v}_P = 3\left(\dfrac{7\sqrt{3}}{2} - 1\right)\mathbf{I} + \left(\dfrac{15\sqrt{3}}{2} - 2\right)\mathbf{J} + \left(\dfrac{15}{2} + 2\sqrt{3}\right)\mathbf{K}$ m/sec.

4.3. (a) $\dot{\mathbf{x}} = 36\mathbf{i} - 12\mathbf{j} - 27\mathbf{k}$ m/sec; (c) $\mathbf{a}_O = -56\mathbf{i} + 78\mathbf{j} - 47\mathbf{k}$ m/sec^2.
4.5. $\dot{\mathbf{x}} = 58\mathbf{i} - 62\mathbf{j} - 14\mathbf{k}$ ft/sec; $\ddot{\mathbf{x}} = -734\mathbf{i} - 539\mathbf{j} - 596\mathbf{k}$ ft/sec^2, no.
4.7. (a) $\boldsymbol{\omega}_{21} = \dfrac{v_A(y^2\mathbf{i} - ay\mathbf{j})}{(a^2+y^2)(l^2-a^2-y^2)^{1/2}}$;

(c) $\boldsymbol{\omega}_{20} = \dfrac{v_A}{(a^2+y^2)^{1/2}}\left[\dfrac{y\boldsymbol{\beta}}{(l^2-a^2-y^2)^{1/2}} + \dfrac{a\mathbf{k}}{[a^2+y^2]^{1/2}}\right]$.

4.9. $\mathbf{v}_B = 750\mathbf{j} + 600\mathbf{i}$ cm/sec, $\boldsymbol{\omega}_{21} = -90\mathbf{k}$ rad/sec.
4.11. $\dot{\boldsymbol{\omega}}_{30} = 2\mathbf{i} + 4\mathbf{j} + 45\mathbf{k}$ rad/sec^2.
4.13. (b) $\boldsymbol{\omega}_{10} = -\omega \tan\alpha\,\mathbf{K}$.
4.17. $\dot{\boldsymbol{\omega}}_{30} = -149\mathbf{I} + 15\mathbf{J} + 90\mathbf{K}$ rad/sec^2.
4.19. $\boldsymbol{\omega}_{30} = 20\mathbf{i} + 2\mathbf{j} + 2\mathbf{k}$ rad/sec, $\dot{\boldsymbol{\omega}}_{30} = 6\mathbf{i} + 39.4\mathbf{j} - 39.8\mathbf{k}$ rad/sec^2.
4.21. $\dot{\boldsymbol{\omega}}_{30} = 0.02\mathbf{i} + 0.21\mathbf{j} + 0.1\mathbf{k}$ rad/sec^2.
4.23. (a) $\dot{\boldsymbol{\omega}}_{30} = -(2 + 4500\sqrt{2})\,\mathbf{I} + 150(1 - 5\sqrt{2})\,\mathbf{J} + 750\sqrt{2}\,\mathbf{K}$ rpm^2;
(c) $\dot{\boldsymbol{\omega}}_{31} = -2\mathbf{I} + 750\sqrt{2}\,(\mathbf{K} - \mathbf{J})$ rpm^2.
4.25. (b) $\boldsymbol{\omega}_{30} = \omega_1[\mathbf{i} - \alpha^2 \cos\theta \cos\phi \sin\phi(\cos\theta\,\mathbf{j} + \sin\theta\,\mathbf{k})$;
(c) $\mathbf{v}_A = r\alpha\omega_1(\cos\phi\,\boldsymbol{\gamma}_2 + \sin\phi \sin\theta\,\boldsymbol{\gamma}_3)$,
$\mathbf{v}_D = -r\alpha\omega_1 \cos\phi(\boldsymbol{\gamma}_1 + \alpha \sin\phi \cos\theta\,\boldsymbol{\gamma}_3)$.

4.27. (a) $\mathbf{a}_P = 6\mathbf{I}$, (b) $\mathbf{v}_P = 6t(\cos\theta\,\mathbf{i} - \sin\theta\,\mathbf{j})$.

4.29. $\mathbf{v}_{WG} = \dfrac{V\sqrt{2}}{6}(-\mathbf{I} + 3\mathbf{J})$.

4.31. $\mathbf{v}_{QF} = -a\omega\dfrac{\cos(\theta - \psi)}{\sin\psi}\mathbf{i}$, $\mathbf{a}_{QF} = a\omega^2\dfrac{\sin(\theta - \psi)}{\sin\psi}\mathbf{i}$.

4.33. $\mathbf{a}_{AF} = -(9\sqrt{6}/4)\,\mathbf{j}$ ft/sec^2, $\mathbf{v}_{AB} = -3\sqrt{6}\,\mathbf{j}$ ft/sec.

4.35. $R = 125$ cm.

4.37. $t = 2.5$ min.

4.39. $\mathbf{a}_{MF} = -(bp^2\sin pt + r\omega^2\cos\omega t)\,\mathbf{i} - r\omega^2\sin\omega t\,\mathbf{j}$.

4.41. (a) $\mathbf{v}_{AF} = \dfrac{3\sqrt{3}}{2}(\sqrt{3}\,\mathbf{i} - \mathbf{j})$ cm/sec, (b) $\mathbf{a}_{AB} = -\dfrac{12\sqrt{3}}{5}\mathbf{j}$ cm/sec^2.

4.43. $a = 2b = rV/v$.

4.45. (a) $\ddot{\mathbf{r}} = 100a(-\mathbf{i} + \sqrt{3}\,\mathbf{j})$, (c) $\boldsymbol{\omega}_{AB} = \left[10 + \dfrac{\cos pt}{(3 + \cos^2 pt)^{1/2}}\right]\mathbf{k}$.

4.47. $\boldsymbol{\omega}_3 = 7\mathbf{k}$ rad/sec.

4.49. $\mathbf{a}(P, t) = (2v^2/m)(-\sin\phi\,\mathbf{e}_r + \cos\phi\,\mathbf{e}_\phi)$.

4.53. $\mathbf{v}_R = (5\pi/2)(-2\mathbf{e}_r + 3\mathbf{e}_\phi)$ cm/sec, $\mathbf{a}_R = -(5\pi^2/8)(3\mathbf{e}_r + 4\mathbf{e}_\phi)$ cm/sec^2.

4.57. $\mathbf{a}_S = -4b\omega^2(\cos\phi\,\mathbf{e}_r + \sin\phi\,\mathbf{e}_\phi)$.

4.59. $\mathbf{a}_P = -rb^2\sin^2 qt\,\mathbf{e}_r + rbq\cos qt\,\mathbf{e}_\phi - ap^2\sin pt\,\mathbf{e}_z$.

4.61. $\phi_{\max} = 60°$, $\mathbf{a}_P = -11.65\mathbf{e}_r + 38.64\mathbf{e}_\phi$ ft/sec^2.

4.63. $\mathbf{v}_P = \dot{a}\mathbf{e}_r + a\omega\mathbf{e}_\phi$.

4.65. $\mathbf{a}_P = (6\pi - 135\pi^2 - 1/15)\,\mathbf{e}_r + 30\pi\,\mathbf{e}_\phi + 2\mathbf{e}_z$ cm/sec^2.

4.67. $\mathbf{a} = (c^2/2)[\sin\phi\,\mathbf{e}_\phi - 2(1 - \cos\phi)\,\mathbf{e}_r]$.

4.69. $\mathbf{v}_E = c[2a/p]^{1/2}\cos\psi(\mathbf{e}_\phi - \dfrac{p}{c}\mathbf{e}_z)$, $\mathbf{a}_E = \dfrac{ac}{p}\cos^2\psi\left(-4\pi\mathbf{e}_r + \mathbf{e}_\phi - \dfrac{p}{c}\mathbf{e}_z\right)$.

4.71. $\boldsymbol{\omega}_{SG} = \dfrac{r}{2L}(\dot{\beta} - \dot{\alpha})\,\mathbf{e}_z$, $\mathbf{a}_O = \dfrac{r^2}{4L}(\dot{\beta}^2 - \dot{\alpha}^2)\,\mathbf{e}_r - \dfrac{r}{2}(\ddot{\alpha} + \ddot{\beta})\,\mathbf{e}_\phi$.

4.75. $\mathbf{a}_P = -a\omega^2\sin\theta(\sin\theta\,\mathbf{i} + \cos\theta\,\mathbf{j}) + 2v\omega\sin\theta\,\mathbf{k}$.

4.77. $\mathbf{v}_B = -100\mathbf{e}_\phi + 40\mathbf{e}_\theta$ cm/sec,
$\mathbf{a}_B = -580\mathbf{e}_r - 500\sqrt{3}\,\mathbf{e}_\theta - 400\sqrt{3}\,\mathbf{e}_\phi$ cm/sec^2.

4.79. $\mathbf{v}_B = l\dot{\phi}(\cos\phi\,\mathbf{i} + \sin\phi\,\mathbf{j}) - l\omega\sin\phi\,\mathbf{k}$.

4.81. (a) $\dfrac{\delta\mathbf{x}}{\delta t} = 20(-3\mathbf{j} + 4\mathbf{k})$ cm/sec; (c) $\mathbf{a}_M = 1130\mathbf{i} - 3000\mathbf{j} - 1200\mathbf{k}$ cm/sec^2.

4.83. $\dot{\boldsymbol{\omega}}_{30} = -500\mathbf{j} + 5\mathbf{k}$ rad/sec^2.

4.85. $\mathbf{a}_P = -\dfrac{\pi^2}{12}[6\sqrt{2}\,\mathbf{i} + (25 + 6\sqrt{2})\,\mathbf{j}]$ m/sec^2.

4.87. $\dot{\boldsymbol{\omega}}_f \times \mathbf{x} = 3\mathbf{i} - 2\mathbf{j} - 4\mathbf{k}$ cm/sec, $\mathbf{a}_R = -17\mathbf{i} + 7\mathbf{j}$ cm/sec^2.

4.89. $\mathbf{v}_R = b\omega\mathbf{j}$, $\dot{\boldsymbol{\Omega}} = -\dfrac{\sqrt{3}}{3b}(\ddot{a} - b\omega^2)\,\mathbf{k}$.

4.91. (i) $\mathbf{a}_P = -(l\omega_1^2 + a\dot{\omega}_1\cos\phi - 2a\omega_1\omega_2\sin\phi)\,\mathbf{i}$
$+ [l\dot{\omega}_1 - a\cos\phi(\omega_1^2 + \omega_2^2)]\,\mathbf{j} - a\omega_2^2\sin\phi\,\mathbf{k}$.

4.93. $\mathbf{v}_A = -\omega d\mathbf{i}$, $\mathbf{a}_B = -4a\Omega\omega\mathbf{i} + (2a\dot{\Omega} - d\omega^2)\,\mathbf{j} - a\Omega^2\,\mathbf{k}$.

4.95. $\mathbf{a}_P = -1525\mathbf{i} + (120 - 625\sqrt{3})\,\mathbf{j} - (100 + 1500\sqrt{3})\,\mathbf{k}$ cm/sec^2.

4.97. $\boldsymbol{\omega} = 6\mathbf{k}$ rad/sec, $\dot{\boldsymbol{\omega}} = -27\mathbf{k}$ rad/sec^2.

4.99. (b) $\mathbf{v}_P = a\omega_1 \sin\theta\, \mathbf{e}_3$, $\mathbf{a}_P = -a\omega_1^2 \sin\theta(\sin\theta\, \mathbf{e}_1 + \cos\theta\, \mathbf{e}_2)$;

(d) $\mathbf{a}_P = -\left(\dfrac{v^2}{a} + a\omega_2^2 \sin^2\theta\right)\mathbf{e}_1 - a\omega_1^2 \sin\theta\cos\theta\, \mathbf{e}_2 + 2v\omega_1 \cos\theta\, \mathbf{e}_3$.

4.101. (a) $\boldsymbol{\Omega} = 16\sqrt{5}/3\mathbf{k}'$ rad/sec, $\dot{\boldsymbol{\Omega}} = -(320/3)\,\mathbf{i}'$ rad/sec^2.

4.103. (i) $\mathbf{v}_P = r(\omega_1 + \omega_2)(-\sin\theta\, \mathbf{i} + \cos\theta\, \mathbf{j})$, (ii) $\mathbf{a}_P = -r\dot{\omega}_2\mathbf{e}_1 - r(\omega_1 + \omega_2)^2\, \mathbf{e}_2$.

4.105. (i) $\mathbf{v}_B = l\dot{\psi}(-\sin\psi\, \mathbf{i} + \cos\psi\, \mathbf{j}) - \omega(a + l\sin\psi)\, \mathbf{k}$,

(ii) $\mathbf{v}_B = l\dot{\psi}\mathbf{e}_\theta - \omega(a + l\sin\psi)\, \mathbf{e}_\phi$.

4.107. $\mathbf{a}_B = 3h\omega^2\mathbf{i} - h\dot{\omega}\mathbf{j}$.

4.109. $\mathbf{v}_P = -30\mathbf{i}$ ft/sec, $\mathbf{a}_P = -450\mathbf{j} - 1200\mathbf{k}$ ft/sec^2.

4.111. (a) $\mathbf{v}_S = -40\mathbf{i} - 20\mathbf{j} + 2\mathbf{k}$ ft/sec; (b) $\mathbf{a}_S = -60\mathbf{i} - 30\mathbf{j} - 999\mathbf{k}$ ft/sec^2.

4.113. $\mathbf{a} = 0.0557(\mathbf{e}_r + \mathbf{e}_\theta)$ ft/sec^2.

4.115. $N = -\omega(2\dot{x} + \omega\sqrt{3})$.

Appendix A

A.1. (a) $d(\mathbf{u} \pm \mathbf{v})/ds = 12s^2\mathbf{j} + (\pm 12s - 5)\,\mathbf{k}$; (b) $d(\mathbf{u}\cdot\mathbf{v})/ds = -90s^2$,

$d(\mathbf{u} \times \mathbf{v})/ds = 24s(5s^3 + 3)\,\mathbf{i} + 15\mathbf{j} + 36s^2\,\mathbf{k}$.

A.7. $\mathbf{x}(P, t) = \mathbf{x}_0 + \mathbf{v}_0 t + \frac{1}{2}\mathbf{a}t^2$.

Appendix B

B.1. $\alpha = 16$, $\beta = -32$.

B.3. Form AB, BC, and ABC

$$(AB)^T = \begin{bmatrix} 13 & 1 & -11 & 6 \\ -6 & 0 & 9 & 3 \end{bmatrix}, \quad ABC = \begin{bmatrix} 1 & -32 \\ 1 & -2 \\ 7 & 31 \\ 12 & -9 \end{bmatrix}.$$

B.7. (a) $S = \frac{1}{2}(M + M^T)$, $A = \frac{1}{2}(M - M^T)$.

(c)

$$B_S = \frac{1}{2}\begin{bmatrix} 4 & 0 & 7 \\ 0 & -2 & 2 \\ 7 & 2 & 2 \end{bmatrix}, \quad B_A = \frac{1}{2}\begin{bmatrix} 0 & -2 & -1 \\ 2 & 0 & -2 \\ 1 & 2 & 0 \end{bmatrix}.$$

B.9 tr $M = 2$, det $M = -1$.

B.15. $\det(PQ) = 208$.

B.17. $x = 1$, $y = 2$, $z = -1$.

B.21. There are no antisymmetric, orthogonal 3×3 matrices.

B.23. (b) A and C are not orthogonal, B is proper, D is improper.

Index

CORRIGENDA: PRINCIPLES OF ENGINEERING MECHANICS
Volume 1: Kinematics - The Geometry of Motion
Millard F. Beatty, Jr.

Chapter 1.

p.4, line 5 from bottom; read "consumed"

p.7, line 2 below (1.2); read "discard"

p.8, line 3 from bottom; read "therefore,"

p.10, line 2 from bottom; change $0z$ to Oz

p.18, mid-page; change - "path equation (1.32)" - to "last equation in (1.32)"

p.19, line 2 below (1.37); read "m,"

p.23, line 5; *for all times* in italics

p.24, line 3, equation; replace the comma with a period

p.45, line 1 below (1.114); change "half of" to read: "equation in"

p.72, line 5; 15° should read 1.5°; however, it is best to update the statement as follows. Line 3 from the top: Replace two sentences "Then.... Run... V vs X. Identify..." with "Then plot computer graphs of the normalized (nondimensional) motion $X \equiv x/a$ and the normalized velocity $V \equiv \dot{x}/a\omega$ of S in terms of the crank angle θ, and plot the so-called phase plane graph of V vs X. Identify..."

p.74, Problem 1.39, reverse the direction of $\mathbf{v}_A$ in the figure

p.75, Problem 1.43, last line; insert before the period: "and $D = |\mathbf{D}|$"

p.77, Problem 1.52, read $\mathbf{v} = \dot{x}\mathbf{i} + \dot{y}\mathbf{j}$.

p.78, Problem 1.59, line 2; read: "move from $y = 0$ with..."

p.81, Problem 1.81, line 3, equation; read: "$-\tau$"

Chapter 2.

p.86, line 7 from bottom; change "the same as" to read: "indistinguishable from"

p.89, line 2 below (2.11); *point* in italics

p.90, **Solution** line 1; read: "shown in Fig. 2.5a."

p.93, paragraph 2, midpage, line 2; read: "$\hat{\mathbf{x}} = \mathbf{x}$ for each P."

p.95, line 1 below (2.18); read: "$\boldsymbol{\Delta}\theta$ is called the"

p.99, line 1 below (2.27); read: "$\mathbf{v}_O = \mathbf{v}(O, t)$"

p.104, line 3 above (2.39); remove minus sign, read: "$a\omega^2\mathbf{e}_n$"

p.105, line 1 after (2.44); no bold for components ω_1 and ω_2

p.122, line 1 below (2.110); read: "from O to"

p.124, last line; read: "about an axis"

p.132, Problem 2.20; read: "A rigid rod"

p.136, Problem 2.33; line 3; change "cosθ" to "sinθ"

p.141, Problem 2.52; line 5; read: "after 2 sec when the system is in the position shown."

p.142, Problem 2.55; line 3; read: "Determine for the instant shown"

p.142, Problem 2.57; line 2; replace "side" with "sense"

Chapter 3.

p.160, subscripts on $\mathbf{e}_{ij}$ in (3.30) are italics, not bold

p.162, equation above (3.40); replace "$\mathbf{e}_1$" with "$\mathbf{e}_l$"

p.165, second relation in (3.56); replace bold **1** with 1

p.165, line 4 below (3.58); remove one comma after (3.6)

p.166, line 3 above subsubsection 3.3.4.5.; replace italics v, read: "$\boldsymbol{\Omega} \times \mathbf{v}$"

p.168, line 2 above (3.69); revise to read: "in (3.57) write the..."

p.169, equation (3.78b); read: "ε_{ijk}"

p.174, line 9; insert **R** to read: "$(\mathbf{R}^T\mathbf{R} - \mathbf{1})\mathbf{x} = \mathbf{0}$"

p.176, line 2 below (3.97); replace "$\mathbf{e}_i^{'}$" with "$\mathbf{e}_1^{'}$"

p.178, before the first paragraph starting with: Another interesting..., insert the subsubheading "*3.6.1.2. The Basis Transformation Tensor*"

p.178, last paragraph, first word; read: "Finally,"

p.179, equation (3.108b); insert prime to read: "$QT^{'}Q^T$"

p.181, change subsubsection "*3.6.2.1. Invariant...*" to subsection "**3.6.3. Invariant Properties of Tensors**"; and note the entry in the Table of Contents, p. xvi.

p.184, line 5; read: "we know that"

p.184, last line; read: "coincide initially"

p.190, The vector "$\mathbf{i}_1'$" nearest the bottom of the page in Figure 3.7 should read "$\mathbf{i}_1$"

p.203, line 4 below (3.147b); revise sentence to read: "The resultant axis and angle of rotation generally will depend on the order of rotations in (3.147b). For two rotations, however, it follows..."

p.203, In the same paragraph below (3.148), add the sentence: "Note, however, that $\mathrm{tr}(\mathbf{R}_3\mathbf{R}_2\mathbf{R}_1) = \mathrm{tr}(\mathbf{R}_1\mathbf{R}_3\mathbf{R}_2) \neq \mathrm{tr}(\mathbf{R}_1\mathbf{R}_2\mathbf{R}_3)$.

p.217, Problem 3.18, equation; read numerator: "$2[T^2_{[12]} + T^2_{[23]} + T^2_{[31]}]^{1/2}$"

p.218, Problem 3.23; read: "$-\sqrt{2}/10$" (see revised problem solution p.383 below)

p.218, Problem 3.24; restore the upper left-hand corner of brackets $\lceil$

p.220, Problem 3.34, line 5; read: "position vectors in Φ"

Chapter 4.

p.237, line below (4.16c); replace the sentence to read: "This is the total time rate of change in frame Φ of the relative angular velocity vector ω_2 of the arm OA."

p.238, equation (4.17c); replace italics δx with bold $\delta\mathbf{x}$

p.242, amend line 2 below (4.22) to read: "from (4.21) or (4.22)"

p.243, line below equation (4.25a); change (4.25) to (4.25a)

p.252, line 2 after (4.33); read: "Then..."

p.254, **Example 4.6**; revise last sentence to read: "What is the time derivative of this result in frame 0?"

p.254, line 2 below (4.39b); add the sentence: "The time derivative of this result in frame 0 is determined next."

p.255, subsubsection *4.5.2.1*, lines 1 and 2; read: "elec-tromechanical"

p.264, line 1 above (4.51); read: "*rules*"

p.268, line 3 from the end of the central paragraph; read: "the cylindrical coordinate system is..."

p.269, last equalty in line 3 above (4.60); read: "$2\dot{r}\dot{\phi}\mathbf{e}_\phi$."

p.271, **Solution (ii)**, line 2; change (4.47) to (4.48)

p.272, line 3 after (4.65d); add to the last line: "Alternatively, one may begin with $\mathbf{x}(P,t) = r\mathbf{e}_r$."

p. 274, line 2 above (4.66g) read: "...use of (4.66e) or (4.66f) delivers..."

p.276, remove the minus sign in equation (4.67c)

p.282, amend (4.76b) to read:

$$"\omega_f = \omega_{20} = \omega\mathbf{e}_\theta + \Omega\mathbf{e}_\phi, \quad \dot{\omega}_f = \dot{\omega}_{20} = -\Omega\omega\mathbf{e}_r"$$

p.282, line below (4.76b); replace (4.14) with (4.39a)

p.282, insert comma after $\mathbf{e}_r$ in (4.76d) and append the equation $\dot{\omega}_f \times \mathbf{x} = \mathbf{0}$.

p.282, line below (4.76d); delete the phrase: "the second equation in (4.67b),"

p.283, last relation in (4.77b); insert R to read: "$-R(1-\cos\psi)\mathbf{n}$"

p.284, equation (4.77f); replace r with R (italics) in the first term; change bold face $\mathbf{R}$ to R (italics) in the second

p.288, **Example 4.17**, line 9; read: "acceleration"

p.290, line 1 of (4.83b); in the $\mathbf{j}'$ component insert $\cos\phi$ to read: "$\omega_2 \cos\phi \cos pt$"

p.293, lines 1 and 2 below (4.87c); replace c with q (2 places)

p.293, Figure 4.29; the axis at θ from the vertical is $\mathbf{k}$

p.297, line 7; replace "the" with "we"

p.298, line 4; replace "closely approximated" with "approximate"

p.298, line 5; insert after the period the following addendum: "A variety of factors that affect this estimate, including the relative motion of the Earth's atmosphere, a dominant influence that causes changes in the Earth's angular speed, have been ignored. Nevertheless, the approximation suffices to show that time variations in the Earth's rotational rate are extremely small. Therefore, ..."

p.301, line 3 below (4.93c); read: "Hence, relative to C in the preferred frame Φ, the..."

p.301, line 1 below (4.93d); after "Coriolis acceleration" insert "relative to C in Φ"

p.301, line 2 below (4.93d); replace the sentence beginning with "Therefore," to read: "Therefore, in its flow southward in the preferred frame relative to C, the river particle has a Coriolis acceleration directed eastward in the northern hemisphere where $\cos\theta > 0$, westward in the southern hemisphere where $\cos\theta < 0$.

p.309, line 1 below (4.105); read: "$\boldsymbol{\Omega}$=(det $\mathbf{Q}$)$\mathbf{Q}\boldsymbol{\Omega}^*$"

p.316, line 3 below (4.138b); replace bold $\mathbf{t}_\Phi$ with t_Φ

p.331, Problem 4.40; the respective $\mathbf{E}, \mathbf{N}$ directions in the figure are the x, y axes

p.342, Problem 4.84, last line; read: "$\psi = \{Q; \mathbf{i}_k\}$"

p.343, Problem 4.88, line 2; change 4.12 to 4.10

Appendix B.

p.375, line 1 after (B.24); read: "The transposed matrix C^T..."

p.376, line 1 delete - "adjoint"

p.379, Problem B.16, lines 1 and 2; change C to C^T

p.380, Problem B.25, line 1; change C to C^T

Answers to Selected Problems.

p.381, answer to 1.21; amend to read: "$\dot{s}(2) = \frac{\pi}{4}\sqrt{5}$ cm/sec, $\mathbf{x}(P, 4) = 2\mathbf{J}$ cm."

p.383, answer to 3.23; amend to read: "(a) $\boldsymbol{\alpha} = -\frac{\sqrt{7(4+\sqrt{2})}}{14}[\mathbf{i} + (1-\sqrt{2})\mathbf{j} - 2\mathbf{k}]$,

(b) $\mathbf{d}(P) = -3\mathbf{i} + (1+\sqrt{2})\mathbf{j} - 2\mathbf{k}$."

p.384, answer to 4.21; amend to read: "$\dot{\omega}_{30} = 0.014\mathbf{I} + 0.1\mathbf{J} - 0.21\mathbf{K}$ rad/sec^2"

9 780306 421310